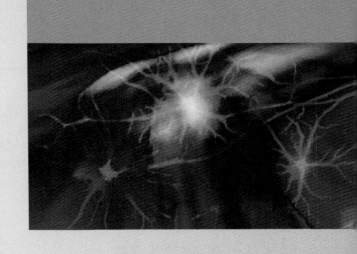

In Situ Molecular Pathology and Co-Expression Analyses

ELSEVIER
science & technology books

Companion Web Site:

http://booksite.elsevier.com/9780124159440

In Situ Molecular Pathology and Co-Expression Analyses
Gerard J. Nuovo

Resources for available:

- All figures from the book available as both Power Point slides and .jpeg files

TOOLS FOR ALL YOUR TEACHING NEEDS
textbooks.elsevier.com

Ⓐ ACADEMIC PRESS

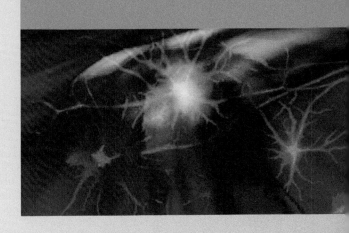

In Situ Molecular Pathology and Co-Expression Analyses

Gerard J. Nuovo MD

Ohio State University Comprehensive Cancer Center, Columbus, OH, USA

AMSTERDAM • BOSTON • HEIDELBERG • LONDON
NEW YORK • OXFORD • PARIS • SAN DIEGO
SAN FRANCISCO • SINGAPORE • SYDNEY • TOKYO

Academic Press is an imprint of Elsevier

ELSEVIER

AcademicP ressisan imp rinto fE lsevier
32 Jamestown Road, London NW1 7BY, UK
525 B Street, Suite 1800, San Diego, CA 92101-4495, USA

Medicine is an ever-changing field. Standard safety precautions must be followed, but as new research and clinical experience broaden our knowledge, changes in treatment and drug therapy may become necessary or appropriate. Readers are advised to check the most current product information provided by the manufacturer of each drug to be administered to verify the recommended dose, the method and duration of administrations, and contraindications. It is the responsibility of the treating physician, relying on experience and knowledge of the patient, to determine dosages and the best treatment for each individual patient. Neither the publisher nor the authors assume any liability for any injury and/or damage to persons or property arising from this publication.

Notice

No responsibility is assumed by the publisher for any injury and/or damage to persons or property as a matter of products liability, negligence or otherwise, or from any use or operation of any methods, products, instructions or ideas contained in the material herein. Because of rapid advances in the medical sciences, in particular, independent verification of diagnoses and drug dosages should be made

British Library Cataloguing-in-Publication Data
A catalogue record for this book is available from the British Library

Library of Congress Cataloging-in-Publication Data
A catalog record for this book is available from the Library of Congress

ISBN:978-0-12-415944-0

For information on all Academic Press publications
visito urwe bsiteat elsevierdirect.com

Typeset by MPS Limited, Chennai, India
www.adi-mps.com

Printed and bound by CPI Group (UK) Ltd, Croydon, CR0 4YY

Contents

Acknowledgments

There are many people whom I have met in my career that, if I had not known them, this book may never have been written. First is Dr. Jackson Clemmons. He is a true gentleman who encouraged me both by example and by his support to become a pathologist. Next are the many pathologists at Columbia Presbyterian College of Physicians and Surgeons with whom I had the distinct pleasure of working during my residency in the 1980s. Of course, I did not realize it at the time, but I now clearly see that the Pathology Department of Columbia Presbyterian set the standard for both patient care and integrity that would be unmatched in the rest of my career. It also had the premier surgical pathologists of the time. Anyone who was fortunate to be trained by Drs. Karl Perzin and Raphael Lattes would have to have a good foundation in Anatomic Pathology. Add to that people such as Drs. John Fenoglio, Daniel Knowles, Jay Lefkowitz, and David Silvers, and it is easy to see why anyone leaving such a program would be committed to the art and science of surgical pathology. I also had the great fortune to work with Drs. Ralph Richart and Chris Crum, who were the best possible role models for me to become an academic pathologist interested in HPV disease. Dr. Richart, despite being a very busy person, taught me to always put the female patient first, and it was a lesson that has stayed with me during my career. I am so appreciative to Drs. Crum and Saul Silverstein, who first introduced me to the world of academic research pathology. Any pathologist reading this list, especially if he or she is around my age, will quickly understand the basis of whatever success I have had in the field is due primarily to the influence, example, and teachings of those who trained me.

I also want to thank my colleagues who have had a large part of my later career. This includes Dr. Roy Steigbigel and so many people at the Ohio State University Comprehensive Cancer Center. What a privilege it has been, and continues to be, working with people such as Drs. Clay Marsh, Mike Ostrowski, Michael Caligiuri, Nino Chiocca, and Carlo Croce! Each of these senior investigators has a team that reflects his high ethics and dedication to good patient-driven research. I must add a special thanks to Dr. Carlo Croce, whose entire large team consists of highly intelligent, motivated, good people. Working with someone as brilliant and "cutting edge" as Dr. Croce makes me feel like a kid in a candy store, because it is so exciting to go to work and do the *in situ* based experiments that can supplement his superior work. I also want to thank others such as Drs. Saul Suster, Deborah Bartholomew, Lauren Ackermann, and Cynthia Magro, who are superior examples of the type of people I first met at Columbia Presbyterian—excellent medical doctors who put patient care first and foremost. I continue to be surprised by how poorly known the simple fact is that it is the pathologist who determines treatment with his or her diagnosis. I am so grateful to have worked with so many medical doctors who understand this and

base their careers on this point. I also want to thank Dr. Elazar Rabbani and Mr. Shahram Rabbani. They are true gentlemen, and their laboratory, Enzo Laboratory, reflects their intelligence, hard work, and dedication. It is truly an honor to be part of their team. In a similar light, I would also like to thank Dr. Adel Mikhail for his collaborations over the years. He demonstrates so well that hard work and success can go hand-in-hand with integrity and his team at Phylogeny is superior. I also want to thank Ms. Kathleen Sergott and Mr. JD Kimple for their superior assistance over a long period of time with the Ventana Benchmark. Much of the data seen in this book is the result of the excellent Benchmark system.

Finally, very special thanks go to several people. First, I want to thank several people who did a lot of the work for this book. My niece Elana Nuovo edited the English. I don't mind saying that although I was born in New York and have lived in the United States my entire life, I once had a paper accepted only after revision of the English that had to be "done by someone whose primary language is English, which does not appear to be the case of the primary author"! I also thank my niece Allison Nuovo who helped me with some of the experiments. It has been a real pleasure working with her as she starts her career in medicine at OSU. I thank my sister-in-law Karen who did the excellent cover illustration for the book. I want to also deeply thank Sandra Sue Otera who did all the graphics in the book; her professionalism and talent seem to be infinite. Sandra Sue is a very special person whose courage and humanism matches her amazing talent and ability. I feel very fortunate to work with someone who is so dear to me, and who I respect so much. All the photomicroscopy/images were done by Dr. Margaret Nuovo. I thank her for not becoming violent with me (and she would have been acquitted by any jury if she did!) during all the revisions!

I also want to thank my good friends Bruce Parizo and Joey Alfieri. I also want to thank the many friends I made during my part time employment in college and Medical School such as Jim, Country, Martha, Jim, and Brewster. These people showed me better than any classroom lecture why medicine can be such a noble profession. The next people to thank are no longer with us. My dear sister Jan, my cousin Gloria, my Uncle Jimmy, and my parents will always have a special place in my heart, and their spirit is in this book.

The next special person to thank is Mr. Salim B. Lewis, President of the Lewis Foundation. He is that very rare combination of intelligence, integrity, and toughness. His financial and personal support of my work has been critical for more than 22 years.

Finally, the last person to thank is the one to whom I dedicate this book. It is to my best friend and wife, Dr. Margaret Nuovo. More than anyone, Margaret is responsible for my love of both my work and my life. I am truly a fortunate man to have Margaret as my wife and soul-mate!

Introduction

1

One of the most dramatic changes in my 30+ year career has been the explosion of the field of molecular pathology. Technological changes have resulted in the evolution of a field that basically did not exist 20 years ago, to the point that it is now a dominant player in both research and clinical medicine. There are two in situ-based tests: in situ hybridization and immunohistochemistry. Indeed, in clinical pathology, many diagnoses are dependent on performing an in situ-based test. Further, clinicians often depend on specific in situ-based results to make the definitive decision on how to treat a patient's disease. For example, the her-2-neu immunohistochemical test is routinely done to decide whether a woman's breast cancer will show reduced growth if treated with a specific drug called herceptin that has minimal side effects.

The critical need for immunohistochemistry and in situ hybridization tests in the diagnostic and research arenas has led to the involvement of major biotechnical companies into this area. Large companies such as Enzo Biochem, Roche, Ventana Medical Systems, Dako, Leica, and others offer many products for in situ-based molecular pathology including automated platforms, and they, and many other large companies, market reagents for such tests. This has led to basically all diagnostic and most research laboratories using these automated in situ hybridization- and immunohistochemical-based systems. I can attest that 25 years ago the idea that automated machines could do in situ hybridization and immunohistochemistry was almost in the realm of science fiction!

As a result of these advances in the field of in situ-based molecular pathology, many more laboratories are either using these tests for their diagnostics or research, or wanting to incorporate them into their work. Hence, the purpose of this textbook. It is my hope that this book can make in situ-based molecular pathology more accessible and understandable to both the research and diagnostic laboratory. I hope to do this by focusing on two key goals: (1) to explain the theory and foundation of immunohistochemistry and in situ hybridization, and (2) to present simplified protocols that are easy to follow for the different in situ-based protocols. I also include protocols for the identification of two or more DNA/RNA/protein targets in a given tissue.

This textbook has been written assuming minimal prior knowledge of the topics of molecular pathology in general and in situ-based molecular pathology in particular.

The first chapters focus more on the biochemistry of the processes inherent in any molecular pathology-based method, including the polymerase chain reaction (PCR) and hybrid capture solution phase detection of DNA or RNA, as well as Western blot detection of proteins. The biochemistry part, though, strongly emphasizes just the key parts you must understand to be able to "visualize" what is actually happening inside the intact cell when doing either immunohistochemistry or in situ hybridization. Since all such methods, of course, use intact tissues, I also include a chapter to assist you in being better able to determine the cell type(s) that contain the target sequence of interest. Specifically, I include a chapter that is meant to teach the basics of histopathology to the nonpathologist. After this basic introduction to these key topics, we move on to the practical applications of in situ hybridization, immunohistochemistry, and co-expression analyses.

Thus, it is certain that all readers will be able to either just breeze through or skip certain sections, depending on your training. It is my strong hope that all readers, after finishing this book, will not only want to try their hand at in situ-based molecular pathology, but also have the confidence that they will be able to reason out the best way to solve the problems that arise when using any such methodology. The end result, I hope, will be well worth the effort. For one, the power of the in situ-based molecular pathology tests is extraordinary. By knowing the cell type or types that contain the target of interest, you typically get tremendous insight into the role of the target that simply cannot be achieved by PCR, Western blots, or any of the other solution-based methods, since each of the latter tests requires the pulverization of the tissue as a prerequisite to doing the test. Also, with these methods, you can get the true pleasure of looking under a microscope and often seeing for the first time data that no one before has seen, especially when working with novel DNA/RNA or protein sequences. Thus, we can appreciate the wonder and excitement of Van Leeuwenhoek when he first examined microbes under the microscope. The fun and enjoyment of doing this is why I enjoy in situ hybridization and immunohistochemistry today every bit as much as, if not more so, when I started 30 years ago!

When I started writing this book, I realized that I had certain preconceived notions about in situ hybridization **1**

DOI: http://dx.doi.org/10.1016/B978-0-12-415944-0.00001-2

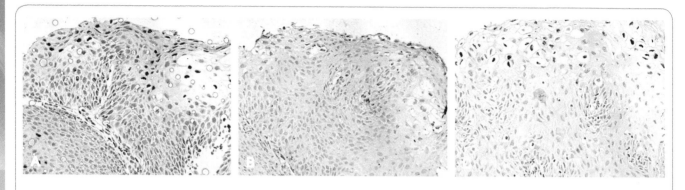

Figure1-1 Effect of the tissue block age on the HPV in situ hybridization signal and its "rejuvenation." Panel A shows the intense signal for HPV in situ hybridization in this cervical intraepithelial lesion grade 1 (CIN 1) obtained in 1992. Serial section slides were saved for 20 years. In 2012, when the serial section was tested for the same HPV type (HPV 51), the signal was lost (panel B). However, when another serial section was treated with a series of reagents meant to regenerate the signal, the intense HPV 51 signal returned and was, thus, rescued (panel C). We discuss in detail the "regeneration of the signal" concept in subsequent chapters. But, for now, these data show that such aged blocks still can be useful for in situ hybridization-based research.

and immunohistochemistry. It seemed that the format of writing a book in this field was the perfect time to test such preconceived notions. For example, I had assumed for my entire career that if I was unable to get a good signal for either immunohistochemistry or in situ hybridization with an older block (usually defined as at least 10 years old), the target DNA/RNA/protein had simply degraded and that was that. I was trained (and it made perfect sense) to simply avoid such blocks of tissue, as they probably were not fixed properly at the time of biopsy and, more importantly, nothing could be done to "rejuvenate" the signal. Similarly, I was trained to assume that any RNA would quickly degrade in the tissue sections, either from just time-related degradation and/or RNase activity in the tissue/in situ solutions and, thus, to only use recently done formalin-fixed biopsies for RNA in situ hybridization analysis, and to also use strict "RNase-free" protocols. Although I could give you many such other examples, let me end with just one more. I was trained to rely primarily on one method to "expose the target" when doing in situ hybridization or immunohistochemistry. This method has many names, including *antigen retrieval, cell conditioning*, and *liquid-based denaturation*. Again, this made perfect sense since it was well documented that formalin fixation cross-linked cellular proteins to each other and to RNA/DNA. The logic went that this extensive cross-linking created many small pores that needed to be opened for the DNA/RNA probe or primary antibody and ancillary reagents to enter the cell and access the target. Of course, this theory became very popular when antigen retrieval first came on the scene about 25 years ago, and many proteins that were otherwise undetectable with immunohistochemistry became evident. Certainly, I clearly remember the importance of antigen retrieval to the anatomic pathologist in breast cancer, since the ER/PR and her-2-neu testing required this pretreatment to get an accurate idea of the signals which, in turn, had important implications for the treatment of the woman.

An important focus of this book is that all the preconceived notions noted in the preceding paragraph, despite making sense, are simply wrong! Look at Figure 1-1. This is the result of in situ hybridization for HPV DNA in a tissue sample over 20 years old. When I first tested the block in 1992 (just when HPV could be successfully detected in situ), it produced an intense signal, as seen in panel A. When I tested the same tissue in 2012, basically no signal was evident (panel B). Again, I just simply assumed that the HPV DNA had degraded over time and probably simply diffused out of the cell. I also assumed that this block was therefore worthless for any further DNA or RNA testing, and probably for any protein testing by immunohistochemistry as well. But look at panel C. This is a serial section of the same tissue. I treated the tissue with a "rejuvenating" agent and then did the in situ hybridization. The signal was beautifully regenerated! Figure 1-2 shows the exact same situation for a protein (cytokeratin AE1/3) in a block of tissue 20 years old. Figure 1-3 shows the same result for an RNA, in this case microRNA-let-7c. Clearly, the idea that DNA, proteins and, especially, RNA degrade over time in formalin-fixed tissue and that this per se precludes their detection by in situ hybridization or immunohistochemistry is simply wrong! And, again, this is what I was taught and I certainly believed for many years.

We spend several of the next chapters on understanding the reason that you can use old tissue blocks and slides with immunohistochemistry and in situ hybridization and, by understanding what is happening to the tissues and macromolecules over time, regenerate the signal and basically make the tissue not only as good as new but, in most cases, "better than new." This leads to a fundamental part of this book. Recipes (often called cookbook recipes) for in situ hybridization and immunohistochemistry are helpful; indeed, they are essential because they serve as a starting point for these methods. However, I do *not* want this book to only give you such recipes. I think it is essential that we all, to the best of

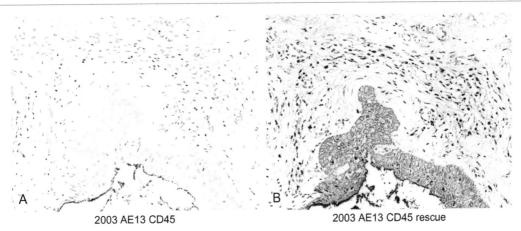

2003 AE13 CD45

2003 AE13 CD45 rescue

Figure 1-2 Effect of the tissue block age on the immunohistochemistry signal and its "rejuvenation." Panel A shows results of immunohistochemistry for two proteins, cytokeratin AE 1/3 and CD45, in a skin biopsy of a patient with nonspecific dermatitis. The biopsy was done in 2003, and serial sections saved for the last 9 years. Note the very weak signal for the cytokeratin and the lack of a signal for CD45 in the lymphocytes that are present in the dermis. Both proteins should yield intense signals in such a biopsy. An additional serial section slide was treated with the same series of reagents meant to regenerate the signal as used for the HPV test in Figure 1-1. Note that the intense signals for each cytokeratin and CD45 are now evident (panel B). We discuss in detail the "regeneration of the signal" concept in subsequent chapters. But, for now, these data show that such aged blocks still can be useful for immunohistochemistry research.

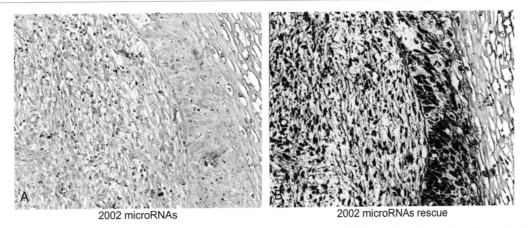

2002 microRNAs

2002 microRNAs rescue

Figure 1-3 Effect of the tissue block age on the in situ hybridization signal for microRNAs and its "rejuvenation." Panel A shows a cervical biopsy with nonspecific inflammation taken in 2002. Unstained slides were stored for 10 years. The tissue was tested for miR-31 and miR-let-7c. These miRNAs should be present in high copy number in the cervix in the stromal inflammatory cells and basal epithelial cells, respectively. However, no signal was noted. An additional serial section slide was treated with the same series of reagents meant to regenerate the signal as used for the HPV test in Figure 1-1 and the immunohistochemistry test in Figure 1-2. Note that the intense signals present in the submucosal inflammatory cells and basal epithelial cells are now evident (panel B).

our ability, understand the biochemistry of each step of immunohistochemistry and in situ hybridization. This requires an in-depth knowledge of what actually happens at a biochemistry level inside the intact cells when we use cryostat sections versus denaturing fixatives (such as ethanol, acetic acid, and alcohol) versus the most common fixative, 10% neutral buffered formalin. This knowledge will be *by far* the most important tool you will have to troubleshoot when you are experiencing problems with immunohistochemistry and in situ hybridization.

Now look at Figure 1-4. These are all images of HPV in situ hybridization. Note that in some of the tissues the optimal signal requires DNA retrieval. By "DNA retrieval," I mean exposing the tissue to 95°C in an aqueous solution prior to in situ hybridization (like antigen retrieval for proteins with immunohistochemistry). However, you will see tissues that are histologically equivalent such that to maximize the HPV in situ hybridization signal you need to use protease digestion. You will also see, as illustrated in panels C and D, cases in which

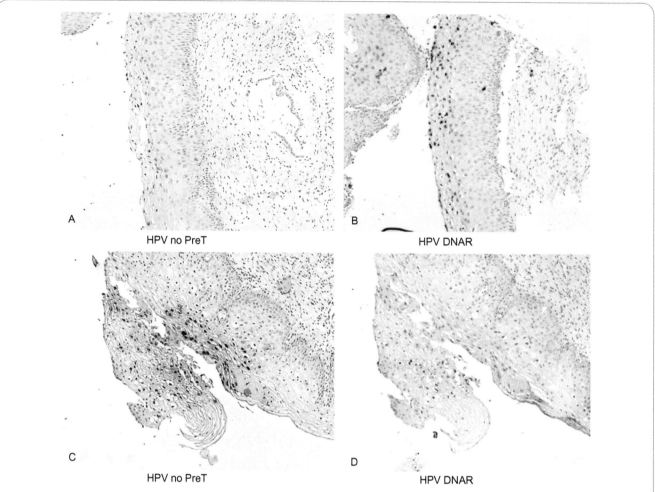

Figure 1-4   Different optimal protocols for HPV in situ hybridization for different CIN tissues. Each tissue was diagnosed as CIN, and each was obtained in 2011 or 2012. Note that one of the biopsies showed a very weak signal with no pretreatment (panel A), but when a serial section was incubated at 95°C for 30 minutes in an EDTA solution, the signal became much stronger (panel B). However, note that a different CIN tissue, which looked equivalent to the CIN shown in panels A and B, yielded the exact opposite results. Specifically, there was a strong signal with no pretreatment (panel C) that was much reduced when the serial section was incubated at 95°C for 30 minutes in an EDTA solution (panel D). These data underscore the important point that tissues from the same site with the same diagnosis may well require different pretreatment conditions when doing in situ hybridization for RNA or DNA.

tissues with the same histologic diagnosis require *no* pretreatment to get the best signal. In yet other tissues, antigen retrieval *plus* protease digestion gives the best signal! I can assure you that it is impossible to predict which HPV-infected formalin-fixed, paraffin-embedded tissue will require no pretreatment, antigen retrieval, protease, or a combination of the last two pretreatment regimes. It is important to stress that if you do not use the right pretreatment regime for that given tissue, then you might not see *any* signal. Why does this happen? What is the biochemical basis of this observation? We discuss this topic at length in the book and, I hope, by the time you reach the end with Chapter 10 you have a solid understanding of this and related phenomena.

The preceding paragraph, where it is clear that one pretreatment regime may be perfect for one tissue and give no signal at all for another tissue with the same

pathologic diagnosis, may be a bit disheartening to the beginner. Not to worry! We discuss in detail the biochemical basis for this observation and learn how to use it to our advantage when we devise our in situ hybridization and immunohistochemical-based protocols. So let's begin with a discussion of some of the basic concepts of molecular biology in Chapter 2.

Before we move to Chapter 2, let's take a quantitative look at how the fields of in situ hybridization and immunohistochemistry have grown over the past three decades. The suggested readings show the number of publications produced in 1975 on the topic of in situ hybridization or immunohistochemistry; note that there were eight such papers [1–8]. Compare this to the list of publications produced in 1980 on in situ hybridization or immunohistochemistry; note that there were 34 [9–42]. These references are included to give homage to the pioneers in these two fields.

Let's now look at the number of peer review references on either in situ hybridization or immunohistochemistry over the past 35 years. These data are presented in the following table.

As is evident, an explosion of papers on in situ-based molecular pathology were published between 1985 and 1995. In 1995, the field became firmly established in biomedical research and diagnostics.

Year	Number of Peer Review Papers on Either In Situ Hybridization or Immunohistochemistry
1975	8
1980	34
1985	150
1990	5,030
1995	11,023
2000	12,216
2005	16,610
2010	13,337

SUGGESTED READINGS

Publications on in situ hybridization and immunohistochemistry in 1975

[1] Boorsma DM, Nieboer C, Kalsbeek GL. Cutaneous immunohistochemistry. The direct immunoperoxidase and immunoglobulin-enzyme bridge methods compared with the immunofluorescenceme thodin d ermatology. J CutanP athol 1975;2:294–301.

[2] Brandtzaeg P. Rhodaminec onjugates: specific and non specific binding properties in immunohistochemistry. AnnNYAc adS ci 1975;254:35–54.

[3] Hokfelt T, Fuxe K, Goldstein M. Applications of immunohistochemistry to studies on monoamine cell systems with special reference ton ervoustissu es. AnnNYAc adS ci 1975;254:407–32.

[4] Hokfelt T, Fuxe K, Johansson O, Jeffcoats S, White W. Distributiono ft hyrotropin-releasing hormone (TRH) in the central nervous system asre vealedwith immu nohistochemistry. EurJ Pharmacol 1975;34:389–92.

[5] Kemler R, Mossmann H, Strohmaier U, Kickhofen B, Hammer DK. Invit ros tudies on the selective binding of IgG from different species to tissue sections of the bovine mammaryglan d. EurJ I mmunol 1975;5:603–8.

[6] Martinez-Hernandez A, Merrill DA, Naughton MA, Geczy C. Letter:ac rylamideaf finity chromatography for immunohistochemistry. Purificationo fs pecifican tibodies. JHis tochem Cytochem 1975;23:146–8.

[7] Pearse AG, Polak JM. Bifunctionalre agents as vapour- and liquid-phase fixatives for immunohistochemistry. HistochemJ 1975;7:179–86.

[8] Pickel VM, Joh TH, Field PM, Becker CG, Reis DJ. Cellular localization of tyrosine hydroxylase byimmu nohistochemistry. JHis tochem Cytochem 1975;23:1–12.

Publications on in situ hybridization and immunohistochemistry in 1980

[9] Bauman JG, Wiegant J, Borst P, vanDu ijn P. A new method for fluorescence microscopical localization of specific DNA sequences by in situ hybridization of fluorochrome-labelled RNA. ExpCe ll Res 1980;128:485–90.

[10] Buffa R, Crivelli O, Lavarini C, et al. Immunohistochemistry of brain 5-hydroxytryptamine. Histochemistry 1980;68:9–15.

[11] Cote BD, Uhlenbeck OC, Steffensen DM. Quantitation of in situ hybridization of ribosomal ribonucleic acids to human diploid cells. Chromosoma 1980;80:349–67.

[12] Cumming R, Dickinson S, Arbuthnott G. Cyclic nucleotide losses during tissue processing forimmu nohistochemistry(letter). JHis tochem Cytochem 1980;28:54–5.

[13] de Armond SJ, Eng LF, Rubinstein LJ. The application of glial fibrillary acidic (GFA) protein immunohistochemistry in neurooncology.Ap rogressr eport. PatholRe s Pract 1980;168:374–94.

[14] Debbage PL, O'Dell DS, Fraser D, James DW. Tubulinimmu nohistochemistry.F ixation methods affect the response of spinal cord cells invit ro. Histochemistry 1980;68:183–95.

[15] Doerr-Schott J. Immunohistochemistry of the adenohypophysis of non-mammalian vertebrates. ActaHis tochemS uppl 1980;22:185–223.

[16] Dube D, Kelly PA, Pelletiar G. Comparative localization of prolactin-binding sites in different rat tissue by immunohistochemistry, radioautography,an dr adioreceptoras say. Mol CellE ndocrinal 1980;18:109–22.

[17] Engel A. The immunopathology of myasthenia gravis. IntJ Ne urol 1980;14:35–46.

[18] Gasc JM, Sar M, Stumpf WE. Immunocharacteristics of oestrogen and androgen target cells in the anterior pituitary gland of the chick embryo as demonstrated by a combined method of autoradiography andimmu nohistochemistry. JE ndocrinol 1980;86:245–50.

[19] Hata S, Endo H, Yabuuchi H. Incontinentia pigmentiac hromians(Ito). JDe rmatol 1980;7:49–54.

[20] Hokfelt T, Skirboll L, Rehfeld JF, et al. A subpopulation of mesencephalic dopamine neurons projecting to limbic areas contains a cholecystokinin-like peptide: evidence from immunohistochemistry combined with retrograde tracing. Neuroscience 1980;5:2093–124.

[21] Ibata Y, Watanabe K, Kinoshita H, etal. Dopaminean dalp ha-endorphinar e contained in different neurons of the arcuate nucleus of hypothalamus as revealed by combined fluorescence histochemistry and immunohistochemistry. NeurosciL ett 1980;17:185–9.

[22] Johansen P, Jensen MK. Enzymecytochemistry and immunohistochemistry in monoclonal gammapathyan dr eactivep lasmacytosis. Acta PatholM icrobiolS candA 1980;88:377–82.

[23] Kameya T, Tsumuraya M, Adachi I, et al. Ultrastructure, immunohistochemistry and hormone release of pituitary adenomas in relationt op rolactinp roduction. VirchowsAr ch AP atholAn atHis tol 1980;387:31–46.

[24] Kimora H, McGeer PL, Peng F, McGeer EG. Choline acetyltransferase containing neurons in rodent brain demonstrated by immunohistochemistry. Science 1980;208:1057–9.

[25] Kimura S. Pseudopyogenic granuloma: effects ofc orticosteroido nn ewly-formedve ssels. J Dermatol 1980;7:29–35.

[26] Klein C, Van Noorden S. Pancreatic polypeptide (PP)- and glucagon cells in the pancreatic islet of Xiphophorus helleri H. (Telecostei). Correlative immunohistochemistry ande lectronmic roscopy. CellT issueRe s 1980;205:187–98.

[27] Korsrud FR, Brandtzaeg P. Quantitative immunohistochemistry of immunoglobulin- and J-chain-producing cells in human parotid and submandibulars alivaryglan ds. Immunology 1980;39:129–40.

[28] Leathem A, Atkins N. Fixation and immunohistochemistryo flymp hoidt issue. J ClinP athol 1980;33:1010–2.

[29] Murao S, Horita Y, Maeda S, Sugiyama T. Genemap pingb yin s itumo lecular hybridization(author'st ransl). Tanpakushitsu KakusanKo so 1980;25:178–91.

[30] Omata M, Liew CT, Ashcavai M, Peters RL. Nonimmunologicb inding of horseradish peroxidase to hepatitis B surface antigen. A possible source of error inimmu nohistochemistry. AmJ Clin P athol 1980;73:626–32.

[31] Osamura RY, Hata J, Watanabe K, Shimizu K, Akatsuka Y. Applicationo fe nzyme histochemistry and immunohistochemistry to cytologicd iagnosis(author'st ransl). Rinsho Byori 1980;28:50–4.

[32] Rognum TO, Brandtzaeg P, Orjasaeter H, Fausa O. Immunohistochemistryo fe pithelial

cell markers in normal and pathological colon mucosa. Comparison of results based on routine formalin-an dc olde thanol-fixationme thods. Histochemistry 1980;67:7–21.

[33] Rosekrans PC, Meijer CJ, Cornelisse CJ, vand erW al AM, Lindeman J. Useo f morphometry and immunohistochemistry of small intestinal biopsy specimens in the diagnosiso ffo odalle rgy. JClin P athol 1980;33:125–30.

[34] Sar M, Stumpf WE. Simultaneous localization of [3H] estradiol and neurophysin I or arginine vasopressin in hypothalamic neurons demonstrated by a combined technique of dry-mount autoradiography andimmu nohistochemistry. NeurosciL ett 1980;17:179–84.

[35] Schauenstein K, Wick G. Quantitative immunohistochemistry. ActaHis tochemS uppl 1980;22:101–10.

[36] Scheider HM, Storkel FS, Will W. The influence of insulin on local amyloidosis of the islets of Langerhansan din sulinoma. PatholRe sP ract 1980;170:180–91.

[37] Tubbs RR, Sheibani K, Sebek BA, Weiss RA. Immunohistochemistryve rsus immunofluorescence for non-Hodgkin's lymphomas[le tter]. AmJ Clin P athol 1980;73:144–5.

[38] Tubbs RR, Sheibani K, Weiss RA, et al. Immunohistochemistryo fW arthin's tumor. Am JClin P athol 1980;74:795–7.

[39] Verhofstad AA, Steinbusch HW, Penke B, Varga J, Joosten HW. Useo fan tibodies to norepinephrine and epinephrine in immunohistochemistry. AdvB iochem Psychopharmacol 1980;25:185–93.

[40] Watanabe K, Fujisawa H, Oda S, Kameyama Y. Organc ulturean dimmu nohistochemistry of the genetically malformed lens, in eye lens obsolescence,E lo,o ft hemo use. ExpE yeRe s 1980;31:683–9.

[41] Watts G, Leathem AJ. A non-immunoglobulin link reagent for use in the unlabelled antibody method(pap)o fimmu nohistochemistry. Med LabS ci 1980;37:359–60.

[42] Yasuno H, Maeda M, Sato M, et al. Organ cultureo fad ulth umans kin. JD ermatol 1980;7:37–47.

The Basics of Molecular Pathology

INTRODUCTION

It is very helpful if you can "with your mind's eye" visualize what is actually happening inside a cell when you are doing in situ hybridization and immunohistochemistry. To be able to do that, you must first have a basic understanding of the structure of DNA, RNA, and proteins. This understanding does not have to be the comprehensive understanding you would expect in a high-level organic biochemistry course. Rather, it simply needs to focus on the components of nucleic acid's and protein's structure that impact your ability to detect it within intact cells.

THE STRUCTURE OF DNA AND RNA

The nucleotide is the basic unit of DNA and RNA. It consists of the 5 carbon ribose sugar (RNA) or 2′ deoxyribose sugar (DNA) (Figure 2-1), the base, and the triphosphate group. There are five bases: two purines (adenine and guanine, Figure 2-2) and three pyrimidines (cytosine, thymine for DNA, and, for RNA, uracil, Figure 2-3). The base is linked to carbon 1 of the ribose sugar, and the triphosphate group is linked to carbon 5 (Figure 2-3). Note that in the large DNA molecule on either strand, there will only be one free 3-OH group and one free triphosphate group (Figure 2-4). By convention, the free triphosphate group marks the 5′ end of the DNA molecule, whereas the free 3-OH group delineates the 3′ end of the molecule. When DNA polymerase synthesizes DNA, the single 3′OH group of the larger molecule links to the triphosphate group of the single nucleotide (Figure 2-4). Thus, DNA synthesis always occurs in a 5′ to 3′ direction that, stated another way, simply means that the triphosphate group of the single nucleotide is always added to the only free OH group of the larger macromolecule. Of course, the DNA molecule is double-stranded, with one strand located in the 5′ to 3′ orientation, and the other strand in the "antiparallel" direction (3′ to 5′). This results in the phosphate and ribose sugar serving as the outer backbone of the double-stranded structure with the bases facing inward, where they can link with each other via hydrogen bonds. As we discuss, the hydrogen bonds are the key concept with regards to DNA and RNA in situ hybridization.

As we will learn, hydrogen bonding is also important in understanding the interaction of proteins with each other and with other types of macromolecules. However, since proteins have many side chains that can be ionically charged, whereas DNA and RNA have only two and one such ionically charged units, respectively, on the entire macromolecule, the ionic potential will be much greater for proteins than for nucleic acids. Similarly, proteins will be able to form hydrophobic and hydrophilic pockets in regions of their sequence, which DNA and RNA macromolecules cannot do. The bottom line is that two related phenomenon will dictate the strength of the hybridization between DNA/DNA and cDNA/RNA:

1. Hydrogenb onds
2. Breathabilityo fth emac romolecule

We discuss the concept of breathability in much more detail later in this book. For now, let's simply understand the term as a way to denote the fact that hybridized macromolecules will have a tendency to separate and reattach. When in this state, they are more susceptible to complete separation. This tendency of RNA/cDNA and DNA/DNA hybrids to separate and reattach (and separate and reattach again) we call "breathability."

THE BASIC UNITS OF DNA AND RNA

Let's look at how an understanding of the structure of DNA and RNA can help us when doing in situ hybridization. First, let's discuss the phosphodiester bond. Although the phosphodiester bond is strong, high temperatures can cause a small percentage of these bonds to break in formalin-fixed, paraffin-embedded tissues. During the paraffin embedding process, prolonged exposure to temperatures at 65°C are used. If you extract the DNA from such samples and compare it to fresh, unfixed samples of the same tissues, it will be evident that the high temperatures have induced rare nicks among the double-stranded human DNA or, as shown by multiple groups, in the doubled-stranded HPV DNA. The term "nick" simply refers to breakage of the phosphodiester bond. Since the two DNA strands are perfectly matched strands in parallel and antiparallel orientation, they certainly would be able to stay hybridized if a small percentage of the phosphodiester bonds were broken, much the same way that a zipper

DOI: http://dx.doi.org/10.1016/B978-0-12-415944-0.00002-4

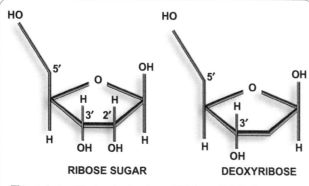

Figure2-1   The basic structure of DNA and RNA: the deoxyribose and ribose sugar. The basic "backbone" of RNA and DNA are the ribose and deoxyribose sugar, respectively. The triphosphate group and bases attach to the ribose/deoxyribosesug ars.

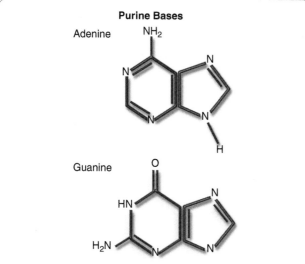

Figure2-2   The purine bases of RNA and DNA. The structures of the two purine bases found in both DNA and RNA, adenine and guanine, are depicted.

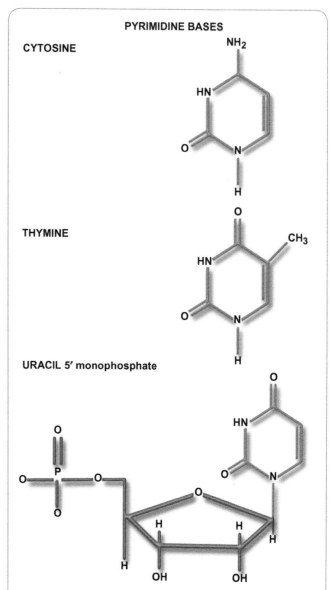

Figure2-3   The pyrimidine bases of RNA and DNA. The structures of the three pyrimidine bases found in DNA and RNA—cytosine, thymine for DNA, and, for RNA, uracil—are depicted. The latter shows the nucleotide where the base is added to carbon 1 of the ribose sugar and the triphosphate to carbon5.

would stay attached if one or two "teeth" of a large zipper had broken.

Do these nicks have any consequences for our in situ hybridization results? The answer is *no* if you are doing in situ hybridization, and *yes* if you are doing reverse transcriptase (RT) in situ PCR. In the latter process, the DNA polymerase (called rTth) is added to the amplifying solution with the unlabeled and labeled nucleotides. The polymerase will see these DNA nicks and it will "repair them" by using them as a starting point for synthesizing a new strand of DNA via its endonuclease activity. Thus, you get primer-independent DNA synthesis, which will generate a nonspecific signal. We can easily eliminate this DNA synthesis pathway by performing an RNase-free DNase incubation step prior to the RT in situ PCR. Of course, we discuss this in much more detail in Chapter 10, which deals with RT in situ PCR.

As indicated previously, the bases are *the* key component when doing in situ hybridization. They are the key component because of the ability of A = T (or for RNA A = U) and G = C to form hydrogen bonds (Figure 2-5). This is the glue that will keep our DNA probe hybridized to our DNA (or RNA) target and allow us to visualize it when we perform in situ hybridization.

For DNA, the two bases A and T will attach to each other, as will the base pair G and C (Figure 2-5). I do not know how many of you readers are (like me) old enough to remember such things as building a model train set, as well as using the building tools that attach together (such as LEGO blocks). Such sets at times included linking together units in a long chain and then using that to attach

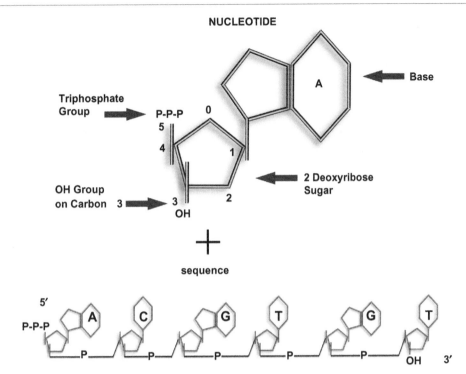

Figure2-4 The single-stranded DNA molecule. The structure of the nucleotide with a focus on its key components is depicted, where the purine or pyrimidine base is linked to carbon 1 of the deoxyribose sugar and the triphosphate group is linked to carbon 5. These nucleotides are joined via phosphodiester bonds between the free triphosphate group and the 3′ OH group to form the single-stranded DNAmo lecule.

to a similarly sized long chain with complementary pieces (Figure 2-5). In this way, you could make a much stronger "dual-linked chain" than if you used just one of the chains. This simple concept is a key part of in situ hybridization. By having a DNA or RNA molecule in the intact tissue and then introducing the complementary labeled sequence, you can create a strong double-stranded sequence that will have a strong proclivity to remain attached. And, again, we are only going to have to be concerned with two main topics when determining the forces that will allow our DNA probe and DNA or RNA target to remain attached:

1. The number of hydrogen bonds between the two strands
2. The breathability of the DNA/DNA or cDNA/RNA molecule

Although the latter is clearly affected by the number of hydrogen bonds, it will also be affected by other factors that are more dependent on the interactions of the DNA/RNA and surrounding proteins. Breathability of DNA/DNA and DNA/RNA hybrids is also much affected by the fact that the ribose (deoxyribose) sugar backbone has a strong propensity to rotate in three-dimensional space. The propensity is strong enough to overcome the forces of hydrogen bonding. Again, let's hold this discussion for later, after we review these basics of molecular biology. For now, let's realize that DNA–DNA and cDNA–RNA hybrids will be pretty easy to understand because so few forces can affect them, with hydrogen bonding and breathability being the key concepts.

HOMOLOGY

If you are like me (and old enough to remember these toys whose units joined together in long chains), you probably had quite a few of them break over time. Specifically, the small plastic knobs of the units of one chain that attached to the small plastic holes of the complementary chain would break. Little did we realize that we were working with an excellent model of the concept of *homology* that is so important for in situ hybridization. Let's assume that we had two chains that had 100 such subunits that linked together. When the set arrived, all the pieces had the knobs and the complementary small holes that were used to attach the two chains together. Over time, some knobs broke. In each case, 10/100 of the knobs are still intact. However, in one set, the 10 intact knobs are all together, whereas in the other one, they are dispersed evenly throughout the 100, with one connected link per 10 in the chain. Which of the two pairs of chains will be more difficult to pull apart? Clearly, it will be the first pair, where the 10 paired units are together, one after the other.

Figure 2-6 shows the difference between *strong* homology and *weak* homology with hybridized nucleic acids. Of course, DNA strands with 100% homology will have a strong propensity to stay attached, no matter what the stringency conditions. At 50% homology, the hydrogen bonds in the hybridized complex would still be able to keep the hybridized complex together, though nowhere near as strong as the perfectly matched two chains. However, it would matter a great deal if the base

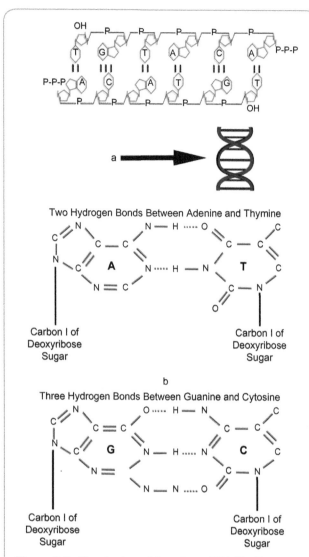

Figure 2-5   The structure of the polymer DNA. The figure depicts the DNA macromolecule, made from the joining of multiple nucleotides together in sequence, that then hybridize together in complementary chains via the matching hydrogen bonds between A and T plus G and C. Phosphodiester bonds, though strong, can be broken by physical conditions such as the high temperatures generated (60°C) during paraffin embedding, as well as certain chemicals, notably picric acid, that may be used in the pathology laboratory. Note that in the large DNA molecule on either strand, there will only be one free 3-OH group and one free triphosphate group. Thus, DNA and RNA have much less ionic potential than proteins, which typically have large numbers of positively and negatively charged side chains. By convention, the free triphosphate group marks the 5′ end of the DNA molecule, whereas the free 3-OH group delineates the 3′ end of the molecule. Also note that the free triphosphate of the macromolecule is never used for DNA or RNA synthesis; rather, only the free 3-OH group of the RNA or DNA macromolecule is used in the elongation of the molecule. This is why modified nucleotides that lack the 3-OH group (dideoxy nucleotides) can stop the synthesis of DNA or RNA. A nucleotide that lacked the 5′ triphosphate group could not stop the synthesis of DNA or RNA, as it could not participate in the phosphodiester bond. Also note that the alternating phosphate groups/ribose sugar serve as the backbone of the DNA molecule, allowing the bases to point inward and, thus, participate via hydrogen bonding with their matched nucleotide (A and T or U plus G and C).

pairs with the 50% homology were together or dispersed throughout the chain. Why in nature would matched base pairs tend to be together? The most likely reason is that we are looking at consensus sequences between two similar organisms (or viruses). One simple example would be comparing two HPV types. By definition, two distinct HPV types must have less than 50% homology. However, consensus regions within the 8000 base pair sequence between two different HPV types may show much stronger homology, and this would typically manifest itself as complementary base pairs that cluster together. For example, if you look at the entire 8000 base pair sequence, then HPV 16 and HPV 51 share only about 27% homology. However, in certain regions of their genome, such as in the E6/7 region, the homology can be over 75%. We can use this simple fact to use a probe against one specific DNA or RNA sequence to detect similar related DNA or RNA sequences. Finally, if there is poor homology between two different DNA or RNA sequences, represented here by 10% homology, then it is very unlikely that we will ever be able to keep them attached during the in situ hybridization process.

KEY TERMS IN MOLECULAR BIOLOGY

THE HYDROGEN BOND

The physical force that keeps our knob/hole chains linked together is simply the resistance/friction of the small knob as it sits in the correctly sized hole. If we apply a larger counter force, the bond is broken. The equivalent force that holds together matched nucleotides is the hydrogen bond. The nucleotide match A–T has two hydrogen bonds (or A–U for RNA), whereas G–C bonds are 50% stronger, since they form three hydrogen bonds. What forces will tend to break hydrogen bonds? One is simple heat. Certain chemicals can displace the hydrogen bond between the base pairs, for the simple reason that they can form hydrogen bonds and thus displace the relatively weaker G = C and A + T hydrogen bonds. Two common examples of such chemicals include formamide and urea.

Thus, one way to separate hybridized nucleic acids is to use a wash (or hybridization) solution that tends to disrupt hydrogen bonds. However, there is still one more force that helps keep hybridized nucleic acids together that we can manipulate to our advantage. This is based in the fact that nucleic acids are weak acids, with a negative charge. Higher salt concentrations tend to "cloak" these negative charges that want to repel the two hybridized strands. Hence, if we lower the concentration of the salt in the solution, we tend to make hybridized nucleic acids less likely to stay together. Understand that this force is much weaker than the cumulative force of the hydrogen bonds. We can use this to our advantage when the hybridized molecules are the probe and a sequence of DNA/RNA that has poor homology by simply lowering the concentration of salt in the hybridization solution. Still, although lowering the salt concentration can help denature the probe from nontarget DNA or RNA sequences, let's remember that the *primary* force that keeps DNA probe/DNA target and DNA probe/RNA target molecules together is the hydrogen bond.

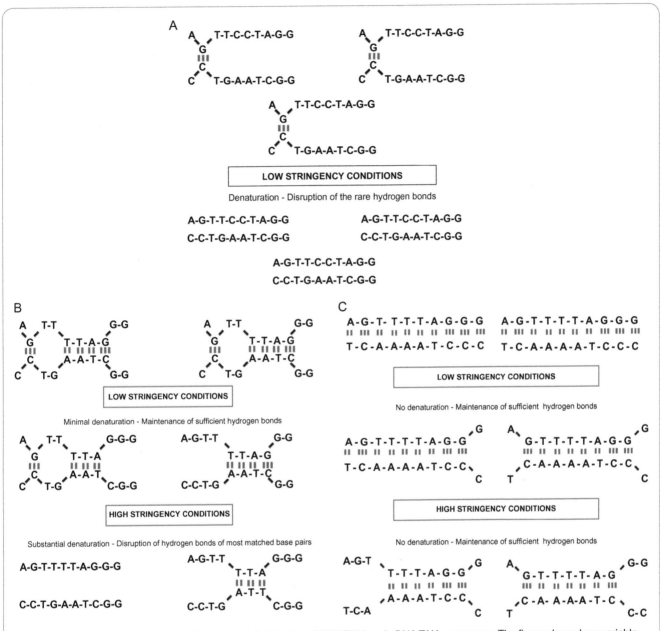

Figure2-6 The importance of homology to the hybridization of DNA/DNA and cDNA/RNA sequences. The figure shows how variable homology will affect the ability to hybridize under different stringency conditions. Note that even hybrids with 100% homology will show disruption of matched base pair during high stringency. However, 100% homology will allow the remaining hybridized base pair to keep the macromolecule together at high stringency, whereas 50% homology would only allow for persistent hybridization due to the remaining base pairs that are still attached at low stringency conditions. DNA molecules that share 10% homology would not stay hybridized even atl owst ringencyc onditions.

We also need to remember that we are talking about a DNA–DNA hybridization reaction where the DNA molecules are surrounded by a three-dimensional protein/protein cross-linked network. This means that other forces will come into play. For example, several protein side chains can form hydrogen bonds with the DNA or RNA target and the DNA/cDNA probe. Several protein side chains can also be positively or negatively charged, and thus either attract or repel the DNA/RNA macromolecule. Further, even in fixed tissues, DNA molecules will have a tendency to separate due to the strong propensity of the deoxyribose sugar to rotate in three-dimensional space. This leads us to a discussion of the "breathability" of the macromolecules.

THE "BREATHABILITY" OF THE HYBRIDIZED COMPLEXES

As an anatomic pathologist, I have the occasion to look at bone biopsies from time to time. I remember thinking how static bony tissue appears to be. Nothing could be further from the truth. Bone tissue constantly remodels itself, which is quickly evident if a person is not getting sufficient calcium or Vitamin D in his diet.

DNA–DNA and DNA–RNA hybridized molecules are the same way. They are not like LEGO blocks that remain static in space and simply bind to each other if there is sufficient homology. DNA–DNA and RNA–DNA hybrids breathe! That is, in the normal process of in situ hybridization, the DNA probe and DNA target, even if there is 100% homology, can partly open and close, much like moving a zipper on your coat up and down. And, like the zipper on your coat, if you move it too far, the coat can open. In the same way, DNA–DNA complexes with strong homology can separate even if the wash conditions are not too stringent.

Let's put on our "mad scientist" hats for a while and think of ways we might be able to block such "unzippering" of the DNA probe and target during in situ hybridization. One way would be to surround the DNA target with a protein "coat" that would not allow the DNA strand to move in space. This probably happens to some degree with formalin fixation. Of course, one runs the risk with the three-dimensional protein–DNA cross-linked network after formalin fixation of creating such a tight web around the DNA target that our probe (and other in situ hybridization reagents) cannot access the DNA target molecule. This may necessitate protease digestion or "DNA retrieval."

In nature, of course, the ability of the DNA–DNA double-stranded molecule to breathe is built into the structure of the DNA. It is essential that this can happen when the DNA needs to be synthesized when the cell divides or, of course, when a region of the DNA has to be made into RNA. Stated another way, the backbone of DNA and RNA, the ribose sugar, is able to move in three-dimensional space very well. We can think of the ribose/deoxyribose sugar backbone of RNA/DNA much like a hinged joint, capable of easily rotating in space when the conditions are favorable for this. Thus, if we modify the ribose/deoxyribose sugar, we probably would make the sugar much less able to rotate in three-dimensional space and have the effect of "locking the joint" so it simply can't move. One way to do this is to attach a bridge between carbon 4 and OH-2. This bridge has the effect of locking the nucleotide in place, so the probe made from such nucleotides that contains these modified nucleotides is rigidly held in place and cannot "unzip." We have just described locked nucleic acid probes (LNA probes), which will become a key part of our strategy for performing in situ hybridization. Their success reminds us just how important the breathability of DNA–DNA and RNA–DNA hybrids is to success with in situ hybridization!

DENATURING AND ANNEALING: SIGNAL AND BACKGROUND

Let's summarize what we have discussed thus far and introduce a few new terms. The two complementary DNA strands will be in different orientations. If we are working with RNA, then the probe will be generated to be complementary to the RNA molecule. Since in most cases the probe is a DNA sequence, it is referred to as

a *complementary DNA molecule*, or, more commonly, cDNA. The 5' end and 3' end represent these mirror image strands. Newly synthesized DNA or RNA is made from the 3' end. Stated another way, in newly synthesized DNA or RNA, the triphosphate group of the single nucleotide is added to the free 3-OH group on the much larger macromolecule. This is why it is so easy to terminate DNA or RNA synthesis. One simply needs to use a nucleotide (such as a dideoxynucleotide that lacks the 3-OH group) and, when this is added, no more nucleotides can be added to the DNA molecule. The two complementary nucleic acids will have their bases pointing in the same direction, toward the center of the two strands. In this way, base pairs with A–T (or A–U) will attach, as will G–C base pairs. The ribose/deoxyribose sugar that serves as the backbone of the macromolecule is capable of freely and profoundly moving in three-dimensional space, which no doubt is a critical evolutionary-based ability as this is essential to both DNA replication and RNA synthesis. The percentage of base pair matches in the two strands is referred to as *homology*. If there is sufficient homology, the two strands will have a strong propensity to remain attached, due to the strength of the large number of hydrogen bonds. This is especially true if the matched base pair are together as compared to being dispersed throughout the hybrid. The attachment of two complementary strands of DNA (or RNA and cDNA) is more commonly known as either *annealing* or *hybridization*. If we apply forces that tend to disrupt hydrogen bonds (or make the weak negative charge of each strand want to repel each other), then we are increasing the *stringency* of the reaction. Under high stringency conditions, only hybridized pairs with strong homology will remain hybridized. The hybridized strands with poor homology will separate, or, as it is more commonly called, *denature*.

We can introduce a known strand of DNA or RNA in order to determine if the complementary strand is in the tissue of interest. We usually will label the known strand with tagged nucleotides. The most common tags are biotin, digoxigenin, and fluorescein. At this stage, most are visualized via a colorimetric reaction. Of course, we can visualize fluorescein directly using a specialized (and expensive) microscope called a dark field microscope. However, I prefer to use an anti-fluorescein-alkaline phosphatase (or peroxidase) conjugated complex when using fluorescein tagged probes, which will lead to a color-based signal that can be seen with the standard light field microscope. The labeled DNA/RNA sequence is called the *probe*, whereas the complementary sequence in the tissue is called the *target*. DNA is double-stranded, and RNA is single-stranded. Hence, we must denature the DNA target in the tissue first if we wish to hybridize it to the complementary strand in our probe. Of course, the ultimate goal with in situ hybridization is to visualize the probe/target complex. The color produced from the probe/target complex is referred to as the *signal*, whereas any color produced from the probe and nontarget molecules is referred to as *background*.

THE MELTING TEMPERATURE CURVE

It is time to get a bit more quantitative. Let's assume we have 100 DNA–DNA hybrids, and that they have 100%

homology. For purposes of this illustration, let us assume that each hybrid is 100 nucleotides in length. We decide to do a simple experiment whereby we slowly increase the temperature of the hybridization reaction and measure the number of DNA–DNA hybrids that have denatured. A representative set of data is provided in Figure 2-7. Note that at 70°C one-half of the hybrids have denatured. Thus, by definition, the other one-half are still hybridized. We refer to this graph as a *melting curve*. The temperature at which one-half of the hybrids are denatured is called the *melting temperature*, or Tm for short. Clearly, the more strongly attached two hybridized DNA or RNA/DNA complexes are, the "further to the right" will be the graph of the melting temperature.

For hybridized DNA/DNA hybrids, what will be the main determinant of the melting temperature? Let's keep this discussion simple, because, at its heart, the ideas are pretty straightforward (even if the formulae are rather complex!). Clearly, the key issue will be the number of hydrogen bonds and whether they are located close to each other, or randomly dispersed throughout the hybridized complex. Of course, if the DNA hybrids are the same size, then the degree of homology will be the key factor in determining the Tm value, as seen in Figure 2-7. But are there other factors we can use to increase the likelihood that DNA/DNA and DNA/RNA hybrids stay annealed? The answer is *yes*, and the key concept is breathability!

Remember, even 100% homologous base pairs will tend to breathe and spontaneously denature due, in part, to the deoxyribose's sugar propensity to move in three-dimensional space. If the hybridized complex was 1000 base pairs, then the odds that the hybridized complex would completely denature are remote. But what if the DNA–DNA complex was only 20 base pairs? Then it would be easy to imagine that the breathability of the hybridized complex could well allow it to denature, which, in turn, would decrease the melting temperature of the complex. This is illustrated in Figure 2-8, where the Tm of a 20 base pair hybridized complex is only 29°C.

We now need to remember an important practical issue regarding DNA and cDNA probes. They tend to be small, usually around 20 nucleotides in size. Certainly, this was the case 10–20 years ago. It is true that in the past few years it has become simpler to make larger probes of up to 100 nucleotides, but these probes are more expensive. Further, as we will see in more detail in Chapter 10 on in situ hybridization, 100 nucleotides can be a problem for DNA and cDNA probes, as this large size can make it physically more difficult for probes to enter the cell and get to the (typically) nuclear target. However, as anyone who did in situ hybridization 10–20 years ago can attest, DNA probes that are 20 nucleotides in size gave poor signal and high background. The reason they gave poor signal is clear in Figure 2-8. With a melting temperature typically in the 20°C–30°C range, the probe–target complex tended to denature even at room temperature. The high background typical of these unmodified oligoprobes simply reflects the fact that to counteract the low signal, we were often forced to use high concentrations of the probe, which, in turn, is strongly correlated to background.

Enter the LNA modification. The importance of the breathability of DNA–DNA and DNA–RNA hybrids is

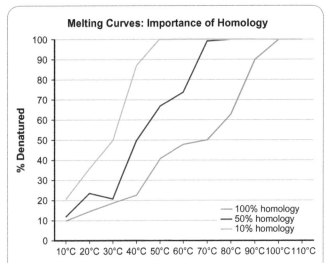

Figure 2-7 The melting curve for hybridized DNA: Part one—the importance of homology. In this standard melting curve plot, the percentage of hybridized DNA complexes were plotted versus temperature for DNA hybrids with varying homology. Note the "S-shaped" curves where the temperature at which half the DNA strands are separated (denatured) shows a much higher number for the 100% homology situation (70°C) as compared to the 50% homology situation and the 10% homologyhyb rids.

well underscored by just how effective the LNA modification is in "shifting the melting curve" to the right. By bridging the 2-O and the 4-C molecules, we do the equivalent to a DNA–DNA or DNA–RNA hybrid of putting a "block" on the wheel of a car if it is parked illegally. The hybridized complex starts to become locked in place, unable to move in three-dimensional space. Here is a key point. If we add six such LNA modified nucleotides to a 20-nucleotide probe, then the melting temperature of the probe/target complex will increase by about 45°C! This is illustrated in Figure 2-8. Note the key end result, which is so important for when we do in situ hybridization. The Tm of the 20 base pairs complex, which includes at least six dispersed LNA modified nucleotides, is equivalent to a 1000 base pairs complex with 100% homology! This will make it relatively easy for us to generate a strong signal using a low concentration probe that, in turn, will result in low background.

PROBE CONCENTRATION AND PROBE SIZE

There are many formulae that can be used to calculate the melting temperature of the DNA–DNA and DNA–RNA complexes. One reason there are many formulae is that the concept is quite complicated, with variable thermodynamic states that exist that are different when comparing completely denatured, partly denatured, and completely hybridized complexes. Following is a formula that I have found useful in my work for Tm:

$$Tm = 8\,15°C + 16.6\log[Na]\,(as\ molarity)$$
$$+ 0.41(\%\,G + C\,content) - 0.61(\%\,formamide)$$
$$- 500/\#\,base\,pairs\,in\,the\,complex$$

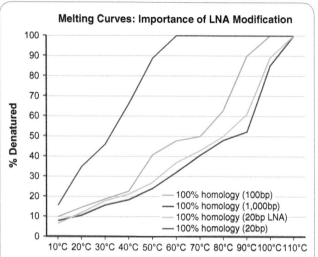

Figure2-8 The melting curve for hybridized DNA: Part two—the importance of the size of the hybrids and the LNA modification. In these experiments, the Tm is calculated for DNA hybrids that all show 100% homology but vary in size. As expected, since the larger DNA hybrids have far greater numbers of hydrogen bonds, they show a higher melting temperature. Note the very low Tm for the 20 base pairs hybrid despite its 100% homology. This reflects, in part, the ability of the ribose sugar to move in three-dimensional space. If we modify 6 of the 20 nucleotides with the LNA modification and then redo the experiment, it is evident that the LNA-modified nucleotides in the 20 base pairs hybrid dramatically increase the Tm of the reaction to the equivalent of the 1000 base pairs hybrid that does not have the LNA modification. This allows us to use these very small oligoprobes (20 nucleotides) yet still have the very high Tm needed for optimal detection of the target with in situ hybridization.

First, it is clear why both the G/C content and the percentage of formamide as well as sodium concentration are part of the equation. As we know, the G–C base pairs have 50% stronger hydrogen bonds than the A–T base pairs. Thus, DNA/DNA complexes with 60% G–C base pairs will have a higher melting temperature when compared to the same size annealed complex with a 40% G–C content. Indeed, when we do the math and make certain assumptions (no formamide, low salt, 100 base pair complex), then the annealed complex with the 60% G–C content will have a melting temperature that is 9°C higher than the one with the 40% G–C complex. Although DNA and RNA sequences can vary a lot in their G–C content, in most cases the end result is a G–C content of about 50%. Thus, this tends not to be an important factor for probes, but does tend to be a factor for the small oligoprobes and oligoprimers that we routinely use for in situ hybridization (with the LNA modification) or in situ PCR.

Since formamide can avidly break hydrogen bonds, it again makes sense that formamide in the solution will decrease the melting temperature. Specifically, if we use a stringent wash solution (or a probe cocktail) with 50% formamide, then the melting temperature will decrease by 30°C. I do not like to use formamide in either the stringent wash or the probe cocktail because: (1) it is a

hazardous chemical; (2) it is relatively expensive; and (3) although its use tends to reduce background a lot, it also tends to have a strong negative effect on the signal.

Let's get back to the probe size. The result on the melting temperature of increasing the probe/target complex from 100 base pairs to 1000 base pairs is seen in Figure 2-8. The increase in the Tm can be calculated from the last part of the equation (assuming the same G–C content) and, of course, that the other conditions such as sodium and formamide concentration are the same. As you see, the increase from a 100 to a 1000 base pair complex increases the melting temperature by only 4.5°C! When we add to that calculation that 1000 base pair probes would have a very difficult time getting into the cytoplasm or nucleus of the cell, it becomes clear why we are going to focus our attention on improving the melting curve features of small probes and not worry about larger probes. From 1990 through 2005 or so, small probes meant around 50 base pairs. Today, small probes mean 20 nucleotides in size, thanks to the LNA modification. It is not possible in this book to overstress the importance of the LNA modification of the DNA (or cDNA) probe to our results with in situ hybridization. So let's look at this point in another way to highlight this key concept. Going back to our melting curve formula, what is the change in the Tm if we use a probe that is 20 nucleotides is size? Whereas the change from a 1000 base pair complex to a 100 base pair complex decreased the Tm by only 4.5°C, the reduction from a 100 base pair to a 20 base pair complex decreases the Tm by 25°C! Such a change typically means a loss of signal. This is illustrated in Figure 2-9, where I compare the data for the in situ hybridization analysis of miR-221 using a 20 nucleotide probe that is labeled with digoxigenin versus the same size probe with the same 5′ digoxigenin label, but where 6 of the 20 nucleotides have had the LNA modification. The tissue in panels A–C is a leiomyosarcoma. It contains a much increased amount of miR-221 when compared to either normal smooth muscle of the uterus or the common benign smooth muscle of the uterus, called a leiomyoma. If the LNA modification is not used, no signal is evident (panel C). This is clearly a false-negative reaction, since we know that miR-221 is present in high copy number in these tumor cells. The LNA modified miR-221 probe is associated with a strong signal in this malignant tumor (panel A). This figure also introduces us to some other concepts and observations with in situ hybridization that are discussed in length in other chapters:

1. The importance of pretreatment with in situ hybridization. The miR-221 signal required strong protease digestion prior to the in situ hybridization. If the protease digestion step was not done, then no signal was seen (panel B).
2. Most tissues show an internal control. The internal control can be a positive or negative control. These are very important in the interpretation of the data. Note in panel A that the signal is strongly positive in the cancer cells (large arrow) but absent in the surrounding benign stromal cells (small arrow). This gives us more confidence in the specificity of the reaction. A strong foundation in histopathology

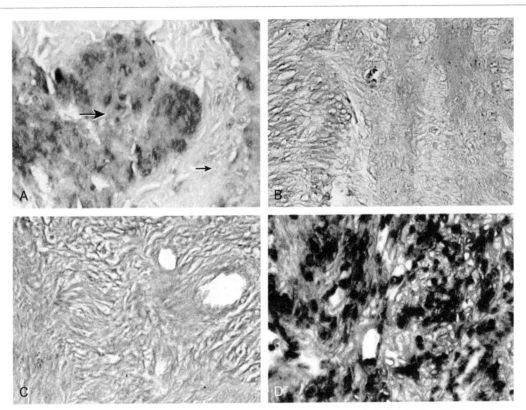

Figure2-9 The importance of the LNA modification to the oligonucleotide probe for in situ hybridization. The tissue is a leiomyosarcoma that shows a strong upregulation of the oncomiRNA-221. Note the strong signal with this LNA modified probe and how it localizes to the cytoplasm and nuclei of the malignant cells (large arrow) and not the adjoining benign cells (small arrow). The signal is lost if either the protease digestion is omitted (B) or the probe does not have the LNA modified nucleotides (C). (D) The results for a benign leiomyoma, in which miR-221 was not upregulated, where the probe concentration was increased 50-fold. This is background as defined by the strong colorimetric reaction in cells that are known to not have the miR upregulated. Also note the primarily nuclear-based localization, which is another clue that this is background, as miR-221 typically localizes to the cytoplasm and nucleus (A).

interpretation will help a great deal in finding these internalc ontrols.

3. Signal and background are very much dependent on the probe concentration. Note that in panel D we can generate a signal with the miR-221 probe in the benign leiomyoma, even if we know it does not contain much of this oncomiRNA if we dramatically increase the concentration from 1 pmole/microliter to 50 pmoles/microliter.

Of course, we cannot explain all the results with in situ hybridization by relying on just the melting temperature curves. Ultimately, the signal with in situ hybridization reflects the *density* of the labeled nucleotide over the target sequence of interest. Thus, if we can increase the amount of the labeled (often called the reporter) nucleotide in the probe/target complex, the signal will increase even if the melting curve has not changed. Of course, one way to do this is to simply increase the concentration of the probe. However, as we just saw in Figure 2-9, increase the probe concentration also tends to increase background. One way many companies have tried to get around increasing the probe concentration yet still increase the density of the label in the probe/target complex is to use what is sometimes referred to as in

situ hybridization or immunohistochemistry "branching trees." These are methods whereby we can, via biochemical steps, add many more (for example) biotin molecules per probe–target complex surface area. Perhaps the best known of such methods is the TSA.

Tyramide Signal Amplification (TSA) is based on the observation that, in the presence of hydrogen peroxide, the peroxidase that is conjugated to the streptavidin, which, in turn, is connected to the biotin (that, in turn, is attached to the DNA or RNA probe—hence, the many branches of a tree), converts the chemical tyramide into a highly reactive intermediate that, in turn, binds to the streptavidin (or protein epitope in immunohistochemistry). So we get many more streptavidin/peroxidase conjugates in the "branching tree." The tyramide also contains a dye that is converted by the peroxidase to a dark precipitate that we can see under the microscope. The large amplification of signal, unfortunately, has associated with it a high propensity for background. Attempts to increase the signal in either immunohistochemistry or in situ hybridization by increasing the density of the reporter nucleotide or detection molecule are often associated with background. This may be due to the very nature of the amplification step, since if it occurs away from the probe–target complex, it will, by definition, produce

15

a strong color. It is typically a more effective strategy to increase the dynamics of the binding of the probe to the target complex when trying to increase the signal-to-background ratio.

Let's go back to the melting curve again. It will be useful to look at the effects of varying homology on the melting curve, as this is important in our understanding of the theoretical basis of the stringent wash. Let's use a 100 base pair hybrid that has only 10% homology. How will this affect the melting curve? Clearly, since the homology is so much less than 100%, the hybrids will be much more easily coaxed to denature at lower temperatures. So, the melting curve will "shift to the left." This is evident in Figure 2-7. It is clear from the figure that you would have to do the in situ hybridization reaction at very low temperatures to see a signal if there was only 10% homology. You can take advantage of this fact if you are looking to detect a novel DNA sequence that you know is related to a known DNA sequence for which you have the probe. A good example of this is human papillomavirus (HPV). By definition, any new HPV type can share no more than 50% homology with all the other known already-characterized types of HPV. However, the less than 50% homology applies to the entire genome. Within that genome, you will find regions of relatively high homology that will be over 50%. Specific examples of this are HPV 16 and HPV 51. Thus, by purposely using low stringency conditions and an HPV 16 probe, you should be able to detect the HPV 51 genome in the infected cells. However, if you go to a high stringency wash, then the signal is lost. Thus, here we have a practical example where you can use varying stringencies to identify novel DNA or RNA sequences using in situ hybridization by using a probe that shares some homology with the novel sequence. From the melting curve, it is clear that you would probably need at least 25% homology to have a good chance of visualizing the novel sequence.

We can also use the variable hybridization with different stringencies to determine the type of HPV in a specific lesion using in situ hybridization. Figure 2-10, panel A, shows a CIN 1 lesion (cervical intraepithelial lesion) that was strongly positive for HPV DNA using a probe cocktail that can detect HPVs 31, 33, 35, and 51 at low stringency; this cocktail is made by Enzo Life Sciences. Note the results if we test serial sections of the CIN 1 lesion at high stringency for HPVs 31, 33, and 35 (panels B, C, and D, respectively). Only the HPV 31 signal gives an intense signal equivalent to that seen with the probe cocktail at low stringency (no signal was evident with the HPV 51 probe as well). Hence, the lesion is clearly HPV 31 positive. Similarly, a signal was seen with both the HPV 16/18 and HPV 31/33/35/51 probe cocktails from Enzo Life Sciences at low stringency in a vulvar intraepithelial grade 2 lesion (data not shown). When serial sections for this lesion were analyzed for HPV 16, 18, 31, 33, 35, and 51 at high stringency, a signal was seen only with the HPV 16 probe (panel E); panel F show the negative results typical of the other individual probes at high stringency, in this case for HPV 31.

Let's now shift to a more practical consideration of the melting curve, and that is background. We will use our HPV example again. What would we predict would be the degree of homology between HPV DNA and human DNA? Evolutionarily speaking, it would be difficult to consider two sequences of DNA more likely to be disparate. This is the case the degree of homology between HPV DNA and the human genomic sequence, on average, is less than 1%. Going to our melting curve, we can easily see that 1% homology would not be capable of causing a signal with in situ hybridization. Why, then, do we sometimes see background when using an HPV probe and analyzing human biopsies by in situ hybridization? This is an important practical question for in situ hybridization, because it gets to the heart of an issue that we must control if we are to get easy-to-interpret and good data. There are two possible considerations:

1. The problem may be based in the melting curve in the sense that the background is due to HPV DNA probe binding to human DNA targets that share about 1% homology. That is, even though the homology may be very low at 1%, the fact that there are billions more of the human sequences (using an average of 8000 base pair to directly compare it to the HPV genome of this size) still will produce enough HPV DNA–human DNA hybrids to see a signal at temperatures between room temperature and 60°C.
2. Background per se may have nothing to do with the unwanted hybridization between HPV DNA and human DNA. Rather, it may reflect unwanted nonspecific binding of the HPV DNA with other macromolecules in the tissue.

Let's see what the data say about point 1 versus point 2. If point 1 were correct, then simply raising the temperature of the stringent wash would eliminate the background. As an aside, you could raise the temperature of the hybridization step. I prefer to keep the hybridization parameters constant and alter the stringency conditions. Decreasing the concentration of the probe may also help reduce background in this situation. However, altering the pretreatment conditions should have no effect on the background if the primary cause is the binding of HPV DNA with human DNA sequences with which it shares 1% homology.

If point 2 were correct, then altering the stringent conditions or probe concentration probably would not affect background. If the problem reflected nonspecific "stickiness" of the probe to the tissues, macromolecules, then they probably would involve other forces, such as hydrophobicity, that would not respond to stringent condition variation. Rather, it may respond to using pretreatments such as proteinase digestion or antigen retrieval since they can, of course, alter the physical characteristic of the macromolecules within a cell and make them less likely to bind nonspecifically to themselves. Of course, a similar discussion could be made about immunohistochemistry since the nonspecific binding of the primary (or secondary) antibody to the tissues macromolecules could also create background.

Figure 2-11 shows an example of background being dramatically altered by keeping all experimental conditions constant (e.g., primary antibody concentration, hybridization time, stringent wash conditions, etc.) and varying only the pretreatment conditions. Panels A and B

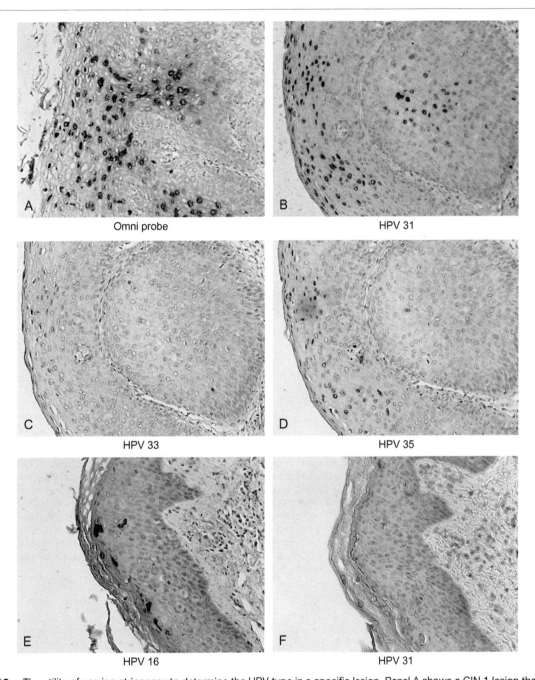

Figure2-10 The utility of varying stringency to determine the HPV type in a specific lesion. Panel A shows a CIN 1 lesion that was strongly positive for HPV DNA using a probe cocktail that can detect HPVs 31, 33, 35, and 51 at low stringency. Note the results if we test serial sections of the CIN 1 lesion at high stringency for HPVs 31, 33, and 35 (panels B, C, and D, respectively). Only the HPV 31 signal gives an intense signal equivalent to that seen with the probe cocktail at low stringency (no signal was evident with the HPV 51 probe as well). Hence, the lesion contains HPV 31. Similarly, a signal was seen with both the HPV 16/18 and HPV 31/33/35/51 probe cocktails from Enzo Life Sciences at low stringency in a vulvar intraepithelial grade 2 lesion (data not shown). When serial sections for this lesion were analyzed for HPV 16, 18, 31, 33, 35, and 51 at high stringency, a signal was seen only with the HPV 16 probe (panel E); panel F shows the negative results typical of the other individual probes at high stringency, in this case for HPV 31.

show the results with immunohistochemistry for the epitope to Iba-1. Iba-1 is an actin-binding protein that is important in macrophage activation and function. It plays other roles in inflammation, such as in the proliferation of blood vessels and lymphocyte migration. It is a cytoplasmic protein and, based on its functions, is easily found in an inflamed tonsil. Note the strong cytoplasmic

signal for Iba-1 after immunohistochemistry with no pretreatment (panel A). If you do a 4-minute protease digestion prior to immunohistochemistry, the signal is drowned in a sea of background (panel B); note the high background in the nuclei. Panels C and D show the results for immunohistochemistry analyzing for CD68, which is the classic marker for activated macrophages.

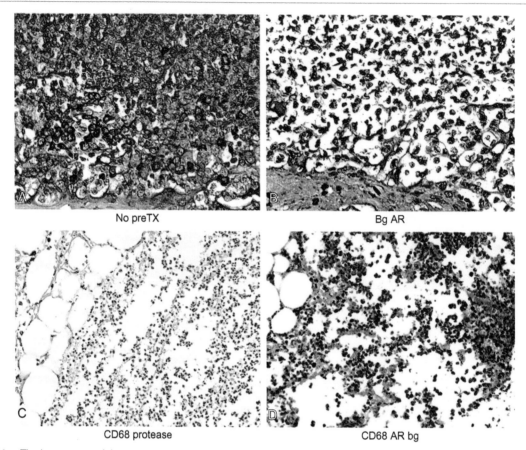

Figure2-11   The importance of the pretreatment conditions to background with immunohistochemistry and in situ hybridization. Although, on a theoretical basis, hybridization of the probe to nontarget nucleic acid sequences through weak base pair matching and the primary antibody to nonepitope protein targets can cause background, other factors likely play a much more important role. Panels A and B show the results for immunohistochemistry in a tonsil that should contain many cells that express the cytoplasmic protein Iba-1. Note that no pretreatment leads to an intense signal and no background (panel A). However, simply adjusting the pretreatment conditions to 4 minutes of protease digestion and keeping all other conditions constant, including the stringent wash, led to very high background (panel B). Similarly, the lesion in panel C contains mostly neutrophils and lymphocytes with a few scattered macrophages in the upper left area that show a strong signal for the macrophage marker CD68 after a protease digestion (panel C). If you change the pretreatment from protease digestion to antigen retrieval (panel D), then intense background is evident.

As we will see in Chapter 6, we can differentiate macrophages from other inflammatory cells such as lymphocytes and neutrophils based on their larger size, ample cytoplasm, and bean-shaped nucleus. Panel C shows a few such cells toward the upper left part of the panel that show the intense red stain indicative of the CD68 protein. Most of the cells in this biopsy are lymphocytes and neutrophils that do not show a signal for CD68. Panel C was the result after protease digestion. Note the dramatic increase in background when all conditions are kept the same except for the pretreatment condition that was changed from protease digestion to antigen retrieval (panel D). Thus, here are two good examples where background can be much affected by conditions that most likely have little to no effect on the forces that keep the primary antibody hybridized to the epitope (that is, stringency conditions).

This is *not* to say that altering the probe concentration or stringency conditions can affect background. However, the fact is that altering the stringency conditions, in my experience, for both immunohistochemistry and in situ hybridization, although it can affect the signal, *rarely* will affect the background.

There is another very simple experiment you can do to get more data to support the notion that background with in situ hybridization and immunohistochemistry is due primarily to the nonspecific binding of the probe to macromolecules in the tissue. This is an important point, as we use it later in the book in the sections on in situ hybridization and immunohistochemistry to get rid of the background. The key point is based on the fact that the albumin we use in the stringent wash has the ability to "stick" nonspecifically to the macromolecules in the tissue and, thus, displace any probe that may be nonspecifically bound to these cellular components. Let's do a reaction whereby we omit the albumin (used in the form of bovine serum albumin, or BSA for short) from the wash solution and see if background is affected. As seen in Figure 2-12, the omission of BSA from the stringent wash solution causes an incredible increase in the background in the in

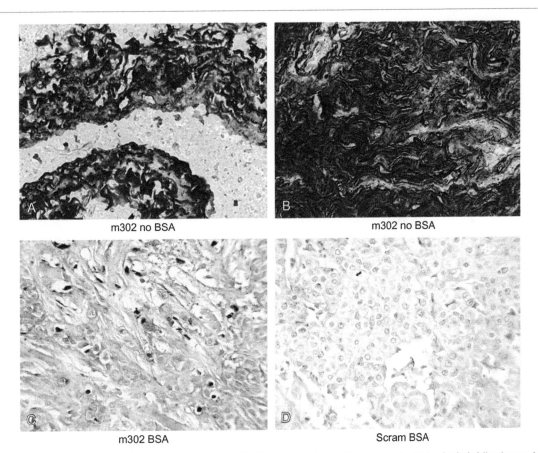

m302 no BSA

m302 no BSA

m302 BSA

Scram BSA

Figure2-12 The importance of the bovine serum albumin (BSA) to the blocking of background with in situ hybridization and immunohistochemistry. In this simple set of experiments, a mouse embryo was tested for the stem cell related microRNA-302. Embryonic tissues should show occasional cells expressing this stem cell miRNA. In one set of slides, the stringent wash contained BSA, whereas the conditions for the other set of slides were identical except that the BSA was omitted from the stringent wash. Note the strong background evident if the BSA was omitted from the stringent wash (panels A and B). The strong intensity of the colorimetric reaction and the fact that it involved all the different cell types is certainly consistent with background. In comparison, if the same tissue was also tested for microRNA-302 (panel C) but where BSA was included in the wash, then the background is gone and a few cells show an intense target-specific signal. The specificity of the reaction is evident from panel D, where no cells were positive if the scrambled probe was used in conjunction with a BSA containing wash.

situ hybridization reaction. Here is a good example of the theoretical helping us with the practical. The preceding data clearly show us that we should attack background from a nonspecific adsorption of the probe with tissue macromolecules, and not a probe-non-DNA target or primary antibody-nonepitope target hybridization event.

THE HYBRIDIZATION SIGNAL WITH IN SITU HYBRIDIZATION IS MORE THAN THE MELTING CURVE: THE EFFECT OF TARGET COPY NUMBER

Before we leave our discussion of the melting curve and our DNA–DNA as well as cDNA–RNA hybrids with in situ hybridization, let's remind ourselves that this theoretical consideration explains *part* but not *all* of the signal we see with our experiments. Rather, the key to the signal with in situ hybridization is the *density* of the reporter nucleotide per probe–target complex. This can be condensed to several key points that are illustrated in Figure 2-13. There are five key components that relate to the density (or, if

you prefer, the total numbers of the reporter nucleotide per probe–target complex):

1. The concentration of the target (or, as it is sometimes called, the copy number of the target, which we cannot vary unless we do in situ amplification)
2. The concentration of the probe (which we, of course, canvary)
3. The density of the reporter nucleotide on the probe (which, for a full-length probe, is set at about 1 label per 20 nucleotides and which, for an oligoprobe, is typically 1 or 2 tags per probe)
4. The accessibility of the probe to the target (which will be dependent, in turn, on the pretreatment and the fixation conditions)
5. The strength of the forces between the probe and target (which will clearly relate to factors such as homology, LNA modification, and stringent conditions)

Before we examine these components in detail, let's realize that ultimately it is the density of our reporter *enzyme* that determines the strength of the in situ

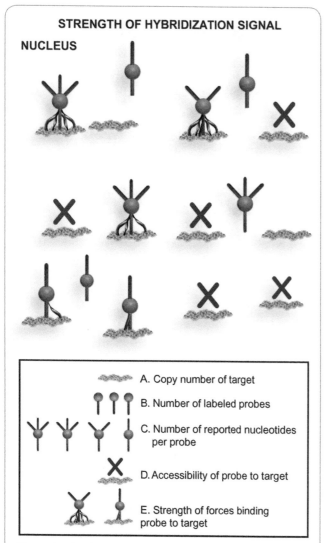

STRENGTH OF HYBRIDIZATION SIGNAL

NUCLEUS

A. Copy number of target

B. Number of labeled probes

C. Number of reported nucleotides per probe

D. Accessibility of probe to target

E. Strength of forces binding probe to target

Figure 2-13 Schematic representation of the key variables that determine the strength of the hybridization signal with in situ hybridization. The model system is HPV; hence, the signal is seen in the nucleus. Signal intensity can be affected by the HPV DNA copy number, the concentration of the labeled probe, the number of biotin or digoxigenin molecules per HPV genomic probe, the accessibility of the probe to the target, and the strength (that is, numbers) of hydrogen bonds between the HPV probe and target.

signal. In molecular pathology jargon, the amount of the target is usually referred to as the *copy number* of the target.

Figure 2-14 shows the importance of target copy number and probe concentration, respectively, to the strength of the signal with in situ hybridization. Indeed, panels A and B that each represent the detection of HPV DNA by in situ hybridization make us aware that basically every HPV in situ hybridization experiment reminds us of the importance of target copy number to the signal. The reason is that HPV DNA shows a dramatic variability in copy number in the upper part of the lesion. In brief, HPV DNA starts its infection as a very low target copy event in the basal cells, especially in the area of the transformation zone of the cervix. This is why the transformation zone is the site of about 95% of CIN lesions and cervical cancer. The very low copy HPV DNA in the basal cells requires PCR in situ-based methods for routine detection. As the viral infected cells move toward the surface, they mature into keratin-making cells. As they do this, in some cells the viral DNA proliferates from one copy to a few copies. In other cells, especially those with perinuclear halos, the viral DNA can proliferate into thousands of copies. This is why, when you examine panels A and B, it is evident that some cells show an intense signal, whereas other adjacent cells show a very weak signal. In each image it is clear that the cells toward the base of the CIN tend to have much lower copy numbers of HPV DNA. This certainly makes sense, given that HPV is transmitted via sexual intercourse, as the cells producing the largest amounts of infectious virus are on the surface. Panels C and D show the importance of probe concentration to the signal of in situ hybridization. The section is from the spinal cord of a paralyzed mouse. Several miRNAs are increased in this situation. One of these miRNAs (miR-681) that was strongly upregulated according to the solution phase qRTPCR data was not evident when in situ hybridization was done (panel D); the probe concentration was 1 femtomole/microliter. Note the difference when the probe concentration was increased to 100 femtomole/microliter (panel C). A signal is now evident, and it localizes to cells in the anterior and posterior horns, which is the location of the cells (alpha motor neurons) that would be involved with the paralysis.

THE CONCENTRATION OF SALT IN THE STRINGENT WASH

Since DNA and RNA are weak acids, their hybrids are influenced by the concentration of salt in the hybridization and wash solutions. Higher salts tend to "mask" the opposing negative charges of the nucleic acid complexes, which makes them less likely to denature. Similarly, if we decrease the salt concentration of the solution, then the opposing negative charges are more exposed and, like two negative poles of adjoining magnets, can cause the hybridized strands to denature.

Let's recall the formula we have been using for the melting temperature:

$$T_m = 8\,15°C + 16.6\log[Na]\,(as\,molarity)$$
$$+ 0.41(\%\,G + C\,content) - 0.61(\%\,formamide)$$
$$- 500/\#\,base\,pairs\,in\,the\,complex$$

hybridization signal. Of course, the two common reporter enzymes we use are peroxidase and alkaline phosphatase. However, at this stage of in situ hybridization and immunohistochemistry, the different commercial products have pretty much equalized and maximized the number of reporter *enzymes* we can get per probe–target complex. Hence, there is no need to discuss this at length.

The concentration of the probe and target clearly affect the ultimate density of the reporter nucleotide on the probe–target complex. The reason is that when we do in situ hybridization, we're basically running a simple diffusion-based "competitive binding" experiment or, if you prefer, a "lock and key experiment," where probe and target concentration determine the ultimate strength of the

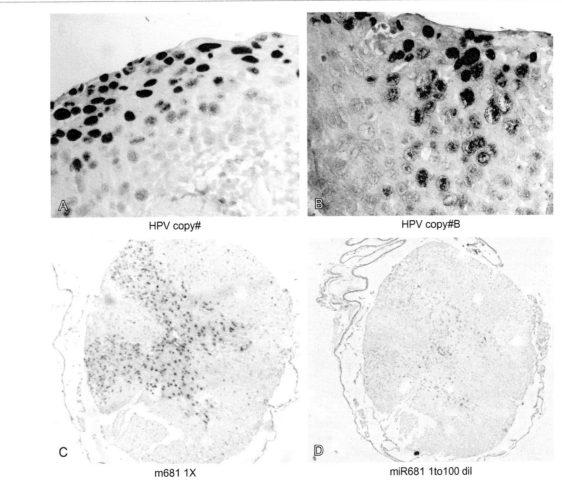

Figure2-14 Relationship of the signal intensity to the target's copy number per cell and the probe concentration. The HPV positive CIN is an excellent model to see the influence on the target copy number and the strength of the hybridization signal. The reason is that the amount of HPV varies from one copy (in the basal cells) to thousands of copies in the cells toward the surface. As is evident from panels A and B, the intensity of the HPV in situ hybridization signal varies dramatically from the surface to the base of the lesion, paralleling the great variability in the viral copy number. Panels C and D show the importance of the probe concentration to the signal for a microRNA. Note the strong signal in panel C for miRNA-681, with a probe concentration of 100 femtomoles/microliter. However, if the probe concentration is decreased to 1 femtomole/microliter, then the signal is lost (panel D).

As is evident from the formula, if you change the sodium concentration in the stringent wash from 150 mM to 15 mM, the end result is less than a degree change in the Tm. These are the usual concentrations of salt you would employ in the stringent wash. Thus, I like to keep the salt concentration constant for the hybridization solution and wash. More specifically, I will keep it low (around 15 mM salt). However, as you can see, altering the sodium concentration will not have much of an effect on the in situ hybridization results. As indicated previously, I also prefer not to use formamide in the wash. It is carcinogenic, toxic to the skin, and relatively expensive. Thus, I prefer to use temperature changes as my primary mode to obtain a maximum signal-to-background ratio with in situ hybridization. Still, I should add that if you use a commercially available probe diluent, it likely does contain formamide. The reason is that the formamide not only protects the solution from bacterial degradation, but also, of course, raises the Tm of the reaction and in this way helps reduce background. However, to stress again,

probe–nontarget DNA binding due to weak homology is *not* the primary determinant of background with in situ hybridization. Thus, altering the formamide concentration in the probe cocktail and/or the wash will usually not be an effective way to reduce background.

THE STRUCTURE OF PROTEINS

There are some obvious similarities when we compare DNA–DNA and RNA–RNA hybridization to protein–protein hybridization. For one, protein–protein hybridization does involve hydrogen bonds, though not to the same order of predominance as with DNA–DNA or RNA–DNA hybrids. Of course, protein–protein hybridization with immunohistochemistry involves the joining of the primary (often monoclonal) antibody against the epitope of interest. Antibody–antigen hybridization is equivalent to the DNA–DNA hybridization process, as it is a "lock and key" process where antibody and target

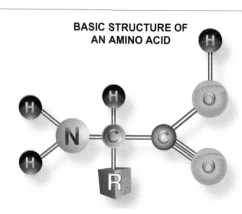

Figure 2-15 The basic structure of the amino acid. The amino acid's versatility in molecular biology is based in the different functional class of the R side chains that can confer to the protein ionic potential, hydrogen bond capability, as well as hydrophobica ndhyd rophilicp ockets.

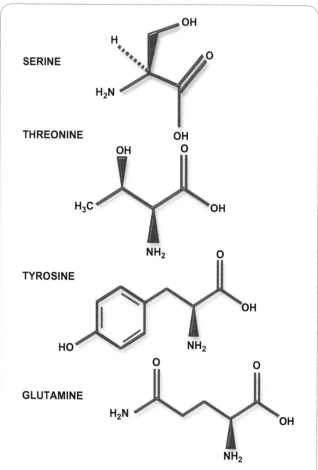

Figure 2-16 The amino acids capable of forming hydrogen bonds with other proteins as well as with RNA and DNA. The serine, threonine, tyrosine, and glutamine side chains in proteins play key roles during in situ hybridization and immunohistochemical, due to their ability to interact with the target and the key reagents used in the process due to their ability to form hydrogen bonds.

concentration will be key variables to consider when trying to maximize our results. So, to better understand the key features of primary antibody–antigen hybridization, we need to examine in some detail what forces are in play that allow an antibody to bind to its epitope.

Before we get into that discussion, I think it would be helpful to examine the basics of protein structure. Of course, proteins are made up of individual amino acids. Figure 2-15 shows the basic structure of an amino acid. Note the COOH and NH_2 groups. More importantly, note the R side chain. Whereas DNA and RNA are relatively (biochemically speaking) homogenous molecules, proteins are far more heterogeneous, and they owe this exclusively to the marked variability present biochemically on their R side chains. The side chains give the protein the capability to exhibit ionic charges (negative or positive), to form hydrogen bonds, to form hydrophobic or hydrophilic pockets, and to cross-link with other proteins or nucleic acids in the presence of formalin. Each and every one of these biochemical abilities will be a very key part in understanding how immunohistochemistry and in situ hybridization work, as well as manipulating the system to our advantage.

Figures 2-16–2-19 provide the structures of most of the different amino acids found in human proteins. I have divided the different amino acids into four functional groups. By "functional," I mean different amino acid groups that have specific functions for us in the in situ hybridization and immunohistochemistry process:

1. Figure 2-16—Amino acids that can form hydrogen bonds (important in epitope–antibody hybridization and in the function of the three-dimensional formalin-induced cross-linked macromolecular cellular cage): *serine, threonine, tyrosine, and glutamine.*
2. Figure 2-17—Amino acids that are positively or negatively charged (important in epitope–antibody hybridization and in the function of the three-dimensional formalin-induced cross-linked macromolecular cellular cage): *aspartic acid and glutamic acid. The positively charged amino acids include arginine, histidine, and lysine.*

3. Figure 2-18—Amino acids that are able to cross-link with other proteins/DNA/RNA molecules (key to the formation of the three-dimensional formalin-induced cross-linked macromolecular cellular cage): *tryptophan, arginine, asparagine, glutamine, and histidine.*
4. Figure 2-19–Those amino acids capable of forming hydrophobic pockets in the protein molecule: *valine, isoleucine, methionine, phenylalanine, tryptophan, cysteine, arginine, lysine, and proline.*

I hope that adopting a KISS ("keeping it simple is smart") philosophy and organizing the amino acids into groups that have specific functions that can determine the success (or failure) of in situ hybridization or immunohistochemistry will help you use this theoretical construct to better understand the molecular bases of the in situ-based reactions.

THE HYBRIDIZATION OF A PRIMARY ANTIBODY AND ITS EPITOPE

The interaction between primary antibodies and epitopes is relatively straightforward. There are basically two types

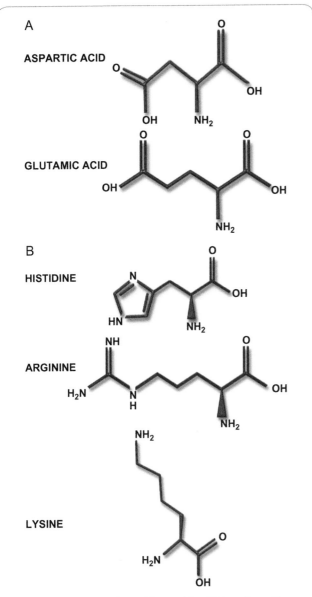

Figure2-17 The amino acids capable of interacting with key proteins such as peroxidase, streptavidin, and alkaline phosphatase due to their ionically charged side chains. The negatively charged amino acid side chains from aspartic acid and glutamic acid as well as the positively charged amino acids including arginine, histidine, and lysine play key roles in the three-dimensional macromolecule cross-linked cage with the hydrogen bonding R side chains via their ability to serve as "gatekeepers" for the key reagents used with in situ hybridizationa ndi mmunohistochemistry.

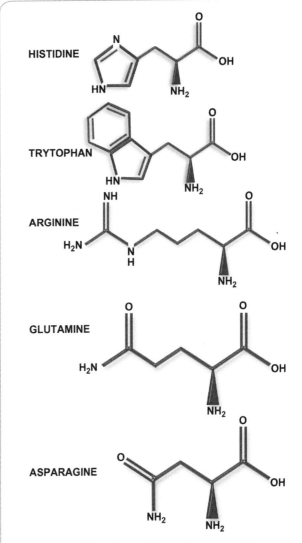

Figure2-18 The amino acids capable of cross-linking with other proteins as well as DNA and RNA molecules. The basic structure of the three-dimensional macromolecule cross-linked cage that surrounds each DNA and RNA target and is intimately associated with every epitope is created due to the R side chains of the amino acids that actually form these cross-links and include tryptophan, arginine, asparagine, glutamine, and histidine. We discuss in subsequent chapters how the three-dimensional macromolecule cross-linked network is likely the key variable in determining success or failure with immunohistochemistry and in situ hybridization.

of epitopes: two-dimensional and three-dimensional. Two-dimensional epitopes are, as they imply, consecutive sequences of amino acids that are able to be "recognized" by the antibody. The recognition involves the same forces that we have discussed for DNA and RNA: hydrogen bonds and ionic charges. However, for proteins, we need to add hydrophobic and hydrophilic forces, since some side chains are strongly hydrophobic and other amino acid side chains are strongly hydrophilic.

Three-dimensional epitopes simply reflect the fact that the protein must alter its conformation in such a way that nonconsecutive amino acids are recognized by the antibody. The same forces just described apply in these interactions.

The number of amino acids required to make an epitope varies a great deal. It can be as few as six to eight, or as large as several hundred. Certainly, epitopes that consist of longer amino acid sequences likely will be more strongly attached to their antibody than epitopes that contain only a few amino acids. Of course, even though the large antibody has only a relatively small region that

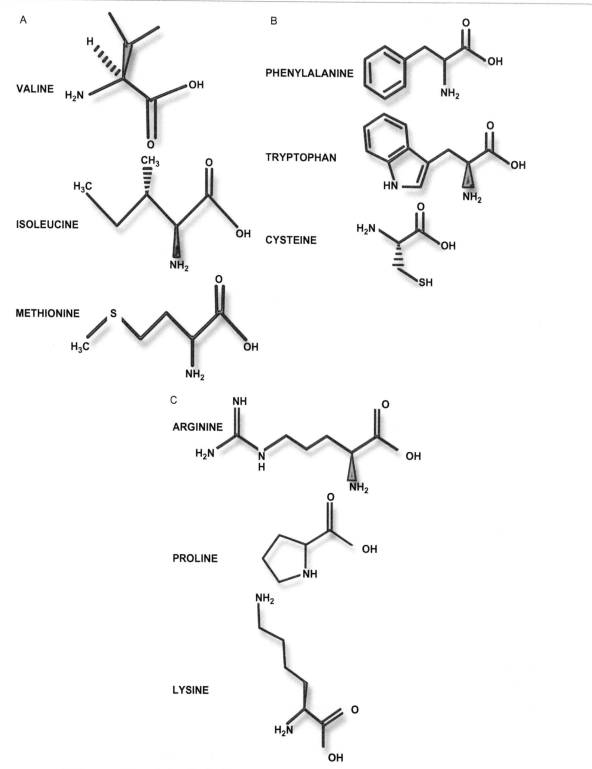

Figure2-19   The amino acids capable of forming hydrophobic pockets. Hydrophobic and hydrophilic pockets play key roles in the three-dimensional conformation of proteins. The amino acids that can aid in the formation of hydrophobic areas in the protein include valine, isoleucine, methionine, phenylalanine, tryptophan, cysteine, arginine, lysine, and proline.

attaches to the epitope, it does not preclude other parts of the primary antibody being able to form hydrogen/ionic or hydrophobic/hydrophilic bonds with the nonepitope region of the protein of interest. These forces would serve to basically "dock" the epitope in place in three-dimensional space, such that it can interact with the antibody, which, in turn, will trigger the molecular signaling events that ultimately result in the cells' response to the protein.

The signaling events can include activation of natural killer cells, being engulfed by macrophages, activation of the complement cascade, induction of migration of inflammatory cells to the region, and so on.

Now that we have discussed the forces in play that allow primary antibodies to bind to their specific epitopes, let's look at some more practical considerations as they relate to in situ hybridization and immunohistochemistry.

DNA and RNA molecules, and the target sequences that are derived from such, tend to be very long. Most genes are many hundreds to thousands of bases long (and longer), as are many mRNAs. Thus, we have the option of making many smaller probes to hybridize with a greater percentage of the DNA or RNA target. This simple way of increasing the total number of reporter nucleotides per target sequence will by itself increase the signal intensity. As a reminder, we are obliged to the synthesis of large numbers of small probes that are no greater than 100 base pairs in size in this situation. The reason is simple: if we make a probe too large (greater than 150 base pairs), then it has difficulty getting into the intact cell due to the constraints placed on it by the three-dimensional formalin-induced cross-linked macromolecular cellular cage. In comparison, multiple probes that cover the same DNA or RNA sequence of between 50 and 100 base pairs in size will readily enter the cell and avidly bind to the homologous target sequence, assuming that we have optimized the conditions via our pretreatment regime.

However, with protein–protein hybrids, the area of the target protein's epitope that is accessible (homologous) to the matching antibody probe protein is fixed and relatively small. This target protein sequence is commonly referred to as the *epitope*. Of course, the probe that is directed against it is derived in an animal by exposing it to the epitope after which specific *antibodies* are formed. These antibodies are purified, concentrated, and become the primary probe against the target of interest. A simple way to appreciate this point is to realize that every three nucleotides translates into one amino acid when the mRNA is transcribed to the protein. Plus, that said, only a relatively small part of the protein in the form of the epitope is available for hybridization with the primary antibody. As discussed previously, this may be as few as 10 amino acids, or even fewer. However, it is important to remember that while this small-sized region may be responsible for the physiologic effects of antibody–epitope binding, it is highly likely that other parts of the antibody are interacting with other regions of the protein besides the epitope, which, in turn, stabilizes the antibody–antigen complex. Indeed, we will devote a large part of several subsequent chapters to describing how the interactions between the three-dimensional protein/protein cross-linked cage induced by the formalin fixation *and* the primary antibody/probe and other key reagents with in situ hybridization and immunohistochemistry are *critical* in achieving success with these molecular assays.

Before we discuss some data that relate to epitope/primary antibody hybridization and denaturation, let's see what our knowledge of biochemistry might predict about such an event. You probably would predict that nucleic acid complexes tend to be much stronger than epitope–antibody complexes. That is to say, they should show

quicker and stronger hybridization dynamics and also be more difficult to denature. This is based on the forces that keep these different molecules together. As we have discussed, DNA–DNA and RNA–RNA complexes are kept together primarily by hydrogen bonds. These are relatively strong bonds. Their strength is reinforced by the long stretches of matched base pairs in a DNA–DNA double-stranded complex. Thus, much like the links in a chain, whereas each link per se may be strong, their strength is greatly amplified when they form long chains. Clearly, teleologically speaking, it was highly advantageous to have a genetic library via the chromosomes that under normal conditions was held tightly together. In comparison, the forces that hold epitopes and antibodies together appear to be relatively weaker. For one, there are no long chains of hydrogen bonds. Another point is that the number of amino acids that comprises an epitope tends to be far less than the number of nucleotides that can base pair with in situ hybridization for RNA–cDNA or DNA–DNA, although this, of course is variable. This may be a reflection of the fact that in nature antibodies do not have to be solidly attached to their epitope to exert their neutralizing effects. Rather, antigen–antibody complexes often serve as a primary catalyst to other events (such as the activation of the complement cascade, T-cell, or macrophage activation). Once these other events are activated, the antigen–antibody complex does not have to continue to be attached to one another. Of course, the cell has multiple mechanisms whereby the double-stranded DNA can be partly denatured to allow RNA transcription, but these events are both very selective and involve a very small percentage of a given chromosome.

DENATURING PROTEIN–PROTEIN COMPLEXES

Let's now look at some actual data and see what they tell us regarding the relative strength of the forces that keep DNA–DNA and RNA–cDNA complexes together versus epitope–primary antibody complexes. Here are the two possibilities:

1. The forces that allow DNA and DNA plus RNA and cDNA to hybridize and remain annealed (primarily hydrogen bonds) are *much stronger* than the forces that allow epitope–primary antibodies (a mixture of variable forces) to hybridize and to resist denaturation.
2. The forces that allow DNA and DNA plus RNA and cDNA to hybridize and remain annealed (primarily hydrogen bonds) are *equivalent* to the forces that allow epitope–primary antibodies (a mixture of variable forces) to hybridize and to resist denaturation.

Let's see what hypothesis point 1 would predict versus point 2. If point 1 were correct, then DNA–DNA hybrids would develop with in situ hybridization much more quickly than epitope–primary antibody hybrids. The reason is that the rate of hybridization for the DNA and protein complexes would be dependent on:

A. The concentration of the probe/primary antibody andtarge t/epitope

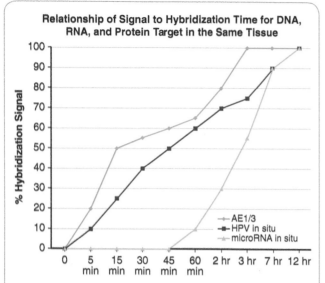

Figure 2-20 The effect of time of incubation on the strength of the hybridization signal for DNA, RNA, and a protein in the same tissue. The same CIN 1 lesion was tested for the cytoplasmic cytokeratin AE 1/3, HPV DNA, and microRNA let-7c and -29a with the signal scored at successive time intervals starting at 5 minutes and ending at 12 hours. Note the markedly different "diffusion kinetics" for the cytokeratin (the fastest of the three) when compared to HPV DNA and, especially, the microRNAs, which are by far the slowest over time in reaching maximum signal. This suggests that the three-dimensional macromolecule cross-linked cage surrounding microRNAs may be greatly restrictive to diffusion relative to the cytokeratin and viralDN A.

B. The ability (or rate of diffusion) of the probe/primary antibody to diffuse through the cell to find its target

C. The strength of the forces that allow the probe-target and epitope/primary antibody to initially bind together and then stay hybridized

Let me clarify the point that I am not a physicist. However, very knowledgeable physicists and biochemists with whom I have discussed this issue have assured me that if you use a high concentration of probe and primary antibody (1 microgm/ml of each) and use a probe of 100 nucleotides and a primary antibody of about 50 to 100 kDaltons (standard sizes), and analyze a high copy number target, then the strength of the forces allowing the hybridization and persistent annealing will be the key variable that determines the *rate* at which the respective complexes form with in situ hybridization and immunohistochemistry, respectively,

Thus, I ran a series of time course experiments using the high copy cytokeratin AE1/3 for the immunohistochemistry and HPV 16 DNA in a CIN 1 lesion for the in situ hybridization. Since the latter is nuclear based, and the AE1/3 is present in the cytoplasm. I also analyzed a very high copy microRNA present in the cervical epithelia—microRNA-let-7c. The results are presented in

graphic form in Figure 2-20, with some of the actual data shownin Figure2-21 .

As you can see, the data show that, if anything, the rate at which the immunohistochemistry reaction reaches maximum intensity (or 3+, as we pathologists call it) is *faster* than the rate for either HPV DNA or microRNA. I have done many such experiments with equivalent results assuming, of course, that the conditions described previously were met and that the pretreatment conditions were also optimal.

If you have biochemical expertise, you may be thinking, "Wait a moment; those data do not make sense. It does not seem possible that DNA–DNA complexes should show much more rapid hybridization kinetics compared to microRNAs in the same cells." Indeed, if we did the reactions of DNA–DNA hybridization and microRNA–cDNA hybridization in an aqueous media (especially since we are using LNA probes for the microRNA), as compared to cells in tissues fixed in formalin, then we would assume that the RNA–cDNA hybridization kinetics would be much faster than the HPV DNA probe/target complex. For one, the microRNA is cytoplasmic and the HPV DNA is nuclear. Also, microRNAs tend to be present in much higher copy numbers than HPV DNA. But that is the key point. We are *not* doing in situ hybridization or immunohistochemistry in a simple aqueous solution. Rather, we are doing the reaction in an aqueous solution that is surrounded by a very complex three-dimensional formalin-induced cross-linked macromolecular cellular cage. As we progress through this book, I continue to share data with you that strongly suggests that this three-dimensional formalin-induced cross-linked macromolecular cellular cage plays a critical role in both signal and background in general, and in such key events as the rate and strength of the hybridization and in background.

Before we leave this topic, let's look at the data from one more set of experiments. We will compare the denaturation of probe–target and epitope–primary antibody complexes that were generated via in situ hybridization and immunohistochemistry, respectively. We will do the experiments under the same equivalent conditions noted previously with one exception. We will markedly increase the stringency of the immunohistochemistry reaction. As you will see in Chapters 7 and 8, I recommend a relatively high stringent wash for in situ hybridization and a relatively low stringent wash for immunohistochemistry. This reflects my own experience over the past 30 years, and the actuality of many of the published protocols for these two in situ-based methods. But, for the purposes of this discussion, let's go to a much more stringent wash for the immunohistochemistry. These data are presented in Figure 2-22. Note that the AE 1/3 immunohistochemistry signal stays strong even after a very stringent wash. It is my hypothesis that this observation does not just reflect the forces that tend to directly link the epitope and primary antibody. This observation also reflects the *indirect* forces that are part of the three-dimensional formalin-induced cross-linked macromolecular cellular cage. These forces, which can include ionic charges, hydrogen bonds, hydrophilic/hydrophobic, and physical constraint (or "anti-breathability") forces, may well play a very

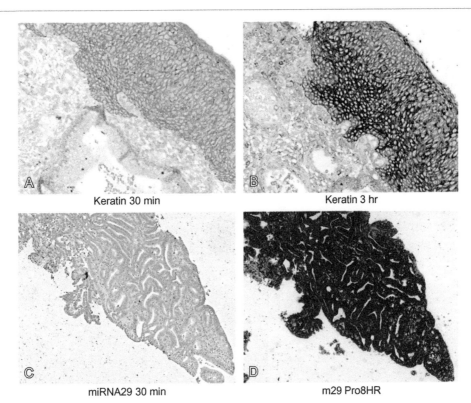

Figure 2-21 Representative examples of the microRNA and cytokeratin signal in the same tissue at different time points. The signal for the cytokeratin was evident after 30 minutes of incubation (panel A) and was maximum at 3 hours (panel B). No signal was evident for the microRNAs after 30 minutes of incubation (panel C) and required 8 hours to maximize (panel D).

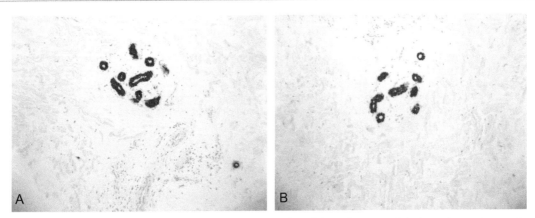

Figure 2-22 The signal for cytokeratin AE1/3 is strong in this breast lobule if the stringent wash is done at room temperature (panel A) or at 45°C (panel B).

important role in our in situ hybridization and immuno-histochemistry results.

DETECTION SYSTEMS FOR NUCLEIC ACID–NUCLEIC ACID AND PROTEIN–PROTEIN HYBRIDS

Brief mention should be made of the biochemistry involved in the detection of DNA/RNA or DNA/DNA hybrids as compared to epitope–antibody complexes. For nucleic acid hybrids detected by in situ hybridization, the process is straightforward. We typically place a labeled nucleotide(s) in the probe. This is usually digoxigenin or biotin. Digoxigenin is favored when doing in situ hybridization with mammalian tissues, since this compound is not found naturally in mammals. Other choices of tagged nucleotides include fluorescent molecules and radioactive isotopes. We focus our attention on the colorimetric and not radioactive-based signals. Of course, radioactive isotopes are more expensive, much shorter lasting, and

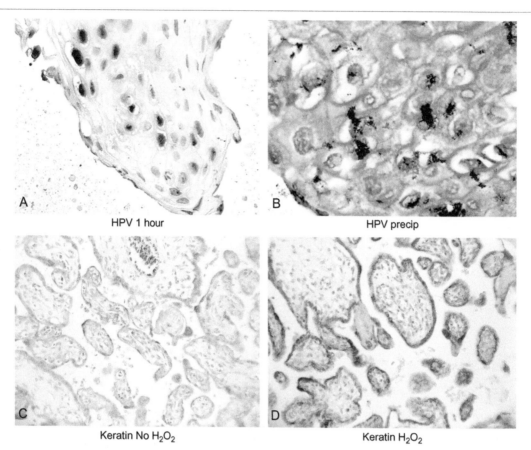

Figure 2-23   Different sources of background with in situ hybridization and immunohistochemistry. Panel A shows the sharp nuclear-based signal that is diagnostic of a target-specific signal for HPV in this CIN lesion. Compare this to panel B, where the blue precipitate is not nuclear based but, rather, in a plane just above the tissue. The latter is characteristic of NBT/BCIP crystallization when using an alkaline phosphatase system and can be eliminated by reducing the concentration of the NBT/BCIP, heating the solution to 37°C prior to adding the chromogen, and by vigorously moving the slide up and down immediately after adding the slides to the chromogen. Peroxidase-based systems (and alkaline phosphatase systems) often show background in red blood cells, perhaps owing to their very high concentration of proteins. Panel C shows a placenta tested by immunohistochemistry for cytokeratin where the tissue was treated only with protease digestion prior to immunohistochemistry. Note the strong nonspecific colorization of the red blood cells and lack of any signal in the trophoblasts, which are rich in cytokeratin. The same results were obtained if the tissue was treated with the protease, then with 10 buffered formalin for 15 minutes, then tested by immunohistochemistry for cytokeratin. However, if the tissue was incubated in H_2O_2 after the protease digestion step, then no background in the red blood cells was evident, and target-specific signal was now present int het rophoblasts.

carry with them serious safety issues. Further, they are no longer any more sensitive than the color-based systems. When using one of the colorimetric-based tagged nucleotides, we will use a secondary compound that will specifically target the probe's tag. For biotin, this will be streptavidin. For digoxigenin and fluorescein, we can use an antibody that is directed against the tag that now is functioning as an epitope. An enzyme is conjugated to the streptavidin or antibody. The enzyme is usually alkaline phosphatase or peroxidase. The next and final step is simple. The target-probe-secondary molecule with the alkaline phosphatase or peroxidase that we generated while doing in situ hybridization is then incubated in a solution that contains the chromogen (NBT/BCIP for alkaline phosphatase or DAB for peroxidase). The enzyme catalyzes a reaction that results in an insoluble precipitate (blue or brown, respectively).

For epitope–antibody reactions, it is very rare for the primary antibody to contain a tag. The key is to realize that the primary antibody was made in some animal. The most common animals used to generate primary antibodies are mice and rabbits. However, other animals can be used; the next most common animals are goats and rats. By knowing the source animal of the primary antibody, we can easily obtain a secondary antibody directed against the epitopes (including primary antibodies) raised in that animal. These secondary antibodies are at times called *consensus* antibodies or *multilink* antibodies. These names connote that these secondary antibodies can recognize any primary antibody directed in the animal of interest. Most secondary antibodies are made in rabbits and mice. There is one more key feature of the secondary antibody, which makes them equivalent to the first detection step of in situ hybridization. That is, they are

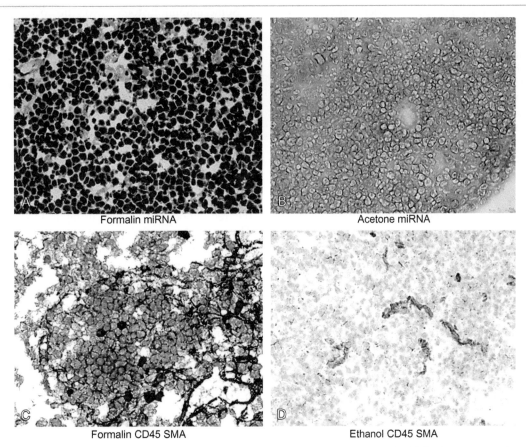

Formalin miRNA

Acetone miRNA

Formalin CD45 SMA

Ethanol CD45 SMA

Figure 2-24   Formalin fixation allows for the sharp cellular localization of the target after in situ hybridization, whereas denaturing fixatives are associated with much diffusion of the signal. Panel A shows the sharp lymphocyte-based signal for two miRNAs (miRNA-31 and 328) in formalin-fixed tissues. Compare this to panel B, where the acetone-fixed tissue causes the signal to, in large part, diffuse out of the cell. Similarly, the immunohistochemical signal for CD45 and smooth muscle actin (SMA) are sharply defined to the cell membrane of the lymphocytes and cytoplasm of the blood vessels, respectively, if the tissue is fixed in formalin (panel C). The same tissue fixed in 70% ethanol showed marked *diffusion* of the CD45 and SMA signals (panel D). This and related data serve as a springboard to discuss the biochemistry of formalin fixation, especially as it relates to the three-dimensional macromolecule cross-linked cage that surrounds the targetso fi nterest.

conjugated to either alkaline phosphatase or peroxidase. At this stage, the in situ hybridization and immunohistochemistry reactions become identical. We simply then add the chromogen, which generates the target-specific precipitate; do the counterstain; coverslip the slide; and examine the result.

THE BIOCHEMICAL BASIS OF SIGNAL VERSUS BACKGROUND

First, let us recall the meaning of the terms "signal" and "background." Signal is easy. It simply represents the colored precipitate that forms due to the enzyme, which is conjugated to the molecule (either secondary antibody for immunohistochemistry or streptavidin/antidigoxigenin antibody for in situ hybridization) that, in turn, is attached to the tagged probe or primary antibody that, finally, attaches to the target DNA/RNA sequence or the epitope of interest.

The biochemical basis of background is more complicated. I think the best definition of background is any

colorimetric precipitate that does not correspond to the target DNA/RNA sequence (for in situ hybridization) or the target epitope (for immunohistochemistry). Thus, to understand the biochemical basis of background, you need to consider each possible source of nonspecific precipitation of the chromogen.

One source is the chromogen itself. It is not too uncommon for either NBT/BCIP or DAB to crystallize and precipitate during the reaction. This is usually seen as small aggregates of amorphous material; for NBT/BCIP it can also take the form of needle-like crystals. The precipitation is taking place in the solution and not in the cells per se. Thus, usually you can identify this type of background because the colored precipitate is over the glass or in a different plane from the tissue (Figure 2-23). The latter can be seen simply by adjusting the fine focus. However, sometimes the precipitate can appear on the cell. It can be recognized by doing the in situ hybridization or immunohistochemistry reaction with no probe and no detection steps; you only add the chromogen to the slide and look for the precipitate. Thus, if you are concerned that you are getting nonspecific precipitation of the chromogen, one way to determine if this is the case is simply to apply the chromogen after the pretreatment step and look under the microscope. If you see the precipitate, this is obviously background, and the question then becomes how do you get rid of it? There are two ways. The simplest and most obvious is to alter the conditions that favor the precipitation in solution. This means using a lower concentration of the chromogen. It also means using solutions that are preheated and not just out of the refrigerator, since lower temperatures typically favor chromogen precipitation prior to actually doing the detection step. Another trick is to agitate the slide (move it up and down) while it is in the chromogen, as precipitation is usually favored by a nonmoving solution. This can be accomplished by placing the slides in the chromogen solution on a shaker, which, in turn, is run at high speed. A simple alternative is to move the slide up and down about 10 times soon after you add the slides to the chromogen solution. In my experience, NBT/BCIP or DAB nonspecific precipitation, when it happens, occurs within the first 5 minutes of incubating the slides in the chromogen. If some precipitate is still left, then a trick of the trade is to put the slide in acetone for 1–2 minutes after the reaction is complete. The signal-based precipitate is much more resistant to acetone than any precipitate that is present in the solution over the tissue, since the latter is free floating and the actual signal is being "protected" by the three-dimensional formalin-induced cross-linked macromolecular cellular cage that surrounds it.

The next possible source of background would be the enzyme that catalyzes the chromogen precipitation—that is, either alkaline phosphatase or peroxidase. Many people, including myself, have assumed that the source of the alkaline phosphatase or peroxidase that was responsible for the background was endogenous. Certainly, some tissues and cells are rich in these enzymes. Thus, if the tissue has not been fixed too strongly, it is possible that some residual activity of these enzymes will remain. The classic example of this is the large amount of alkaline phosphatase that is present in the surface of the bowel epithelia, the apical surface of the proximal convoluted tubules of the kidney, or the surface of the trophoblasts of the placenta. Another classic site of background, especially with a peroxidase-based system, is the red blood cells.

The best and simplest way to reduce background from endogenous peroxidase activity is to incubate the tissue prior to the primary antibody/probe hybridization with the substrate of this enzyme, which is hydrogen peroxide. This is why the protocols in this book for any peroxidase-based assay include the step of incubating the tissue in a 3% solution of hydrogen peroxide for 10 minutes prior to doing the initial denaturation (for in situ hybridization) or adding the primary antibody (for immunohistochemistry). This simple step is effective. A trick of the trade I use to combat background when using DAB as the chromogen is to add 3% hydrogen peroxide to each of the reagents in the reaction, including the primary and secondary antibody. For alkaline phosphatase-based systems that use NBT/BCIP, the chemical levamisole is the standard way to reduce the endogenous activity of the alkaline phosphatase.

Some immunohistochemistry protocols use a post-fixation step in 10% buffered formalin prior to the immunohistochemistry or in situ hybridization reaction. The logic is that the post-fixation may inactivate residual enzyme activity in the tissue that may be contributing to background. This makes sense, since formalin does not penetrate tissue very well when dealing with the standard 5–15mm biopsy samples. However, once cut and placed onto glass slides, the tissue sections are only 4 microns thick and, thus, the formalin should readily cross-link all enzymes in the tissue section if applied to the tissue for a reasonable period of time.

I did a series of experiments to test this hypothesis. Placenta tissues tend to show background, especially in the red blood cells, when using a DAB-based system. Thus, they are a good tissue source to compare the effect of pre- and post-treatments on the signal-to-background ratio. For signal, I tested for the cytokeratin AE 1/3, as this is present in abundance in the placenta, localizing to the trophoblasts and the decidual cells. Representative data from such a series of experiments are shown in Figure 2-23. Note that if you use protease digestion, but no other pre- or post-treatment including no wash in hydrogen peroxide, then there is a strong background in the red blood cells and an associated very weak signal (panel C). As a reminder, the optimal conditions for AE 1/3 include protease digestion. If the background is due to residual peroxidase activity in the red blood cells, then it is reasonable to wonder why the protease digestion did not inactive the peroxidase. The background persisted when the tissue was post-fixed in 10% buffered formalin for 15 minutes after the protease digestion. Surely, the protease and post-fixation in formalin combined would be sufficient to destroy any enzyme still operative in the tissue section. However, if you simply incubate the tissue in 3% H_2O_2 for 10 minutes after the protease digestion and then do the immunohistochemistry, the background is gone and the signal is evident (panel D)! This brings two basic and important points for those learning to do immunohistochemistry. First, high background usually will rob some, or all, of the signal. Perhaps the reason is that the chromogen is being used at such a high rate in the generation of the background (which tends to be intense) that the remaining amount of free chromogen is too low to generate the optimal signal. Second, a simple incubation with a 3% H_2O_2 solution for 10 minutes is sufficient to eliminate the background with a peroxidase-based immunohistochemistry assay system. How does the H_2O_2 eliminate the background? It does not appear by "exhausting" the residual peroxidase activity. Rather, the hydrogen peroxide may be altering chemicals present in certain cell types (such as red blood cells) such that the peroxidase you add when doing the immunohistochemistry reaction can no longer catalyze the precipitation of the DAB. Whatever the source of the DAB-associated background in red blood cells, it does not appear that the major cause is residual peroxidase activity in the tissue.

Another source of background is when the probe or antibody (either primary or secondary) nonspecifically adsorbs to either the glass slide or to the proteins/nucleic acids/carbohydrates/fat or some other common component of the tissue. Since the probe has the tagged nucleotide, and the primary/secondary antibody complex contains the reporter enzyme, nonspecific background is sure to follow. In my experience, the more common scenario from this type of background is nonspecific binding of the probe or primary/secondary antibody to some tissue component. We can exploit this to solve the problem, as we discussed previously in this chapter. Basically, we can say to the probe or primary/secondary antibody, "If you want to bind nonspecifically to macromolecules in the tissue, let me put so much protein in the solution, so you can bind to that, and then we can wash these nonspecific complexes away." Thus, the simple act of treating the slide with a solution that contains 2% bovine serum albumin prior to adding the reporter enzyme (peroxidase or alkaline phosphatase) will greatly reduce background, as shown in Figure 2-12. It is my opinion that nonspecific binding of the tagged probe (for in situ hybridization) or the primary/secondary antibody (for immunohistochemistry) is the major cause of background.

THE MOLECULAR-BASED ASSAYS

Of course, we are focusing our attention in this book on the two in situ-based molecular pathology assays: in situ hybridization and immunohistochemistry. It would be useful to describe some of the other commonly used molecular assays to show the points they have in common with in situ hybridization and immunohistochemistry.

THE POLYMERASE CHAIN REACTION

The polymerase chain reaction (PCR) has become the most commonly used molecular assay. It shares one key component with in situ hybridization: they are both based on the ability of a known DNA sequence to hybridize with its complementary target sequence, which, of course, is the DNA or RNA molecule we are interested in detecting. Thus, if you have a good knowledge of the basics of in situ hybridization, then PCR will not be difficult to understand. With PCR, we use a short DNA sequence, typically about 20 nucleotides, to prime the extension of the target sequence of interest. Logically, this is why these short DNA sequences are called primers. We will learn with in situ hybridization that the hybridization between a 20-nucleotide probe and target is not optimal for getting a reproducibly strong signal. Stated another way, the melting curve of a 20-nucleotide probe/target complex, even at room temperature, shows too strong a propensity toward denaturation to be able to provide a strong signal-to-background ratio with in situ hybridization. Of course, as seen in Figure 2-8, we can solve this problem with LNA-modified 20-nucleotide primers. However, a regular 20-nucleotide primer costs about $20, whereas an LNA-modified 20-nucleotide primer costs about $500! However, for PCR, we are not asking the primer to stay bound to the target sequence through the entire molecular assay, which is typically several hours long. We are only asking the primer to be bound long enough to allow the DNA polymerase (or rTth polymerase if one is doing RTPCR) to extend the primer using the target sequence as the template. This process typically takes less than one minute! Indeed, after the primer extension occurs, we need the primer and target sequence to denature in order to allow the geometric amplification of the extended primer that is mandatory for a successful PCR reaction. Hence, there is no need for an LNA-modified primer for PCR or in situ PCR. Indeed, we could argue that it probably would inhibit the reaction, given its strong propensity to remain hybridized throughout the reaction.

We have already discussed the key step whereby the primer hybridizes to the target sequence, which then allows the DNA polymerase to extend the primer using the target sequence as the template. Of course, as we discussed previously under "The Structure of DNA and RNA," the DNA polymerase extends the primer at its 3′ end (or, as more commonly stated, the DNA is synthesized in the 5′ to 3′ direction). The next key step to realize is that this is occurring at the same time with two different primers. The two primers are made such that they will induce the synthesis of two DNA sequences that will overlap with each other. Remember that, by convention, DNA sequences are written in a 5′ to 3′ direction. With this in mind, the top (or positive) strand is called the *sense strand*. The bottom (or negative) strand is called the *antisense strand*. Since DNA is a negatively charged acid, the designation of positive and negative strands may be a bit confusing. The sense and antisense designations do make more logical sense. When RNA is made from the DNA template, the top strand is used as the template, and thus, it is the "sense strand" because it is used as the template from which the mRNA is made. Of course, neither life nor DNA–RNA is that simple, as it is now clear that the antisense strand can be used by a cell to generate active RNA molecules, including microRNAs.

So to summarize, here are the key steps with PCR:

1. You obtain the double-stranded sequence of the DNA or cDNA sequence of the RNA target molecule that you are interested in detecting. Remember that the upper strand will be in the 5′ to 3′ direction and the lower strand will be in the 3′ to 5′d irection.
2. Remember that for the detection of mRNA by RTPCR, the cDNA sequence will have as its top strand the mirror image of the lower strand of the DNA sequence and as its lower strand will be the mirror image of the top or sense strand of the DNA molecule.
3. You then generate two primers that are each about 20 nucleotides long. Remember that primers are always designated in the 5′ to 3′ direction. Hence, primer *one* will be made from the sequence that will be complementary to a sequence near the 3′ end of the *lower* strand, and primer *two* will be complementary to a sequence near the 3′ end of the *upper* strand. *Stated another, more simple way, choose a region of the upper strand toward its 5′ region as primer one and choose a region of the low strand towards its 5′ region as primer two.*
4. In my experience, I have found you can choose a wide range of nucleotides between primer one and primer two and have successful PCR. Sometimes primers can be attached to each other (primer-dimers), which, in turn, can serve as an initiation point of DNA synthesis in PCR. Such nonspecific

PCR products typically are from 40 to 200 base pairs in size. Hence, I prefer to choose an amplicon of between 300 and 600 base pairs in size. In this way, when you are analyzing the amplicon on an agarose gel, the size of the product you see can be used as one indicator of the specificity of the reaction.

5. For the amplification of mRNA, you need to first synthesize the corresponding cDNA. This can be done at 65°C for 30 minutes. The primer/RNA hybrids tend to have a high melting temperature, so you can actually use a cDNA synthesis temperature of 60–70°C. As you know, the higher temperatures discourage the nonspecific binding of the primer to other RNA or DNA molecules. I strongly recommend you use a reverse transcriptase enzyme for the cDNA synthesis that also has the capacity to perform DNA synthesis. This way, you can do both the RT and PCR step with one enzyme. One such enzyme is called rTth. The other advantage of using just one enzyme for both the RT and PCR steps is that you can also use the same buffer/nucleotide solution, which makes for an overall much simpler process.

6. For DNA (or cDNA) amplification, the first step will be to denature the double-stranded target molecule. This can be done at temperatures between 90°C and 96°C. Since the temperature of the solution with in situ PCR tends to be a few degrees lower than that recorded by the thermal cycler, I prefer 95°C–96°C for the denaturing step of in situ PCR. At these high temperatures, the DNA/cDNA tends to denature rapidly, and, thus, 3–5 minutes should be sufficient to make the target DNA/cDNA into the single-stranded molecules that are the essential starting point of PCR/RTPCR. Although, of course, the Taq polymerase (or equivalent polymerase such as rTth) can withstand these denaturing temperatures, it does not mean that these polymerases are forever impervious to these very high reaction temperatures. The polymerase can lose a few percentage points of activity with each cycle, depending in part on the denaturing temperatures used and the amount of time they are employed This may become an important factor in the latter cycles. Thus, the goal is to use the denaturing temperatures for the shortest amount of time so as to be able to still have sufficient polymerase activity toward, for example, cycles 25 to 35.

7. After the first denaturing step, you allow the primers to bind to the target sequences. Just as importantly, you need to ensure that the primers do not bind to their two other nonspecific options. Two nonspecific DNA synthesis pathways can be operative during PCR:
   A. The primers can bind to themselves, creating a single-stranded region that leads to a double-stranded region and, thus, can lead to DNA synthesis (called primer-dimers).
   B. The primers can bind to nontarget RNA or single-stranded DNA molecules that, of course, can lead to DNA synthesis (this is called *mispriming*).

8. It is useful to remember that the concentration of the primers is far greater than the concentration of the target. Also, the concentration of the nontarget RNA and DNA will be many log-fold higher than

the target DNA or cDNA. Of course, this is the exact same problem we face with in situ hybridization, where the concentration of the labeled probe is much higher than the target, and where the number of nontarget DNA/RNA molecules is always millions to billions times greater than the comparably sized target DNA or RNA. How do we get around this problem? There are two simple ways:
A. Hot Start PCR. In hot start PCR, you take advantage of the simple fact that temperatures of around 60°C–70°C will permit hybridization of the primers to their targets, but *not* allow the persistent annealing of the primers to themselves or to nontarget RNA or single-stranded DNA molecules. So, if you do not add the DNA polymerase until the reaction is at 60°C–70°C, then primer-dimers and primer/nontarget binding cannot serve as pathways of nonspecific DNA synthesis during PCR. There are two ways to do hot start PCR. One is the so-called manual hot start PCR in which you physically withhold the polymerase until the thermal cycler is at 60°C. This is the method I prefer. The second way is to add chemicals that will not allow the polymerase to function until the reaction temperature is at least 60°C. One simple way to achieve the latter "chemical" hot start PCR is to add an antibody directed against the DNA polymerase. This antibody, like most antibodies, will denature and no longer function at temperatures that are 60°C or higher.
B. For RTPCR, you can render the DNA in the cell incapable of supporting DNA synthesis by extensively degrading it with RNase-free DNase. As we will see, RNase-free DNase digestion is an essential component of RT in situ PCR.

9. The other point to remember about the nonspecific pathways that can operate during PCR and RTPCR is identical to what we have discussed with in situ hybridization and immunohistochemistry. That point is that the operation of nonspecific pathways will invariably tend to *weaken* or *eliminate* the target-specific signal.

10. Once the primers anneal to the DNA/cDNA target, the polymerase rapidly synthesizes DNA using the target DNA/cDNA as the template. Rapidly means *really* rapidly! Within just 1–2 minutes, the majority of DNA that can be synthesized from the primer/targets will have been made! Hence, you may use annealing times of 1–2 minutes with temperatures of 60°C–65°C after the hot start PCR/ RTPCR beginning to ensure an initial robust ampliconamp lification.

11. After the initial DNA synthesis step, you must then denature the newly synthesized amplicons. This way, the double-stranded DNA synthesized using primer one as the template will now be able to bind to primer two and vice versa. As is evident from what we just discussed, this can be accomplished with a short incubation (1 minute) at high temperatures(94°C–96°C).

12. Next, you bring the reaction back to 60°C–70°C to allow the primers to bind to their respective

locations on the newly synthesized target-specific DNA templates. By keeping the reaction mixture at 60°C–70°C for 1–2 minutes, you can synthesize the first double-stranded amplicons that have the base pair size you set when you created and ordered the two primers.

13. The next steps are clear: you cycle the machine between denaturation and annealing/DNA synthesis for 25–40 cycles. When the reaction is over, many prefer to use an "extension step" of 70°C–75°C for 10–20 minutes. The idea here is that the polymerase may not have completely synthesized all of the amplicons but, rather, may have left many amplicons many nucleotides too "short." The extension step will rectify this problem. I have not seen much utility in this step. The key would be when you run your amplicon on the agarose gel. Say the amplicon should be 400 nucleotides in size. If you see a band at 400 nucleotides and a conspicuous "tail" below the band (that is, toward the smaller-sized DNA molecules), then you have evidence of incomplete extension of many of the amplicons. If you do not see this, then I recommend that you go from the last cycle directly to 4°C, which, of course, will basically stop the activity of whatever polymerase has survived the repeated denaturation/annealing steps so as to preclude any primer-dimers or primer/nontarget RNA or DNA based synthesis.

There are a few other "tricks of the trade" to discuss before finishing this discussion of PCR. First, let's think of some simple, inexpensive ways we may increase the efficiency of our PCR reactions. We have already discussed the most important: hot start PCR. Another simple trick is to take advantage of the high specificity of the primers, especially at temperatures between 60°C and 70°C. How about we have primer *one* and primer *two* make a real large amplicon, say of 600 nucleotides? Then it would be very inexpensive and simple for us to create two more primers that we will call primer *three* and primer *four*. These will anneal *inside* the 600-nucleotide amplicon. Since the amplicons are smaller, they will have a large advantage over the large amplicon during PCR. Many investigators wait until the first 25–40-cycle reaction is over and then do a separate PCR reaction with only primers three and four. The bottom line is that by using additional primers we are making it much less likely that the resultant band evident on the agarose gel came from the nonspecific DNA synthesis pathways of primer-dimers or primer-nontarget nucleic acid annealing. This strategy of using several primers pairs is called *nested PCR*. Another trick of the trade is to realize that Taq polymerase (and the other polymerase that we can use for PCR) has a tendency to adsorb to either the plastic used with PCR or the glass slides used with in situ PCR. We will learn later that nonspecific adsorption of alkaline phosphatase and peroxidase conjugates can be a big problem with in situ hybridization, and that the problem is easily resolved by adding a protein to the solution that basically adsorbs so efficiently to the glass/plastic that there is no "space" left over for our conjugates to adsorb

to. The protein I prefer for this step is bovine serum albumin (BSA). A small amount of BSA added to the PCR or in situ PCR solution is a very simple way to block adsorption of the polymerase to the plastic and glass. This raises the effective concentration of the polymerase in the reaction solution. This also allows us to use much less of the polymerase, which typically is the most expensive reagent used in either PCR or in situ PCR.

SOUTHERN BLOT, NORTHERN BLOT, AND DOT BLOT HYBRIDIZATION

Several older methods that predated PCR were used for the analysis of DNA and RNA that were extracted from cells or tissue. These methods are called Southern blot hybridization (for DNA), Northern blot hybridization (for RNA), and dot blot hybridization (for DNA or RNA). Dot blot hybridization has many different names that include perhaps the most commercially common name, hybrid capture hybridization (for the HPV hybrid capture test that is extensively used in diagnostic pathology).

Since you already have a good knowledge of the basics of DNA–DNA and RNA–DNA hybridization, you will see that Southern, Northern, and dot blot hybridization are easy to understand both theoretically and as a practical matter of actually either doing or interpreting data from these tests.

The first step with Southern, Northern, or dot blot hybridization is to extract either the DNA or RNA. About 20 years ago, this process involved rather laborious protocols that included phenol, chloroform, high concentration sodium acetate, and isopropanol or cold ethanol. Nowadays, there are many commercially available kits on the market that allow very simple extraction and purification of DNA or RNA from tissue or cell preparations. Optimally, you use unfixed fresh (or frozen) cells or tissue for such analyses. But it is uncommon for fresh unfixed tissue to be available for nucleic acid extraction, especially with regards to human tissues. Is there a way around this? This leads me to an important trick of the trade—specifically, the use of formalin-fixed (and typically paraffin-embedded) tissues for PCR or Southern, Northern, and dot blot hybridization.

FORMALIN FIXATION AND MOLECULAR HYBRIDIZATION

Since the topic of formalin fixation is perhaps the most important topic with in situ hybridization and immunohistochemistry, and is important for PCR/Southern, Northern, or dot blot hybridization, let's examine formalin fixation in detail.

Living cells contain many molecules that are capable of either killing the cells or quickly changing their metabolic fingerprint in a short period of time. These include DNases, RNases, apoptotic factors (such as caspase-3), and so on. If you want to capture the molecular footprint of the cell at a given moment in time, you must either rapidly freeze it or "fix" it. This way, when you extract the

RNA/DNA/protein, you can be sure that cellular digestive enzymes and other processes have not altered the profile of proteins/DNA/RNA characteristic of the cell. This is clear to any pathologist who has viewed tissues from autopsies. Many times, because 10–30 hours often have elapsed from time of death to the autopsy, the cells of the tissues literally start to fall apart in a process called *autolysis.* You can easily imagine the change in the cells' proteins/DNA/RNA profile that must be occurring when the cells are basically falling apart.

There are two basic ways to fix tissues. One is to use an agent that denatures the proteins, which, of course, will include the enzymes that do most of the mayhem to a cell once a biopsy is removed from a living patient. The most common protein denaturing agents are alcohol (e.g., methanol or ethanol) and acetone. The second way to fix tissue is to make the proteins cross-link with each other, as well as with DNA and RNA. These fixatives include formalin, paraformaldehyde, and glutaraldehyde. These are often called the cross-linking fixatives. Thus, denaturing fixatives basically take the proteins in a cell and change their conformation so they no longer function. Cross-linking fixatives are more like "construction workers" because they create extensive cross-linked "cages" in the cytoplasm and nucleus involving all the macromolecules in the cell. Imagine how complex the three-dimensional spider-web-like labyrinth must be that is present within a cell after fixation with 10% neutral-buffered formalin, which is by far the most commonly used of the cross-linking fixatives! I will offer this sentence now as "food for thought" for later chapters: it is my contention that the data will show that the understanding of the biochemistry of the three-dimensional macromolecule cross-linked cage surrounding each DNA/RNA/epitope target is the key to successful in situ hybridization and immunohistochemistry! For now, this is just food for thought. We will discuss this topic and the associated data in detail later.

There is one more set of molecules to discuss before leaving this basic discussion of the different types of fixatives available to the molecular biologist. These are the molecules used by pathologists to enhance viewing the stained nucleus under the microscope. There are two classes of these chemicals: the heavy metals and picric acid. Both heavy metals (e.g., zinc and mercury) and picric acid have been added to solutions that contain formalin to enhance the cytologic appearance of the cells in the tissue. The most common such solution that has a heavy metal is probably Zenker's solution, and Bouin's solution is certainly the most common of the formalin/picric acids mixes.

An important biochemical-related question is this: do the cell fixatives alter the ability to detect the RNA or DNA in the cell? (We will leave out proteins for a while, since we are discussing Southern, Northern, or dot blot hybridization, but will come back to this issue with immunohistochemistry.) Well, denaturing fixatives do nothing to DNA or RNA that would affect our ability to detect it after we do the necessary extraction procedures. Thus, when we examined human warts either unfixed or frozen, and compared their DNA to that from tissues fixed in ethanol, we found no differences whatsoever

regarding the structure of their HPV DNA. That is, in each case, the HPV DNA was about 8000 base pairs long and was just as easy to detect by Southern blot or dot blot hybridization after extraction. However, using either picric acid or a heavy metal quickly degraded the DNA and RNA. The degradation was so quick and so complete that only a few hours of fixation with picric acid or a heavy metal rendered the DNA or RNA undetectable by Southern, Northern, or dot blot hybridization. Formalin (again, the standard 10% neutral-buffered formalin) also did not alter the HPV DNA in any way if it was used as a fixative *without* paraffin embedding. However, since the standard is, of course, to take formalin-fixed tissue and embed it in paraffin, it was evident that the process of paraffin embedding does indeed induce rare nicks in the DNA. We will discuss this point in detail later in the book. I should add that unbuffered formalin did allow for the DNA to be degraded via nicks and, thus, was less than optimal for Southern, PCR, or dot blot hybridization, but since unbuffered formalin is so rarely used, there is no reason to stress this point.

The key points I want to stress regarding formalin-fixed, paraffin-embedded tissues and their DNA and RNA are two fold:

1. The DNA and RNA are well preserved, but due to the extensive cross-links with the cellular proteins that comprise the three-dimensional formalin-induced cross-linked macromolecular cellular cage, you must do a *robust protease digestion step* prior to DNA/RNAe xtraction.
2. The DNA and RNA (and cellular proteins) will degrade slowly over time. This key point is discussed in detail in the next chapter. The practical point regarding this second point is that the DNA and RNA extracted from aged blocks (more than 5 years old) are often too degraded for Southern blot/Northern blot or PCR/RTPCR analysis. As you will see in the following chapters, the best (indeed only) way to detect the DNA and RNA in these aged blocks will be with in situ hybridization after you "regenerate" the tissue.

Let me define what I mean by "robust protease digestion" in point 1. When I extract RNA or DNA from formalin-fixed, paraffin-embedded tissues, I first incubate the tissues in 1 mg/ml of proteinase K at 60°C overnight, followed by an additional protease digestion in fresh 1 mg/ml of proteinase K at 60°C for 5 hours! Then I proceed to the RNA or DNA extraction step.

After RNA or DNA extraction, the next step for Northern and Southern blot hybridization is electrophoretic separation of the respective nucleic acid. The RNA and DNA molecules in the cell will vary in size and separate accordingly via electrophoresis. This gives us a bit more specificity to the reaction, as we are removing a lot of nontarget RNA and DNA from our nucleic acid target simply by doing the electrophoresis. Of course, we already have removed another potential source of background, nonspecific binding of our DNA or RNA probe to cellular proteins, via the robust protease digestion step. With dot blot hybridization, we are basically saying that we have so much confidence in our ability to reduce background by using the correct probe concentration and stringent conditions that we do not need electrophoretic separation, so we are just going to analyze the extracted DNA or RNA directly.

Thus, after electrophoresis (or directly with dot blot hybridization), we need to place our DNA or RNA sample on some solid support that will allow us to do the hybridization step. For Southern, Northern, or dot blot hybridization, this typically involves a filter, often made of nitrocellulose. Again, all the filter needs to do are two things: (1) allow the DNA or RNA to strongly bind to it; and (2) to remain intact during the hybridization and detection steps. With Southern and Northern blot hybridization, you simply take the agarose gel after electrophoretic separation, place the filter over it, add some weight, and incubate the entire apparatus in a solution that facilitates transfer of the nucleic acids to the filter. You can also use vacuums and DNA/RNA cross-linkers to speed up the transfer of the nucleic acids to the filter and to ensure that the maximum amount of nucleic acid stays attached to the filter. Then you only have to do the hybridization step (usually for 2–5 hours), followed by the stringent wash (often 30 minutes) and detection step and read the results. The theory and, indeed, practice of the hybridization, stringency wash, and detection steps with Southern, Northern, or dot blot hybridization are very similar to in situ hybridization. Specifically, you want to use a probe of optimal size and concentration, a wash stringent enough to remove all probe–nontarget hybrids, and a detection step in which you have optimized the signal-to-background ratio. Twenty years ago, basically all Southern/Northern/dot blots were done with radioactive probes. Thus, you would take the filter after hybridization and the stringent wash, place it in the dark with autoradiograph film, and then develop the image in a darkroom using the appropriate photographic chemicals. Today, you typically use other, safer detection methods, including chemiluminescence and colorimetric-based assays.

Although this book is devoted to in situ hybridization and immunohistochemistry with a focus on co-expression analyses, in this chapter on basic molecular pathology, let's focus on the many similarities between in situ hybridization/immunohistochemistry and the solution-phase-based methods of molecular biology, including PCR, Southern/Northern/dot blot hybridization, and Western blot hybridization.

A simple but important question regarding in situ hybridization is this: compared to the other DNA/RNA methods of PCR, Southern/Northern/dot blot hybridization, what is(are) the best fixative(s)? We have already established with PCR, Southern/Northern/dot blot hybridization that we can use unfixed tissue or tissue fixed in denaturing fixatives (such as ethanol) with equally good results, since the latter does very little to "alter or hide" the nucleic acids. We also established that formalin fixation, due to the extensive three-dimensional labyrinth it creates around each DNA and RNA molecule, requires a robust protease digestion prior to DNA/RNA extraction, after which PCR, Southern/Northern/dot blot hybridization could be done with excellent results since the formalin fixation does not alter the DNA or RNA.

Indeed, the formalin fixation protects the DNA and RNA from alterations such as conformational change by wrapping it in this cross-linked protein spider web.

So let's try to answer the question for in situ hybridization: how do unfixed tissues versus tissues fixed in denaturing fixatives versus formalin-fixed, paraffin-embedded tissues compare? And do heavy metals and picric acid influence the results with in situ hybridization? Before we try to answer the question, we need to discuss one more topic: *diffusion*. Of course, with PCR, Southern/Northern/dot blot hybridization diffusion is not an issue. The point is to extract the DNA and RNA for analyses as completely as possible. With in situ hybridization, the point is the exact opposite. We want the DNA and RNA to stay exactly where it is in the cell without moving one angstrom! It stands to reason that a DNA or RNA molecule within an unfixed cell probably would move a lot if we do in situ hybridization, considering that we are going to have to heat the tissue to 65°C–100°C, expose it to a stringent wash, and to a detection step. Will the denaturing fixatives keep the RNA and DNA from moving in the cell with in situ hybridization? Not likely, because the proteins have not been "strengthened" with the fixation, just altered in their conformation. This is easy to prove. Do PCR (or in situ hybridization) with unfixed tissue or tissue fixed in ethanol and then test the amplifying solution for the amplicon. There will be a lot that has diffused out of the cells into the solution. But how about in situ hybridization with formalin-fixed, paraffin-embedded tissues? The extensive three-dimensional protein cross-linked network serves as the equivalent of an ion-exchange resin or, if you will, a labyrinth. Have you ever tried to find your way out of a labyrinth? Or perhaps you heard a recent news article about a couple who were in a labyrinth in an amusement part and needed to dial 911 because they could not find their way out! Well, it is the same for DNA and RNA molecules. They tend to *not* diffuse with in situ hybridization when using formalin-fixed, paraffin-embedded tissues and, as a corollary, if you do PCR (or in situ hybridization) with formalin-fixed, paraffin-embedded tissues and extract the DNA from the amplifying solution, no amplicons will be detected. This point is illustrated in Figure 2-24. The tissue is a lymph node, and it was fixed in 10% buffered formalin (panel A) or acetone (panel B). The tissues were tested for two microRNAs that are abundant in lymphocytes and tend to produce a nuclear and cytoplasmic signal—miRNA-31 and -328—by in situ hybridization. Note the intense signal in panel A that is very tightly confined to the cell. Note that the formalin fixation allows for a very sharp localization of the signal to the lymphocytes. Compare this to the signal in the acetone-fixed tissue (panel B), where it is clear that most of the signal left the cell and is either in the stroma or was lost in the RNA in situ hybridization solution.

So, this simple idea of a three-dimensional cross-linked protein network with formalin-fixed, paraffin-embedded tissues explains why, indeed, it is true that such tissues are optimal for in situ hybridization! But can you add heavy metals or picric acid to the formalin and still get good in situ hybridization data? The answer in general is no. Remember, only a few hours of fixation in heavy metal or picric acid are enough to degrade the RNA or HPV DNA to the point that the fragments are so small that they cannot be detected after electrophoresis. Such small fragments (say, fewer than 50 nucleotides in size) are also too small to be detected by in situ hybridization.

IMMUNOHISTOCHEMISTRY

Let's go a bit more into the specifics of immuno-histochemistry. First of all, what fixation is best for immunohistochemistry? If you have been doing immunohistochemistry for at least 10 years, you probably have done the method using either unfixed (frozen) sections or cells fixed in ethanol/methanol or acetone. But are these the best fixatives? The answer really lies within the biochemistry of the cell and the reactions you use for immunohistochemistry. If the tissue is unfixed, or if it's fixed with a denaturing fixative, then you face the same problem with immunohistochemistry as you did with in situ hybridization, namely diffusion. Since the epitope has nothing to anchor it down within the cell, it is certainly possible that either it may detach during the immunohistochemistry process and no longer be detected, or one of the many molecules used in the detection process may do the same, leading to a negative result even though the epitope was present in the cells of the tissue. But if you use formalin-fixed, paraffin-embedded tissues for immunohistochemistry, then the three-dimensional protein-protein-macromolecule network can serve the very important function of physically and chemically (through biochemical bonds including the ionic and hydrogen bonds) keeping the epitope-primary antibody-secondary antibody-chromogen "tree" in place. Also, denaturing fixatives, by changing the conformation of the proteins, can render an epitope unavailable to the primary antibody. Once that happens, there is nothing you can do chemically or physically to the tissue to expose the antigen. This is true primarily because the tissue that is unfixed or fixed in alcohol or acetone has no internal rigid framework to speak of. If you do protease digestion or antigen retrieval (i.e., heating the tissue at 95°C), then the cell may be destroyed. Formalin does not tend to change the confirmation of proteins as much as simply "hiding" the epitope. Also, since it provides an extensive cross-linked network throughout the cell, you can heat the tissue to 95°C for 60 minutes *and* expose it to protease, yet the morphology will remain as clear and intact as the original hematoxylin and eosin stained slide assuming, of course, that the time the tissue was incubated in formalin was sufficient.

As you may gather, I recommend that you use only formalin-fixed, paraffin-embedded tissues for immunohistochemistry. There are several reasons for this: (1) you will get the best possible morphology; (2) formalin-fixed, paraffin-embedded tissues are mandatory if you wish to do co-expression analyses with a RNA/DNA/microRNA and a protein of interest; and (3) in my experience, if you use a simple optimizing protocol for immunohistochemistry, then over 90% of primary antibodies will generate a good signal with immunohistochemistry using formalin-fixed, paraffin-embedded tissues. Let's examine some data to underscore this point. Figure 2-24 shows some

data from the same lymph node, but where the test is now immunohistochemistry for two proteins: CD45 (present in most lymphocytes) and smooth muscle actin (present in the larger blood vessel wall). Compare the data for the formalin-fixed tissue (panel C) to that for the tissue fixed in 70% ethanol (panel D). The CD45 and smooth muscle actin signals in the formalin-fixed tissue are very sharply defined, as if they had been drawn with a sharp pen. They are also continuous around the cells, as they should be. The equivalent signal in the ethanol-fixed tissue is *diffuse* and not continuous. It is reasonable to speculate that the three-dimensional macromolecule cross-linked cage around these two epitopes in the formalin-fixed tissue kept the signal sharply localized in the cell to the point where these proteins are actually located. The ethanol fixation does not produce this three-dimensional macromolecule cross-linked cage and, as a consequence, the signal is *not* held rigidly in the place where the epitope is located but, rather, *diffuses* out of the cell.

There is another area where in situ hybridization and immunohistochemistry share a commonality that may not be evident on first glance. That is in the area of degradation of the target (DNA, RNA, or protein epitope) in formalin-fixed, paraffin-embedded tissues over time. Our discussions of formalin-fixed, paraffin-embedded tissues suggests that the macromolecules become very stable with the fixation. This is relatively true but *only* if you define the conditions. Specifically, expose DNA, RNA, or protein epitopes in formalin-fixed, paraffin-embedded tissues to *dry* heat, and they will quickly be undetectable by in situ hybridization or immunohistochemistry, respectively, even if they are in high copy numbers. Allow the paraffin blocks (or unstained slides) to sit for many years and then analyze them, and you can certainly see reduced signals with both DNA/RNA and protein targets. We go into the details of this observation in detail, and it reinforces the point that immunohistochemistry and in situ hybridization can be viewed as one basic method—in situ-based molecular pathology—with many features in common.

Let's now make brief mention of Western blot analysis. With this method, you extract the proteins from unfixed cells and run them on an electrophoretic gel typically using polyacrylamide to separate them based on size, conformation, and charge. You then transfer the proteins to a filter (typically nitrocellulose or polyvinylidene fluoride) as we did for Southern and Northern blot hybridization using capillary action, although it is typical to include either a vacuum or an electric current to speed up the transfer from the gel to the filter. You can do the equivalent of a dot blot hybridization and place the protein extract directly on a filter; this is called the ELISA method. With Western blots, you often denature the proteins with a strong detergent such as SDS to allow better separation of the proteins on the gel, thus making it more likely that the epitope will be available for the detection step. The SDS also confers a negative charge on the denatured proteins and, hence, the primary variable that is used to separate them is their size. Either way, you detect the protein epitope of interest using the specific antibody of interest. I routinely use the same primary antibodies with immunohistochemistry and Western blot analysis. One key similarity between Western blotting/ELISA and immunohistochemistry is the need for a protein block to prevent nonspecific adsorption of the antibodies/detection enzymes onto the filter/intact cells. In each case, you typically use bovine serum albumin. With Western blotting/ELISA methods, chemiluminescence is commonly used for the detection step, although colorimetric assays may also be used.

The suggested readings [1–252] cover a broad range of papers that deal primarily with the biochemistry and structure of macromolecules, especially in tissue sections.

These papers emphasize that both immunohistochemistry and in situ hybridization are primarily biochemical reactions in which the structure of RNA, DNA, and proteins is key to understanding what actually happens at the macromolecule level when we do any in situ-based methodology. I have also included references that detail the three-dimensional macromolecule structure of the intact cell, because this is such a key component of in situ hybridization or immunohistochemistry.

SUGGESTED READINGS

Structural aspects of DNA, RNA, and proteins

[1] Abdalla MY, Britigan BE, Wen F, et al. Down-regulation of heme oxygenase-1 by hepatitis C virus infection in vivo and by the in vitro expression of hepatitis C core protein. J Infect Dis 2004;190:1109–18.

[2] Akari H, Terao K, Nam KH, Adachi A, Yoshikawa Y. Comparative analysis of human and macaque monkey CD4: differences in formaldehyde lability and conformation. Exp Anim 1998;47:23–7.

[3] al-Shawi R, Burke J, Wallace H, et al. The herpes simplex virus type 1 thymidine kinase is expressed in the testes of transgenic mice under the control of a cryptic promoter. Mol Cell Biol 1991;11:4207–16.

[4] Alterio V, Aurilia V, Romanelli A, et al. Crystal structure of an S-formylglutathione hydrolase from Pseudoalteromoas haloplanktis TAC125. Biopolymers 2010;93:669–77.

[5] Anderson K, Karle E, Reiner A. A pre-embedding triple-label electron microscopic immunohistochemical method as applied to the study of multiple inputs to defined tegmental neurons. J Histochem Cytochem 1994;42:49–52.

[6] Ash W, Stockner T, MacCallum J, Tieleman D. Computer modeling of polyleucine-based coiled coil dimers in a realistic membrane environment: insight into helix-helix interactions in membrane proteins. Biochemistry 2004;43:9050–60.

[7] Baev AS, Lyubchenko Yu L, Lazurkin Yu S, Trifonov EN, Frank-Kamenetskii MD. Study of low-melting segments in T2 phage DNA by electron microscopy and the kinetic formaldehyde method. Mol Biol 1972;6:612–7.

[8] Banci L, Bertini I, D'Amelio N, et al. Fully metallated S134N Cu,Zn-superoxide dismutase displays abnormal mobility and intermolecular contacts in solution. J Biol Chem 2005;280:35815–21.

[9] Bielka H, Schneiders I, Henske A. [Studies on the structure of ribosomal RNA from normal and tumor tissues. I. Thermal denaturation, reaction with formaldehyde and base composition] Untersuchungen zur struktur ribosomaler RNS aus normal und tumorgeweben. I. Thermische denaturierung, reakton mit formaldehyd und basenzusammensetzung. Acta Biol Med Ger 1964;13:13–23.

[10] Bishop B, Dasgupta J, Klein M, et al. Crystal structures of four types of human papillomavirus L1 capsid proteins: understanding the specificity of neutralizing monoclonal antibodies. J Biol Chem 2007;282:31803–11.

[11] Bizzzini B, Raynaud M. [Detoxification of protein toxins by formol: supposed mechanisms and new developments] La detoxication des toxines proteiques par le formol: mecanismes supposes at nouveaux developpements. Biochimie 1974;56:297–303.

[12] Blank M, Tang Y, Yamashita M, et al. A tumor suppressor function of Smurf2 associated with controlling chromatin landscape and genome stability through RNF20. Nat Med 2012;18:227–34.

[13] Boedtker H, Kelling DG. The ordered structure of 5S RNA. Biochem Biophys Res Commun 1967;29:758–66.

[14] Boehmer T, Enninga J, Dales S, Blobel G, Zhong H. Depletion of a single nucleoporin, Nup107, prevents the assembly of a subset of nucleoporins into the nuclear pore complex. Proc Natl Acad Sci U S A 2003;100:981–5.

[15] Bohne S, Sletten K, Menard R, et al. Cleavage of AL amyloid proteins and AL amyloid deposits by cathepsins B, K, and L. J Pathol 2004;203:528–37.

[16] Boyle MP, Enke RA, Mogayzel Jr. PJ, et al. Effect of adeno-associated virus-specific immunoglobulin G in human amniotic fluid on gene transfer. Hum Gene Ther 2003;14:365–73.

[17] Boyoglu S, Vig K, Pillai S, et al. Enhanced delivery and expression of a nanoencapsulated DNA vaccine vector for respiratory syncytial virus. Nanomedicine 2009;5:463–72.

[18] Braconi C, Valeri N, Gasparini P, et al. Hepatitis C virus proteins modulate microRNA expression and chemosensitivity in malignant hepatocytes. Clin Cancer Res 2010;16:957–66.

[19] Brandtzaeg PS. Rhodamine conjugates: specific and nonspecific binding properties in immunohistochemistry. Ann N Y Acad Sci 1975;254:35–54.

[20] Braun V, Herrmann C. Docking of the periplasmic FecB binding protein to the FecCD transmembrane proteins in the ferric citrate transport system of Escherichia coli. J Bacteriol 2007;189:6913–8.

[21] Bredemeyer AL, Huang CY, Walker LM, Bassing CH, Sleckman BP. Aberrant V (D) J recombination in ataxia telangiectasia mutated-deficient lymphocytes is dependent on nonhomologous DNA end joining. J Immunol 2008;181:2620–5.

[22] Brentani M, Kubota M, Brentani R. Studies on the secondary structure of nuclear ribonucleic acids. Biochem J 1972;130:11–17.

[23] Brodolin KL, Studitskii VM, Mirzabekov AD. [Study of the structure of Escherichia coli RNA polymerase and its complex with the lacUV5-promotor using protein-protein and DNA-protein crosslinks, formed by formaldehyde] Issledovanie struktury RNK-polimerazy Escherichia coli i ee komplekxa s lacUV5-promotorom s pomoshch'iu belok-belkovykh i DNK-belkovykh sshivok, obrazuemykh formal'degidom. Mol Biol (Mosk) 1993;27:1085–93.

[24] Brookes SM, Hyatt AD, Wise T, Parkhouse RM. Intracellular virus DNA distribution and the acquisition of the nucleoprotein core during African swine fever virus particle assembly: ultrastructural in situ hybridisation and DNase-gold labelling. Virology 1998;249:175–88.

[25] Brown JA, Fowler JD, Suo Z. Kinetic basis of nucleotide selection employed by a protein template-dependent DNA polymerase. Biochemistry 2010;49:5504–10.

[26] Brown LR, Bradbury JH. Proton-magnetic-resonance studies of the lysine residues of ribonuclease A. Eur J Biochem 1975;54:219–27.

[27] Brunton H, Goodarzi AA, Noon AT, et al. Analysis of human syndromes with disordered chromatin reveals the impact of heterochromatin on the efficacy of ATM-dependent G2/M checkpoint arrest. Mol Cell Biol 2011;31:4022–35.

[28] Cardinale A, Racaniello M, Saladini S, et al. Sublethal doses of β-amyloid peptide abrogate DNA-dependent protein kinase activity. J Biol Chem 2012;287:2618–31.

[29] Castel M, Belenky M, Cohen S, Ottersen OP, Storm-Mathisen J. Glutamate-like immunoreactivity in retinal terminals of the mouse suprachiasmatic nucleus. Eur J Neurosci 1993;5:368–81.

[30] Ceci P, Ilari A, Falvo E, Giangiacomo L, Chiancone E. Reassessment of protein stability, DNA binding, and protection of Mycobacterium smegmatis Dps. J Biol Chem 2005;280:34776–85.

[31] Cepica A, Beauregard M, Qian B. Fluorescence spectroscopy monitoring of the conformational restraint of formaldehyde- and glutaraldehyde-treated infectious bursal disease virus proteins. Vaccine 1998;16:1957–61.

[32] Chahwan R, van Oers JM, Avdievich E, et al. The ATPase activity of MLH1 is required to orchestrate DNA double-strand breaks and end processing during class switch recombination. J Exp Med 2012;209:671–8.

[33] Chima SC, Agostini HT, Ryschkewitsch CF, Lucas SB, Stoner GL. Progressive multifocal leukoencephalopathy and JC virus genotypes in West African patients with acquired immunodeficiency syndrome: a pathologic

and DNA sequence analysis of 4 cases. Arch Pathol Lab Med 1999;123:395–403.

[34] Cho MY, Kim TH, Yi SY, Jung WH, Park KH. Relationship between Epstein-Barr virus-encoded RNA expression, apoptosis and lymphocytic infiltration in gastric carcinoma with lymphoid-rich stroma. Med Princ Pract 2004;13:353–60.

[35] Choi YS, Lee JS, Choi YK, Kim HS, Kim CJ. Expression of angiostatin using DNA-based semliki forest virus replicon. J Vet Sci 2002;3:41–5.

[36] Cooper S, Khatib F, Treuille A, et al. Predicting protein structures with a multiplayer online game. Nature 2010;466:756–60.

[37] Cox RA, Bonanou SA. A possible structure of the rabbit reticulocyte ribosome. An exercise in model building. Biochem J 1969;114:769–74.

[38] Coy JF, Wiemann S, Bechmann I, et al. Pore membrane and/or filament interacting like protein 1 (POMFIL1) is predominantly expressed in the nervous system and encodes different protein isoforms. Gene 2002;290:73–94.

[39] Daniel DC, Kinoshita Y, Khan MA, et al. Internalization of exogenous human immunodeficiency virus-1 protein, Tat, by KG-1 oligodendroglioma cells followed by stimulation of DNA replication initiated at the JC virus origin. DNA Cell Biol 2004;23:858–67.

[40] Datta RK, Ghosh JJ. Effect of strychnine sulphate and nialamide on hydrogen bonded structure of ribonucleic acid of brain cortex ribosomes. J Neurochem 1964;11:357–66.

[41] Datta RK, Ghosh JJ. Mescaline-induced changes of brain-cortex ribosomes. Effect of mescaline on the hydrogen-bonded structure of ribonucleic acid of brain-cortex ribosomes. Biochem J 1970;117:969–80.

[42] de Graaf B, Clore A, McCullough AK. Cellular pathways for DNA repair and damage tolerance of formaldehyde-induced DNA-protein crosslinks. DNA Repair (Amst) 2009;8:1207–14.

[43] De la Llosa P, Durosay M, Tertrin-Clary C, Jutisz M. Chemical modification of lysine residues in ovine luteinizing hormone. Effect on biological activity. Biochem Biophys Acta 1974;342:97–104.

[44] Deckelbaum RA, Majithia A, Booker T, Henderson JE, Loomis CA. The homeoprotein engrailed 1 has pleiotropic functions in calvarial intramembranous bone formation and remodeling. Development 2006;133:63–74.

[45] Dekker J, Rippe K, Dekker M, Kleckner N. Capturing chromosome conformation. Science 2002;295:1306–11.

[46] Deussing JM, Kühne C, Pütz B, et al. Expression profiling identifies the CRH/CRH-RI system as a modulator of neurovascular gene activity. J Cereb Blood Flow Metab 2007;27:1476–95.

[47] Dias-Gunasekara S, Gubbens J, van Lith M, et al. Tissue-specific expression and dimerization of the endoplasmic reticulum oxidoreductase Ero1beta. J Biol Chem 2005;280:33066–75.

[48] Dixon DP, Cummins L, Cole DJ, Edwards R. Glutathione-mediated detoxification systems in plants. Curr Opin Plant Biol 1998;1:258–66.

[49] Doane KJ, Yang G, Birk DE. Corneal cell-matrix interactions: type VI collagen promotes adhesion and spreading of corneal fibroblasts. Exp Cell Res 1992;200:490–9.

[50] Dobrov EN, Kust SV. [A spectrophotometric study of the structure of RNA in particles of tobacco mosaic virus] Spektrofotometricheskoe issldovanie struktury RNK v chastitsakh virusa tabachnoi mozaiki. Vopr Virusol 1971;16:366–71.

[51] Dobrov EN, Kust SV, Mazhul LA, Tikhonenko TI. [Study of the RNA structure in the particles of the rod-like cucumber virus] Izuchenie struktury RNK v chastitsakh palochkovidnogo virusa ogurtsov. Nauchnye Doki Vyss Shkoly Biol Nauki 1974:96–102.

[52] Dobrov EN, Kust SV, Tikchonenko TI. The structure of single-stranded virus RNA in situ. A study of absorption spectra and optical rotatory dispersion of tobacco mosaic virus and potato virus X preparations. J Gen Virol 1972;16:161–72.

[53] Dolgova EV, Proskurina AS, Nikolin VP, et al. "Delayed death" phenomenon: a synergistic action of cyclophosphamide and exogenous DNA. Gene 2012;495:134–45.

[54] Dragan AI, Li Z, Makeyeva EN, et al. Forces driving the binding of homeodomains to DNA. Biochemistry 2006;45:141–51.

[55] Ebrahimi B, Allsopp TE, Fazakerley JK, Harkiss GD. Phenotypic characterisation and infection of ovine microglial cells with Maedi-Visna virus. J Neurovirol 2000;6:320–8.

[56] Eccleston J, Schrader CE, Yuan K, Stavnezer J, Selsing E. Class switch recombination efficiency and junction microhomology patterns in Msh2-M1h1-, and Exo1-deficient mice depend on the presence of mu switch region tandem repeats. J Immunol 2009;183:1222–8.

[57] Eccleston J, Yan C, Yuan K, Alt FW, Selsing E. Mismatch repair proteins MSH2, MLH1 and EXO1 are important for class-switch recombination events occurring in B cells that lack nonhomologous end joining. J Immunol 2011;186:2336–43.

[58] Endt H, Sprung CN, Keller U, et al. Detailed analysis of DNA repair and senescence marker kinetics over the life span of a human fibroblast cell line. J Gerontol A Biol Sci Med Sci 2011;66:367–75.

[59] Eslami MH, Gangadharan SP, Sui X, et al. Gene delivery to in situ veins: differential effects of adenovirus and adeno-associated viral vectors. J Vasc Surg 2000;31:1149–59.

[60] Faedo M, Ford CE, Mehta R, Blazek K, Rawlinson WD. Mouse mammary tumor-like virus is associated with p53 nuclear accumulation and progesterone receptor positivity but not estrogen positivity in human female breast cancer. Clin Cancer Res 2004;10:4417–9.

[61] Fan J, Li L, Small D, Rassool F. Cells expressing FLT3/ITD mutations exhibit elevated repair errors generated through alternative NHEJ pathways: implications for genomic instability and therapy. Blood 2010;116:1737–46.

[62] Fleckenstein J, Kuhne M, Seegmuller K, et al. The impact of individual in vivo repair of DNA double-strand breaks on oral mucositis in adjuvant radiotherapy of head-and-neck cancer. Int J Radiat Oncol Biol Phys 2011;81:1465–72.

[63] Fleishman SJ, Whitehead TA, Ekiert DC, et al. Computational design of proteins targeting the conserved stem region of influenza hemagglutinin. Science 2011;332:816–21.

[64] Fowler CB, Evers DL, O'Leary TJ, Mason JT. Antigen retrieval causes protein unfolding: evidence for a linear epitope model of recovered immunoreactivity. J Histochem Cytochem 2011;59:366–81.

[65] Fowler S, Byron O, Jumel K, et al. Novel configurations of high molecular weight species of the pertussis toxin vaccine component. Vaccine 2003;21:2678–88.

[66] Frelin L, Alheim M, Chen A, et al. Low dose and gene gun immunization with a hepatitis C virus nonstructural (NS) 3 DNA-based vaccine containing NS4A inhibit NS3/4A-expressing tumors in vivo. Gene Ther 2003;10:686–99.

[67] Friedrich RE, Bartel-Friedrich S, Lobeck H, Niedobitek G, Arps H. Epstein-Barr virus DNA, intermediate filaments and epithelial membrane antigen in nasopharyngeal carcinoma. Anticancer Res 2000;20:4909–16.

[68] Frischholz S, Beier F, Girkontaite I, et al. Characterization of human type X procollagen and its NC-1 domain expressed as recombinant proteins in HEK293 cells. J Biol Chem 1998;273:4547–55.

[69] Galazka M, Tang M, DeBari VA, et al. Modification of beta 2 glycoprotein I by glutardialdehyde. Conformational changes and aggregation accompany exposure of the cryptic autoepitope. Appl Biochem Biotechnol 1999;76:1–13.

[70] Gapud EJ, Lee BS, Mahowald GK, Bassing CH, Sleckman BP. Repair of chromosomal RAG-mediated DNA breaks by mutant RAG proteins lacking phosphatidylinositol 3-like kinase consensus phosphorylation sites. J Immunol 2011;187:1826–34.

[71] Garcia V, Phelps SE, Gray S, Neale MJ. Bidirectional resection of DNA double-strand breaks by Mre11 and Exo1. Nature 2011;479:241–4.

[72] Garcia-Martin A, Kwa LG, Strohmann B, et al. Structural role of bacteriochlorophyll ligated in the energetically unfavorable beta-position. J Biol Chem 2006;281:10626–34.

[73] Garel A, Kovacs AM, Champagne M, Daune M. Comparison between histones F Van d F2a2 of chicken erythrocyte. II. Interaction with homologous DNA. Biochem Biophys Acta 1975;395:16–27.

[74] Ghosh SK, Chaudhuri S, Roy J, Sinha NK, Sen A. Physicochemical investigations on buffalo-lactoglobulin. Studies on sedimentation, diffusion and hydrogen ion titration. Arch Biochem Biophys 1971;144:6–15.

[75] Giladi H, Ketzinel-Gilad M, Rivkin L, et al. Small interfering RNA inhibits hepatitis B virus replication in mice. Mol Ther 2003;8:769–76.

[76] Gluck M, Sweeney WV. 13C-NMR of Clostridium pasteurianum ferredoxin after reductive methylation of the amines using (13C) formaldehyde. Biochim Biophys Acta 1990;1038:146–51.

[77] Gnatenko D, Arnold TE, Zolotukhin S, et al. Characterization of recombinant adeno-associated virus-2 as a vehicle for gene delivery and expression into vascular cells. J Investig Med 1997;45:887–98.

[78] Goemaere EL, Cascales E, Lloubes R. Mutational analyses define helix organization and key residues of a bacterial membrane energy-transducing complex. J Mol Biol 2007;366:1424–36.

[79] Gonzalez CF, Proudfoot M, Brown G, et al. Molecular basis of formaldehyde detoxification. Characterization of two

S-formylglutathione hydrolases from Escherichia coli, FrmB and YeiG. J Biol Chem 2006;281:14514–22.

[80] Gosert R, Kanjanahaluethai A, Egger D, Bienz K, Baker SC. RNA replication of mouse hepatitis virus takes place at double-membrane vesicles. J Virol 2002;76:3697–708.

[81] Gospodinov A, Vaissiere T, Krastev DB, et al. Mammalian Ino80 mediates double-strand break repair through its role in DNA end strand resection. Mol Cell Biol 2011;31:4735–45.

[82] Goto M, Nagatomo Y, Hasui K, et al. Chromaticity analysis of immunostained tumor specimens. Pathol Res Pract 1992;188:433–7.

[83] Gouk SS, Lim TM, Teoh SH, Sun WQ. Alterations of human acellular tissue matrix by gamma irradiation: histology, biomechanical property, stability, in vitro cell repopulation, and remodeling. J Biomed Mater Res B Appl Biomater 2008;84:205–17.

[84] Gray SJ, Blake BL, Criswell HE, et al. Directed evolution of a novel adeno-associated virus (AAV) vector that crosses the seizure-compromised blood-brain barrier (BBB). Mol Ther 2010;18:570–8.

[85] Grayling RA, Becktel WJ, Reeve JN. Structure and stability of histone HMf from the hyperthermophilic archaeon Methanothermus fervidus. Biochemistry 1995;34:8441–8.

[86] Guikema JE, Schrader CE, Brodsky MH, et al. p53 represses class switch recombination to IgG2a through its antioxidant function. J Immunol 2010;184:6177–87.

[87] Gunn A, Bennardo N, Cheng A, Stark JM. Correct end use during endjoining of multiple chromosomal double strand breaks is influenced by repair protein RAD50, DNA-dependent protein kinase DNA-PKcs, and transcription. J Biol Chem 2011;286:42470–82.

[88] Guo H, Lokko K, Zhang Y, et al. Overexpression and characterization of Wzz of Escherichia coli 086:H2. Protein Expr Purif 2006;48:49–55.

[89] Gwak GY, Lee DH, Moon TG, et al. The correlation of hepatitis B virus pre-S mutation with cellular oxidative DNA damage in hepatocellular carcinoma. Hepatogastroenterology 2008;55:2028–32.

[90] Hauser N, Paulsson M. Native cartilage matrix protein (CMP). A compact trimer of subunits assembled via a coiled-coil alpha-helix. J Biol Chem 1994;269:25747–53.

[91] Hawkins AJ, Dodd PR. Localisation of GABA (A) receptor subunits in the CNS using RT-PCR. Brain Res Brain Res Photoc 2000;6:47–52.

[92] Herpers B, Xanthakis D, Rabouille C. ISH-IEM: a sensitive method to detect endogenous mRNAs at the ultrastructural level. Nat Protoc 2010;5:678–87.

[93] Hershman JM, Okunyan A, Rivina Y, Cannon S, Hogen V. Prevention of DNA double-strand breaks induced by radioiodide-(131) I in FRTL-5 thyroid cells. Endocrinology 2011;152:1130–5.

[94] Hisatomi T, Sueoka-Aragane N, Sato A, et al. NK314 potentiates antitumor activity with adult T-cell leukemia-lymphoma cells by inhibition of dual targets on topoisomerase II{alpha} and DNA-dependent protein kinase. Blood 2011;117:3575–84.

[95] Hollander AP, Heathfield TF, Liu JJ, et al. Enhanced denaturation of the alpha (II) chains of type-II collagen in normal adult human intervertebral discs compared with femoral articular cartilage. J Orthop Res 1996;14:61–6.

[96] Holmes DS, Bonner J. Preparation, molecular weight, base composition, and secondary structure of giant nuclear ribonucleic acid. Biochemistry 1973;12:2330–8.

[97] Hu HQ, Sun YN, Luo SP, et al. Generation of a mouse monoclonal antibody recognizing both the native and denatured forms of human VEGF. Hybridoma (Larchmt) 2009;28:51–7.

[98] Hu Y, Faham S, Roy R, Adams MW, Rees DC. Formaldehyde ferredoxin oxidoreductase from Pyrococcus furiosus: the 1.85 A resolution crystal structure and its mechanistic implications. J Mol Biol 1999;286:899–914.

[99] Hubert Jr. L, Lin Y, Dion V, Wilson JH. Topoisomerase 1 and single-strand break repair modulate transcription-induced CAG repeat contraction in human cells. Mol Cell Biol 2011;31:3105–12.

[100] Hurley TD, Yang Z, Bosron WF, Weiner H. Crystallization and preliminary X-ray analysis of bovine mitochondrial aldehyde dehydrogenase and human glutathione-dependent formaldehyde dehydrogenase. Adv Exp Med Biol 1993;328:245–50.

[101] Inbal A, Loscalzo J. Glycocalicin binding to von Willebrand factor adsorbed onto collagen-coated or polystyrene surfaces. Thromb Res 1989;56:347–57.

[102] Iwata M, Carlson SS. A large chondroitin sulfate proteoglycan has the characteristics of a general extracellular matrix component of adult brain. J Neurosci 1993;13:195–207.

[103] Jani D, Lutz S, Marshall NJ, et al. Sus1, Cdc31, and the Sac3 CID region form a conserved interaction platform that promotes nuclear pore association and mRNA export. Mol Cell 2009;33:727–37.

[104] Jekely G, Arendt D. Cellular resolution expression profiling using confocal detection of NBT/BCIP precipitate by reflection microscopy. Biotechniques 2007;42:751–5.

[105] Jinn TL, Chiu CC, Song WW, Chen YM, Lin CY. Azetidine-induced accumulation of class I small heat shock proteins in the soluble fraction provides thermotolerance in soybean seedlings. Plant Cell Physiol 2004;45:1759–67.

[106] Jo N, Ju M, Nishijima K, et al. Inhibitory effect of an antibody to cryptic collagen type IV epitopes on choroidal neovascularization. Mol Vis 2006;12:1243–9.

[107] Joh NH, Min A, Faham S, et al. Modest stabilization by most hydrogen-bonded side-chain interactions in membrane proteins. Nature 2008;453:1266–70.

[108] Joosten EA, Reshilov LN, Gispen WH, Bar PR. Embryonic form of N-CAM and development of the rat corticospinal tract; immuno-electron microscopical localization during spinal white matter ingrowth. Brain Res Dev Brain Res 1996;94:99–105.

[109] Jorgensen E, Stinson A, Shan L, et al. Cigarette smoke induces endoplasmic reticulum stress and the unfolded protein response in normal and malignant human lung cells. BMC Cancer 2008;8:229.

[110] Journet L, Bouveret E, Rigal A, et al. Import of colicins across the outer membrane of Escherichia coli involves multiple protein interactions in the periplasm. Mol Microbiol 2001;42:331–44.

[111] Kanai R, Rabkin SD, Yip S, et al. Oncolytic virus-mediated manipulation of DNA damage response: synergy with chemotherapy in killing glioblastoma stem cells. J Natl Cancer Inst 2012;104:42–55.

[112] Keimling M, Volcic M, Csernok A, et al. Functional characterization connects individual patient mutations in ataxia telangiectasia mutated (ATM) with dysfunction of specific DNA double-strand break-repair signaling pathways. FASEB J 2011;25:3849–60.

[113] Kemler R, Mossmann H, Strohmaier U, Kickhofen B, Hammer DK. In vitro studies on the selective binding of IgG from different species to tissue sections of the bovine mammary gland. Eur J Immunol 1975;5:603–8.

[114] Kendall T, Mukai L, Jannuzi AL, Bunch TA. Identification of integrin beta subunit mutations that alter affinity for extracellular matrix ligand. J Biol Chem 2011;286:30981–93.

[115] Khoury M, Courties G, Fabre S, et al. Adeno-associated virus type 5-mediated intraarticular administration of tumor necrosis factor small interfering RNA improves collagen-induced arthritis. ArthritisRh eum 2010;62:765–70.

[116] Kim Han JS, O'Malley KL. Cell stress induced by the parkinsonian mimetic, 6-hydroxydopamine, is concurrent with oxidation of the chaperone, ERp57 and aggresome formation. AntoxidRe doxS ignal 2007;9:2255–64.

[117] Kiselev LL, Frolova L, Borisova OF, Kukhanova MK. [Ont hes econdarys tructure of transfer RNA determined from data of the formaldehyde reaction and ribonuclease hydrolysis.] O vtorichno I strukture transportnykh rnk po dannym reaktsii s formal'degidom i gidroliza ribonukleazoi. Biokhimiia 1964:116–25.

[118] Klein CR, Kesseler FP, Perrei C, et al. A novel dye-linked formaldehyde dehydrogenase with some properties indicating the presence of a protein-bound redox-active quinone cofactor. BiochemJ 1994;301:289–95.

[119] Klingen AR, Ullmann GM. Negatively charged residues and hydrogen bonds tune the ligand histidine pKa values of Rieske iron-sulfur proteins. Biochemistry 2004;43:12383–89.

[120] Koana T, Takahashi T, Tsujimura H. Reduction of spontaneous somatic mutation frequency by a low-dose X irradiation of Drosophila larvae and possible involvement of DNA single-strand damage repair. RadiatRe s 2012;177:265–71.

[121] Koike M, Yutoku Y, Koike A. KARP-1 works as a heterodimer with Ku70, but the function of KARP-1 cannot perfectly replace that of Ku80 in DSB repair. ExpCe llRe s 2011;317:2267–75.

[122] Komura J, Ikehata H, Mori T, Ono T. Fully functional global genome repair of (6-4) photoproducts and compromised transcription-coupled repair of cyclobutane pyrimidine dimers in condensed mitotic chromatin. ExpCe llRe s 2012;318:623–31.

[123] Kooijman EE, Tieleman DP, Testerink C, et al. Ane lectrostatic/hydrogenb ond switch as the basis for the specific interaction of phosphatidic acid with proteins. JB iolCh em 2007;282:11356–64.

[124] Kramer HS. Preliminary studies on the clinical application of antigenically altered collagen as a biomaterial in oral surgery. TransI ntCo nf Oral Surg 1973;4:223–8.

[125] Kubic VL, Brunning RD. Immunohistochemical evaluation of neoplasms in bone marrow biopsies using

monoclonal antibodies reactive in paraffin-embedded tissue. Mod Pathol 1989;2:618–29.

[126] Kust SV, Dobrov EN, Tikhonenko TI. NA structure in potato virus X particles. Mol Biol 1972;6:33–40.

[127] LaPres JJ, Glover E, Dunham EE, Bunger MK, Bradfield CA. ARA9 modifies agonist signaling through an increase in cytosolic aryl hydrocarbon receptor. J Biol Chem 2000;275:6153–9.

[128] Larsen RA, Foster-Hartnett D, McIntosh MA, Postle K. Regions of Escherichia coli TonB and FepA proteins essential for in vivo physical interactions. J Bacteriol 1997;179:3213–21.

[129] LaVail JH, Tauscher AN, Sucher A, Harrabi O, Brandimarti R. Viral regulation of the long distance axonal transport of herpes simplex virus nucleocapsid. Neuroscience 2007;146:974–85.

[130] Le L, Ayer S, Place AR, Benyajati C. Analysis of formaldehyde-induced Adh mutations in Drosophila by RNA structure mapping and direct sequencing of PCR-amplified genomic DNA. Biochemical Genetics 1990;28:367–87.

[131] Lensink MF, Govaerts C, Ruysschaert JM. Identification of specific lipid-binding sites in integral membrane proteins. J Biol Chem 2010;285:10519–26.

[132] Leonardi R, Almeida LE, Loreto C. Lubricin immunohistochemical expression in human temporomandibular joint disc with internal derangement. J Oral Pathol Med 2011;40:587–92.

[133] Li H, Haurigot V, Doyon Y, et al. In vivo genome editing restores haemostasis in a mouse model of haemophilia. Nature 2011;475:217–21.

[134] Lickert H, Cox B, Wehrle C, et al. Dissecting Wnt/beta-catenin signaling during gastrulation using RNA interference in mouse embryos. Development 2005;132:2599–609.

[135] Liu X, Wang Y, Benaissa S, et al. Homologous recombination as a resistance mechanism to replication-induced double-strand breaks caused by the antileukemia agent CNDAC. Blood 2010;116(10):1737–46.

[136] Liu Y, Liu R, Mou Y, Zhou G. Spectroscopic identification of interactions of formaldehyde with bovine serum albumin. J Biochem Mol Toxicol 2011;25:95–100.

[137] Lodish HF. Secondary structure of bacteriophage f2 ribonucleic acid and the initiation of in vitro protein biosynthesis. J Mol Biol 1970;50:689–702.

[138] Looger LL, Dwyer MA, Smith JJ, Hellinga HW. Computational design of receptor and sensor proteins with novel functions. Nature 2003;423:185–90.

[139] Lopez JA, Li CQ, Weisman S, Chambers M. The glycoprotein Ib-IX complex-specific monoclonal antibody SZ1 binds to a conformation-sensitive epitope on glycoprotein IX: implications for the target antigen of quinine/quinidine-dependent autoantibodies. Blood 1995;85:1254–8.

[140] Lubin M, Erisir A, Aoki C. Ultrastructural immunolocalization of the alpha 7 nAChR subunit in guinea pig medial prefrontal cortex. Ann N Y Acad Sci 1999;868:628–32.

[141] Machida T, Kameyama K, Onda M, Naito Z, Kumazaki Characteristic localisation of denatured high-density lipoprotein (HDL) at the periphery of a lipid core in human atherosclerotic lesions. Pathology 2005;37:32–8.

[142] Magea J, Furukawa C, Ogata H. Dose-rate effect on proliferation suppression in human cell lines continuously exposed to γ rays. Radiat Res 2011;176:447–58.

[143] Malewicz M, Kadkhodaei B, Kee N, et al. Essential role for DNA-PK-mediated phosphorylation of NR4A nuclear orphan receptors in DNA double-strand break repair. Genes Dev 2011;25:2031–40.

[144] Mandal SM, Hegde ML, Chatterjee A, et al. Role of human DNA glycosylase Nei-like 2 (NEIL2) and single strand break repair protein polynucleotide kinase 3′-phosphatase in maintenance of mitochondrial genome. J Biol Chem 2012;287:2819–29.

[145] Mani R, Tang M, Wu X, et al. Membrane-bound dimer structure of a beta-hairpin antimicrobial peptide from rotational-echo double-resonance solid-state NMR. Biochemistry 2006;45:8341–9.

[146] Männich M, Hess I, Wiest W, et al. Developing T lymphocytes are uniquely sensitive to a lack of topoisomerase III alpha. Eur J Immunol 2010;40:2379–84.

[147] Martánez MC, Anchkor H, Persson B, et al. Arabidopsis formaldehyde dehydrogenase. Molecular properties of plant class III alcohol dehydrogenase provide further insights into the origins, structure and function of plant class p and liver class I alcohol dehydrogenases. Eur J Biochem 1996;241:849–57.

[148] Martán-Folgar R, Lorenzo G, Boshra H, et al. Development and characterization of monoclonal antibodies against Rift Valley fever virus nucleocapsid protein generated by DNA immunization. MAbs 2010;2:275–84.

[149] Mason JT, O'Leary TJ. Effects of formaldehyde fixation on protein secondary structure: a calorimetric and infrared spectroscopic investigation. J Histochem Cytochem 1991;39:225–9.

[150] Mason RP, Sanders JK. In vivo enzymology: a deuterium NMR study of formaldehyde dismutase in Pseudomonas putida F61a and Staphylococcus aureus. Biochemistry 1989;28:2160–8.

[151] McFarland JM, Joshi NS, Francis MB. Characterization of a three-component coupling reaction on proteins by isotopic labeling and nuclear magnetic resonance spectroscopy. J Am Chem Soc 2008;130:7639–44.

[152] Metz B, Jiskoot W, Hennink WE, Crommelin DJ, Kersten GF. Physicochemical and immunochemical techniques predict the quality of diphtheria toxoid vaccines. Vaccine 2003;22:156–67.

[153] Hammel M, Rey M, Yu Y, et al. XRCC4 protein interactions with XRCC4-like factor (XLF) create an extended grooved scaffold for DNA ligation and double strand break repair. J Biol Chem 2011;286:32638–50.

[154] Miller DW, Vosseler S, Mirancea N, et al. Rapid vessel regression, protease inhibition, and stromal normalization upon short-term vascular endothelial growth factor receptor 2 inhibition in skin carcinoma heterotransplants. Am J Pathol 2005;167:1389–403.

[155] Miller MM, Zhu L. Aging changes in the beta-endorphin neuronal system in the preoptic area of the C57BL/6J mouse: ultrastructural analysis. Neurobiol Aging 1992;13:773–81.

[156] Miura K. Specificity in the structure of transfer RNA. Prog Nucleic Acid Res Mol Biol 1967;39–82.

[157] Momma K, Mishima Y, Hashimoto W, Mikami B, Murata K. Direct evidence for Sphingomonas sp. A1 periplasmic proteins as macromolecule-binding proteins associated with the ABC transporter: molecular insights into alginate transport in the periplasm. Biochemistry 2005;44:5053–64.

[158] Monti S, Bronco S, Cappelli C. Toward the supramolecular structure of collagen: a molecular dynamics approach. J Phys Chem B 2005;109:11389–98.

[159] Moore G, Crichton RR. Reductive alkylation of ribosomes as a probe to the topography of ribosomal proteins. Biochem J 1974;143:607–12.

[160] Nakano T, Katafuchi A, Matsubara M, et al. Homologous recombination but not nucleotide excision repair plays a pivotal role in tolerance of DNA-protein cross-links in mammalian cells. J Biol Chem 2009;284:27065–76.

[161] Nasirizadeh N, Zare HR, Pournaghi-Azar MH, Hejazi MS. Introductiono fh ematoxylin as an electroactive label for DNA biosensors and its employment in detection of target DNA sequence and single-base mismatch in human papilloma virus corresponding to oligonucleotide. BiosensB ioelectron 2011;26:2638–44.

[162] Neculai AM, Neculai D, Griesinger C, Vorholt JA, Becker S. Ad ynamicz incr edoxs witch. J Biol Chem 2005;280:2826–30.

[163] Nie CL, Zhang W, Zhang D, He RQ. Changes in conformation of human neuronal tau during denaturation in formaldehyde solution. ProteinP eptL ett 2005;12:75–8.

[164] Nielsen BS, Sehested M, Duun S, et al. Urokinase plasminogen activator is localized in stromal cells in ductal breast cancer. Lab Invest 2001;81:1485–501.

[165] Nienhaus K, Renzi F, Vallone B, Wiedenmann J, Nienhaus GU. Exploringc hromophore-protein interactions in fluorescent protein cmFP512 from Cerianthus membranaceus: X-ray structure analysis and optical spectroscopy. Biochemistry 2006;45:12942–53.

[166] Nitta K, Suzuki N, Honma D, Kaneko Y, Nakamoto H. Ultrastructurals tabilityu nder high temperature or intensive light stress conferred by a small heat shock protein in cyanobacteria. FEBSL ett 2005;579:1235–42.

[167] Norin A, Van Ophem RW, Piersma B, Duine JA, Jornvall H. Mycothiol-dependent formaldehyde dehydrogenase, a prokaryotic medium-chain dehydrogenase/reductase, phylogenetically links different eukaryotic alcohol dehydrogenases—primary structure, conformational modelling and functional correlations. EurJ B iochem 1997;248:282–9.

[168] Nozawa S, Yuhima M, Kojima K, et al. Tumor-associated mucin-type glycoprotein (CA54/61) defined by two monoclonal antibodies (MA54 and MA61) in ovarian cancers. CancerRe s 1989;49:493–8.

[169] Okamura-Ikeda K, Hosaka H, Yoshimura M, et al. Crystal structureo fh umanT -protein of glycine cleavage system at 2.0 A resolution and its implication for understanding non-ketotic hyperglycinemia. JM olB iol 2005;351:1146–59.

[170] Ono-Nita SK, Kato N, Shiratori Y, et al. YMDD motif in hepatitis B virus DNA polymerase influences on replication and lamivudine resistance: a study by in vitro full-length viral DNA transfection. Hepatology 1999;29:939–45.

[171] Ozvegy-Laczka C, Laczká R, Hegedus C, et al. Interaction with the 5D3 monoclonal antibody is regulated by intramolecular rearrangements but not by covalent dimer formation of the human ABCG2 multidrug transporter. J Biol Chem 2008;283:26059–70.

[172] Paliwal R, London E. Comparison of the conformation, hydrophobicity, and model membrane interactions of diphtheria toxin to those of formaldehyde-treated toxin (diphtheria toxoid): formaldehyde stabilization of the native conformation inhibits changes that allow membrane insertion. Biochemistry 1996;35:2374–9.

[173] Pan WH, Xin P, Morrey JD, Clawson GA. A self-processing ribozyme cassette: utility against human papillomavirus 11 E6/E7 mRNA and hepatitis B virus. Mol Ther 2004;9:596–606.

[174] Patriksson A, Marklund E, van der Spoel D. Protein structures under electrospray conditions. Biochemistry 2007;46:933–45.

[175] Petrovic A, Abramovic M, Mihailovic D, et al. Multicolor counterstaining for immunohistochemistry—a modified Movat's pentachrome. Biotech Histochem 2011;86:429–35.

[176] Pettigrew JD, Kirsch AW. Base-compositional biases and the bat problem. I. DNA-hybridization melting curves based on AT- and GC-enriched tracers. Philos Trans R Soc Lond B Biol Sci 1998;353(1367):36–79.

[177] Philip V, Harris J, Adams R, et al. A survey of aspartate-phenylalanine and glutamate-phenylalanine interactions in the protein data bank: searching for anion-π pairs. Biochemistry 2011;50:2939–50.

[178] Pleshko NL, Boskey AL, Mendelsohn R. An FT-IR microscopic investigation of the effects of tissue preservation on bone. Calcif Tissue Int 1992;51:72–7.

[179] Preiss S, Argentaro A, Clayton A, et al. Compound effects of point mutations causing campomelic dysplasia/autosomal sex reversal upon SOX9 structure, nuclear transport, DNA binding, and transcriptional activation. J Biol Chem 2001;276:27864–72.

[180] Printz MP, Gounaris AD. Substrate- and inhibitor-induced conformational changes in enzymes measured by tritium-hydrogen exchange. II. Yeast pyruvate decarboxylase. J Biol Chem 1972;247:7109–15.

[181] Prives CL, Silverman PM. Replication of RNA viruses: structure of a 6 s RNA synthesized by the Q RNA polymerase. J Mol Biol 1972;71:657–70.

[182] Prokova V, Mavridou S, Papakosta P, Petratos K, Kardassis D. Novel mutations in Smad proteins that inhibit signaling by the transforming growth factor beta in mammalian cells. Biochemistry 2007;46:13775–86.

[183] Prud'homme J, Jolivet A, Pichon M, Savouret J, Milgrom Monoclonal antibodies against native ant denatured forms of estrogen-induced breast cancer protein (BCEI/pS2) obtained by expression in Escherichia coli. Cancer Res 1990;50:2390–6.

[184] Quadt I, Gunther AK, Voss D, Schelhaas M, Knebel-Morsdorf D. TATA-binding protein and TBP-associated factors during herpes simplex virus type 1 infection: localization at viral DNA replication sites. Virus Res 2006;115:207–13.

[185] Radhakrishnan S, Gordon J, Del Valle L, Cui J, Khalili K. Intracellular approach for blocking JC virus gene expression by using RNA interference during viral infections. J Virol 2004;78:7264–9.

[186] Ray S, Grove A. Interactions of Saccharomyces cerevisiae HMO2 domains with distorted DNA. Biochemistry 2012;51:1825–35.

[187] Reichmann D, Phillip Y, Carmi A, Schreiber G. On the contribution of water-mediated interactions to protein-complex stability. Biochemistry 2008;47:1051–60.

[188] Ressler N. A systematic procedure for the determination of the heterogeneity and nature of multiple electrophoretic bands. Anal Biochem 1973;51:589–610.

[189] Ritter E, Stehfest K, Berndt A, Hegemann P, Bartl FJ. Monitoring light-induced structural changes of Channel rhodopsin-2 by UV-visible and Fourier transform infrared spectroscopy. J Biol Chem 2008;283:35033–41.

[190] Roberts WK. Use of benzoylated cellulose columns for the isolation of poly(adenylic acid) containing RNA and other polynucleotides with little secondary structure. Biochemistry 1974;13(18):3677–82.

[191] Rodriguez RR, Seegmiller RE, Stark MR, Bridgewater LC. A type XI collagen mutation leads to increased degradation of type II collagen in articular cartilage. Osteoarthritis Cartilage 2004;12:314–20.

[192] Rosu S, Libuda DE, Villeneuve AM. Robust crossover assurance and regulated interhomolog access maintain meiotic crossover number. Science 2011;334:1286–9.

[193] Ruggeri G, Santambrogio P, Bonfiglio F, et al. Antibodies for denatured human H-ferritin stain only reticuloendothelial cells within the bone marrow. Br J Haematol 1992;81:118–24.

[194] Ruifrok AC, Johnston DA. Quantification of histochemical staining by color deconvolution. Anal Quant Cytol Histol 2001;23:291–9.

[195] Sacho EJ, Maizels N. DNA repair factor MRE11/RAD50 cleaves 3′-phosphotyrosyl bonds and resects DNA to repair damage caused by topoisomerase 1 poisons. J Biol Chem 2011;286:44945–51.

[196] Sallmyr A, Tomkinson AE, Rassool FV. Up-regulation of WRN and DNA ligase IIIalpha in chronic myeloid leukemia: consequences for the repair of DNA double-strand breaks. Blood 2008;112:1413–23.

[197] Sanghani PC, Bosron WF, Hurley TD. Human glutathione-dependent formaldehyde dehydrogenase. Structural changes associated with ternary complex formation. Biochemistry 2002;41:15189–94.

[198] Sanghani PC, Davis WI, Zhai L, Robinson H. Structure-function relationships in human glutathione-dependent formaldehyde dehydrogenase. Role of Glu-67 and Arg-368 in the catalytic mechanism. Biochemistry 2006;45:4819–30.

[199] Sanghani PC, Robinson H, Bennett-Lovsey R, Hurley TD, Bosron WF. Structure-function relationships in human Class III alcohol dehydrogenase (formaldehyde dehydrogenase). Chem Biol Interact 2003;143-144:195–200.

[200] Sanghani PC, Robinson H, Bosron WF, Hurley TD. Human glutathione-dependent formaldehyde dehydrogenase. Structures of apo, binary, and inhibitory ternary complexes. Biochemistry 2002;41:10778–86.

[201] Saribasak H, Maul RW, Cao Z, et al. XRCC1 suppresses somatic hypermutation and promotes alternative nonhomologous end joining in Igh genes. J Exp Med 2011;208:2209–16.

[202] Schneider EL, Marletta MA. Heme binding to the histidine-rich protein II from Plasmodium falciparum. Biochemistry 2005;44:979–86.

[203] Schott A, Ravaud S, Keller S, et al. Arabidopsis stromal-derived Factor2 (SDF2) is a crucial target of the unfolded protein response in the endoplasmic reticulum. J Biol Chem 2010;285:18113–21.

[204] Schubassi G, Robert T, Vanoli F, Minucci S, Foiani M. Acetylation: a novel link between double-strand break repair and autophagy. Cancer Res 2012;72:1332–5.

[205] Scott KA, Daggett V. Folding mechanisms of proteins with high sequence identity but different folds. Biochemistry 2007;46:1545–56.

[206] Seeliger MA, Spichty M, Kelly SE, et al. Role of conformational heterogeneity in domain swapping and adapter function of the Cks proteins. JB iolCh em 2005;280:30448–59.

[207] Senior MB, Olins DE. Effect of formaldehyde on the circular dichroism of chicken erythrocyte chromatin. Biochemistry 1975;14:3332–7.

[208] Sfeir A, de Lange T. Removal of shelterin reveals the telomere end-protection problem. Science 2012;336:593–7.

[209] Shaheen M, Allen C, Nickoloff JA, Hromas R. Synthetic lethality: exploiting the addiction of cancer to DNA repair. Blood 2011;117:6074–82.

[210] Shamanna RA, Hoque M, Lewis-Antes A, et al. TheNF 90/NF45c omplexp articipates in DNA break repair via nonhomologous end joining. MolCe llB iol 2011;31:4832–43.

[211] Shanmugam M, Zhang B, McNaughton RL, et al. Thes tructureo ff ormaldehyde-inhibited xanthine oxidase determined by 25 GHz 2H ENDOR spectroscopy. JAmCh emS oc 2010;132:14015–17.

[212] Shin HJ, Baek KH, Jeon AH, et al. Dual roles of human BubR1, a mitotic checkpoint kinase, in the monitoring of chromosomal instability. CancerCe ll 2003;4:483–97.

[213] Siegel JB, Zanghellini A, Lovick HM, et al. Computational design of an enzyme catalyst for a stereoselective bimolecular Diels-Alder reaction. Science 2010;329:309–13.

[214] Singh VR, Kopka M, Chen Y, Wedemeyer WJ, Lapidus LJ. Dynamics imilarityo f the unfolded states of proteins L and G. Biochemistry 2007;46:10046–54.

[215] Slegers H, Fiers W. Studies on the bacteriophage MS2. 23. Fixation of the MS2 RNA acid structure by formaldehyde. Biopolymers 1973;12:2023–31.

[216] Somyajit K, Subramanya S, Nagaraju G. Distinctr oleso fF ANCO/RAD51C protein in DNA damage signaling and repair: implications for Fanconi anemia and breast cancer susceptibility. JB iolCh em 2012;287:3366–80.

[217] Stefanou H, Woodward AE, Morrow D. Relaxationb ehavioro fc ollagen. BiophysJ 1973;13:772–9.

[218] Strom R, Mondovi B. Effect of alkaline pH on the optical properties of native and modified erythrocyte membranes. Biochemistry 1972;11:1908–15.

[219] Sun TT, Traut RR, Kahan L. Protein-protein proximity in the association of ribosomal subunit of Escherichia coli: crosslinking of 30 S protein S16 to 50 S proteins by glutaraldehyde or formaldehyde. JM olB iol 1974;87:509–22.

[220] Takehara T, Suzuki T, Ohkawa K, et al. Viral covalently closed circular DNA in a non-

transgenic mouse model for chronic hepatitis B virusre plication. JHe patol 2006;44:267–74.

[221] Tanaka N, Kusakabe Y, Ito K, Yoshimoto T, Nakamura KT. Crystalstru cture of formaldehyde dehydrogenase from Pseudomonas putida: the structural origin of the tightly bound cofactor in nicotinoprotein dehydrogenases. JM olB iol 2002;324:519–33.

[222] Tanaka Y, Maniwa Y, Bermudez VP, etal. Nonsynonymoussin glen ucleotide polymorphisms in DNA damage repair pathwaysan dlu ngc ancerrisk . Cancer 2010;116:896–902.

[223] Tann AW, Boldogh I, Meiss G, et al. Apoptosis induced by persistent single-strand breaks in mitochondrial genome: critical role of EXOG (5'-EXO/endonuclease) in their repair. JB iolCh em 2011;286:31975–83.

[224] Taskent-Sezgin H, Chung J, Patsalo V, et al. Interpretation of p-cyanophenylalanine fluorescence in proteins in terms of solvent exposure and contribution of side-chain quenchers: a combined fluorescence, IR and moleculard ynamicsstu dy. Biochemistry 2009;48:9040–6.

[225] Tehei M, Madern D, Franzetti B, Zaccai G. Neutron scattering reveals the dynamic basis ofp roteinad aptationto e xtremete mperature. JB iolCh em 2005;280:40974–79.

[226] Telclemariam-Mesbah R, Wortel J, Romijn HJ, Buijs RM. Asimp lesilve r-gold intensification procedure for double DAM labelingstu diesin e lectronmic roscopy. J HistochemCyto chem 1997;45:619–21.

[227] Tesmer VM, Kawano T, Shankaranarayanan A, Kozasa T, Tesmer JJ. Snapshoto fac tivated G proteins at the membrane: the Galphaq-GRK2-Gbetagammac omplex. Science 2005;310:1686–90.

[228] Tsai CS, Tsai YH, Lauzon G, Cheng ST. Structure and activity of methylated horse liveralc ohold ehydrogenase. Biochemistry 1974;13:440–3.

[229] Usha R, Ramasami T. Structure and conformation of intramolecularly cross-linkedc ollagen. ColloidsS urfB B iointerfaces 2005;41:21–4.

[230] Vagenende V, Yap MG, Trout BL. Mechanisms of protein stabilization and preventiono fp roteinaggre gationb yglyc erol. Biochemistry 2009;48:11084–96.

[231] Van Regenmortal MH, Lelarge N. The antigenic specificity of different states of

aggregationo ft obaccomo saicvir usp rotein. Virology 1973;52:89–104.

[232] Walczak A, Rusin P, Kziki L, et al. Evaluation of DNA double strand breaks repair efficiency inh eadan dn eckc ancer. DNACe llB iol 2012;31:298–305.

[233] Wang AT, Sengerová B, Cattell E, et al. Human SNM1A and XPF-EROC1 collaborate to initiate DNA interstrand cross-linkr epair. GenesDe v 2011;25:1859–70.

[234] Weintraub H, Palter K, Van Lente F. Histones H2a, H2b, H3, and H4 form a tetrameric complexin s olutiono fh ighs alt. Cell 1975;6:85–110.

[235] Weirich CS, Erzberger JP, Berger JM, Weis K. TheN- terminald omaino fNu p159f ormsa beta-propeller that functions in mRNA export by tethering the helicase Dbp5 to the nuclear pore. MolCe ll 2004;16:749–60.

[236] Wellauer PK, Dawid IB. Secondary structure maps of ribosomal RNA and its precursors as determinedb ye lectronmic roscopy. Cold Spring Harb Symp Quant Biol 1974;38:525–35.

[237] Wheatcroft AC, Hollander AP, Croucher LJ, eta l. Evidenceo fin s itus tabilityo ft het ypeI V collagen triple helix in human inflammatory bowel disease using a denaturation specific epitopean tibody. MatrixB iol 1999;18:361–72.

[238] Whiley DM, Sloots TP, Whiley DM, Sloots TP. Meltingc urvean alysisu singh ybridisation probes: limitations in microbial molecular diagnostics. Pathology 2005;37:254–6.

[239] Winder DM, Pett MR, Foster N, et al. An increase in DNA double-strand breaks, induced by Ku70 depletion, is associated with human papillomavirus 16 episome loss and den ovovir alin tegrratione vents. JP athol 2007;213:27–34.

[240] Wu BK, Li CC, Chen HJ, et al. Blocking of G1/S transition and cell death in the regenerating live of Hepatitis B virus X proteint ransgenicmic e. BiochemB iophysRe s Commun 2006;340:916–28.

[241] Wu L, McElheny D, Takekiyo T, Keiderling TA. Geometryan de fficacyo fc ross-strand Trp/Trp, Trp/Tyr, and Tyr/Tyr aromatic interactionin ab eta-hairpinp eptide. Biochemistry 2010;49:4705–14.

[242] Wu XH, Chen RC, Gao Y, Wu YD. The effect of Asp-His-Ser/Thr-Trp tetrad on the thermostabilityo fW D40-repeatp roteins. Biochemistry 2010;49:10237–45.

[243] Yahr TL, Wickner WT. Evaluating the oligomeric state of SecYEG in preprotein translocase. EMBOJ 2000;19:4393–401.

[244] Yang F, Wang Y, Zhang Z, et al. The study of abnormal bone development in the Apert syndrome Fgfr2+/S252W mouse using a 3D hydrogelc ulturemo del. Bone 2008;43:3282–8756.

[245] Yang H, Magpayo N, Rusek A, et al. Effects of very low fluences of high-energy protons orir onio nso nir radiatedan db ystanderc ells. RadiatRe s 2011;176:695–705.

[246] Yelina NE, Erokhina TN, Lukhovitskaya NI,e tal. Localizationo fP oas emilatent virus cysteine-rich protein in peroxisomes is dispensable for its ability to suppress RNA silencing. JGe nVir ol 2005;86:479–89.

[247] Yew WS, Wise EL, Rayment I, Gerlt JA. Evolution of enzymatic activities in the orotidine 5'-monophosphate decarboxylase suprafamily: mechanistic evidence for a proton relay system in the active site of 3-keto-L-gulonate6- phosphated ecarboxylase. Biochemistry 2004;43:6427–37.

[248] Zhang JH, Cerretti DP, Yu T, Flanagan JG, Zhou R. Detectiono fli ga ndsin regions anatomically connected to neurons expressing the Eph receptor Bsk: potential rolesin n euron-targetin teraction. JNe urosci 1996;16:7182–92.

[249] Zhang R, Kang KA, Piao MJ, et al. Triphlorethol-a improves the non-homologous end joining and base-excision repair capacity impairedb yf ormaldehyde. JT oxicolE nviron HealthA 2011;74:811–21.

[250] Zhao QY, Chen Q, Yang DJ, et al. Endomorphin 1[psi] and endomorphin 2[psi], endomorphins analogues containing a reduced (CH2NH) amide bond between Tyr1 and Pro2, display partial agonist potency buts ignificantan tinociception. LifeS ci 2005;77:1155–65.

[251] Zheng J, Zanuy D, Haspel N, et al. Nanostructure design using protein building blocks enhanced by conformationally constraineds ntheticr esidues. Biochemistry 2007;46:1205–18.

[252] Zimowska G, Aris JP, Paddy MR. A Drosophila Tpr protein homolog is localized both in the extrachromosomal channel networkan dt on uclearp orec omplexes. JCe ll Sci 1997;110:927–44.

The Biochemical Basis of In Situ Hybridization and Immunohistochemistry

3

INTRODUCTION

Although we have just finished an in-depth look at the foundation of molecular biology, it will be useful if we devote one more chapter to this type of basic information. In this chapter, we focus on the biochemical reactions that occur within the cells and tissues during in situ hybridization and immunohistochemistry. As you will see, I am not referring just to the biochemical reaction that gives us our signal. Although we discuss that, it is a minor part of this chapter. Rather, this chapter gives us the foundation to understand at a biochemical level what actually is happening inside the cells from start to finish when we do in situ hybridization or immunohistochemistry. Given that by far and away the most common type of cells/tissue we will use is formalin-fixed and paraffin-embedded, most of our attention is directed to these types of specimens.

THE BIOCHEMICAL EFFECTS OF FIXING CELLS IN FORMALIN

A common mistake many people make when either doing or interpreting data from in situ hybridization is to assume that all formalin-fixed tissues behave pretty much the same. This implies that formalin fixation basically affects all tissues the same way. Nothing could be further from the actual reality. If you want further evidence, simply ask any researchers who have performed in situ hybridization for at least several years about this theory. They likely will tell you that tissues vary widely in the optimal conditions required for a strong signal with minimal background. Some examples of how variable formalin-fixed, paraffin-embedded tissues can behave with in situ hybridization are shown in Figure 3-1. Note that panels A–C are serial sections from a vulvar condyloma. They were each tested for HPV DNA using no pretreatment (panel A), protease digestion (panel B), or DNA retrieval (panel C). Note that you get a robust and equivalent signal in this tissue for HPV in situ hybridization regardless of the pretreatment. In comparison, look at the data presented in panels D–F. These are also serial section analyses of a different vulvar condyloma. The tissues came from the same laboratory within 2 weeks of each other. The women's age was only 4 years different.

Histologically, the tissues looked very similar. However, note that a good signal was evident in panel D with no pretreatment, but that the signal was weaker if DNA retrieval was done (panel E). Vulvar condyloma are excellent positive controls for HPV in situ hybridization, because they typically contain very high copy numbers of the virus. This should translate into an intense signal. Note the intense signal that was evident if DNA retrieval was done with protease digestion (panel F).

Now compare these data to what is shown in Figure 3-2. Panel A shows the HPV in situ hybridization result in a cervical intraepithelial grade 1 lesion (CIN 1). Like their vulvar condyloma counterparts, the lesions should be intensely HPV positive. Panel A (no pretreatment) shows a good signal, but many of the koilocytes are negative, and these are the cells that typically contain the highest copy numbers of HPV. Can this signal be increased? Is there some biochemical barrier to the signal induced in the formalin-fixation process? Clearly, the signal can be augmented because the serial section, which was predigested in protease, shows a marked increase in the intensity of the HPV in situ hybridization signal and the number of positive cells (panel B). Then compare these data to that derived from immunohistochemistry in another vulvar lesion. This case showed Paget's disease of the vulva. This is where an adenocarcinoma invades the epithelia. Since the cancer cells are adenocarcinoma based, they express a lot of the cytokeratin AE 1/3. Note the complete absence of a signal in the vulvar tissue if no pretreatment was done (panel C). However, note the intense signal in the serial section if 4 minutes of protease digestion was performed before adding the primary antibody (panel D). What is the biochemical basis in these formalin-fixed, paraffin-embedded tissues that explains why protease digestion should lead to an intense signal in one tissue, yet in a different tissue have either no effect on the signal or lead to the diminution of the signal? Surely, the explanation must include an understanding that different epitopes/DNA/RNA targets can show very variable biochemical microenvironments in different formalin-fixed, paraffin-embedded tissues. Indeed, we will see many examples in this book where the same formalin-fixed, paraffin-embedded tissue shows very different optimization profiles for different DNA, RNA, or protein targets! Let me illustrate one such example now.

43

DOI: http://dx.doi.org/10.1016/B978-0-12-415944-0.00003-6

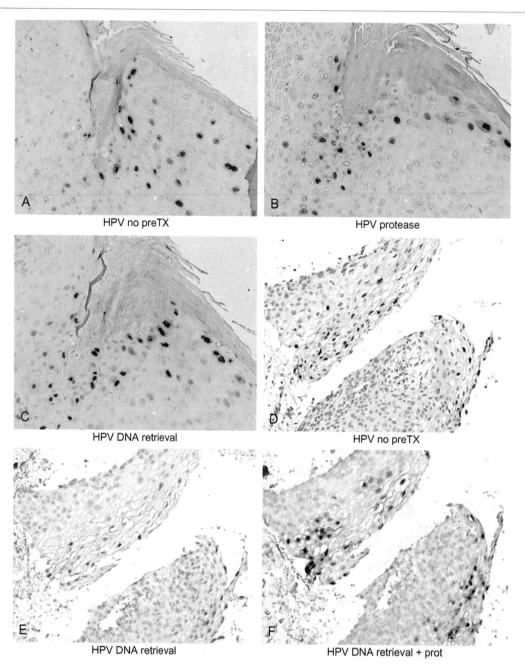

Figure3-1   Histologically equivalent vulvar condyloma require different optimal pretreatment conditions for in situ hybridization. Serial sections of the vulvar genital wart shown in panels A–C showed similarly strong HPV in situ hybridization signals if no pretreatment (panel A), protease digestion (panel B), or DNA retrieval (panel C) was used. DNA retrieval involved pretreating the tissues at 95°C for 30 minutes in an EDTA solution. Another vulvar wart likewise showed a strong signal with no pretreatment (panel D), but a much reduced signal with DNA retrieval (panel E) and a strong signal if DNA retrieval and protease digestion were done (panel F). Such differences in optimal pretreatment conditions among lesions that are histologically equivalent is very common with in situ hybridization.

Figure 3-3 shows a formalin-fixed, paraffin-embedded tonsil that was tested for two proteins that should be evident in such a tissue: TNF alpha (TNFα) and Iba-1. Each protein plays a key role in inflammation. Each protein should be produced by macrophages (although Iba-1 can also be made by lymphocytes). Each protein is cytoplasmic. Since we are examining the same tonsil, by definition, the cells' fixation for the two different epitopes, of course, must be the same. Yet, note in Figure 3-3 that the

TNFα signal is not evident without pretreatment (panel A), but is intense after using 4 minutes of protease digestion prior to incubating the primary antibody (panel B). In comparison, the Iba-1 signal is strong with no pretreatment (panel C) and is lost with protease digestion, where background is now evident in place of this signal (panel D)! Let's use this chapter to better grasp the biochemical bases of these simple, yet very important observations because in them lies the true key to our understanding

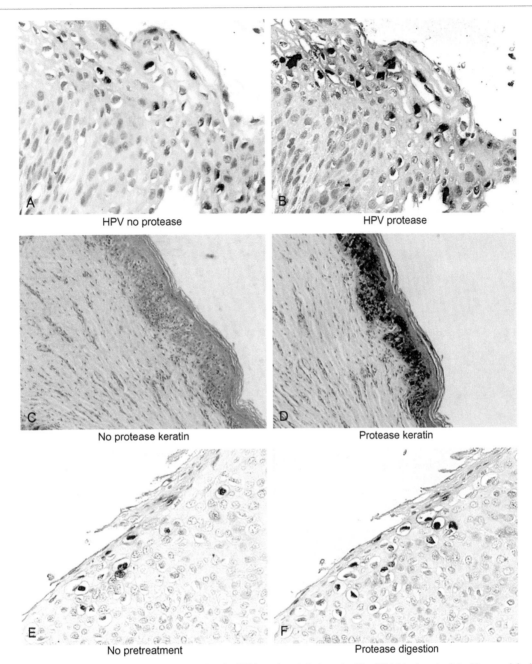

Figure 3-2 Differential responses to protease digestion for DNA and protein targets. The CIN 1 lesion depicted in panel A showed a good signal for HPV DNA in situ hybridization with no pretreatment. The signal in the serial section was enhanced with 4 minutes of protease digestion (panel B). In comparison, the cytokeratin signal as detected by immunohistochemistry was nonexistent in this vulvar tissue with Paget's disease with no pretreatment (panel C), and became intense after a pretreatment step of 4 minutes of protease digestion (panel D). In comparison, a different CIN 1 lesion showed equivalent results for HPV in situ hybridization with either no pretreatment (panel E) or protease digestion (panel F).

of how to achieve success with in situ hybridization and immunohistochemistry.

CROSS-LINKING OF PROTEIN SIDE CHAINS TO OTHER PROTEINS, DNA, RNA, AND GLYCOPROTEINS

Recall that formaldehyde is the simplest of the aldehydes. Its chemical formula is CH_2O. Formaldehyde is able to cross-link the primary amino groups of proteins with other nitrogen molecules that are present either on other proteins, DNA, or RNA, through a CH2 link. Several amino acids in proteins contain nitrogen molecules as part of their unique molecular structure, or, as it is more commonly referred to, as its side or "R" chain (Figure 2-18). These amino acids include tryptophan, arginine, asparagine, glutamine, and histidine. Further, as we have already discussed, each nucleotide in DNA or RNA has

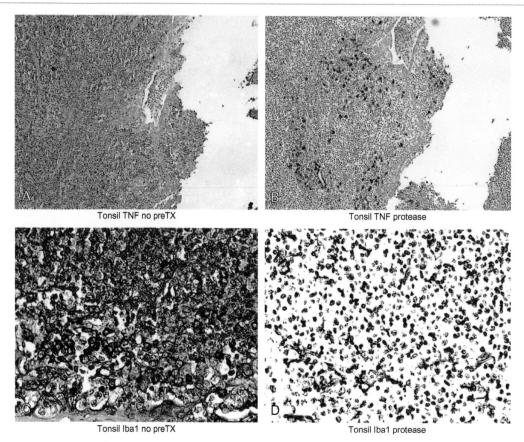

Tonsil TNF no preTX

Tonsil TNF protease

Tonsil Iba1 no preTX

Tonsil Iba1 protease

Figure3-3   Differential responses to no pretreatment and protease digestion of two cytoplasmic proteins both involved in inflammation in the same formalin-fixed, paraffin-embedded tonsil. This chronically inflamed tonsil showed no signal for the common cytokine TNF alpha with no pretreatment (panel A) and an intense signal with a 4-minute pretreatment with protease. In comparison, the same exact lymph node showed an intense signal for the cytoplasmic protein Iba-1 with no pretreatment (panel C), whereas the signal was completely lost with concomitant background if the tissue was pretreated with protease for 4 minutes (panel D). These and similar data suggest a key role for variable three-dimensional macromolecule cross-linked cages associated with any given epitope or DNA/RNA target.

from two nitrogen molecules (thymidine) up to five nitrogen molecules (adenosine and guanosine). Hence, there is ample opportunity for formalin to extensively cross-link proteins to each other and to DNA and RNA.

When a house is built, the many different pieces of wood are attached together in such a way that the final product has enormous structural integrity. Although this description is a bit of an oversimplification, the more wood beams that are attached together, the stronger the house will be to withstand high winds or whatever else Mother Nature throws at it. In some ways, a fence may be a better analogy. Some of the properties in my neighborhood have horses and dogs. Having watched many of the fences built to contain the horses and dogs, I could see the builders start with heavy wood poles. To this, they would attach a chain link fence that had slats small enough that even a toy dog could not get through. They then put large poplar boards across the top that joined each of the heavy wood poles. For those who know about such things, this is called a diamond mesh or diamond link fence. The process of building such a fence is similar to formalin-induced cross-linking of proteins in the cell. With adequate formalin fixation, the final product will be a cell with so much cross-linking that it will have superior structural integrity, able (in most cases) to withstand

any protease or antigen retrieval that you may throw at it. Further, it is easy to envision that as the proteins are cross-linked to one another, you develop a complex three-dimensional structure in which the "pore size" will depend on the overall number of cross-linked macromolecules, especially proteins. Compare this to tissue that has not been fixed adequately in formalin. This would be more like the standard vinyl fence for horses with two or three slats per unit. Such a fence would easily allow dogs (and another large assortment of wild critters) to pass through. Also, if a horse kicks such a fence (as happened to a neighbor), the fence slats may break and the horse may get through. This would be impossible with a diamond mesh fence. As we will see, over time even well-fixed formalin-fixed, paraffin-embedded tissues will show degradation of the three-dimensional protein cross-linked network, so the fence basically changes over time from a diamond mesh fence to a three-slat vinyl fence! This begs the question: can we fix a tissue long enough in 10% buffered formalin that it becomes the equivalent of a brick wall, where nothing can get through? The answer is NO, as defined by the observation that I have yet to see a formalin-fixed tissue (even after years of fixation!) for which we could not get successful in situ hybridization or immunohistochemistry.

However, as is evident if we remember some basics of the biochemistry of amino acids, once the proteins are cross-linked to themselves and to neighboring RNA and DNA molecules, two other forces can come into play regarding the molecular structure of this labyrinth: hydrogen bonds and ionic bonds. Amino acids that are capable of forming hydrogen bonds with other amino acids and nucleic acids include serine, threonine, tyrosine, and glutamine (Figure 2-16). It is clear that most if not all proteins, since they normally consist of hundreds to thousands of amino acids, will have ample supplies of one or more of these "hydrogen bonding amino acids." Of course, DNA molecules will naturally form hydrogen bonds between the two strands, and RNA molecules routinely do the same via three-dimensional conformational changes. However, DNA and RNA molecules cannot contribute to the ionic charge network in the formalin-fixed cell, since they have only one carboxyl group in the entire sequence of a DNA or RNA molecule. Proteins may contain many R side chains capable of creating positive or negative charges (Figure 2-17) that can exert powerful influences, especially if present in groups. Here is where the molecular properties of proteins become a key part of understanding of how formalin fixation "works" and, more importantly, how we can use this to our advantage when doing either in situ hybridization or immunohistochemistry.

At pH 7.0, which is, of course, the pH of most living or fixed cells, amino acids have both negative and counterbalancing positive charges. However, as amino acids are linked during protein synthesis, the negative and positive charges (from the carboxylic acid and amine groups, respectively) are lost, which means that the charge on the amino acid and, thus, protein will be dependent solely on the specific "R" group of that amino acid. It may surprise you, but it is no exaggeration that the simple fact that some R groups are negatively charged and others are positively charged, as well as the fact that others can form hydrogen bonds, may be a critical factor in understanding how in situ hybridization and immunohistochemistry works!

The negatively charged amino acids include aspartic acid and glutamic acid. The positively charged amino acids include arginine, histidine, and lysine (Figure 2-17). Since any protein will have an admixture of these five amino acids, fixed proteins can be viewed as a "skeletal network" of cross-linked proteins/glycoproteins/DNA/RNA that supports an ion exchange and hydrogen-bonding network. The three-dimensional skeletal network will have four different ways it can influence the movement of the chemicals needed for in situ hybridization or immunohistochemistry not only in the cell, but also in the specific cellular compartment from which it originated:

1. Physical constraint (from the size of the molecules, or what is more commonly referred to as the "pore size").
2. Ion exchange (where the positively charged amino acids will restrict the movement of negatively charged in situ and immunohistochemistry molecules, and where the negatively charged amino acids will restrict the movement of the positively charged molecules that we use with in situ hybridization and immunohistochemistry).

3. Hydrogenb onding.
4. Hydrophobic and hydrophilic interactions. The amino acids with the strongest hydrophobic side chains include valine, leucine, isoleucine, methionine, phenylalanine, tryptophan, and cysteine. The amino acids with the strongest hydrophilic side chains include asparagines, arginine, lysine, histidine, and proline.

Let's summarize some of these ideas in graphic form. Formalin fixation will create a three-dimensional protein/protein cross-linked network that will surround every DNA, RNA, and protein target we might be interested in detecting. This cross-linked network, of course, includes the inactivated digestive enzymes (proteases, RNases, DNases, etc.) and, thus, the tissue morphology tends to be both strong and resistant to degradation (Figure 3-4). Figure 3-5 illustrates in graphic form the idea that the number or density of cross-linked amino acids may relate to the pore size that, in turn, may control the flux of some of our key reagents during in situ hybridization and immunohistochemistry. Figure 3-6 shows us the theory that breathability of macromolecules could have a strong effect on our signal with in situ hybridization and immunohistochemistry. Of course, this theory is the basis for the strong positive effect locked nucleic acid probes have on the signal with in situ hybridization. You can envision where the diffusion of key reagents such as streptavidin/peroxidase or streptavidin/alkaline phosphatase conjugates could be influenced by pore size, or may not "dock" well with the biotin-tagged probe or secondary antibody if the three-dimensional protein/protein cross-linked network allows too much breathability of the biotin-tagged macromolecule. Similarly, Figure 3-7 shows in graphic form how ionic pockets (either positively or negatively charged) could attract and allow key reagents to dock in the in situ hybridization or immunohistochemistry process that were either negatively or positively charged, respectively. Of course, most proteins, including the antibodies and streptavidin and reporter enzymes we routinely use with in situ hybridization or immunohistochemistry, will have hydrophobic/hydrophilic regions as well as ionically charged areas in their sequences as well.

Let's discuss the biochemical "ion-exchange resin" and "hydrogen bond resin." For the organic chemistry buffs reading this chapter, you are familiar with the fact that biochemists can fix positively and negatively charged molecules rigidly in place on an immobile matrix, and form a material that can purify organic molecules based on their relative charge. Plus, hydrogen bonding, of course, is the primary force that keeps DNA strands together in the living cell's nucleus. A key point to realize is that these forces are most likely relatively weak in the living cell's cytoplasm. The reason is, in part, that proteins and other macromolecules have, relatively, a great deal of liberty in moving in three-dimensional space in the unfixed cell. But if you fix the cell in formalin, this all changes. You create a rigid, less mobile structure in which the amino acid side chain capable of forming hydrogen or ionic bonds is relatively fixed in space, at least in relation to the living cell. Although this degree of immobility will vary a great deal depending on the type of tissue and the length of time the material was fixed in formalin, it will be present in all

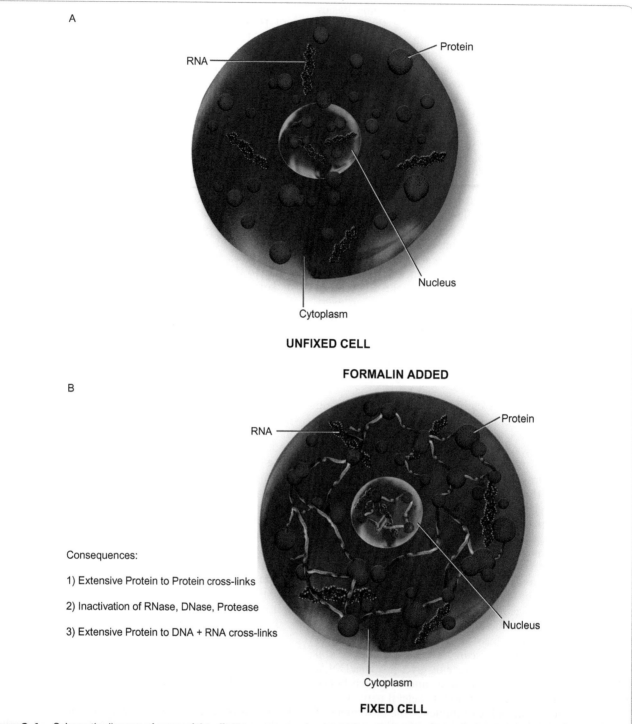

A

RNA

Protein

Nucleus

Cytoplasm

UNFIXED CELL

FORMALIN ADDED

B

RNA

Protein

Nucleus

Cytoplasm

Consequences:

1) Extensive Protein to Protein cross-links

2) Inactivation of RNase, DNase, Protease

3) Extensive Protein to DNA + RNA cross-links

FIXED CELL

Figure3-4   Schematic diagram of some of the effects on macromolecules of formalin fixation. Formalin fixation, via its ability to cross-link the macromolecules of the cell, will strengthen the integrity of the cell such that it can usually resist the high temperatures and other conditions of in situ hybridization and immunohistochemistry. Further, the degradative enzymes of the cell will become inactive due to the extensive cross-linking of their active sites.

formalin-fixed material. How important is this creation of a "three-dimensional ionic bond, hydrogen bond network" to the success of in situ hybridization? The answer is *critical*. Indeed, as alluded to previously, and as we will learn when we discuss the topic, the primary reason that

LNA probes are so effective for in situ hybridization is that they cannot move in three-dimensional space (also referred to as "breathing"), because the probe is "locked" rigidly into shape when it attaches to its target sequence. It is the same with formalin-fixed cells.

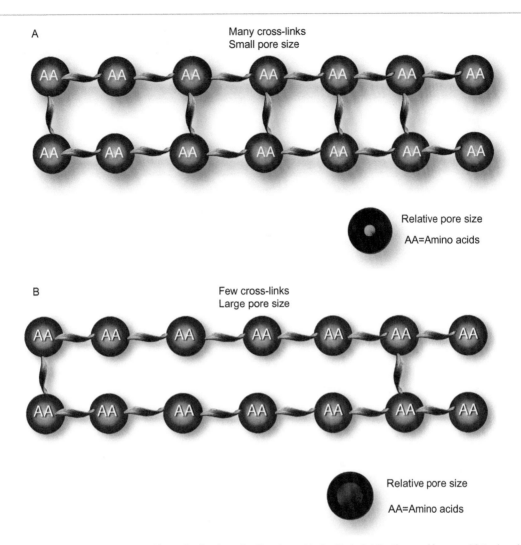

Figure 3-5   Biochemical consequences of formalin fixation of cells relevant to in situ hybridization and immunohistochemistry: pore size. The concept of pore size is important in the living cell. For example, the protein albumin plays a key role in maintaining the blood volume because it is too large to pass through the pore sizes of the cells (called podocytes) that are the gatekeepers of molecular passage into the urine via the glomeruli. However, in a disease called the nephrotic syndrome, the pore size increases due to structural damage to the podocyte's "foot processes," and albumin pours out and is lost in the urine. Pore size, thus, is also important in the cells of formalin-fixed, paraffin-embedded tissues. This graphic representation suggests that pore size around targets may decrease in size with prolonged formalin fixation, due to the greater number of cross-linked molecules per unit volume around the target of interest.

The cells in formalin-fixed, paraffin-embedded tissues have two gatekeepers to the entry of reagents in and out of the cell:

1. Pore siz e.
2. A complex three-dimensional cage of ionic and hydrogen-bonding potential. To be complete, we should also add hydrophobic and hydrophilic potential to the forces of this three-dimensional cross-linked protein cage that can affect the movement and retention of the reagents we use when we do in situ hybridization or immunohistochemistry.

This is the key point of the three-dimensional macromolecule cross-linked network: the movement of the molecules we use with in situ hybridization (probes, antibodies such as antidigoxigenin conjugate, proteins such as streptavidin conjugate, reporter enzymes) will all, by definition, be strongly influenced by the specifics of the pore size and how "active" the R chains of amino acids are in terms of their ionic or hydrogen potential.

Although we cannot actually look inside the cell during in situ hybridization or immunohistochemistry and see which of these forces are predominating in any given experiment, we can do some simple experiments that can give us some important insight into these different biochemical-related aspects of in situ hybridization. It is my contention that a solid understanding and awareness of these key biochemical points will allow you to achieve success routinely with in situ hybridization or immunohistochemistry and, as a corollary, know how to adjust the experimental conditions when in situ hybridization or immunohistochemistry does not work optimally.

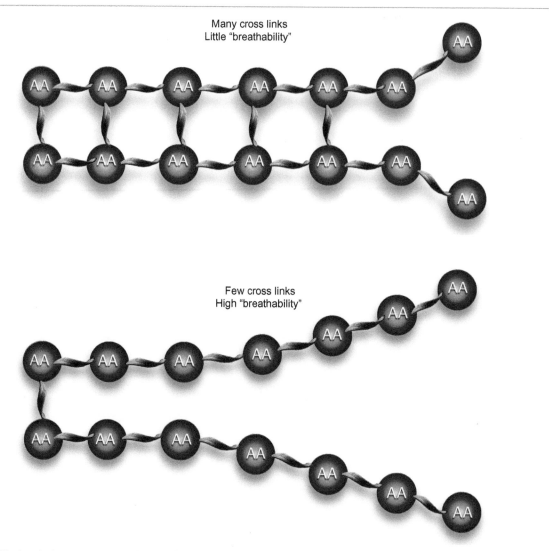

Figure 3-6   Biochemical consequences of formalin fixation of cells relevant to in situ hybridization and immunohistochemistry: breathability of macromolecules. In the living cell, macromolecules need the ability to freely rotate in three-dimensional space. Of course, this is essential to the "unzippering" of DNA during its synthesis, to cite one example. Macromolecules retain this ability to change conformation in three-dimensional space (i.e., breathe) in formalin-fixed, paraffin-embedded tissues. The figure theorizes that this ability may be physically constrained, in part, by the cross-linking induced by formalin and that this, in turn, may relate to the degree of formalin fixation.

Also, realize that it is not possible to determine before doing an in situ hybridization experiment on any tissue what the relative pore sizes are and the relative strength of the "ionic charge and hydrogen bond network." It follows that it is to your advantage to purposely vary the pretreatment conditions when testing a tissue by in situ hybridization.

So, to summarize, we will use this model of a three-dimensional cross-linked network of macromolecules, predominated by proteins, where pore size, ionic charges, and hydrogen bonds that, in turn, can affect the "breathability" of macromolecules including our key in situ hybridization and immunohistochemistry reagents, will be the key variables that will determine our pretreatment regimes and where we can improve the signals we get with both in situ hybridization and immunohistochemistry by manipulating this network. Let's examine

this theory in more detail, and do some experiments in an attempt to test the hypothesis.

PORE SIZE AND BIOCHEMICAL CHARGES

For the purpose of this discussion, let's refer to the ability of fixed cells to either retain or impede the movement of macromolecules based on the relatively rigidity of the three-dimensional network of R chains that can form either hydrogen bonds or ionic bonds as its "ionic and hydrogen bond potential." Tissues that have been fixed for long periods of time in formalin will tend to have a high ionic and hydrogen bond potential. The reason is that the protein side chains are more rigidly held in three-dimensional space due to more cross-links. That is, they will have less breathability. Tissues that have been fixed for a short period of time will have a low ionic and hydrogen bond

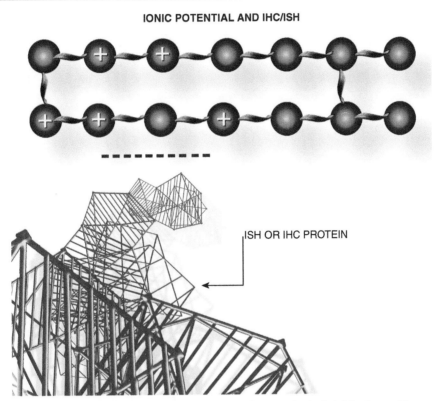

IONIC POTENTIAL AND IHC/ISH

ISH OR IHC PROTEIN

Figure3-7   Biochemical consequences of formalin fixation of cells relevant to in situ hybridization and immunohistochemistry: ionically charged regions. This graphic depiction highlights one role that ionically charged regions of the three-dimensional macromolecule cross-linked cage that surround any DNA/RNA target or epitope may play in the immunohistochemistry and in situ hybridization reaction. The charged region may assist the docking of key molecules such as streptavidin, peroxidase, or alkaline phosphatase to the region of the target of interest, and thus assist in the generation of the in situ hybridization/immunohistochemistry signal. Alternatively, such charges could serve to temporarily hold the probe/primary antibody when surrounding nontarget sequences to facilitate their washing away with the stringent wash. If such regions were not available, then the probe/primary antibody may be more likely to absorb to the macromolecules of the cell, which, in turn, may not be washed away at the stringent wash. The end result may be background.

potential, because the protein side chains are much more capable of moving in three-dimensional space. We are also assuming that such relatively weakly fixed tissues will have larger pore sizes surrounding the target molecules, and show greater breathability of the macromolecules.

Clearly, since the basis of either pore size or the ionic and hydrogen bond potential is cross-linked proteins, either should be able to be altered by protease digestion. When we do protease digestion for either in situ hybridization or immunohistochemistry, we usually use between 5 and 30 minutes of digestion. In more than 99% of such cases, the integrity of the tissue will show no ill effect. This suggests that a three-dimensional macromolecule cross-linked network must allow for superior cytoskeleton integrity, since even 30 minutes of intense protease digestion (in most cases) will not affect it to the point that the cells show poor morphology. But there is no way of knowing how 30 minutes of protease digestion will affect pore size.

This leads to the question: is there any way to differentiate between the pore size (by which I am referring to the physical barrier that will impede the movement of the reagents we use in the in situ hybridization and immunohistochemical processes) and the ionic/hydrogen bond potential of formalin-fixed, paraffin-embedded tissues?

One way we may be able to do this is by using very brief protease digestions. By "very brief," I mean *really* brief! Specifically, only 10 seconds of protease digestion! This seems too short a time for the protease digestion to affect pore size. But it may be enough time to remove some R side changes, such as those of serine and tyrosine as well as glutamic and aspartic acid, that will reduce the ionic and hydrogen bond potential of the fixed tissue.

I did a series of experiments with both in situ hybridization and immunohistochemistry using these very brief (10-second) protease digestion times. I was very surprised to see that, in many cases, even 10 seconds of protease digestion could have a large effect on the signal with either in situ hybridization or immunohistochemistry. Two examples of this are shown in Figure 3-8 (for in situ hybridization) and Figure 3-9 (for immunohistochemistry). Figure 3-8 shows a weak signal in this CIN 1 lesion if no pretreatment, 30 minutes of protease digestion, or DNA retrieval was used. Note the strong enhancement of the signal for HPV DNA in the same cells in the serial section if antigen retrieval (30 minutes at 95°C) pretreatment was followed by a 10-second protease digestion. I will admit to being surprised by this result. But there must be a simple explanation. Note in the figure that

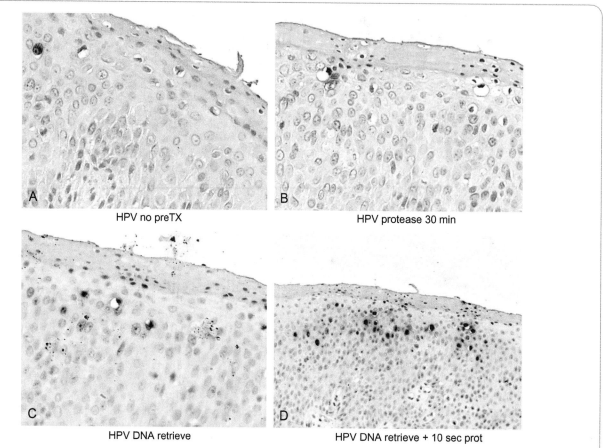

A   HPV no preTX

B   HPV protease 30 min

C   HPV DNA retrieve

D   HPV DNA retrieve + 10 sec prot

Figure3-8   Effect of very short protease digestion on the signal with in situ hybridization in serial sections. This CIN 1 lesion showed a very weak signal for HPV DNA if no pretreatment (panel A), protease digestion (panel B), or DNA retrieval was done (panel C). However, these same cells showed an intense signal if the DNA retrieval was followed by a 10-second protease digestion (panel D).

30-minute protease digestion alone is associated with a weak signal for the HPV DNA. This would imply that the pore size may not be a key issue in this particular tissue with regards to the HPV in situ result, as 30 minutes is a very long time for protease digestion. Also note that DNA retrieval alone has little effect on the DNA signal. DNA retrieval alone should not affect pore size as much as prolonged protease digestion. We can speculate that the DNA retrieval is altering the three-dimensional macromolecule cross-linked network to the point that the R side chains are more accessible to protease digestion, even for only 10 seconds. In Figure 3-9, note how the background is very high with antigen retrieval for parvovirus detection in a muscle biopsy, to the point that no signal can be seen. However, when antigen retrieval was followed by a 10-second protease digestion, the background is eliminated and the signal is evident. By the way, this inverse relationship between signal and background will be something we see and discuss throughout this book. The parvovirus immunohistochemistry result reminds us of a simple but critical point. Our knowledge of surgical pathology can be invaluable for interpreting the immunohistochemical or in situ hybridization results. Specifically, parvovirus (in muscle biopsies) typically localizes to the endothelial cells. The positive cells in this case, as seen in panels C, D, and E, are clearly endothelial cells. This

gives us more confidence as to the specificity of the reaction. Finally, panel F of Figure 3-9 shows another classic finding in immunohistochemistry (also evident in in situ hybridization). It is where a thin rim at the edge of the tissue shows enhanced background. It is, logically, called edge artifact. The basis of this phenomenon may be the relatively slow rate of diffusibility/cross-linking by formalin. The edge of the tissue will tend to be more strongly fixed than the center of the tissue. This will be more likely if the tissue is large and/or it was not fixed for too long a period of time. Since the rim of the tissue may be more strongly fixed, it may tend to show either enhanced signal or background (as seen in panel F) than the center of the tissue. Another explanation for "edge artifact" is that the edge of the tissue would, of course, be more likely to be exposed if one or more of the reagents in the different steps of immunohistochemistry or in situ hybridization dried out during the procedure. Drying of any reagent with immunohistochemistry or in situ hybridization will invariably cause more background.

Whether or not the theoretical construct as described here is the correct explanation for the observation that even 10 seconds of protease digestion can have a strong effect on the in situ hybridization or immunohistochemistry signal is not important. What is important is that you realize that the fixed cells from tissue to tissue (and even

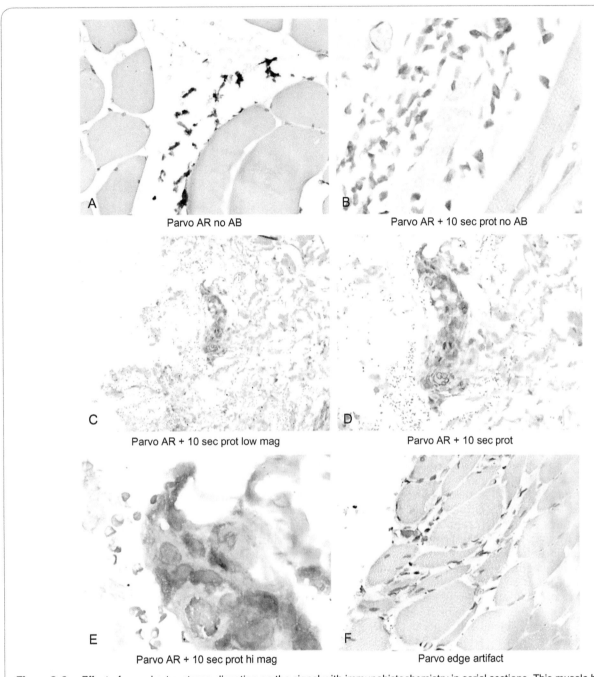

Parvo AR no AB

Parvo AR + 10 sec prot no AB

Parvo AR + 10 sec prot low mag

Parvo AR + 10 sec prot

Parvo AR + 10 sec prot hi mag

Parvo edge artifact

Figure3-9   Effect of very short protease digestion on the signal with immunohistochemistry in serial sections. This muscle biopsy is from a patient with acute parvoviral infection and myositis. If antigen retrieval was done *without* the primary antibody, background colorization was seen (panel A). The background was lost if the reaction was done at the same time, with only one difference: the antigen retrieval was followed by a 10-second protease digestion (panel B). Under these conditions, a target specific signal is seen for the parvoviral epitope (panels C, D, and E) as defined by the presence of the signal in a cell type known to be trophic for the virus: the endothelial cell. Finally, panel F shows the classic "edge artifact" where typically all cell types at the very periphery of the tissue show a precipitate (in this case, DAB based).

within a given tissue) likely vary from sample to sample *and in a given sample* relative to its pore sizes and ionic and hydrogen bond potential! It is my experience that the variation is so profound and impossible to predict that the basic framework (pore size and ionic and hydrogen bond potential) from sample to sample and for any given epitope/DNA/RNA target probably is as distinct for that given tissue or target as a fingerprint is for a person. This is why it is to your advantage when doing in situ hybridization or immunohistochemistry to intentionally vary the pretreatment conditions in such a way as to increase the chances of a successful experiment. From a practical standard, there may not be a more important statement regarding successful in situ hybridization and immunohistochemistry!

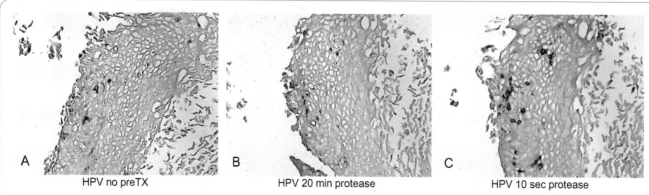

Figure 3-10   Optimal conditions for HPV in situ hybridization in a tissue fixed for 24 hours in 10% buffered formalin. This high-grade SIL contained HPV 16 DNA. It was fixed for exactly 24 hours in formalin. In situ hybridization for HPV DNA showed a weak signal with no pretreatment (panel A) and a similarly weak signal with 20 minutes of protease digestion (panel B). However, if only 10 seconds of protease digestion was done prior to the in situ hybridization, then a strong signal was seen in the same cells of the serial section (panel C).

RELATIONSHIP OF BIOCHEMICAL CHANGES TO FIXATION TIME AND TEMPERATURE

Although the representations of the three-dimensional macromolecule cross-linked network in formalin-fixed, paraffin-embedded tissues in this book are theoretical, there is some supporting data that has been known for many years. Indeed, it was established in the 1990s that the time of formalin fixation has a strong effect on the signal with in situ hybridization. Practicing pathologists know this from the time they begin residency, although not in terms of in situ hybridization but rather on tissue integrity. As residents, we pathologists learn that formalin diffuses into tissue slowly and rather inefficiently. The theory I prefer to accept in this regard is that formalin diffuses relatively quickly into tissues. The rate-limiting event, and why formalin fixation tends to take a long time, is the time needed for cross-linking once formalin actually diffuses into the tissue.

We can be more quantitative when it comes to the theoretical rate of diffusion of formalin when fixing tissues. The diffusion rate of small molecules in water is $10^{-6}\,cm^2/sec$. This translates to a diffusion rate of 10 microns in 16 minutes. Recall that a red blood cell is about 7 microns. Thus, this gives us an idea why a lot of time is needed to adequately fix the standard surgical biopsy, which typically ranges from 5 to 15 mm in size. Of course, formalin diffusion is not the only factor that relates to formalin's ability to generate the three-dimensional macromolecule cross-linked network. When the formalin diffuses into the cell, the reaction whereby this molecule induces cross-links will take additional time and will be facilitated by increased temperatures, as are most chemical reactions. Going back to our house-building analogy, we would not get upset with a master builder if, after being given a large number of wood beams, he or she had not built the house in a few hours! We would expect it to take days of work to attach all the pieces of wood together. Pathology residents learn that they need to use ample formalin (10× the volume of the tissue) and to fix it for an ample period of time (at least 5 hours for a surgical biopsy of less than 1 cm and at least overnight for any tissue greater than 2 cm); otherwise, the tissue will be fragile and soft, and histotechnicians will

complain that they cannot prepare good tissue sections on the slide. Indeed, at least when I was a resident in pathology 30 years ago, we were required to fix whole brains for 2 weeks in a large volume of 10% buffered formalin to assure adequate cross-linking in each cell.

Similarly, pathologists take advantage of the fact that elevated temperatures increase the ability of formalin to adequately fix the cell by using microwave ovens with formalin-fixed material to speed up the fixation process. In this case, the high temperature of the microwave may be doing two separate things to keep the cell preserved: (1) assisting formalin in cross-linking the proteins, including all degradative enzymes, which, in turn, inactivate the tissue digestive enzyme; and (2) heat-denaturing the degradative enzymes (such as RNase, DNase, and proteases) and, thus, preventing the tissue from degrading over time.

It is not possible to know how long a given formalin-fixed, paraffin-embedded tissue has been fixed in formalin. I have taken tissues and fixed them so that I knew exactly how long the tissue was in formalin before being embedded in paraffin wax. To at least have some better idea about the optimal conditions for formalin-fixed, paraffin-embedded tissues that have been fixed for a defined period of time, I took a high-grade intraepithelial neoplasia of the fingernail region that was HPV 16 positive. I fixed the tissue in formalin for either 24 or 48 hours. I then pretreated the tissues with nothing, or with 10 seconds of proteinase K, or 20 minutes with the protease. I then did in situ hybridization for HPV 16. Representative data are provided in Figure 3-10 for tissues fixed for 24 hours in formalin. Note that the protease digestion and 20 minutes of protease digestion gave equivalent results. In each case, a weak signal is noted in a few cells. We have no way of knowing if this represents the strongest possible signal for this tissue, because there is no way to know the copy number of HPV in the infected cells. However, look at the results for the tissue pretreated with 10 seconds of proteinase K. The HPV signal is much stronger! I never cease to be amazed how variable tissues are relative to optimal pretreatment conditions, especially when the tissues have the same histologic diagnosis and are about the same size. Please don't take the data in Figure 3-10 to suggest that most surgical biopsies will show

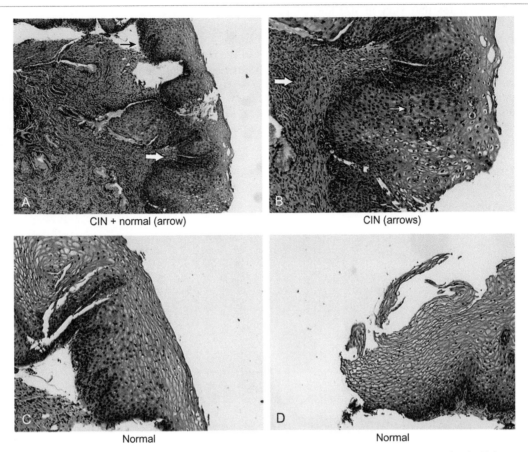

CIN + normal (arrow)

CIN (arrows)

Normal

Normal

Figure 3-11   Variable protein concentrations of cells in a cervical biopsy. This CIN 1 biopsy from the cervix, stained with hematoxylin and eosin, shows the thickened region of the CIN 1 lesion (large arrow) and the adjacent thinner, normal cervical epithelium that was not infected by HPV (small arrow, panel A). A higher magnification of the CIN 1 region (panel B) shows the dark pink of the dysplastic squamous cells (small arrow), indicative of a high protein concentration (primarily keratin). The larger arrow shows the dark blue small cells (lymphocytes) on top of which is a dark pink band that corresponds to the stromal protein collagen. Panels C and D show different parts of the normal squamous epithelia. Note how uniform and well ordered the cells are compared to the more disorganized growth pattern of the cells in the CIN1.

optimal in situ hybridization results with 10 seconds of protease digestion prior to denaturation/hybridization. The actual data show much more variability than that. Finally, these are more data that strongly suggest that pore size is not the key variable for pretreatment when doing in situ hybridization with formalin-fixed, paraffin-embedded tissues. Ten seconds of protease pretreatment could not affect pore size for tissues treated for 24 hours. It is more logical to assume that other factors, such as ionic and hydrogen bond potential, may be the explanation for such a dramatic increase in signal for in situ hybridization with such short protease digestion times.

RELATIONSHIP OF BIOCHEMICAL CHANGES TO CELL TYPE

Cells may vary greatly relative to the concentration of protein, especially in their cytoplasm. Cells also vary tremendously to the extent of their cytoskeleton. Thus, it is easy to predict that the individual cells in a given tissue may show differences in their three-dimensional macromolecule cross-linked network even if, by definition, they have been fixed for the same amount of time in formalin.

To see with your own eyes how the cells in tissues vary a great deal in their protein concentration, you only need to look at the hematoxylin and eosin stain. The hematoxylin stains the nucleic acids in the cell blue. Hence, the hematoxylin stains mostly the nucleus. However, some cells, such as plasma cells, that have abundant mRNAs in the cytoplasm will also stain blue in the cytoplasm. The eosin stains the proteins in the cell pink. Look at Figure 3-11. The biopsy comes from the uterine cervix. Note that the mature squamous cells are dark pink (small arrow, panel B), which reflects the high concentration of the protein keratin present in the cytoplasm. Compare this to the lymphocytes that stain blue and not pink (panel B, large arrow), which reflects the high DNA content of the nucleus and the relatively low protein content of the cytoplasm, since most of these cells are "resting" or, at least, not yet very metabolically active. Also note that the epithelial layer (panel A) varies a great deal over the course of the biopsy. The epithelia toward the upper part of panel A (thin arrow) are thinner and show uniform perinuclear halos and well-organized cells. Compare these to the epithelia toward the base of panel A, which are much thicker and show a disorganized cell growth where the cells show nuclei and halos that vary in size and shape (thicker arrow). The reason that the lower part of the epithelia is thicker is that the cells are growing more rapidly. This is referred to as hyperplasia. Hyperplasia can be a benign process, or a premalignant/malignant process. The way pathologists tell these two key categories apart is simple: benign hyperplasia shows an orderly arrangement of the increased numbers of cells where each cell looks pretty much the same as the next one. In premalignant/malignant

hyperplasia, the cells are growing in a disorganized fashion, and the cells and their nuclei vary in their size, shape, and color. This process is often called dysplasia. In the cervix, dysplasia is always started by the venereally spread virus human papillomavirus (HPV).

Before moving on with the discussion of the biochemistry of formalin-fixed, paraffin-embedded cells, let's look at Figure 3-12, because it illustrates another key point about the interpretation of immunohistochemistry and

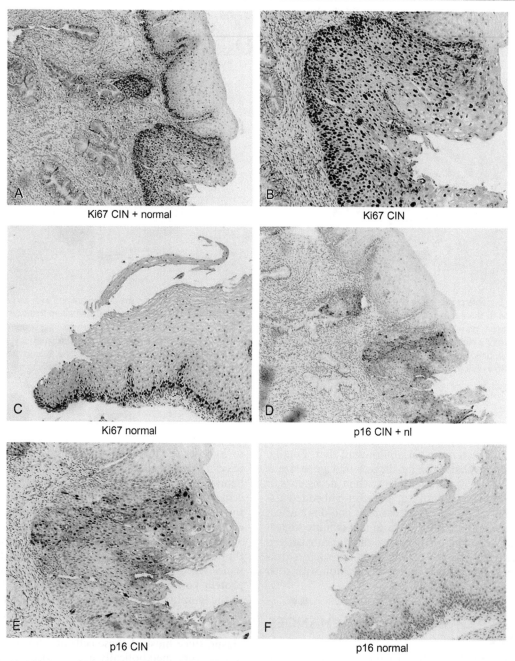

Figure 3-12 Variable immunohistochemical profile in dysplastic versus normal cells. The same CIN 1 lesion shown in Figure 3-11 (indeed, the serial section such that we are examining the same cells) were tested with the mitotic index or proliferative marker Ki-67 and the tumor suppressor gene p16, which tends to be increased in CIN lesions. Note that the Ki-67 positive cells in the CIN 1 lesion (panels A and B) show a disorganized pattern of growth, going from the base of the squamous epithelia to many cells toward the surface. Panel C shows the orderly arrangement of the Ki-67 positive cells in the basal layer of the normal epithelia. Panels D–F show that the signal for p16 is present only in the CIN 1 area. Thus, the normal epithelia serve as an excellent internal negative control for the p16 test in the CIN 1 lesion.

THE BIOCHEMICAL BASIS OF IN SITU HYBRIDIZATION AND IMMUNOHISTOCHEMISTRY

in situ hybridization and how much we can gain from a solid understanding of the histologic findings. Figure 3-12 shows two immunohistochemistry stains on the CIN 1 lesion (and the adjacent normal tissue) shown in Figure 3-11. One immunohistochemistry stain is called the Ki-67 stain. It is one of the most common markers in diagnostic immunohistochemistry. The reason is that it marks the key event of both benign and premalignant/malignant hyperplasia, the *Mitotic Index*. Although pathologists have counted mitoses for decades, there is some subjectivity to this process. The subjectivity is removed by the immunohistochemistry test measuring the percentage of cells that are expressing the Ki-67 protein, which is an excellent marker of a cell's proliferative potential.

Note two observations in panels A–C:

1. There are many more cells that express Ki-67 in the CIN 1 lesion as compared to the adjacent normal epithelia.
2. The Ki-67 positive cells are at the base only in the normal epithelia, and go "way up" the epithelia toward the surface in the CIN part of the lesion.

Point 1 is typical of any hyperplastic process; more cells are able to proliferate. Point 2 is typical of any pre-neoplastic process; the cells are showing a *disorganized* pattern of proliferation. The organized pattern is strictly basal, and the disorganized pattern is one in which the proliferating cells are no longer just in the basal zone but, rather, present in the upper parts of the epithelia. We discuss the histologic findings of tissues in Chapter 6. Suffice it to say for now that if you can differentiate an organized growth pattern from a disorganized growth pattern, you will find that it is relatively easy to recognize dysplasia/cancer in the biopsies you study.

Finally, also look at panels D–F of Figure 3-12. This is for the protein p16, which is widely used as a marker of HPV infection in the cervix. Note that the p16 positive cells are clearly evident in the CIN 1 portion of the lesion and nonexistent in the normal epithelia, which are present less than 1mm away from the CIN 1 lesion. This represents a good example of a so-called internal control for immunohistochemistry. These internal controls (negative cells that are known not to have the target) are invaluable for documenting the specificity of an immunohistochemistry (or in situ hybridization) reaction. Indeed, I think such internal controls are more important in some sense than the more standard controls we run with immunohistochemistry such as omitting the primary antibody. The reason is that when we do immunohistochemistry or in situ hybridization, each cell is being exposed to each reagent. Talk about high-throughput! Any immunohistochemistry or in situ hybridization reaction involves the simultaneous analysis of tens or hundreds of thousands of cells! That is more impressive than a 96-well plate for PCR! Although the omission of a probe or primary antibody gives us some useful information, it is not as direct a proof of the specificity of the reaction as cells known not to have (or, as a positive control, have) the target showing no reaction while cells with the target show a strong signal.

Does the highly variable concentration of proteins in different cell types in a given biopsy have any consequences for our in situ hybridization and immunohistochemistry experiments? Stated another way, if a target is present in two different cell types, or if we are doing co-expression analysis for two different targets in two different cell types, can the variable concentration of proteins in these two cell types affect the overall immunohistochemistry (or in situ hybridization) signal? The answer is yes. I suspect there are at least two different explanations for this phenomenon:

1. Because two different cell types may have very different protein concentrations, they may present very different pore sizes, as well as ionic and hydrogen bond potential, to the respective probe/primary antibodies for a given time of formalin fixation.
2. Because each epitope (or RNA/DNA target) probably has its own unique three-dimensional protein/protein cross-linked cage associated with it, by definition, two different targets probably will have different optimal conditions, regardless of the time of fixation in formalin.

Figure 3-13 shows the results of a simple experiment to illustrate this importance of differential optimal conditions for two different epitopes in a given tissue. The

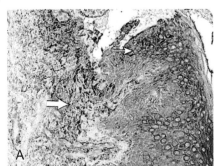 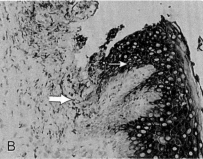 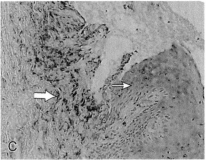

| CD45 AE13 no preTX | CD45 AE13 protease1 | CD45 AE13 protease3 |

Figure 3-13 Variable immunohistochemical profile between different cell types in the same lesion. This cervical biopsy was analyzed at the same time for CD45 and cytokeratin AE 1/3. With no pretreatment, the cytokeratin signal was weak (small arrow), whereas the CD 45 signal was strong (panel A, large arrow). In comparison, if the tissue was pretreated with a highly active protease, then the signal for both CD45 and cytokeratin was strong (panel B). The keratin signal weakened while the CD45 signal was still strong if a weaker protease digestion was used (panel C). This is a good example of how different cells in the same tissue may show different optimal pretreatment conditions.

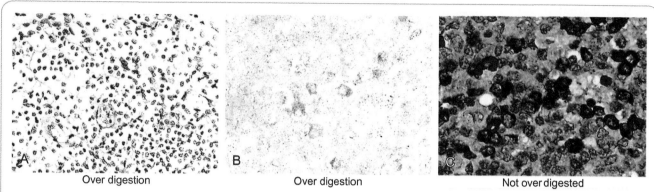

Over digestion   Over digestion   Not over digested

Figure3-14   Overdigestion of tonsillar tissue during immunohistochemistry This tonsil biopsy was tested for the protein TNF alpha. At optimal conditions (panel C) of 4 minutes of protease digestion, a strong signal is seen with the morphology of the cells intact. The cells also take up the hematoxylin counterstain very well. Compare this to panel A, where the tissue was treated with antigen retrieval for 30 minutes. The cell detail is less well defined, and the reticulin fibers are prominent. If the tissue was digested with protease for 30 minutes, the cell morphology is difficult to discern, and the signal is much diminished (panel B).

cervical biopsy was fixed in formalin for an unknown amount of time. I did immunohistochemistry for two different proteins: cytokeratin (specifically, the cytokeratin AE 1/3, which would stain the squamous cells), and CD45, which would stain the lymphocytes. As an aside, this is the first time we have discussed co-labeling. We pay a great deal of attention to the topic of co-labeling in later chapters, especially Chapter 9. For now, let's keep co-labeling as simple as it gets. Specifically, let's simply use two different primary antibodies at the same time, such that they will also be visualized with the same chromogen (in this case, DAB). We will be able to differentiate the two antigens simply by the fact that two very different and easily distinguishable cell types will be positive for the antigens of interest—in this case, CD45 (mononuclear cells) and AE 1/3 (squamous cells).

In the experiment depicted in Figure 3-13, three different cervical biopsies were placed on the same silane-coated (also referred to as PLUS) glass slide. Prior to immunohistochemistry, one tissue was not pretreated, another biopsy received 4 minutes of pretreatment in the strong protease 1 (Ventana Medical Systems), and the final biopsy received 4 minutes of pretreatment in the relatively weak protease Proteinase 3 (Ventana Medical Systems). This is a good example of how placing three sections on a slide (commonly called a paraffin ribbon) allows you to do multiple simple yet very informative experiments with relatively few reagents and little manpower.

Note that the signal for AE 1/3 was poor with no pretreatment, strong with digestion with protease 1 prior to immunohistochemistry, and poor again if the pretreatment was digestion with the weaker protease 3 (all marked by the small arrows). This makes sense in that, in my experience, cytokeratin AE 1/3 usually needs a strong proteinase digestion prior to immunohistochemistry to get a signal. Stated another way, the high concentration of proteins in the squamous cells in the form of the cytokeratin network requires strong protease digestion to allow the epitope to be accessed during the immunohistochemical process. However, look at the CD45 data in

the same sample. Although background may be slightly less in the tissue treated with protease 1, the signal is similar for the no pretreatment and protease 1 or protease 3 digestions. Again, with much less protein per unit volume in the lymphocytes, this cell type may require less proteinase digestion than cytokeratin to generate a strong signal.

There is a corollary to the observation noted in the previous paragraph that mononuclear cells may require less protease digestion than surrounding cells with high protein content such as squamous cells for optimal immunohistochemistry or in situ hybridization. The corollary is that tissues rich in mononuclear cells, such as lymph nodes, tonsil, and spleen, or tissues with relatively low cellular content, such as lung (that consists mostly of air spaces lined by very thin, delicate alveolar walls), tend to be easily overdigested. The ways in which we can identify overdigestion with immunohistochemistry or in situ hybridization are illustrated in Figure 3-14. In panel A, in which this tonsillar tissue was slightly overdigested (by antigen retrieval), note that the nuclear fast counterstain shows the reticulin fibers of the stroma but does not show the cytoplasm of any of the cells. This is a reflection of the fact that reticulin (which lines the blood vessels and the base of epithelia) is much more resistant to protease than the cytoplasm of any individual cell. When overdigestion becomes more pronounced, then two events occur that are usually easily recognizable:

1. The individual cells will not take up the counterstain (Figure 3-14, panel B versus panel C).
2. The morphology of the tissue is no longer recognizable, and there is a concomitant increase in background (panel B).

As we have already discussed, hematoxylin (and Nuclear Fast Red) stain the nucleic acids in the cells. If the tissue was overdigested, the cellular proteins and nucleic acids, for the most part, leak out of the cell into the stroma. This is the reason for the lack of counterstaining of such overdigested tissue and the associated background. The trick of the trade to recognize overdigestion with immunohistochemistry or in situ hybridization is to simply hold the slide up to the light after doing the counterstain. If you cannot see much color, then the tissue was overdigested.

When you are dealing with tissues that are easily over-digested, a trick of the trade is to use relatively weak protease digestion or short-term antigen retrieval. In comparison, tissues that tend to be fixed for a long time and/or have a tight network of cell connections and/or cells with high protein concentration (e.g., brain, skin, heart muscle) usually perform well with in situ hybridization or immunohistochemistry when using long protease time pretreatment or long incubation times with antigen retrieval.

THE BIOCHEMICAL EFFECTS OF EMBEDDING THE TISSUE/CELLS INTO PARAFFIN

After the tissue has been fixed in formalin, the next step is to embed the tissue into paraffin wax. It has been well documented that the DNA or RNA in either unfixed tissue or tissue fixed in formalin is intact. That is to say, there are no detectable alterations in the sequence of the DNA or RNA or, more specifically, no breakage in any of the phosphodiester bonds. However, this all changes when you embed formalin-fixed tissue in paraffin wax. And, of course, you have no choice but to embed formalin-fixed tissues into paraffin wax in order to be able to prepare the unstained slides for immunohistochemistry or in situ hybridization. Although you can use cryostat (frozen sections) for immunohistochemistry or in situ hybridization, as we discuss, this is not optimal for immunohistochemistry or in situ hybridization because the morphology will suffer and, more importantly, the signal will not be as good on a routine basis as for paraffin-embedded, formalin-fixed tissues. The reason that cryostat sections are not as good as formalin-fixed, paraffin-embedded tissues for in situ hybridization or immunohistochemistry is that the former lack the three-dimensional macromolecule cross-linked network that is likely the key feature for optimal in situ detection of any protein, DNA, or RNA sequence. The end result is that tissues fixed with denaturing fixative or no fixative at all may show a signal with immunohistochemistry, *but* the signal will show substantial *diffusion* (Figure 2-23). In comparison, if the same tissue is fixed in formalin, there will be a strong signal with immunohistochemistry, *and* the signal will be sharply localized to the cellular compartment where it is located (same figure). This is an important point, because it underscores a key feature of the three-dimensional protein/protein cross-linked network induced by formalin: *the ability to limit diffusion of the in situ hybridization or immunohistochemistry signal from the subcellular compartment where the target is present.* It is reasonable to say that perhaps the most important point I make in this book regarding in situ hybridization and immunohistochemistry is that the three-dimensional protein/protein cross-linked cage that is associated with any given target is a *primary factor* in determining the success (or failure!) of a given reaction. Given the importance of this point, let's look at some actual data comparing these three-dimensional protein/

protein cross-linked cages created by formalin fixation with that formed by denaturing fixatives (70% ethanol or acetone) or no fixation at all.

In this series of experiments, I obtained serial sections of a cryostat benign lymph node. Some of the sections were fixed in 10% formalin and others in 70% ethanol or acetone for 15 hours. Some of the sections remained unfixed and were stored at −20°C. After fixation, the slides were all tested at the same time for CD45 and smooth muscle actin (SMA) using either no pretreatment, protease digestion, or antigen retrieval. This is the same set of experiments discussed in the previous chapter, but now I want to focus on the more general point of formalin versus denaturing fixatives versus no fixative for immunohistochemistry.

Representative data are shown in Figure 3-15. Note that the signals for CD45 and SMA in the formalin-fixed tissue (panel A and B) are extremely sharply defined (see also Figure 2-23). The CD45 signal is in the cell membrane, and the SMA signal is in the cytoplasm. In each case, it is as if someone drew the signal with a thin pencil, as it is so sharply defined. Compare this to panel C, where the signal for CD45 in acetone-fixed lymph node is present. The signal is poorly defined, and it has diffused to either over the cell or over the stroma. Similarly, the CD45 signal for the ethanol-fixed tissue is not sharply defined but is beaded, and the signal for SMA has diffused markedly from its point of origin. Figure 3-16 shows the comparable data for the unfixed (cryostat section) tissue. It is clear when comparing panels A and B with panel C (the same tissue fixed in formalin) that the key difference is not in the intensity of the hybridization signal, but rather in the fact that there is marked diffusion of the immunohistochemistry signal with the unfixed tissue, and no diffusion with the formalin-fixed, paraffin-embedded tissues. It follows that the three-dimensional protein/protein cross-linked network induced by formalin fixation is doing a superior job in localizing the signal to the specific subcellular compartment to which it belongs. Of course, the latter is a key goal with in situ hybridization and immunohistochemistry, and it is an important reason why I recommend limiting all tissue-based research to formalin-fixed, paraffin-embedded tissues.

Another observation that is evident from Figures 3-15 and 3-16 is that the immunohistochemistry signal for CD45 and smooth muscle actin tend to be stronger for the unfixed tissue (and, of course, the formalin-fixed tissue) when compared to that seen with the denaturing fixatives (ethanol and acetone). The latter two fixatives certainly will affect the three-dimensional protein/protein cross-linked network that exists in the living cell in the form of the complex association of proteins with each epitope, RNA, and DNA molecule. Did denaturing the proteins that form the "protein cage" actually make them less able to assist the key immunohistochemistry reagents in accessing and docking with the epitopes of interest with immunohistochemistry? Is this the reason for the weakened signal evident with the denaturing fixatives when examining the data in Figures 3-15 and 3-16 to either formalin-fixed tissue or, perhaps more to the point, the unfixed samples? Could denaturing the proteins that surround any given RNA or DNA molecule in the fixation

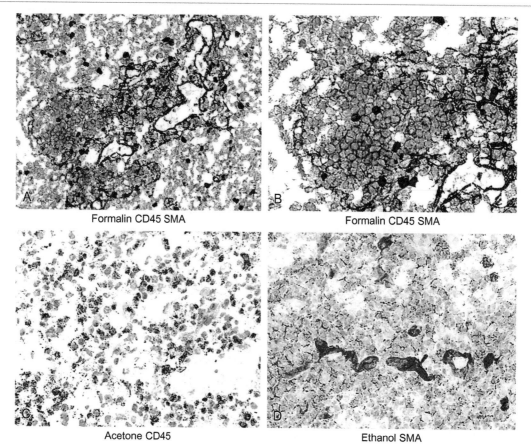

Figure3-15   Formalin fixation allows for a much sharper localization of the immunohistochemical signal when compared to denaturing fixatives and no fixative (cryostat sections): Part A. The immunohistochemical signal for CD45 and smooth muscle actin (SMA) is sharply defined to the cell membrane of the lymphocytes and cytoplasm of the blood vessels, respectively, if the tissue is fixed in formalin (panels A and B). Also note that the signal is continuous. The same tissue fixed in acetone (panel C) or 70% ethanol (panel D) showed marked *diffusion* of the CD45 and SMA signals (panel D) in the form of a beaded pattern and much diffusion throughout the stroma.

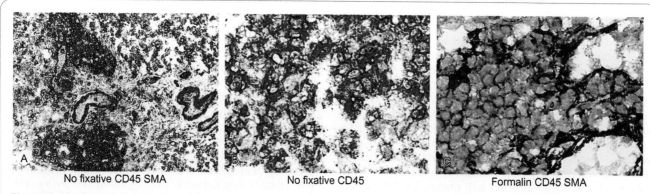

Figure 3-16   Formalin fixation allows for a much sharper localization of the immunohistochemical signal when compared to denaturing fixatives and no fixative (cryostat sections): Part B. The immunohistochemical signal for CD45 and smooth muscle actin (SMA) is sharply defined to the cell membrane of the lymphocytes and cytoplasm of the blood vessels, respectively, if the tissue is fixed in formalin (seen at high magnification in panel C). Compare this to the signal generated in an unfixed cryostat section. Note that the SMA signal (panel A) is very broad and diffuse, whereas the signal for CD45 (panel B) is also very diffuse, especially when compared to that for the formalin-fixed tissue (panel C).

process also diminish the signal compared to not altering them at all?

To address this question, we tested the same lymph node tissue using in situ hybridization for microRNAs 31 and 328. These microRNAs are strongly expressed in lymphocytes, and tend to show a nuclear and cytoplasmic signal. The data are shown in Figures 3-17 and 3-18. Note that the formalin-fixed cells in the lymph node

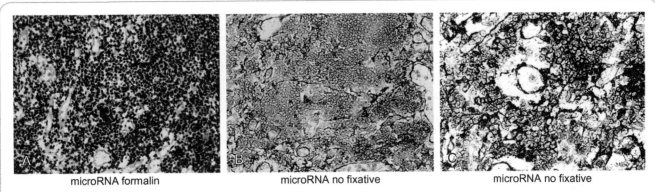

microRNA formalin   microRNA no fixative   microRNA no fixative

Figure3-17 Formalin fixation allows for a much sharper localization of the in situ hybridization signal when compared to denaturing fixatives and no fixative (cryostat sections): Part A. The in situ hybridization signal for microRNAs 31 and 328 is strong and sharply defined in the formalin-fixed tissue (panel A). Although the signal is still relatively strong, it is more diffuse in the cryostat section (panels B and C).

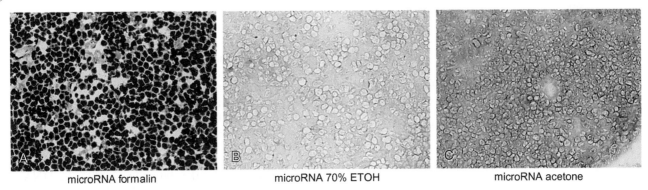

microRNA formalin   microRNA 70% ETOH   microRNA acetone

Figure3-18 Formalin fixation allows for a much sharper localization of the in situ hybridization signal when compared to denaturing fixatives and no fixative (cryostat sections): Part B. The in situ hybridization signal for microRNAs 31 and 328 is strong and sharply defined in the formalin-fixed tissue (panel A). No signal is evident if the same tissue was fixed in 70% ethanol and then tested by in situ hybridization for these microRNAs (panel B), and only a weak, diffuse signal was evident for the acetone-fixed tissue (panel C).

show an intense, well-defined cellular signal (panel A, Figure 3-17). In comparison, the unfixed tissue yielded a good but clearly weaker signal that clearly was, in part, diffusing out of the cell (panels B and C). The acetone- and ethanol-fixed tissues yielded weak, diffuse signals for the microRNAs, despite trying no pretreatment, protease digestion, and RNA retrieval (Figures 3-18 and 3-19). I think the reason was that our key in situ hybridization reagents were simply diffusing away from the miRNA target, since there was no three-dimensional protein/protein cross-linked network to anchor them in place. The unfixed tissue will have the "native" three-dimensional macromolecule network surrounding each epitope; it just won't be cross-linked. The ethanol and acetone fixatives may actually make this network less able to limit diffusion during immunohistochemistry and in situ hybridization by changing the conformation of these proteins by denaturing them. You may recall from the previous chapter that when we did the same experiments with in situ PCR, the amplicons were easily detected in the amplifying solution in the cells fixed in acetone or ethanol, but not in the formalin-fixed cells. So, to summarize the data,

a key feature *unique* to formalin-fixed, paraffin-embedded tissues when doing immunohistochemistry or in situ hybridization is that the signal is sharply localized to the subcellular compartment to where the target is located.

DNA AND RNA NICKS FROM THE PARAFFIN EMBEDDING PROCESS

DNA and RNA do not like being heated to 65°C during the wax embedding process. Specifically, the heating generates nicks into the DNA and RNA. This can be easily demonstrated for the DNA by using a modification of the in situ PCR protocol. The nicks in the DNA will be recognized by the Taq polymerase, which will attempt to "fix" them. This issue is discussed at length in Chapter 10. A representative example of this method and the controls are shown in Figure 3-20 in panel B. The tissue is formalin-fixed, paraffin-embedded brain from a person who died of rabies infection. The rabies virus is an RNA virus that has a tropism to infect neurons. Note that if the in situ PCR reaction is done without a DNase step, then an intense nuclear-based signal is seen in every cell. In short, whenever the Taq polymerase sees a nick in the DNA, it will use this as an initiation point for DNA synthesis. If you incorporate a reporter nucleotide such as digoxigenin or biotin into the newly synthesized DNA and then detect this, you can get an idea of the extent of nicking of the DNA during the paraffin embedding process (Figure 3-20, panel B). If you are familiar with in situ PCR, this DNA synthesis pathway is typically referred to as the primer independent pathway.

When frozen tissue sections (cryostat sections) were fixed in formalin and not heated and then subjected to RT in situ PCR without the primers, no signal was evident (this point is discussed at length in the PCR in situ hybridization book [26]). Since the frozen/fixed tissue was not embedded in paraffin and was never exposed to any temperature above room temperature, the "DNA nick-induced primer-independent DNA synthesis pathway" is not operative. Stated another way, the DNA in the cryostat-sectioned/formalin-fixed tissue is as pristine and intact as the DNA was prior to the biopsy being removed.

Given the obvious fact that it is easy to generate an intense signal with in situ hybridization using a target present in high copy number (such as HPV in genital condyloma or CIN 1 lesions) in formalin-fixed, paraffin-embedded tissues, and the observation that the DNA in such tissue will have a lot of nicks, it is easy to be confident that these nicks in the DNA do not affect our ability to detect such targets. This makes sense when you think about it. Specifically, HPV DNA is 8000 base pairs. Even if there is a nick in the DNA on average after every 1000 nucleotides, the three-dimensional protein/protein cross-linked network should keep the HPV DNA "in place" in the cell. This is one reason I like to refer to the three-dimensional macromolecule cross-linked network as a "cage." Specifically, the data suggest that the DNA and RNA in the cell, even if the phosphodiester bonds are haphazardly broken in the sequence, are still held in place by the three-dimensional macromolecule cross-linked network that surrounds it. Of course, for DNA and RNA (as well as for proteins), the cage is actually physically attached to the nucleic acids. Further, the cross-linked proteins that surround the DNA and RNA in the formalin-fixed, paraffin-embedded tissues are probably able to

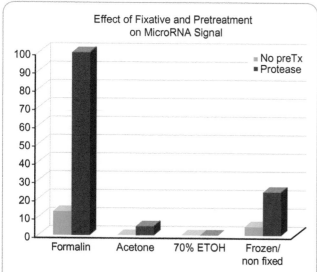

Figure 3-19 A graphic representation of the intensity of the microRNA in situ hybridization signal relative to different fixatives and pretreatment conditions. Note that the microRNA in situ hybridization signal was very weak with the two denaturing fixatives (ethanol and acetone) regardless of whether the tissue was pretreated with protease or was not pretreated. This suggests that the denaturing fixatives, by altering the conformation of the three-dimensional macromolecule cross-linked cage associated with the miRNAs, facilitates diffusion of the signal and/or reduces the ability of the key in situ hybridization reagents to dock with the target. In comparison, formalin-fixed tissue yields the strongest signal that shows no diffusion, but this requires protease digestion. Cryostat sections yield signals stronger than the denaturing fixatives and are slightly increased by protease digestion, yet are not nearly as strong as the signal seen with the formalin-fixed, paraffin-embedded tissues and also show diffusion.

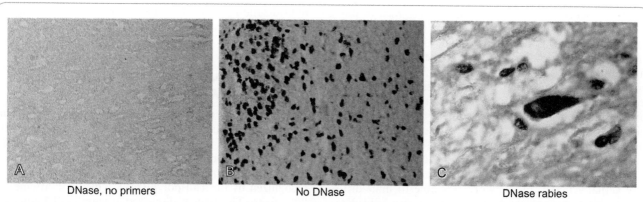

Figure 3-20 RT in situ PCR as a way to demonstrate the nicks in the genomic DNA after paraffin embedding. This brain tissue infected with rabies virus was digested with DNase, and then RT in situ PCR was performed without primers. No signal was evident (panel A). Compare this to panel B, where RT in situ PCR was done on an adjacent section of the same slide, but without DNase digestion and with no primers. Note the strong nuclear-based signal from DNA repair. Thus, the DNase digestion completely eliminated this nonspecific DNA synthesis. Under these conditions, you can do target-specific direct incorporation of the reporter nucleotide. Rabies RNA was detected after RT in situ PCR in cells with the diagnostic feature of rabies infection, the Negri body (panel C).

keep the nucleic acid molecules, even when damaged, in their correct orientation due to the hydrogen bonds and physical restraints that they can impress upon the nucleic acids. In a sense, the three-dimensional protein/protein cross-linked cage acts like a cast or splint for a fractured bone, with the "fractures" being the nicks in the RNA or DNA molecule.

ALTERATIONS IN PROTEINS

It is a routine process to extract DNA and RNA from paraffin-embedded tissues and then to study them by either PCR or dot blot hybridization technologies, such as the hybrid capture method for HPV DNA. Our laboratory (and many other laboratories) have done this and have shown that, if we study formalin-fixed tissues that were recently obtained biopsies that were then paraffin-embedded within the past 6 months, the recovery of DNA or RNA is comparable to frozen, unfixed tissues. Our laboratory has performed these types of experiments for HPV DNA and microRNAs. The logical conclusion is that, despite proving that nucleic acids routinely get nicks in their triphosphate backbone during the paraffin embedding process, the end result is that the size of the nucleic acids is not altered much. Thus, we can successfully extract and precipitate them using the standard protocols for DNA (with ethanol precipitation meant to capture nucleic acids greater than 400 base pairs) and microRNA, and then detect the targets with hybrid capture or real-time qRTPCR, respectively.

It is a different matter to do such experiments with proteins. For DNA and RNA molecules, the extensive protein-nucleic acid cross-linking is easy to solve when we want to extract nucleic acids from formalin-fixed, paraffin-embedded tissues: we do very extensive protease digestion. By "very extensive," I mean anywhere from 8 to 15 hours of protease digestion at the optimal temperature and the salt concentration for the protease! Obviously, this won't work for purifying proteins that have been fixed in formalin.

Though we cannot evaluate formalin-fixed proteins by extraction from the paraffin-embedded block,

purification, and analysis by Western blot or ELISA, we can theorize that these macromolecules behave like DNA and RNA after formalin fixation. That is, the paraffin embedding probably causes a relatively low percentage of the peptide bonds to be disrupted. However, the overall integrity of the protein sequence and the target epitope sequence is intact and, thus, easily detectable with immunohistochemistry. Further, even if there are some peptide bonds that have been broken due to paraffin embedding and the age of the stored block, the epitope is likely being held in place by the protein cage that surrounds it by ionic and hydrogen bonds, direct cross-links with the cage, and by physical constraints that I have referred to in this book as "pore size." As in the previous analogy with DNA and RNA nicks in formalin-fixed, paraffin-embedded tissues, it may help to think of the disrupted peptide bonds with proteins as fractures, and thus to think of the three-dimensional macromolecule cross-linked network as a splint that holds the fractured macromolecules in place so that we can still detect them by immunohistochemistry.

THE BIOCHEMICAL EFFECTS OF PROTEASE DIGESTION (PROTEASE PRETREATMENT)

I suppose there is nothing more obvious than the fact you can modify the three-dimensional protein cross-link generated with formalin fixation by doing a protease digestion. But just how useful is protease digestion for successful in situ hybridization and immunohistochemistry?

The observation that protease digestion can be essential for successful in situ hybridization and immunohistochemistry was illustrated in Figures 3-2 and 3-3. For these particular tissues, where there was abundant cytokeratin AE1/3 and HPV 6 DNA, respectively, our assays would have been negative if we had not done the protease digestion step. However, look at Figure 3-21. For these tissues, which looked identical to the tissue illustrated in Figures 3-1 and 3-2 under the microscope when we did a routine hematoxylin and eosin stain, a strong signal was evident without protease digestion (panel A). Importantly, the

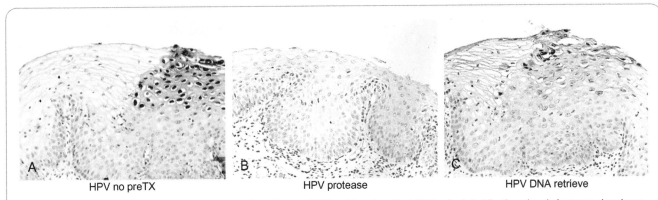

HPV no preTX   HPV protease   HPV DNA retrieve

Figure3-21   The deleterious effect of protease digestion and DNA retrieval on the HPV in situ hybridization signal. A strong signal was evident for this CIN 1 lesion if no pretreatment was done (panel A). If a 4-minute protease digestion preceded the in situ hybridization, then the signal was obliterated (panel B). DNA retrieval weakened the signal relative to no pretreatment (panel C). This biopsy was obtained from the same laboratory and the same year 2012 as the biopsies shown in Figures 3-1 and 3-2.

signal was *lost* if we did the protease digestion step (Figure 3-21, panel B); note that DNA retrieval also greatly reduced the signal compared to no pretreatment (panel C)! Also note the sharp demarcation between the HPV-infected and noninfected cells in panel A, which is typical of this virus. If you are old enough to remember what a broken record sounds like, I need to sound like a broken record. One of the most important concepts to remember with both immunohistochemistry and in situ hybridization is that tissues will vary a great deal in their optimal conditions for these two methods. And, again, these are tissues that may look the same to you under the microscope and may have come from the same laboratory, even collected the same day! This is why it is *not* possible to predict ahead of time what specific pretreatment regime will be needed to obtain optimal immunohistochemistry or in situ hybridization on a given formalin-fixed, paraffin-embedded tissue.

Thus, we need to understand at the biochemical level why protease digestion in some cases is essential for optimal in situ hybridization, yet in other cases is either not needed or actually is deleterious to the DNA- or RNA-based signal. The reasons that protease digestion can, at times, allow us to produce a stronger signal with in situ hybridization are straightforward when we examine our biochemical model of formalin fixation (Figures 3-5 through 3-7). If we assume that the pore size, or breathability, and/or the ionic/hydrogen bond potential surrounding the HPV 6 DNA/protein complexes are so stringent that it will not allow our 60–100 base-pair-sized probes to enter the nucleus, then, of course, we will get no signal without protease digestion. You may recall that if you use HPV probes of from 150 to 200 base pairs in size, you basically *never* will obtain a strong, optimal signal with in situ hybridization. This, by itself, indicates that pore size is likely a factor with the signal we get with in situ hybridization.

Now let's turn our attention to in situ hybridization using LNA probes for microRNA analysis. Here the probes are only 20–23 nucleotides long as compared to the 75–100 base-pair-sized HPV probes. Further, the target is cytoplasmic as compared to the nuclear-based signal of HPV DNA. Thus, diffusion into the cell is probably less of an issue for the in situ detection of microRNAs as compared to HPV DNA. Finally, the microRNA target is also only 20–23 nucleotides long, compared to the 8000 base pair HPV DNA target. Is protease digestion important for microRNA in situ hybridization? The answer is exactly the same answer as with HPV in situ hybridization: at times *yes* and at other times *no*! This is illustrated in Figure 3-22. Note that there is no signal if there is no pretreatment. Only 10 seconds of protease

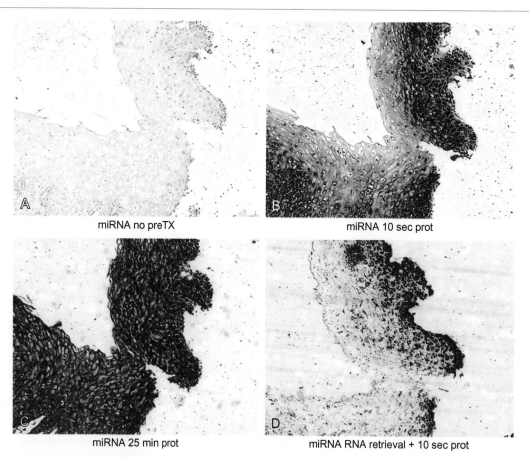

miRNA no preTX

miRNA 10 sec prot

miRNA 25 min prot

miRNA RNA retrieval + 10 sec prot

Figure3-22 The effect of the pretreatment conditions on the in situ hybridization signal for microRNAs. No signal was seen in this cervical biopsy tested for microRNA-let-7c by in situ hybridization if no pretreatment was used (panel A). A 10-second incubation with proteinase K led to some signal (panel B), which, in turn, increased with increased protease digestion time (panel C). The signal was reduced and cell morphology damaged if protease digestion was added to RNA retrieval (panel D).

digestion was enough to generate some signal (panel B) that was much stronger if the protease digestion was 25 minutes in duration (panel C). Whereas robust HPV in situ hybridization signals can be seen relatively commonly with no pretreatment, this is unusual with micro-RNAs where either strong protease or RNA retrieval +/– protease is often needed. This raises the possibility that microRNAs might typically be surrounded by much more dense three-dimensional macromolecule cross-linked cages than HPV and, thus, need protease digestion much more commonly than HPV to generate a signal. Of course, this theory is based on the overlying hypothesis that the biochemical characteristics (including protein density) of the three-dimensional protein cross-linked cage may be the primary factor in determining the optimal pretreatment regime for in situ hybridization. Finally, note that panel D shows a loss of both signal and of cell detail with RNA retrieval and protease; this is typical overdigestion.

Although we leave the details to the next chapter, we will see that the same phenomenon just discussed from protease digestion and in situ hybridization for HPV DNA and microRNA is true for many proteins as analyzed by immunohistochemistry. Look at Figure 3-13 where the signal for the cytokeratin AE 1/3 is very weak with no pretreatment and is intense after 4 minutes of digestion in proteinase K prior to the immunohistochemistry. For the cytokeratin, we can explain the essential need for protease digestion by assuming that the protein-protein cross-linking induced by formalin surrounded the epitope with such a small pore size that the primary antibody was not able to access the epitope. As depicted in Figure 3-6, we can also add to our hypothesis that the cross-linking induced by the formalin fixation will also so tightly encompass the AE 1/3 epitope that it will be incapable of "breathing" (that is, separating itself from the surrounding three-dimensional protein/protein cross-linked network) and that it will be surrounded by such a strong ionic/hydrogen bond network (Figure 3-7) that it would be impossible for the primary antibody to break through this even if the pore size was large enough.

Recall that there will be instances with immunohistochemistry in which protease digestion will *reduce* the signal compared to no pretreatment at all. This is a rare event for immunohistochemistry as compared to in situ hybridization, where it is relatively common. But it does happen. Recall that an example of this was shown in Figure 3-3 for the cell membrane protein Iba1 that is found on macrophages and microglial cells. It makes sense that protease digestion rarely inhibits immunohistochemistry (as compared to in situ hybridization), since proteins absorb the brunt of biochemical changes with formalin fixation and, thus, are much more likely to require protease pretreatment to expose an epitope rendered hidden by the inevitable extensive protein-protein cross-linking caused by formalin.

If we understand that there are logical and practical reasons to use protease digestion before immunohistochemistry or in situ hybridization analysis, the next question is what protease to use? I am old enough to remember when we used saliva, employing the proteases and amylases naturally present in this solution. I do *not*

recommend this now! I do recommend using either proteinase K or pepsin as your major protease. Proteinase K is the stronger of the two and is very effective in destroying the proteins found in mammalian cells. Indeed, the K of proteinase K refers to keratin, as this protease avidly degrades one of the most common of the human proteins, cytokeratin. Proteinase K's activity can be modulated. If you use high temperature (65°C) and salt, then it will show a very high specific activity. Thus, you rarely under-protease-digest a sample with proteinase K under these conditions. However, you also risk seeing overdigested tissue in which the cytoplasm and nuclei basically melt away and, of course, any in situ hybridization or immunohistochemistry data become useless, as we saw in Figures 3-14 and 3-22. That is one advantage of pepsin; it is very unusual to overdigest the sample with pepsin. Also, since pepsin requires a low pH to operate most effectively, simply raising the pH is a simple way to inactivate this protease.

For most of my in situ hybridization (and immunohistochemistry) work, I use proteinase K. This protease comes from a fungus. Its mode of action involves the cleavage of the peptide bond adjacent to the CO groups of aromatic and aliphatic amino acids, which makes it a very broad spectrum protease. As noted, it is called proteinase K for its ability to cleave the peptide bonds in keratin, which gives us a lot of reassurance that it will be able to digest the cytoplasmic proteins required to access both cytoplasmic and nuclear targets. It has a molecular weight of 28.9 k Daltons, making it a relatively small protein. In comparison, the enzyme Taq polymerase, which is able to readily enter the nucleus during in situ PCR, has a molecular weight of 95 k Daltons. As another point of comparison, serum albumin is a protein that can readily diffuse in and out of cells. Indeed, there is a disease called the nephrotic syndrome in which the physical barrier of cells called podocytes that line the glomerular capillaries falls apart and the patient loses massive amounts of serum albumin in the urine. The albumin we will use extensively in our work is bovine serum albumin (BSA). It has a molecular weight of 66.5 k Daltons based on a size of 585 amino acids and dimensions of $140 \times 40 \times 40 \text{Å}^3$. These sizes of different proteins assures us that proteinase K (and pepsin) will readily diffuse into the cell during either in situ hybridization or immunohistochemistry.

A key point we need to keep in the back of our minds when doing either in situ hybridization or immunohistochemistry is diffusion of our reagents into and out of the formalin-fixed cells. Of course, our protocol for in situ hybridization involves re-hydrating the cell such that all of the reactions are done in an aqueous phase. As noted previously, the diffusion rate of small molecules in water such as the proteinase K, albumin, Taq polymerase, and such is $10^{-6} \text{cm}^2/\text{sec}$. Remembering that cells in mammalian tissues typically range from 10 to 100 microns in size, this equals a diffusion rate whereby 16 minutes would be needed for a small molecule such as our LNA probes, 100 bp "genomic probes," various antibodies/conjugates, and chromogens to diffuse throughout a cell with a circumference of 10 microns.

When you are using proteinase K, it is useful to remember that calcium at a concentration of around

5μmoles is able to stabilize the enzyme and allow for more reliable digestion over time, especially if you need to digest the samples for 30 minutes or more. Though detergents such as sodium dodecyl sulphate (SDS) or hydrogen-bond-disrupting agents such as urea do not affect the activity of proteinase K, I do not recommend your using them in the proteinase solution. The reason is that the SDS may have the effect of harming the morphology of the tissue given its ability to digest through lipid-rich membranes. Adding urea can actually enhance the ability of the proteinase to disrupt the three-dimensional protein/protein cross-linked network in the formalin-fixed tissue. However, it tends to be too effective in this regard and may, thus, reduce the signal. Although I would recommend against using SDS, you may wish to experiment with urea if you are consistently getting weak signals yet good morphology. I prefer to alter other conditions when I am getting a weak signal with good morphology, such as either adding antigen retrieval or increasing either the incubation time or concentration of the proteinase K.

So, to summarize, the key features of proteinase K are as follows:

1. It is very efficient in rapidly digesting DNases and RNases (which is more useful for fresh, non-fixed samples).
2. It is an easy matter to increase its specific activity by either increasing the temperature of the reaction from 37°C to 65°C and/or by adding salt (e.g., 1XSSC) to the digestion solution. Therefore, you do not need to add agents such as SDS or urea in an attempt to increase the activity of proteinase K relative to your in situ hybridization signal; you simply need to increase the temperature of the reaction to 65°C and/or add salt to the proteinase K digestion solution.
3. It usually does an excellent job in altering the three-dimensional protein cross-linked network cage around your target by affecting the pore size, ionic and hydrogen bond potential, and breathability of the target and key in situ reagents, which allows you to maximize signal and show little to no diffusion.

With respect to the first point, recall that this is a much more important issue if you are digesting fresh/frozen tissue that is not fixed in formalin. The reason is simply that formalin, by cross-linking enzymes, is very good at inactivating them all by itself and, as we have discussed, is one reason that formalin is so good at preserving cell morphology for decades (indeed, centuries) after the tissue has been fixed.

The last point to discuss about proteinase K is its inactivation. When you are doing in situ hybridization, you can easily inactivate it. Proteinase K can be easily removed by simply holding the slide under running water for DNA or protein targets. For RNA targets, I recommend that you use RNase-free water to rinse the slides. Just a 15- to 30-second rinse should be enough to remove all of the proteinase K from the tissue. There are agents that can be added to the solution to inactivate proteinase K, such as diisopropyl fluorophosphate (DFP). These agents are simply not needed for in situ hybridization, since we have a solid base (the silane-coated glass slide) that allows us to easily and readily rinse away the proteinase.

PORE SIZE

Now that we have discussed the various proteases you can use with in situ hybridization and ways to either enhance their activity or to inactivate them after they have done their job, let's get back to a discussion of the effects of protease digestion on the three-dimensional protein/protein cross-linked network. Of course, the simple and desired effect of treating formalin-fixed, paraffin-embedded tissues with proteinase K is that the treatment will increase the pore size of the three-dimensional macromolecule cross-linked network. The logical question that immediately follows is: how important is the pore size in formalin-fixed, paraffin-embedded tissues when doing in situ hybridization?

Can we isolate the effect of protease digestion on pore size with in situ hybridization? Although it is not possible, of course, to be able to create equal and known pore sizes in formalin-fixed, paraffin-embedded tissues, we can get a clue as to the importance of pore size by doing in situ hybridization on a series of formalin-fixed, paraffin-embedded tissues and compare the results to serial sections in which either no pretreatment was used versus those where either protease digestion or DNA retrieval (95°C for 30 minutes) was done. The rationale of this testing is that protease digestion would directly affect the pore size in the formalin-fixed, paraffin-embedded cells, whereas DNA retrieval, by denaturing the extensively cross-linked proteins, would probably have a lesser-degree effect on pore size. Further, we will assume that the diffusion rate of the HPV DNA probe and other molecules will be related to the pore size. Thus, experiments were done by comparing the different pretreatment regimes in their effect on the intensity of the hybridization signal versus time of hybridization. The time of hybridization was varied from 15 minutes to 12 hours. Finally, we used HPV DNA in CIN 1/condyloma lesions as the model system, since these are associated with high viral copy numbers and, thus, an intense signal when optimal conditions exist. A summary of the data follows in Figure 3-23. It is important to stress that these experiments were done with serial sections, so we are comparing the effects of protease digestion on the same cells in these CIN lesions.

Note that the graph scores on the y-axis the intensity of the HPV 16-specific in situ hybridization signal, and it is plotted versus time on the x-axis (from 15 minutes to 12 hours of hybridization time). CC refers to cell conditioning, which, as you know, is synonymous with antigen retrieval (30 minutes at 95°C). Note that with protease digestion, you reach 50% of the maximum hybridization signal at 45 minutes of hybridization time, and 90% of the signal intensity at around 6 hours of hybridization. In comparison, if you omit the protease (and do no pretreatment at all), then you reach 50% of the maximum hybridization signal at 11 hours of hybridization time and do not reach 100% intensity over the 12 hour time course of the experiment. This marked difference in time (10.25 hours less time needed with protease digestion to reach 50% of the maximum hybridization signal) shows how dramatically the protease digestion can speed up the diffusion of the in situ hybridization reagents during in situ hybridization.

Of course, protease digestion will not just affect pore size. It will also remove protein side chains that can form

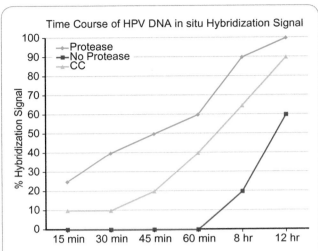

Figure3-23 The effect of the pretreatment conditions on the time kinetics of the HPV in situ hybridization signal. Serial sections of this CIN 1 lesion were subjected to in situ hybridization with hybridization times that ranged from 15 minutes to 12 hours. Some of the sections were not pretreated, whereas others received protease digestion or DNA retrieval (here referred to as CC). Note that the slopes of the three curves are similar, but that protease digestion was associated with a much more rapid generation of the in situ hybridization signal, with DNA retrieval intermediate, and no pretreatment the slowest. These data are consistent with pore sizes *in this particular tissue* that are restrictive to in situ hybridization and require protease digestion. Recall that in many cases formalin-fixed, paraffin-embedded tissues that contain HPV do not require any protease digestion for optimal signal.

hydrogen bonds with the HPV DNA probe. These hydrogen bonds can have a negative effect on our in situ hybridization reaction. The negative effect would result if the hydrogen bonds from the three-dimensional protein/protein cross-linked network so strongly bound to the HPV DNA probe (or target) that it was not available to bind to the target (or probe). The protease digestion may also affect the "breathability" of the double-stranded DNA target. The reduced protein-DNA cross-links may make the double-stranded HPV DNA more accessible by allowing the HPV base pairs more sites to form hydrogen bonds with probe and target, which would also enhance the in situ hybridization signal. However, the hydrogen (and ionic) bonds and "breathability" provided by the three-dimensional protein/protein cross-linked network may also be helpful to the in situ hybridization reaction. For example, ionic/hydrogen bonds between amino acid side chains in our protein/protein cross-linked network may be able to bind to our DNA/cDNA probes and antibodies/conjugates, which, in turn, could facilitate their retention in the part of the cell that contains the target of interest. Of course, we cannot separate out the effects of the proteinase K digestion on the pore size versus these other forces that most likely affect our in situ hybridization signal.

Let's go back to the data in Figure 3-23. It is clear that the protease digestion dramatically increases the rapidity of hybridization between the HPV target and the HPV-labeled probe. We are hypothesizing that this

markedly increased rate of hybridization involves, at least in part, increased diffusion of our different reagents to the nucleus, which is where the episomal HPV DNA is located. Let's return to our discussion of diffusion of macromolecules in the fixed cells and discuss some basic points that relate to the diffusion of our in situ hybridization reagents during the in situ hybridization process.

As noted previously, since macromolecules diffuse through aqueous solutions at a rate of $10^{-2}\,\mu m^2$/sec, this means that a small protein (or our DNA probe or primary antibody) can diffuse through the entire thickness of the tissue (10 microns) in about 16 minutes. Obviously, we are not dealing with a water solution, but rather with water immersed in a complex three-dimensional protein/protein cross-linked network. Any protein or DNA/RNA probe trying to traverse this protein-protein labyrinth will act pretty much like a pinball going through a complex maze; it will be slowed down considerably as it bounces off the many different macromolecules that are standing in its way. What is the effect of the pore size on the diffusion of our protein reagent or DNA probe? The effect of pore size on the diffusion of a macromolecule is dramatic. It is not arithmetic but rather geometric. Let's include some specifics. By increasing the pore size from 0.01 microns to 0.02 microns, we will increase the diffusion rate of a protein (or DNA probe) by 3.5 times. Recall that a typical cell is about 10 microns, so we are referring to an area about 1/1000 the size of a typical cell. If the pore size increases from 0.01 micron to 0.03 microns the diffusion rate increases 9 times. An increase in the pore size from 0.01 micron to 0.04 microns increases the diffusion time by 15 times, and so on. No wonder proteinase K digestion can increase the rate that our hybridization develops during in situ hybridization! And, going back to our example in Figure 3-23, the use of proteinase K decreased the amount of time needed to see 50% of the maximum hybridization signal from 11 hours to 0.75 hours, which is an increase of 15 times! Clearly, since proteinase K can "punch holes" in the three-dimensional protein/protein cross-linked network and, thus, increase pore size, it should be able to increase the diffusion rate of our in situ hybridization reagents in most formalin-fixed, paraffin-embedded tissues. The only times it would not be able to would be if the tissue was inadequately fixed because the pore size was large to begin with and, thus, could not be augmented any more to improve our in situ hybridization signal, or if the tissue had been stored for many years and our three-dimensional protein/protein cross-linked network had degenerated.

Let's go back to Figure 3-23. We have theorized that the 15 times increase in the rate of the hybridization signal with proteinase K digestion likely reflects primarily an increase in pore size of our three-dimensional protein/protein cross-linked network, and we have further theorized that the actual increase in pore size from the proteinase K digestion was four-fold. Now let's compare these data to those derived from cell conditioning (95°C for 30 minutes prior to HPV denaturation/hybridization). First, let's think about the effect of incubating the formalin-fixed, paraffin-embedded tissues at 95°C for 30 minutes prior to our denaturation and hybridization. This high temperature in an aqueous solution probably acts primarily by denaturing the proteins in our three-dimensional protein/protein cross-linked network. Stated another way, denaturing the proteins in the three-dimensional macromolecule cross-linked network will change their conformation. No doubt the change in conformation is associated with some protein-protein linkage destruction (and hence an increase in pore size). However, when you think about it, it is doubtful that increasing pore size is the primary effect of cell conditioning (i.e., DNA retrieval). Indeed, the many cross-links in the three-dimensional protein/protein cross-linked network would tend to minimize the ability of cell conditioning to break the peptide bonds needed to increase the pore size. Rather, the conformational change will tend to change the orientation of proteins relative to each other and, in this way, expose epitopes that otherwise would not have been accessible to our immunohistochemistry or in situ hybridization reagents. Of course, such shifts in the conformation of the proteins in our three-dimensional protein/protein cross-linked network will likely affect the hydrogen/ionic charge potential of this network, as well as the "breathability" of the macromolecules in the cell, especially the RNA and DNA targets.

Based on these suppositions, we would predict that cell conditioning will increase the rapidity of the formation of the hybridization signal compared to no pretreatment in this specific tissue. However, proteinase K digestion will probably have an even a stronger effect on increasing the rapidity of formation of the hybridization signal. If we look at the data again (Figure 3-23), it is clear that our prediction at least for this tissue is correct. Note that the time needed to get 50% of the maximum intensity of the hybridization signal goes from 11 hours (no protease) to 2 hours with cell conditioning, as compared to 45 minutes with protease digestion. These data are consistent with our theory that, at least in some tissues that probably have been very amply fixed in 10% buffered formalin, pore size is a major limiting factor to the success of our in situ hybridization reaction, and that protease digestion can help us a great deal in this regard by punching holes in the three-dimensional protein/protein cross-linked network and, thus, much increasing the pore size.

We need to make an important distinction between *pore size* and *morphology* when discussing in situ hybridization in formalin-fixed, paraffin-embedded tissues. As we have seen in many of the figures in this chapter, pore size can be increased a great deal with no apparent change in tissue morphology. This is best evident in Figure 3-21. Proteinase K digestion eliminated the signal, presumably by increasing the pore size so much that some key reagents simply diffused out of the nucleus. Or stated another way, the key reagents could no longer "dock" with the target and/or probe-target complex. However, the morphology of the tissue is still very good. Of course, if tissue is overdigested to the point that the morphology has been destroyed, then it will be impossible to get a signal with in situ hybridization or immunohistochemistry. But the much more common scenario is that the protease digestion has altered the three-dimensional protein/protein cross-linked network to the point that we cannot generate a signal with in situ hybridization, but the morphology is still very well preserved.

It is logical to assume that there must be a range of pore sizes that allow for the robust diffusion of our reagents that is needed for optimal in situ hybridization. If this is true, it follows that for some tissues it may not matter if we treat the tissue with either a protease, antigen retrieval, or no pretreatment. Protease digestion would increase the pore size, but still within the range to allow for excellent diffusion of our in situ hybridization reagents, while still allowing for excellent retention of the signal. Since cell conditioning alters pore size to a smaller degree than proteinase K digestion, then it should also allow for optimal in situ hybridization under these conditions. We saw an example of such a tissue in Figure 3-1, panels A–C. It is important to stress that I am not implying that pore size is the main determinant for successful in situ hybridization. Far from it. But I do hope that these examples give you the confidence that there are biochemical explanations, probably relative simple, that help explain why seemingly identical-appearing tissues can behave so differently when doing in situ hybridization or immunohistochemistry.

IONIC CHARGE AND HYDROGEN BOND POTENTIAL OF CELL/BREATHABILITY OF THE MACROMOLECULES INSIDE THE FORMALIN-FIXED, PARAFFIN-EMBEDDED TISSUES

As already indicated, there is no way for us to dissect the relative influence of pore size versus other forces that may be operative when doing in situ hybridization. We recognize that the side chains of the amino acids, which, of course, are the main components of our three-dimensional protein/protein cross-linked network, will create the potential for hydrogen and ionic bonding that can obviously impact the movement of the macromolecules that we use when we do in situ hybridization or immunohistochemistry. Further, we realize that the same side chains, as well as the actual N-C-N cross-links, can limit the "breathability" of macromolecules that can also influence our in situ hybridization and immunohistochemistry results. LNA probes are the classic and dramatic example of how important limiting the breathability of the DNA probe and target can be in enhancing in situ hybridization.

Realizing all of the above, are there variables we can adjust to get a better idea of the relative contribution of pore size versus ionic/hydrogen bond potential and breathability to in situ hybridization? One simple way I have tried is very short incubation times with the protease, especially after cell conditioning. As I described in the preceding chapter, my reasoning is that protease digestion times of 10 seconds would be too short to "punch holes" in the three-dimensional protein/protein cross-linked network, but may be sufficient to alter the ionic/hydrogen bond potential and breathability of the network.

I did a series of many HPV in situ hybridization experiments. I subjected the tissues to either no pretreatment, long-standing protease digestion (30 minutes), cell conditioning (30 minutes), short protease digestion (10 seconds) and, finally, cell conditioning (30 minutes) plus 10 seconds of protease. My reasoning was that if the

10 seconds of protease digestion, either alone or combined with cell conditioning, enhanced the signal, this probably would reflect an effect on the ionic/hydrogen bond potential and breathability of the three-dimensional protein/protein cross-linked network and not pore size. And, to review, we are assuming that at times the ionic/hydrogen bond potential and breathability factor may reduce our in situ hybridization signal by making it more difficult for our reagents to access the DNA or RNA target.

You will recall that we have already briefly touched on the simple question: can 10 seconds of protease digestion, either alone or in combination with cell conditioning, enhance our HPV in situ hybridization signal? The answer was *yes*! This point was illustrated in Figures 3-9 and 3-10. Note that the enhancement is dramatic and points to, at least in this tissue, some factor in the three-dimensional protein/protein cross-linked network that is very sensitive to very short protease digestion times. Of course, in many of the tissues the short proteinase K digestion, either alone or with cell conditioning, had no effect on the HPV in situ hybridization signal. As you see, we now have another tool at our disposal in the form of short protease digestion, either alone or in combination with antigen retrieval, which we can use to maximize our signal with in situ hybridization, as well as immunohistochemistry.

THE PRACTICAL SIGNIFICANCE OF THE HIGHLY VARIABLE "FORMALIN-INDUCED THREE-DIMENSIONAL NETWORK" ON OPTIMIZING THE SIGNAL FOR IN SITU HYBRIDIZATION AND IMMUNOHISTOCHEMISTRY

Up until now, we have focused our attention on the theoretical aspects of the biochemical forces operative in the formalin-fixed, paraffin-embedded cell when we do in situ hybridization or immunohistochemistry. I want to end this part by looking at the practical consequences of this to our protocols with immunohistochemistry and in situ hybridization. It is clear that we have at our disposal the following possible pretreatment conditions:

1. Nop retreatment
2. Protease digestion, long-term (from 5 to 30 minutes)
3. Proteased igestion,sh ort-term(10se conds)
4. Antigen retrieval (from 8 to 60 minutes)
5. Antigenre trievalp lusp roteased igestion

It is also clear that we cannot predict which pretreatment regime will lead to the optimal signal when we do immunohistochemistry or in situ hybridization, even if we have controlled the fixation of the tissue ourselves. Thus, the practical consequence of this is that it is to our advantage to try as many of the pretreatment conditions as possible on a given case when doing in situ hybridization or immunohistochemistry. This is why it is so advantageous to have the histotechnologist place three tissue sections per silane-coated slide. This allows you, using just two slides, to compare all five variables listed here. This, in turn, markedly increases the chances that you will find the optimal conditions for detection of the DNA, RNA, or protein target with the first

series of experiments. This rationale will serve as the basis of the in situ hybridization protocol I present in Chapter 7.

THE EFFECT OF AGING OF THE TISSUE/PARAFFIN BLOCK ON THE "FORMALIN-INDUCED THREE-DIMENSIONAL NETWORK" FOR IN SITU HYBRIDIZATION AND IMMUNOHISTOCHEMISTRY

The last topic we discuss in this "basics of biochemistry of in situ hybridization and immunohistochemistry" chapter is the effect of aging of the formalin-fixed, paraffin-embedded tissues on the signal with in situ hybridization and immunohistochemistry. Let me assure you that, if you define "interesting" as unexpected results that shed new light on a topic, this section will be interesting and, as it is to me, quite unexpected!

MOLECULAR BASIS OF THE LOSS OF SIGNAL WITH IN SITU HYBRIDIZATION AND IMMUNOHISTOCHEMISTRY RELATIVE TO THE AGING OF THE SLIDE/PARAFFIN BLOCK

I have on many occasions spoken to investigators who have done immunohistochemistry or in situ hybridization for many years, sometimes for decades. It is not unusual to hear statements like: "We decided not to use any blocks of tissue older than 10 years because the signal was too variable," or "Older blocks are too much of a headache, as the signal can be lost for targets you know have to be in the tissue." What surprises me is that, despite this loss of signal for immunohistochemistry and in situ hybridization with increasing age of the block being understood, there is relatively little information on the cause of the problem.

We undertook a series of experiments to better define the reduction in the signal with in situ hybridization and immunohistochemistry with the increased age of the block of tissue. First, we took a series of CIN 1 or condyloma lesions that were obtained within the past year and compared the data to equivalent biopsies obtained between 10 and 21 years prior to 2012. Further, the biopsies from 1991 through 2002 had one other important feature: I had done HPV in situ hybridization on them when they were initially obtained 10–21 years ago. Further, I selected only cases in which we had identified the HPV type and had demonstrated a strong signal for the virus when the tissue was initially tested 10–21 years ago. I also still had the unstained *plus* slides from these cases, as well as access to the original blocks. In this way, we could easily quantify the reduction in the HPV in situ hybridization signal versus storage of the block for many years. We quantified the score from 0 to 100, with 100 representing an intense signal in the cells (that is koilocytes) known to contain the highest copy number of the virus.

A summary of the data from these experiments is presented in Table 3-1. Note that there was a 74% reduction in the intensity of the HPV signal for those CIN 1/condylomatous lesions where the blocks had been stored from

10 to 21 years. This reduction in signal was evident both when comparing the data to blocks with the same diagnoses obtained within the last 2 years *and* to the signal evident in the same cases when they were initially analyzed 10–21 years ago. A representative example of how the signal for HPV DNA can be reduced after the block has been stored many years is shown in Figure 3-24.

We can address another simple question that may give us some insight into the biochemical basis of the reduction of the HPV signal over time. Since I had both unstained silane-coated slides from the CIN 1/condyloma blocks that were made 10–21 years ago as well as the paraffin blocks, we could see if the degradation of the signal was more dramatic if the sections were cut onto slides 10–21 years ago, or if the sections were prepared right before the HPV in situ hybridization test. In other words, we could test if storing the tissue in the block reduced the degradation of the HPV signal compared to putting the tissue section on a glass slide 10–21 years ago and storing it in light-tight boxes at room temperature for all of these years. These data are presented in Table 3-2. Note that there was a 70% reduction in the intensity of the HPV in situ hybridization signal when comparing unstained slides prepared 10–21 years ago versus the HPV in situ hybridization

data obtained when the biopsy was initially removed all those years ago. However, if we instead went to the paraffin block and made some new recuts in 2012 from the 10–21-year-old cases and then did the in situ hybridization, then the HPV signal intensity still decreased, but only by 48%. Thus, whatever the biochemical basis for the reduction in the HPV in situ hybridization signal, it is enhanced by storing the tissue on a glass slide as compared to storing the tissue in the original paraffin block.

The next obvious question is this: does this reduction of signal with prolonged storage of the blocks also apply to RNA and proteins? This is an easy question to answer. We took a series of aged blocks and did microRNA in situ hybridization on them, and compared these data to those obtained from microRNA in situ hybridization on biopsies obtained within the past 2 years. Of course, microRNA in situ hybridization was not possible 10–21 years ago, as this was way before LNA probes became commercially available. But, by studying

Table 3-1   Degradation of HPV in situ hybridization signal over time

Age of slide	Signal* (mean)	SE	pvalue
10–21 yrs old	25.9	4.5	
0–2 yrs old	84.1	2.0	<0.001

SE refers to Standard Error of the mean.
*Scored blindly as intensity of signal and % of HPV+ cells n = 44 for each group

Table 3-2   Degradation of HPV in situ hybridization signal over time

Age of slide	Signal* (mean)	SE	pvalue
10–21 yrs old (original)	89.4	3.6	
10–21 yrs old (recut – 2012)	41.5	8.3	<0.001
10–21 yrs old (recut >10 yrs ago)	19.4	2.9	0.016

*Scored blindly as intensity of signal and % of HPV+ cells N = 17 for each group

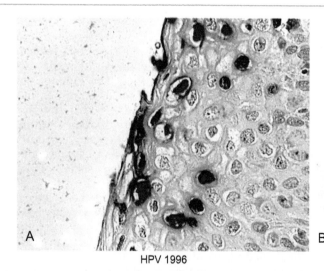
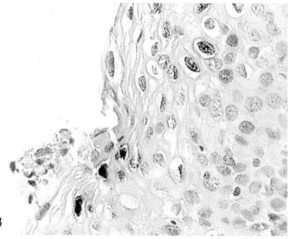

A   HPV 1996   B   HPV redone 2012

Figure 3-24   The effect of the age of the block on the HPV in situ hybridization signal. Panel A shows the HPV in situ hybridization signal that was generated in 1996 for a CIN 1 lesion that contained HPV 51. The intense signal reflects the high copy number of HPV 51 typical of such lesions. Unstained slides from this case were stored for 16 years. After this time (in 2012), the HPV in situ hybridization was repeated using the HPV 51 probe and other reagents from the same source as used in 1996 (Enzo Life Sciences), including the exact same pretreatment conditions. Note that in the serial section, the signal is now much attenuated (panel B).

microRNAs that should be present in high copy number in CIN 1/condylomatous lesions, we could still generate useful comparative data. We also compared the data from unstained silane-coated slides with tissues cut 10–21 years ago with tissue sections from these aged blocks that were cut in 2012, right before we did the experiments. The data are presented in Table 3-3. Note that there was a 62% reduction in the intensity of the micro-RNA in situ hybridization signal if tissue slides prepared 10–21 years ago were tested when compared to the data for the same histologic diagnosis on cases that were 1 to 2 years old. Further, note that if the slide was prepared from the aged block immediately prior to the in situ hybridization, then the reduction in the microRNA in situ hybridization signal was less dramatic (24% reduction). Thus, whatever the biochemical reasons for the degradation of the signal, it is operating on both DNA and RNA.

The next question is obvious: does the signal degradation also apply to proteins? We did the equivalent experiments for two proteins that are present in abundance in CIN 1 lesions: p16 and cytokeratin AE 1/3. The data are presented in Table 3-4. Note the dramatic reduction of 82% in the intensity of the signal when the AE 1/3 and p16 data were analyzed for new versus aged CIN 1 blocks of tissue. We tested just p16 when we compared the unstained slides of the aged blocks made 10–21 years ago with the recuts made just prior to testing this year from the aged paraffin block. These data are presented in Table 3-5. Again, note that making the unstained slide 10–21 years ago and storing it all of these years enhanced the degradation of the immunohistochemistry signal compared to making the unstained slide from the aged block this year and testing it (Table 3-5). Still, even preparing the unstained slide from the aged block this year immediately before testing was associated with a marked reduction of signal intensity (about 30%) when compared to CIN 1 lesions that were obtained within the past 2 years. Let me also stress that for all of these experiments, I used the same pretreatment protocol for the recuts (either original recuts or newly prepared recuts) that I did for the original slide, so as not to make the pretreatment conditions a confounding variable with the interpretation of the data.

What is the biochemical basis for this broad-based reduction in the hybridization signal for DNA, RNA, and proteins with increasing age of the paraffin block? We are assuming that it probably relates to the initial damage that DNA, RNA, and protein molecules obtain when the tissue is embedded in paraffin. The high temperature of the paraffin embedding process induces nicks in the DNA, RNA, and proteins, probably by the breakage of random and relatively few phosphodiester/peptide bonds. We assume that this process will slowly continue over time. We also know that the process is enhanced if the tissue section is cut at 4 microns and placed on a glass slide and stored for many years, as compared to the paraffin block per se. We may speculate that this simply reflects the fact that the section on the unstained plus slide is much thinner at 4 microns than the tissue that stays in the paraffin block, which usually is between 5 and 10 mm in size. The thinner section may make it more susceptible to oxidizing agents that can continue to degrade the macromolecules. Let's use HPV as our model system to get more insight into the biochemical basis of the reduction in our signal for in situ hybridization and immunohistochemistry with storage of the paraffin blocks.

It is routine to extract HPV DNA from formalin-fixed, paraffin-embedded tissues. As noted previously, the key is to incubate the sample in the protease solution for sufficient time (at least 24 hours), and to maximize the activity of the proteinase K by using high temperature and a high sodium concentration. After extraction, we can then analyze the HPV DNA using the very sensitive hybrid capture method, which, in turn, will quantify the amount of HPV DNA present in the sample. The final salient point to realize is that the extraction process we use for precipitating the HPV DNA from the formalin-fixed, paraffin-embedded tissues is based on cold ethanol/sodium acetate precipitation, which will require the HPV DNA sequences to be at least 250 base pairs in size for successful precipitation.

Table 3-3 microRNA in situ hybridization (miRNA let 7c, 205)

Age of block	Signal* (mean)	SE	p value
10–21 yrs old (recuts >10 yrs old)	27.3	6.2	
10–21 yrs old (recuts made in 2012)	65.6	5.9	0.001
1–2 yrs old	89.7	4.0	<0.001 (vs 1)
			0.003 (vs 2)

*Scored blindly as intensity of signal and % of microRNA + cells N = 15 for each group

Table 3-4 Immunohistochemistry for AE1/3 and p16

Age of block	Signal* (mean)	SE	p value
10–21 yrs old (recuts >10 yrs old)	10.9	1.7	
1–2 yrs old	92.9	1.9	0.001

*Scored blindly as intensity of signal and % of p16 or cytokeratin + cells N = 17 for each group

Table 3-5 Immunohistochemistry for p16

Age of block	Signal* (mean)	SE	p value
10–21 yrs old (recuts >10 yrs old)	10.0	2.6	
10–21 yrs old (recuts made in 2012)	63.3	4.2	0.001

*Scored blindly as intensity of signal and % of p16 + cells N = 6 for each group

For about 20 years I had saved paraffin ribbons of CIN 1 lesions in Eppendorf tubes (yes, it's true; pathologists tend to save materials for a long time!). The reason was so that I could see the effect of paraffin block storage on the integrity of the HPV DNA. I also had the original paraffin blocks. Thus, it was simple to obtain additional paraffin ribbons from these blocks for HPV DNA extraction. I prepared an additional hematoxylin and eosin stain after obtaining the paraffin ribbon to make sure that the CIN 1 lesion was still present. Recall that we would predict that the HPV DNA signal after DNA extraction should be greater for the paraffin ribbon cut in 2012 from these aged blocks compared to the ribbon obtained about 20 years ago when the biopsy was first obtained. This, of course, assumes that the reduction of signal with in situ hybridization is due to the HPV degradation over time. I also got some paraffin ribbons of CIN 1 lesions from biopsies that were obtained within the past year, as the positive control. The HPV DNA was extracted from each lesion and analyzed in a blinded fashion with the hybrid capture method. Let's review the two possibilities:

Hypothesis 1: The reduced HPV DNA signal with in situ hybridization over time is due to the degradation of the HPV DNA in the aged blocks.

Hypothesis 2: The reduced HPV DNA signal with in situ hybridization is *not* due to the degradation of the HPV DNA in the aged blocks, but rather due to degradation of the protein "cage" surrounding the HPV DNA.

The data are presented in Table 3-6. It is dramatic. There is a marked degeneration of HPV DNA over time in the aged formalin-fixed, paraffin-embedded tissues. Again, the degradation is such that most of the HPV DNA has been degraded from its original 8000 base pairs size to <250 base pairs in size. However, two observations suggest that hypothesis #2 above is the explanation for the concomitant reduction in the in situ hybridization signal:

One: The amount of HPV degradation is the same for the paraffin ribbons cut about 20 years ago as compared to the paraffin ribbons from the same block prepared just before the extraction. This despite the data that show that the HPV in situ hybridization signal is much stronger for the latter.

Two: Despite the marked degradation of the HPV DNA as measured by hybrid capture, there are still aged blocks in which we can generate a good HPV in situ hybridization signal, albeit nowhere as consistently as with CIN 1 lesions obtained in 2011 or 2012.

Let's now move to a series of experiments to further test hypothesis #2.

Table3- 6	Hybridc apturea nalysis		
Age of block	Signal* (mean)	SE	pval ue
10–21 yrso ld (ribbons >10 yrso ld)	190.5	108.1	
10–21 yrso ld (ribbons made in 2012)	211.5	122.4	No difference
1–2 yrso ld	13,341.0	436.1	<0.001

*Valuesa reRLU (Digenet est)n = 4 for each group

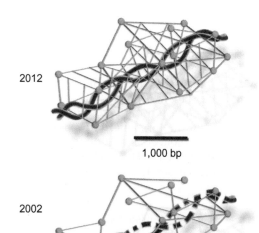

1. Degradation of DNA and RNA molecules

2002

2012

1,000 bp

2. Degradation of the three-dimensional protein/protein cross-link skeleton

A) Increased pore size

B) Increased "breathability" of macromolecules

C) Decreased ionic and hydrogen bond potential

COMBINED EFFECT–DECREASED SIGNAL

Figure3-25 Theoretical construct to explain the reduction of the HPV in situ hybridization signal over time in aged blocks. This graph attempts to explain the data presented in Tables 3-1 through 3-9 in this chapter by stressing that it is the degradation of the three-dimensional protein cross-linked cage that is the key event in the reduced signal over time.

"RECHARGING" THE AGED TISSUE TO ALLOW FOR ENHANCED IN SITU HYBRIDIZATION AND IMMUNOHISTOCHEMISTRY

Our working theory is that the marked reduction in signal for in situ hybridization and immunohistochemistry is due primarily to degradation of the "protein cross-linked cage" that surrounds the DNA/RNA/protein targets, as depicted in Figure 3-25. If this theory is correct, can we biochemically rejuvenate this cage and enhance our in situ hybridization or immunohistochemical signal? The answer is yes. I did this using proprietary compounds that, as of the writing of this book, are part of a provisional patent. Unfortunately, this does not allow me to list the reagents. However, this does not subtract from the point of this section. Suffice it to say that you can treat aged blocks with chemicals that have the end result of "rejuvenating" the "protein cage" that surrounds every DNA, RNA, and protein epitope in the formalin-fixed,

paraffin-embedded tissues. What is the effect of such rejuvenation on the signal?

Let's use HPV in situ hybridization as our model system. Indeed, since I had unstained plus slides from these CIN 1/condylomatous lesions that were stored for 10–21 years, we simply used them to see whether we could regenerate the HPV in situ hybridization signal. These data are presented in Table 3-7. The proprietary chemicals that regenerate the protein cage (called in the table ISHPerfect) clearly and dramatically increased the HPV DNA in situ hybridization signal in these aged tissues. We know from Table 3-6 that the HPV DNA in these same aged samples was much degraded. This provides further evidence that it is not the HPV DNA degradation per se that is responsible for the reduction in the HPV in situ hybridization signal. Rather, it is the associated degeneration of the three-dimensional macromolecule cross-linked network that surrounds our degraded HPV DNA that is responsible for the weakened in situ hybridization signal. But we can regenerate the protein cage that surrounds them, which, as described previously, serves much like a splint or cast for a broken bone and keeps the DNA/RNA/protein in its original order/sequence so we can still detect it by either in situ hybridization or immunohistochemistry. A graphic representation of this "three-dimensional macromolecule cross-linked cage regeneration process" is provided in Figure 3-26 and some representative examples in Figure 3-27.

Can we also augment the signal for RNA and protein using this "cage-rejuvenating compound?" The answer is yes, and the data are provided in Tables 3-8 and 3-9, respectively. Actual photomicrographs showing the enhanced signal for HPV DNA, microRNA, and proteins by in situ hybridization and immunohistochemistry

Table 3-7   HPV DNA

Age of block	Signal* (mean)	SE	pvalue
10–21 yrs old	14.0	6.9	
PreTx–ISH Perfect – analyze SERIAL SECTIONS of same tissues			
10–21 yrs old	64.0	12.7	0.036

*Scored blindly as intensity of signal and % of HPV+ cells N = 5 for each group

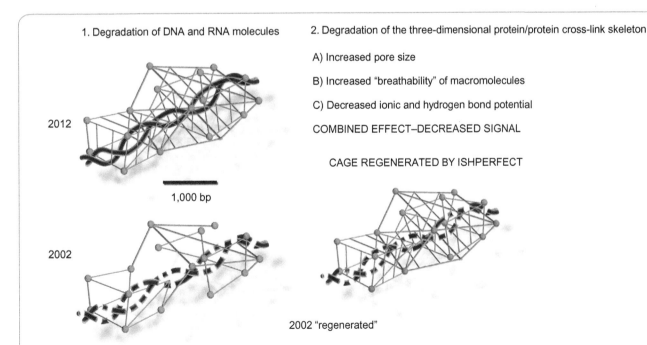

1. Degradation of DNA and RNA molecules

2012

1,000 bp

2002

2. Degradation of the three-dimensional protein/protein cross-link skeleton

A) Increased pore size

B) Increased "breathability" of macromolecules

C) Decreased ionic and hydrogen bond potential

COMBINED EFFECT–DECREASED SIGNAL

CAGE REGENERATED BY ISHPERFECT

2002 "regenerated"

Figure 3-26   Theoretical construct to explain the improved in situ hybridization (and immunohistochemistry) signal in aged blocks after "regeneration" of the three-dimensional macromolecule cross-linked cage. This graph illustrates that by regenerating the cross-linked protein cage that surrounds the HPV DNA target, we can rescue the weakened signal and make it as strong as the original signal.

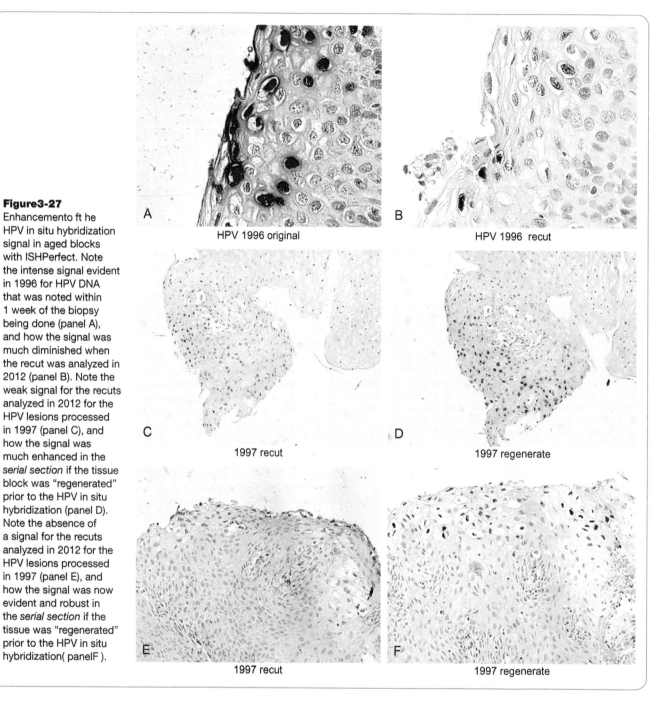

Figure3-27

Enhancemento ft he HPV in situ hybridization signal in aged blocks with ISHPerfect. Note the intense signal evident in 1996 for HPV DNA that was noted within 1 week of the biopsy being done (panel A), and how the signal was much diminished when the recut was analyzed in 2012 (panel B). Note the weak signal for the recuts analyzed in 2012 for the HPV lesions processed in 1997 (panel C), and how the signal was much enhanced in the *serial section* if the tissue block was "regenerated" prior to the HPV in situ hybridization (panel D). Note the absence of a signal for the recuts analyzed in 2012 for the HPV lesions processed in 1997 (panel E), and how the signal was now evident and robust in the *serial section* if the tissue was "regenerated" prior to the HPV in situ hybridization(panelF).

HPV 1996 original

HPV 1996 recut

1997 recut

1997 regenerate

1997 recut

1997 regenerate

Table3- 8 microRNAle t-7d

Ageo fbl ock	Signal*(mean)	SE	pval ue
10–21 yrso ld	5.0	2.5	
PreTx–ISH Perfect – analyze SERIAL SECTIONS of same tissues			
10–21 yrso ld	95.0	2.5	0.001

*Scored blindly as intensity of signal and % of microRNA+ cells N = 4 fore achg roup

Table3-9 Immunohistochemistryfor AE1/ 3a ndp16

Ageo fbl ock	Signal*(mean)	SE	pval ue
10–21 yrso ld	11.1	4.5	
PreTx–IHC Perfect – Score SERIAL SECTIONS of same tissues			
10–21 yrso ld	87.7	1.8	<0.001

*Scored blindly as intensity of signal and % of CD45 or cytokeratin + cells N = 9 for each group

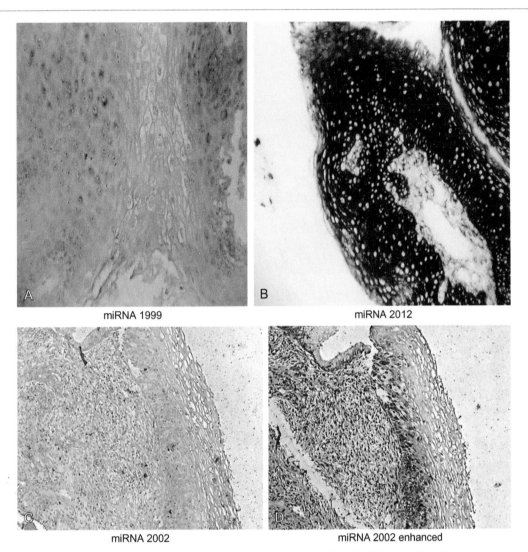

miRNA 1999

miRNA 2012

miRNA 2002

miRNA 2002 enhanced

Figure 3-28 Enhancement of the microRNA in situ hybridization signal in aged blocks with ISHPerfect. Note the absence of a signal from the cervical biopsy obtained in 1999 (panel A) for microRNA let-7c and the intense signal for a tissue obtained in 2012 that was analyzed at the same time as the much older block (panel B). Note the weak signal for the recuts analyzed in 2012 cervical tissue processed in 2002 for several microRNAs (panel C), and how the signal was much enhanced in the *serial section* if the tissue was "regenerated" prior to the microRNA in situ hybridization (panel D). Note that the microRNAs generated a signal in both the stroma and epithelia (shown are microRNAs 31 and let-7d).

with the regenerating agents are shown in Figures 3-27 through3-29 ,r espectively.No teth efo llowing:

Figure 3-27: Note the intense signal evident in 1996 for HPV DNA that was noted within 1 week of the biopsy being done (panel A), and how the signal was much diminished when the recut was analyzed in 2012 (panel B). Note the weak signal for the recuts analyzed in 2012 for the HPV lesions processed in 1997 (panel C), and how the signal was much enhanced in the *serial section* if the tissue block was "regenerated" prior to the HPV in situ hybridization (panel D). Note the absence of a signal for the recuts analyzed in 2012 for the HPV lesions processed in 1997 (panel E), and how the signal was now evident and robust in the *serial section* if the tissue was "regenerated" prior to the HPV in situ hybridization (panel F).

Figure 3-28: Note the absence of a signal from the cervical biopsy obtained in 1999 (panel A) for microRNA

let-7c and the intense signal for a tissue obtained in 2012 that was analyzed at the same time as the much older block (panel B). Note the weak signal for the recuts analyzed in 2012 cervical tissue processed in 2002 for several microRNAs (panel C), and how the signal was much enhanced in the *serial section* if the tissue was "regenerated" prior to the microRNA in situ hybridization (panel D). Note that the microRNAs generated a signal in both the stroma and epithelia (shown are microRNAs 31 and let-7d).

Figure 3-29: Note the absence of a signal from the vulvar biopsy obtained in 2003 (panel A) for CD45 and cytokeratin AE 1/3, and the intense signal for each epitope in the SERIAL SECTION that was "rejuvenated" prior to the immunohistochemistry (panel B). Note the weak signal for the cervical biopsy analyzed for AE 1/3 and CD45 in 2012 from a biopsy done in 2007 (panel C), and how the signal was much enhanced

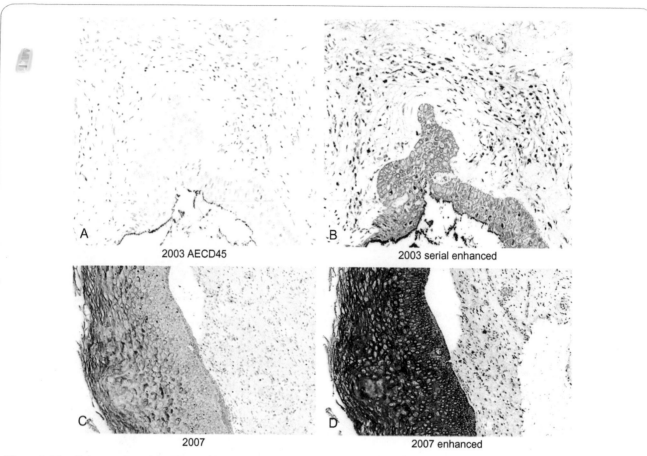

A 2003 AECD45

B 2003 serial enhanced

C 2007

D 2007 enhanced

Figure3-29 Enhancement of the immunohistochemistry signal in aged blocks with IHCPerfect. Note the absence of a signal from the vulvar biopsy obtained in 2003 (panel A) for CD45 and cytokeratin AE 1/3, and the intense signal for each epitope in the *serial section* that was "rejuvenated" prior to the immunohistochemistry (panel B). Note the weak signal for the cervical biopsy analyzed for AE 1/3 and CD45 in 2012 from a biopsy done in 2007 (panel C), and how the signal was much enhanced in the *serial section* if the tissue was "regenerated" prior to the immunohistochemistry (panel D).

in the SERIAL SECTION if the tissue was "regenerated" prior to the immunohistochemistry (panel D).

Needless to say, this regenerative process will open up a treasure trove of old formalin-fixed, paraffin-embedded tissues for research purposes. The importance of this point is underscored by the fact that formalin-fixed, paraffin-embedded tissues at least 5 years old will, of course, have very detailed clinical follow-up data with them that is essential in better understanding the in situ molecular results that are generated from the embedded tissues.

Let me provide one last bit of data in support of the model presented in Figure 3-26. I extracted the microRNAs from the aged blocks and new tissues that were tested for these in situ hybridization experiments. I theorized that the much smaller size of the microRNAs and the fact that they are so strongly encompassed by various proteins may have protected them from degradation. Although there was some variation (I tested five different microRNAs), in general the microRNAs did *not* show increased degradation when comparing the 15- to 20-year-old blocks of tissue to the tissues obtained in the last year (Figure 3-30). Clearly, the reduced microRNA in situ hybridization signal in the aged blocks seen in Figure 3-28 cannot reflect degradation of the target but, rather, must

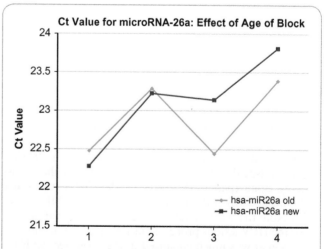

Ct Value for microRNA-26a: Effect of Age of Block

— hsa-miR26a old
— hsa-miR26a new

Figure 3-30 Lack of degradation of microRNAs in aged blocks. microRNA-26a was extracted from four aged cervical biopsies (15–20 years old) and four matching cervical biopsies (with the same diagnosis) obtained this year (2012). The microRNA level was then determined by qRTPCR. Note that there is no statistical difference in the copy number of the microRNA in the aged versus new blocks, despite the dramatic difference in the in situ hybridization signal these disparate blocks can generate (Figure 3-28).

reflect degradation of the three-dimensional macromolecule cross-linked network that surrounds the miRNA.

SUMMARY

I feel very strongly that you cannot maximize your ability to do good in situ hybridization or immunohistochemistry if you just follow recipes. Don't get me wrong, these protocols are very important, and I devote several chapters to them. I hope this chapter has provided the foundation to a better understanding of the biochemical dynamics that are operative every time you do an experiment using in situ hybridization or immunohistochemistry. It is easy to consider formalin-fixed, paraffin-embedded tissues as static tissues basically frozen in time since the formalin was added to the tissue. Clearly, this could not be further from the truth. A better

understanding of the biochemistry of in situ hybridization and immunohistochemistry not only serves as your foundation for practical protocols to improve your chances of success, but also allows you to manipulate the tissues to enhance the results, even with blocks that are probably older than some of the people reading this book!

The suggested readings focus on several aspects of the biochemistry of in situ hybridization and immunohistochemistry. Specifically, references [1–39] deal with the effects of formalin and other fixatives on tissue macromolecules in general and in situ hybridization/immunohistochemistry in general. References [40–74] are a compilation of papers on the biochemical and "practical" properties of different chromogens for in situ hybridization and immunohistochemistry. Finally, references [75–114] describe the biochemical and practical properties of peroxidase and other reporter systems for in situ hybridization and immunohistochemistry.

SUGGESTED READINGS

Effects of formalin and other fixatives on tissue macromolecules

[1] Akao I, Sato Y, Mukai K,e tal. Detection of Epstein-Barr virus DNA in formalin-fixed paraffin-embedded tissue of nasopharyngeal carcinoma using polymerase chain reaction andin situ h ybridization. Laryngoscope 1991;101:279–83.

[2] Ballester M, Galindo-Cardiel I, Gallardo C, etal. Intranucleard etectiono fAfric answin e fever virus DNA in several cell types from formalin-fixed and paraffin-embedded tissues usingan ewin situ h ybridisationp rotocol. JViro lM ethods 2010;168:38–43.

[3] Bejui-Thivolet F, Liagre N, Patricot L, Chardonnet Y, Chignol M. Human papillomavirus DNA in cervix. In-situ hybridization with biotinylated probes on Bouin'sfi xedp araffine mbeddedsp ecimens. PatholRe sP ract 1992;188:67–73.

[4] Blough RI, Smolarek TA, Ulbright TM, Heerema NA. Bicolorfl uorescencein situ hybridization on nuclei from formalin-fixed, paraffin-embedded testicular germ cell tumors: comparisonwith stan dardme taphasean alysis. CancerGe netCyto genet 1997;94:79–84.

[5] Bonds LA, Barnes P, Foucar K, Sever CE. Acetic acid-zinc-formalin: a safe alternative to B-5 fixative. Am J Clin Pathol 2005;124:205–11.

[6] Buwalda B, Nyakas C, Gast J, Luiten PG, Schmidt HH. Aldehydefi xationd ifferentially affects distribution of diaphorase activity but not of nitric oxide synthase immunoreactivityin r atb rain. Brain ResB ull 1995;38:467–73.

[7] Cardoso TC, Gomes DE, Ferrari HF,e tal. A novel in situ polymerase chain reaction hybridisation assay for the direct detection of bovine herpesvirus type 3 in formalin-fixed, paraffin-embeddedt issues. JVir olM ethods 2010;163:509–12.

[8] Cho WS, Chae C. PCRd etectiono f Actinobacillus pleuropneumoniaeapxIV gene in formalin-fixed, paraffin-embedded lung tissues and comparison with in situh ybridization. LettAp plM icrobiol 2003;37:56–60.

[9] Choi YJ. Ins ituh ybridizationu singa biotinylated DNA probe on formalin-fixed liver biopsies with hepatitis B virus infections: in situ hybridization superior to immunochemistry. ModP athol 1990;3:343–7.

[10] Dapson RW. Macromoleculare hangesc aused byf ormalinfi xationan dan tigenr etrieval. BiotechHis tochem 2007;82:133–40.

[11] Dapson RW. Glyoxalfi xation:h owit wo rks and why it only occasionally needs antigen retrieval. BiotechHis tochem 2007;82:161–6.

[12] Downs-Kelley E, Shehata BM, Lopez-Terrada D,e tal. Theu tilityo fF OXO1fl uorescence in situ hybridization (FISH) in formalin-fixed paraffin-embedded specimens in the diagnosis ofalve olarr habdomyosarcoma. DiagnM ol Pathol 2009;18:138–43.

[13] Fowler CB, O'Leary TJ, Mason JT. Modeling formalin fixation and histological processing with ribonuclease A: effects of ethanol dehydration on reversal of formaldehyde cross-links. LabI nvest 2008;88:785–91.

[14] Grantzdorffer I, Yumlu S, Gioeva Z,e tal. Comparison of different tissue sampling methods for protein extraction from formalin-fixedan dp araffin-embeddedt issues pecimens. ExpM olP athol 2010;88:190–6.

[15] Ha SK, Choi C, Chae C. Developmento f an optimized protocol for the detection of classical swine fever virus in formalin-fixed, paraffin-embedded tissues by seminested reverse transcription-polymerase chain reaction and comparison with in situ hybridization. ResVe tS ci 2004;77:163–9.

[16] Heck AJ, Bonnici PJ, Breukink E, Morris D, Wills M. Modificationan din hibitiono f vancomycin group antibodies by formaldehyde andac etaldehyde. Chemistry 2001;7:910–6.

[17] Henke RT, EunKim S, Maitra A, Paik S, Wellstein A. Expressiona nalysiso fmRNA in formalin-fixed, paraffin-embedded archival tissuesb ymRNAin s ituh ybridization. Methods 2006;38:253–62.

[18] Inagaki H, Nonaka M, Nagaya S,e tal. Monoclonality in gastric lymphoma detected in formalin-fixed paraffin-embedded endoscopic biopsy specimens using

immunohistochemistry, in situ hybridization, andp olymerasec hainr eaction. DiagnM ol Pathol 1995;4:32–8.

[19] Kim J, Chae C. Differentiationo fp orcine circovirus 1 and 2 in formalin-fixed, paraffin-wax-embedded tissues from pigs with postweaning multisystemic wasting syndromeb yin -situh ybridization. ResVe tS ci 2001;70:265–9.

[20] Kim J, Chae C. Optimized protocols for the detection of porcine circovirus 2 DNA from formalin-fixed paraffin-embedded tissues using nested polymerase chain reaction and comparison of nested PCR with in situ hybridization. J Virol Methods 2001;92:105–11.

[21] Lewis FA, Andrew A, Cross D, Quirke P. Comparison of in situ hybridization and in situ PCR for the localization of human papilloma virus in formalin-fixed paraffin embeddedt issue(MeetingAb stract). JP athol 1995;176(suppl): 1995.

[22] Lewis FA, Holt A, Cross D, Quirke P. In situ PCR on formalin-fixed paraffin embeddedt issue(meetingab stract). JP athol 1995;suppl:176.

[23] Liu CQ, Shan L, Balesar R,e tal. A quantitative in situ hybridization protocol for formalin-fixed paraffin-embedded archival post-mortemh umanb raint issue. Methods 2010;52:359–66.

[24] Montone KT, Litzky LA, Feldman MD, etal. Ins ituh ybridizationf orCo ccidioides immitis 5.8s ribosomal RNA sequences in formalin-fixed, paraffin-embedded pulmonary specimens using a locked nucleic acid probe: a rapid means for identification in tissue sections. DiagnM olP athol 2010;19:99–104.

[25] Motoi T, Kumagai A, Tsuji K, Imamura T, Fukusato T. Diagnosticu tilityo fd ual-color break-apart chromogenic in situ hybridization for the detection of rearranged SS18 in formalin-fixed, paraffin-embedded synovial sarcoma. HumP athol 2010;41:1397–404.

[26] Nuovo GJ.I n: PCRin s ituh ybridization: Protocolsan dAp plications, 3[rd]e dition Baltimore,M D: LippincottW illiamsan d Wilkins; 1997.

[27] Ooi A, Kobayashi M, Mai M, Nakanishi I. Amplification of c-erbB-2 in gastric cancer: detection in formalin-fixed, paraffin-embedded tissue by fluorescence in situ hybridization. Lab Invest 1998;78:345–51.

[28] Petersen BL, Sorensen MC, Pedersen S, Rasmussen M. Fluorescence in situ hybridization on formalin-fixed and paraffin-embedded tissue: optimizing the method. Appl Immunohistochem Mol Morphol 2004;12:259–65.

[29] Privat N, Sazdovitch V, Seilhean D, LaPlanche JL, Hauw JJ. PrP immunohistochemistry: different protocols, including a procedure for long formalin fixation, and a proposed schematic classification for deposits in sporadic Creutzfeldt-Jakob disease. Microsc Res Tech 2000;50:26–31.

[30] Qian X, Guerrero RB, Plummer TB, Alves VF, Lloyd RV. Detection of hepatitis C virus RNA in formalin-fixed paraffin-embedded sections with digoxigenin-labeled cRNA probes. Diagn Mol Pathol 2004; 13:9–14.

[31] Rait VK, Xu L, O'Leary TJ, Mason JT. Modeling formalin fixation and antigen retrieval with bovine pancreatic RNase A II. Interrelationship of cross-linking, immunoreactivity, and heat treatment. Lab Invest 2004;84:300–6.

[32] Rognum TO, Brandtzaeg P, Orjasaeter H, Fausa O. Immunohistochemistry of epithelial cell markers in normal and pathological colon mucosa. Comparison of results based on routine formalin- and cold ethanol-fixation methods. Histochemistry 1980;67:7–21.

[33] Suzuki SO, Iwaki T. Non-isotopic in situ hybridization of CD44 transcript in formalin-fixed paraffin-embedded sections. Brain research. Brain Research Protocols 1999;4:29–35.

[34] Thompson CH, Rose BR. Deleterious effects of formalin/acetic acid/alcohol (FAA) fixation on the detection of HPV DNA by in situ hybridization and the polymerase chain reaction. Pathology 1991;23:327–30.

[35] Warner CK, Whitfield SG, Fekadu M, Ho H. Procedures for reproducible detection of rabies virus antigen mRNA and genome in situ in formalin-fixed tissues. J Virol Methods 1997;67:5–12.

[36] Wilkens L, Werner M, Nolte M, et al. Influence of formalin fixation on the detection of cytomegalovirus by polymerase chain reaction in immunocompromised patients and correlation to in situ hybridization, immunohistochemistry, and serological data. Diagn Mol Pathol 1994;3:156–62.

[37] Wong KY, Patel J, Krause MO. RNA synthesis and chromatin structure in mammalian cells. In situ detection of template changes in living and formalin fixed L cell nuclei. Exp Cell Res 1971;69:456–60.

[38] Woolums AR, Gunther RA, McArthur-Vaughan K, et al. Cytotoxic T lymphocyte activity and cytokine expression in calves vaccinated with formalin-inactivated bovine respiratory syncytial virus prior to challenge. Comp Immunol Microbiol Infect Dis 2004;27:57–74.

[39] Yang W, Maqsodi B, Ma Y, et al. Direct quantification of gene expression in homogenates of formalin-fixed, paraffin-embedded tissues. Biotechniques 2006;40:481–6.

Biochemical properties of different chromogens for immunohistochemistry and in situ hybridization

[40] Avivi C, Rosen O, Goldstein RS. New chromogens for alkaline phosphatase histochemistry: salmon and magenta phosphate are useful for single- and double-label immunohistochemistry. J Histochem Cytochem 1994;42:551–4.

[41] Awaya H, Takeshima Y, Furonaka O, Kohno N, Inai K. Gene amplification and protein expression of EGFR and HER2 by chromogenic in situ hybridization and immunohistochemistry in atypical adenomatous hyperplasia and adenocarcinoma of the lung. J Clin Pathol 2005;58:1076–80.

[42] Bilous M, Morey A, Armes J, Cummings M, Francis G. Chromogenic in situ hybridisation testing for HER2 gene amplification in breast cancer produces highly reproducible results concordant with fluorescence in situ hybridisation and immunohistochemistry. Pathology 2006;38:120–4.

[43] Cho EY, Choi YL, Han JJ, Kim KM, Oh YL. Expression and amplification of Her2, EGFR and cyclin D1 in breast cancer: immunohistochemistry and chromogenic in situ hybridization. Pathol Int 2008;58:17–25.

[44] Cho EY, Han JJ, Choi YL, Kim KM, Oh YL. Comparison of Her-2, EGFR and cyclin D1 in primary breast cancer and paired metastatic lymph nodes: an immunohistochemical and chromogenic in situ hybridization study. J Korean Med Sci 2008;23:1053–61.

[45] Dandachi N, Dietze O, Hauser-Kronberger C. Evaluation of the clinical significance of HER2 amplification by chromogenic in situ hybridization in patients with primary breast cancer. Anticancer Res 2004;24:2401–6.

[46] Dedes KJ, Lopez-Garcia MA, Geyer FC, et al. Cortactin gene amplification and expression in breast cancer: a chromogenic in situ hybridization and immunohistochemical study. Breast Cancer Res Treat 2010;124:653–66.

[47] Hwang CC, Pintye M, Chang LC, et al. Dual-colour chromogenix in-situ hybridization is a potential alternative to fluorescence in-situ hybridization in HER2 testing. Histopathology 2011;59:984–92.

[48] Jarvela S, Helin H, Haapasalo J, et al. Amplification of the epidermal growth factor receptor in astrocytic tumours by chromogenic in situ hybridization: association with clinicopathological features and patient survival. Neuropathol Appl Neurobiol 2006;32:441–50.

[49] Javanmard SH, Moeiny A. Quantitative immunohistochemistry by measuring chromogen signal strength using a C# written program. J Res Med Sci 2009;14:201–3.

[50] Kasper M, Noll F. 2-bromo-1-naphthol—a novel chromogen for the immunoperoxidase technique in immunohistochemistry. Acta Histochem Suppl 1988;69–75.

[51] Kim HJ, Yoo TW, Park DI, et al. Gene amplification and protein overexpression of HER-2/neu in human extrahepatic cholangiocarcinoma as detected by chromogenic in situ hybridization and immunohistochemistry: its prognostic implication in node-positive patients. Ann Oncol 2007;18:892–7.

[52] Kumamoto H, Sasano H, Taniguchi T, et al. Chromogenic in situ hybridization analysis of HER-2/neu status in breast carcinoma: application in screening of patients for trastuzumab (Herceptin) therapy. Pathol Int 2001;51:579–84.

[53] Lambros MB, Simpson PT, Jones C, et al. Unlocking pathology archives for molecular genetic studies: a reliable method to generate probes for chromogenic and fluorescent in situ hybridization. Lab Invest 2006;86:398–408.

[54] Lehr HA, van der Loos CM, Teeling P, Gown AM. Complete chromogen separation and analysis in double immunohistochemical stains using Photoshop-based image analysis. J Histochem Cytochem 1999;47:119–26.

[55] Li-Ning TE, Ronchetti R, Torres-Cabala C, Merino MJ. Role of chromogenic in situ hybridization (CISH) in the evaluation of HER2 status in breast carcinoma: comparison with immunohistochemistry and FISH. Int J Surg Pathol 2005;13:343–51.

[56] Marquez A, Wu R, Zhao J, Shi Z. Evaluation of epidermal growth factor receptor (EGFR) by chromogenic in situ hybridization (CISH) and immunohistochemistry (IHC) in archival gliomas using bright-field microscopy. Diagn Mol Pathol 2004;13:1–8.

[57] Mayr D, Heim S, Weyrauch K, et al. Chromogenic in situ hybridization for HER-2/neu-oncogene in breast cancer: comparison of a new dual-colour chromogenic in situ hybridization with immunohistochemistry and fluorescence in situ hybridization. Histopathology 2009;55:716–23.

[58] Motoi T, Kumagai A, Tsuji K, Imamura T, Fukusato T. Diagnosticu tilityo fd ual-color break-apart chromogenic in situ hybridization for the detection of rearranged SS18 in formalin-fixed, paraffin-embedded synovial sarcoma. HumP athol 2010;41:1397–404.

[59] Mullink H, Walboomers JM, Tadema TM, Jansen DJ, Meijer CJ. Combinedimmu no- and no-radioactive hybridocytochemistry on cells and tissue sections: influence of fixation, enzyme pre-treatment, and choice of chromogen on detection of antigen and DNA sequences. JHis tochemCyt ochem 1989;37:603–9.

[60] Norgen RB, Lehman MN. An ewc hromogen for use in HRP-tract tracing and double-label immunocytochemistry. BrainRe sB ull 1990;25:393–6.

[61] Ntoulia M, Kaklamanis L, Valavanis C, et al. Her-2 DNA quantification of paraffin-embedded breast carcinomas with LightCycler real-time PCR in comparison to immunohistochemistry and chromogenic in situ hybridization. Clin Biochem 2006;39:942–6.

[62] Pirnik Z, Kiss A. Detectiono fo xytocin mRNA in hypertonic saline Fos-activated PVN neurons: comparison of chromogens in dual immunocytochemical and in situ hybridization procedure. EndocrRe gul 2002;36:23–39.

[63] Riethdorf S, Hoegel B, John B,e tal. Prospective multi-centre study to validate chromogenic in situ hybridization for the assessment of HER2 gene amplification in specimens from adjuvant and metastatic breast cancer patients. JCan cerRe sC linOn col 2011;137:261–9.

[64] Rummukainen JK, Salminen T, Lundin J, Joensuu H, Isola JJ. Amplification of c-myc oncogene by chromogenic and fluorescence in situ hybridization in archival breast cancer tissue array samples. Lab Invest 2001;81:1545–51.

[65] Sheibani K, Tubbs RR, Gephardt GN, McMahon JT, Valenzuela R. Comparison

of alternative chromogens for renal immunohistochemistry. Hum Pathol 1981;12:349–54.

[66] Simone G, Mangia A, Malfettone A, et al. Chromogenic in situ hybridization to detect EGFR gene copy number in cell blocks from fine-needle aspirates of non small cell lung carcinomas and lung metastases from colo-rectal cancer. J Exp Clin Cancer Res 2010;29:125.

[67] Taiambas E, Alexpopoulou D, Lambropoulou S, et al. Targeting topoisomerase IIa in endometrial adenocarcinoma: a combined chromogenic in situ hybridization and immunohistochemistry study based on tissue microarrays. Int J Gynecol Cancer 2006;16:1424–31.

[68] Tanner M, Gancberg D, Di Leo A, et al. Chromogenic in situ hybridization: a practical alternative for fluorescence in situ hybridization to detect HER-2/neu oncogene amplification in archival breast cancer samples. Am J Pathol 2000;157:1467–72.

[69] Todorovic-Rakovic N, Jovanovic D, Neskovic-Konstantinovic Z, Nikolic-Vukosavljevic D. Prognostic value of HER2 gene amplification detected by chromogenic in situ hybridization (CISH) in metastatic breast cancer. Exp Mol Pathol 2007;82:262–8.

[70] Tsiambas E, Karameris A, Lazaris AC, et al. EGFR alterations in pancreatic ductal adenocarcinoma: a chromogenic in situ hybridization analysis based on tissue microarrays. Hepatogastroenterology 2006;53:452–7.

[71] Tsiambas E, Karameris A, Tiniakos DG, Karakitsos P. Evaluation of topoisomerase IIa expression in pancreatic ductal adenocarcinoma: a pilot study using chromogenic in situ hybridization and immunohistochemistry on tissue microarrays. Pancreatology 2007;7:45–52.

[72] Tubbs RR, Sheibani K. Chromogens for immunohistochemistry. Arch Pathol Lab Med 1982;106:205.

[73] Vocaturo A, Novelli F, Benevolo M, et al. Chromogenic in situ hybridization to detect HE-2/neu gene amplification in histological and ThinPrep-processed breast cancer fine-needle aspirates: a sensitive and practical method in the trastuzumab era. Oncologist 2006;11:878–86.

[74] Wixom CR, Albers EA, Weidner N. Her2 amplification: correlation of chromogenic in situ hybridization with immunohistochemistry and fluorescence in situ hybridization. Appl Immunohistochem Mol Morphol 2004;12:248–51.

Biochemistry of peroxidase and other reporter systems

[75] Avissar N, Eisenmann C, Breen J, et al. Human placenta makes extracellular glutathione peroxidase and secretes it into maternal circulation. Am J Physiol 1994;267:E68–76.

[76] Avissar N, Ornt DB, Yagil Y, et al. Human kidney proximal tubules are the main source of plasma glutathione peroxidase. Am J Physiol 1994;266:C367–375.

[77] Baek IJ, Seo DS, Yon JM, et al. Tissue expression and cellular localization of phospholipid hydroperoxide glutathione peroxidase (PHGPx) mRNA in male mice. J Mol Histol 2007;38:237–44.

[78] Baek IJ, Yon JM, Lee SR, et al. Differential expression of gastrointestinal glutathione peroxidase (GI-GPx) gene during mouse organogenesis. Anat Histol Embryol 2011;40:210–8.

[79] Botella MA, Quesada MA, Kononowicz AK, et al. Characterization and in situ localization of a salt-induced tomato peroxidase mRNA. Plant Mol Biol 1994;25:105–14.

[80] Brey EM, Lalani Z, Johnston C, et al. Automated selection of DAB-labeled tissue for immunohistochemical quantifications. J Histochem Cytochem 2003;51:575–84.

[81] Davey FR, Busch GJ. Immunohistochemistry of glomerulonephritis using horseradish peroxidase and fluorescein-labeled antibody: a comparison of two techniques. Am J Clin Pathol 1970;53:531–6.

[82] Den Herder J, Lievens S, Rombauts S, Holsters M, Goormachtig S. A symbiotic plant peroxidase involved in bacterial invasion of the tropical legume Sesbania rostrata. Plant Physiol 2007;144:717–27.

[83] Do HM, Hong JK, Jung HW, et al. Expression of peroxidase-like genes, H_2O_2 production, and peroxidase activity during the hypersensitive response to Xanthomonas campestris pv. vesicatoria in Capsicum annuum. Mol Plant Microbe Interact 2003;16:196–205.

[84] Fu XH. Electrochemical measurement of DNA hybridization using nanosilver as label and horseradish peroxidase as enhancer. Bioprocess Biosyst Eng 2008;31:69–73.

[85] Gijzen M, Miller SS, Bowman LA, et al. Localization of peroxidase mRNAs in soybean seeds by in situ hybridization. Plant Mol Biol 1999;41:57–63.

[86] Hiruta J, Mazet F, Ogasawara M. Restricted expression of NADPH oxidase/peroxidase gene (Duox) in zone VII of the ascidian endostyle. Cell Tissue Res 2006;326:835–41.

[87] Hoffmeister-Ullerich SA, Herrmann D, Kielholz J, Schweizer M, Schaller HC. Isolation of a putative peroxidase, a target for factors controlling foot-formation in the coelenterate hydra. Eur J Biochem 2002;269:4597–606.

[88] Jung KY, Baek IJ, Yon JM, et al. Developmental expression of plasma glutathione peroxidase during mouse organogenesis. J Mol Histol 2011;42:545–56.

[89] Kang JK, Sul D, Kang JK, et al. Effects of lead exposure on the expression of phospholipid hydroperoxidase glutathione peroxidase mRNA in the rat brain. Toxicol Sci 2004;82:228–36.

[90] Krieg R, Halbhuber KJ. Recent advances in catalytic peroxidase histochemistry. Cell Mol Biol 2003;49:547–63.

[91] Krieg R, Halbhuber KJ. Detection of endogenous and immuno-bound peroxidase—the status quo in histochemistry. Prog Histochem Cytochem 2010;45:81–139.

[92] Kunikowska G, Jenner P. Alterations in m-RNA expression for Cu,Zn-superoxide dismutase and glutathione peroxidase in the basal ganglia of MPTP-treated marmosets and patients with Parkinson's disease. Brain Res 2003;968:206–18.

[93] Lanciego JT, Goede PH, Witter MP, Wouterlood FG. Use of peroxidase substrate Vector VIP for multiple staining in light microscopy. J Neurosci Methods 1997;74:1–7.

[94] Maser RL, Magenheimer BS, Calvet JP. Mouse plasma glutathione peroxidase.

cDNA sequence analysis and renal proximal tubular expression and secretion. J Biol Chem 1994;269:27066–27073.

[95] Missirlis F, Rahlfs S, Dimopoulos N, et al. A putative glutathione peroxidase of Drosophila encodes a thioredoxin peroxidase that provides resistance against oxidative stress but fails to complement a lack of catalase activity. Biol Chem 2003;384:463–72.

[96] Moreno SG, Laux G, Brielmeier M, Bornkamm GW, Conrad M. Testis-specific expression of the nuclear form of phospholipid hydroperoxide glutathione peroxidase (PHGPx). Biol Chem 2003;384:635–43.

[97] Murray GI, Foster CO, Ewen SW. A novel tetrazolium method for peroxidase histochemistry and immunohistochemistry. J Histochem Cytochem 1991;39:541–4.

[98] Omata M, Liew CT, Ashcavai M, Peters RL. Nonimmunologic binding of horseradish peroxidase to hepatitis B surface antigen. A possible source of error in immunohistochemistry. Am J Clin Pathol 1980;73:626–32.

[99] Perez-de-Luque A, Gonzalez-Verdejo CI, Lozano MD, et al. Protein cross-linking, peroxidase and beta-1,3-endoglucanase involved in resistance of pea against Orobanche crenata. J Exp Bot 2006;57: 1461–9.

[100] Petersen KH. Novel horseradish peroxidase substrates for use in immunohistochemistry. JI mmunolM ethods 2009;340:86–9.

[101] Pouchkina NN, Stanchev BS, McQueen-Mason SJ. FromE STs equencet os pider silk spinning: identification and molecular characterisation of Nephila senegalensis major ampullate gland peroxidase NsPox. Insect Biochem Mol Biol 2003;33:229–38.

[102] Romero MI, Romero MA, Smith GM. Visualization of axonally transported horseradish peroxidase using enhanced immunocytochemical detection: a direct comparison with the tetramethylbenzidine method. J Histochem Cytochem 1999;47:265–72.

[103] Roth J, Saremaslani P, Warhol MJ, Heitz PU. Improvedac curacyin d iagnostic immunohistochemistry, lectin histochemistry and in situ hybridization using a gold-labeled horseradish peroxidase antibody and silver intensification. LabI nvest 1992;67:263–9.

[104] Sato Y, Demura T, Yamawaki K, et al. Isolation and characterization of a novel peroxidase gene ZPO-C whose expression and function are closely associated with lignification during tracheary element differentiation. Plant Cell Physiol 2006;47:493–503.

[105] Schipper HM, Mateescu-Cantuniari A. Identification of peroxidase-positive astrocytes by combined histochemical and immunolabeling techniques in situ and in cell culture. J Histochem Cytochem 1991;39:1009–16.

[106] Schneider M, Vogt Weisenhorn DM, Seiler A, et al. Embryonic expression profile of phospholipid hydroperoxide glutathione peroxidase. Gene Expr Patterns 2006;6:489–94.

[107] Simonetti E, Veronico P, Melillo MT, et al. Analysiso fc lassI IIp eroxidasege nes expressed in roots of resistant and susceptible wheat lines infected by Heterodera avenae. MolP lantM icrobeI nteract 2009;22:1081–92.

[108] Smith SJ, Kotecha S, Towers N, Latinkic BV, Mohun TJ. XPOX2-peroxidasee xpression and the XLURP-1 promoter reveal the site of embryonic myeloid cell development in Xenopus. MechDe v 2002;117:173–86.

[109] Sollberg S, Peltonen J, Uitto J. Combined use of in situ hybridization and unlabeled antibody peroxidase anti-peroxidase methods: simultaneous detection of type I procollagen mRNAs and factor VIII-related antigene pitopesin keloidt issue. LabI nvest 1991;64:125–9.

[110] Song Z, Marzilli L, Greenlee BM, et al. Mycobacterial catalase-peroxidase is a tissue antigen and target of the adaptive immune responsein syste mics arcoidosis. JE xpM ed 2005;201:755–67.

[111] Thomas MA, Lemmer B. HistoGreen: a new alternative to 3,3′-diaminobenzidine-tetrahydrochloride-dihydrate (DAB) as a peroxidase substrate in immunohistochemistry? BrainRe sB rainRe s Photoc 2005;14:107–18.

[112] Wu LT, Chu KH. Characterization of an ovary-specific glutathione peroxidase from the shrimp Metapenaeus ensis and its role inc rustaceanr eproduction. CompB iochem PhysiolB B iochemM olB iol 2010;155:26–33.

[113] Yoshimura S, Suemizu H, Taniguchi Y, et al. The human plasma glutathione peroxidase-encoding gene: organization, sequence and localizationt oc hromosome5q 32. Gene 1994;145:293–7.

[114] Zhou M, Grofova I. The use of peroxidase substrate Vector VIP in electron microscopic singlean dd oublean tigenlo calization. J NeurosciM ethods 1995;62:149–58.

The Basics of In Situ Hybridization

INTRODUCTION

Now that we have discussed the basics of molecular pathology in general and the in situ-based molecular pathology methods in particular, as well as the fundamentals of formalin fixation and its biochemical ramifications for in situ hybridization and immunohistochemistry, it is now time to turn our attention to the next part of the book, which deals with the details of each of the basic steps of in situ hybridization. Specifically, we focus our attention in chronological order on each of the steps that are part of the in situ hybridization procedure. We explore the biochemistry of each step and offer some tricks of the trade in order to optimize results with the protocol.

STEP ONE: FIXING THE TISSUE OR CELLS

BUFFERED FORMALIN-FIXED CELLS AND TISSUE

For those of us who are working mostly with human tissues, we have no control over the length of time of formalin fixation. At least we can be reassured that most human biopsies and autopsy materials are fixed in a similar fashion. Specifically, most such tissues are fixed in 10% neutral-buffered formalin using an automated tissue processor. The standard procedure is that the biopsy is immediately placed into a labeled jar that contains formalin and then brought to the pathology department. The histotechnologist in the pathology department then records the salient features of the tissue and places it into a plastic cassette. The cassette is then put into a large vat of formalin, which, in turn, is placed on an automated tissue processor. The latter is usually turned on at around 5 p.m., and it will keep the tissue in formalin for another 4 hours before starting the dehydration process that ultimately lands the tissue in paraffin wax that has been heated to 65°C to keep the wax liquid. Once the block of tissue (that is, the tissue in the plastic cassette that, in turn, is embedded in paraffin wax) is brought to room temperature, the wax hardens, and we have what is commonly called a "tissue block" or, to be more complete, a "formalin-fixed, paraffin-embedded tissue block."

Most surgical biopsies range from 4 mm to 1.5 cm in size. Since these tissues are fixed in formalin for a time period that usually ranges from 6 to 24 hours, most surgical biopsies have had sufficient time to be "adequately fixed" and, thus, to preserve the DNA, RNA, and proteins that we want to detect with either in situ hybridization or immunohistochemistry. I put "adequately fixed" in quotation marks because I am referring to the basics of preserving the morphologic features of the cells. As we have seen, adequately fixed is much more difficult to define with in situ hybridization and immunohistochemistry, at least in terms of determining the optimal pretreatment conditions.

Are there any potential problems with this routine fixing and embedding of tissue? The answer is *yes*. First, if the tissue is very large (such as after the removal of a large tumor), which we will define as >2 cm, then the fact that formalin diffuses poorly into tissue means that the internal part of the tissue may be much less well preserved than the outer part of the tissue. Since, as we discuss in detail, fixation time and the "strength of fixation" can have a strong effect on the results with in situ hybridization and immunohistochemistry, it is easy to see why this would give one variable results that would be hard to interpret. Specifically, if you see that the DNA, RNA, or protein of interest shows a very different pattern in the interior of the tissue as compared to the exterior of the tissue, then this is the red flag that suggests that the tissue was not incubated for a sufficient amount of time in the formalin. Another clue in this regard relates to the fact that the internal, poorly fixed tissue will be much more sensitive to protease digestion than the outer, well-fixed tissue. Recall that formalin fixation basically creates a three-dimensional protein cross-linked cage within the cell that surrounds every epitope, DNA, and RNA sequence. Formalin does the same to the cells' cytoskeleton, which is ultimately responsible for its integrity during life and after fixation. If the tissue is well fixed, the cage will be like the steel latticework of a bridge that will be able to resist whatever protease or heat it may be exposed to during the in situ hybridization of immunohistochemistry process. However, if the tissue is poorly fixed, the three-dimensional cage may have the analogous strength of a spider web and not a steel latticework. Although it may be able to maintain its morphology during immunohistochemistry or in situ hybridization, if

DOI: http://dx.doi.org/10.1016/B978-0-12-415944-0.00004-8

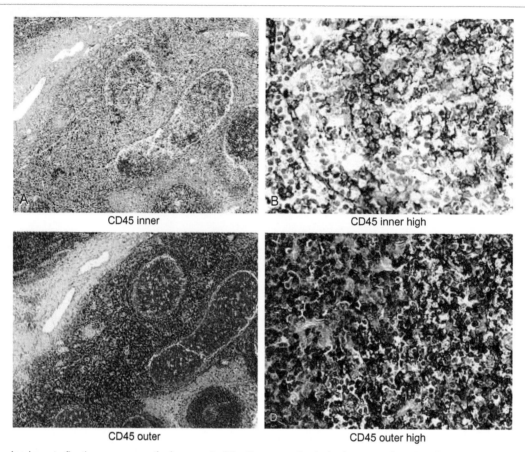

CD45 inner

CD45 inner high

CD45 outer

CD45 outer high

Figure4-1   Inadequate fixation may cause the inner part of the tissue to selectively show overdigestion. This tonsil was very enlarged (3.3 cm) due to chronic inflammation. Most of the cells in the tonsil strongly reacted with the lymphocyte marker CD45. When analyzed for CD45, the inner part of the tissue showed variable signal (panel A) and, at higher magnification, a discontinuous signal with poor morphology typical of overdigestion (panel B). In comparison, the outer part of the tissue, which had far more time to develop the cross-linkage of its cytoskeleton due to the relatively slow action of formalin, showed a continuous and intense signal in the lymphocytes (panels C and, at higher magnification, D).

too much heat and/or protease is used, then it disintegrates and leaves just a few strands as testimony to the tissue destruction. Thus, a very simple and common way to recognize tissue that was inadequately fixed in formalin is to observe that, after immunohistochemistry or in situ hybridization, the inner part of the tissue is overdigested by the protease/heat, whereas the outer part is still well preserved. An example of this is shown in Figure 4-1. The tissue was a hypertrophic tonsil. It was over 3 cm in size. The large size is the red flag that the tissue may not have been fixed adequately. Often there is pressure in the surgical pathology laboratory to get tissues processed as soon as possible. The end result is tissues like the one shown in Figure 4-1, where the outer part of the tissue is relatively well fixed, and the inner part is relatively poorly fixed. The outer, well-fixed part of the tonsil shows an excellent CD45 signal (panels C and D). This is the CD45 distribution pattern and strong signal you would see in any reactive tonsil. However, the inner part of the same tissue, located only 1 cm from the outer surface, shows a poor CD45 signal in the germinal center (panels A and B). Needless to say, this causes a substantial problem when you are trying to interpret the immunohistochemical or in situ hybridization results.

If you are using animal tissues where you are sacrificing the mice, or using cell suspensions that you are growing in cell culture, then you can control the fixation process. For cell suspensions, I prefer to remove the growth media from the cell culture plate right before the cells reach confluence. Then I add about 10 ml of formalin directly to the plate (just enough formalin so that the solution does not overflow). I then wait 4 hours and scrape off the cells using a sterile "rubber policeman." By doing it this way, I know the cells are already very well fixed, and they won't mind at all the pressure exerted upon them by the rubber policeman as they are removed. If you remove the cells first and then fix them, many of the cells will either disintegrate or show many odd forms, all reflecting the damage done by the pressure of the rubber policeman to cells that have essentially no rigid three-dimensional skeleton to protect them from this stress. If you want to see just how important formalin fixation is to preserving the cell's shape (as well as its metabolic machinery), the simplest experiment you can do is to take two identical cell cultures, fix one in formalin, and leave the other unfixed. Then remove the cells and place them side-by-side on a glass slide and do a routine hematoxylin and eosin stain.

The disparity of perfectly well-preserved cells on the one hand and cells with badly damaged morphology on the other will impress you!

THE USE OF FIXATIVES OTHER THAN FORMALIN: CROSS-LINKING FIXATIVES

Up until now we have discussed only 10% neutral-buffered formalin (which I have been calling formalin) for the fixation of either tissue or cell suspensions. The reasons are simple: (1) formalin is by far the most commonly used fixative for human and veterinary biopsies; and (2) formalin does an excellent job in not only immediately freezing the cell's metabolic machinery, but also creating a steel-like latticework in the cytoplasm and nucleus of the cell that allows us to do the pretreatments of heating and protease with no alteration in the cell's morphology. The latter reflects formalin's ability to extensively cross-link cellular proteins with each other and with DNA/RNA, as discussed at length in the preceding chapter.

The next logical question is: can fixatives other than formalin allow for superior immunohistochemistry or in situ hybridization? Clearly, if the fixative is a cross-linking relative of formalin, the answer would be yes! The two most common cross-linking relatives of formalin are paraformaldehyde and glutaraldehyde. The former is commonly used with in situ hybridization and immunohistochemistry. It tends to be more slowly acting than formalin in creating the three-dimensional cross-linked network inside the cell and, thus, you typically need to use it for a longer period of time (such as 4 hours for cell preps and 8 hours for small biopsies). Glutaraldehyde is commonly used in electron microscopy, where perfectly intact cell morphology at the organelle stage is mandatory. If you use either paraformaldehyde or glutaraldehyde as the fixative, and fix the material for an adequate amount of time, then all the tricks of the trade and details described for formalin-fixed tissue in this book would be in effect.

THE USE OF FIXATIVES OTHER THAN FORMALIN: DENATURING FIXATIVES

There is another class of fixatives that do *not* induce cross-links between proteins and other macromolecules. Rather, they denature proteins including, of course, the cell's metabolic enzymes. The most common of these fixatives are the alcohols (ethanol and methanol) and acetone. Some commercial fixatives (histocyte) contain these denaturing fixatives. Since denatured enzymes typically lose their function, the cell's metabolism is preserved because its RNases, DNases, and proteases are all shut down. However, imagine that you made a shed using a wood frame. To support the shed in one case, you linked the different pieces of wood with many steel bars. In the other case, to reinforce the skeleton of the shed, you simply changed the shape of the wood frame. Clearly, if there was a windstorm, your shed with the steel reinforcement bars would likely stay standing, whereas the other shed, like with the wolf in the story of Little Red Riding Hood, would probably blow apart. It is the same way with formalin cross-linking and the denaturing fixatives; the latter simply do not give the cell much of a chance to preserve

its morphology during the various protease and/or heating steps of either immunohistochemistry or in situ hybridization. This is not to say that you cannot use tissues fixed in denaturing fixatives, or not fixed at all, for in situ hybridization or immunohistochemistry. However, the odds of obtaining the maximum strong signal with good preservation of morphology, in my experience, is small, especially when compared to comparable tissues fixed in 10% formalin. You will recall that we discussed this point in some detail with regards to microRNA in situ hybridization and these different fixatives (Figures 3–18 and 3–19). Still, since this is an important point, I want to review it briefly and then address this question: can we "rescue" tissues fixed in acetone or ethanol, or cryostat sections, such that they give us much better results for in situ hybridization?

Although the loss of morphology during in situ hybridization with denaturing fixatives is certainly an issue, the problem literally goes "deeper" than that. Recall from our discussion of the biochemistry of in situ hybridization and immunohistochemistry that a key aspect of success was the ability of formalin fixation to create a complicated three-dimensional network of macromolecules. This complex labyrinth of macromolecules, of course, includes the various proteins involved in the cells' structural integrity and, thus, is one reason we can incubate formalin-fixed tissues at 95°C for 30 minutes and/or treat with highly active proteases and still get superior morphology with our assays. However, the three-dimensional network also supports a complex interlay of amino acid side chains, some of which can form hydrogen bonds with the proteins or nucleic acid probes that we use with in situ hybridization. Further, the amino acid side chains in a given protein also can form ionic bonds with the proteins that we use during the in situ hybridization process, including, of course, the antidigoxigenin antibody, alkaline phosphatase or peroxidase conjugate, or streptavidin. The end result of these hydrogen bonds and ionic bond forces are two-fold: (1) our key reagents are able to stay relatively immobile inside the cell when they come in contact with their target sequence; and (2) not only is the precipitate that marks our in situ hybridization signal (whether it be DAB or NBT/BCIP) present in much higher concentrations inside the subcompartment of the cell where it is formed, *but* it also does not migrate out of the cell into the overlying solution.

If we use a denaturing fixative, this three-dimensional cross-linked, intertwined network of proteins does not form. Rather, the proteins denature and tend to fold onto themselves, which does not allow for optimal exposure of their side chains. The end result is that the hydrogen and ionic bonds needed to maximize our signal *and* keep the key reagents, including the target sequences, from migrating outside the cell are too much reduced to "do their job." Hence, if, for example, you do in situ PCR with cells fixed in formalin versus cells fixed in 95% ethanol or acetone, you will note not only a decreased signal in the denatured fixed cells versus the formalin-fixed cells, but also that the amplicons have left the cell in large numbers and are present in the amplifying solution (Figure 4-2). Also note the relatively weak and more diffuse signal with the ethanol-fixed condyloma as compared to the intense and sharply confined nuclear signal for HPV 6 in the

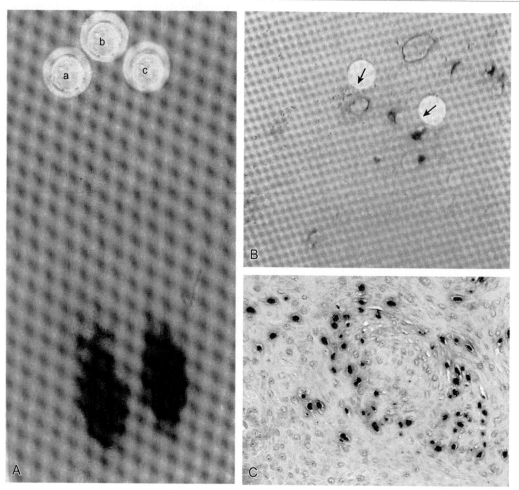

Figure4-2   Migration of the amplicon during PCR in situ hybridization for denaturing fixatives but not for formalin-fixed, paraffin-embedded tissues. Peripheral blood mononuclear cells were fixed in either acetone (lane a), formalin (lane b), or ethanol (lane c), and were then analyzed for the housekeeper bcl-2 mRNA using RT PCR in situ hybridization. After the amplification was done, the supernatant was retrieved, the DNA separated by electrophoresis, and Southern blot hybridization was done using a radioactive bcl-2 probe. The formalin-fixed cells showed a strong in situ signal, whereas the ethanol/acetone-fixed cells showed a very weak signal (not shown, but similar to Figures 3-18 and 3-19, Chapter 3). The reason for these disparate data is clearly evident in the Southern blot (panel A); the amplicons simply diffused out of the cells if they were fixed in acetone or ethanol, but not if they were fixed in formalin. It is highly likely that the same process is operative during in situ hybridization and immunohistochemistry but more easily detected with PCR in situ hybridization due to the amplification part of the process and the small amount of supernatant available in such samples. Panel B shows the weak and diffuse signal for HPV DNA by in situ hybridization in a vulvar wart fixed in ethanol, and panel C shows the very strong and sharply defined signal for a comparable tissue fixed in formalin.

comparable condyloma that was fixed in buffered formalin. Figure 4-2 does remind us that, under optimal conditions, formalin-fixed, paraffin-embedded tissues allow for an incredibly precise localization of RNA and DNA targets by in situ hybridization that simply cannot be afforded by the denaturing fixatives when we are analyzing large numbers of cases. It never ceases to amaze me, after we do such rigorous pretreatments and treatments on formalin-fixed, paraffin-embedded tissues, that the three-dimensional macromolecule protein cross-linked network is still able to do its remarkable job of not only maintaining the integrity of the cell, but also strictly confining the in situ hybridization reagents to the specific cellular compartment of the DNA or RNA target.

To summarize, you can successfully do in situ hybridization with tissues and cells fixed in acetone, ethanol,

methanol, or one of the other denaturing type fixatives. However, if you do a head-to-head comparison, the signal will typically be weaker than for cells/tissue fixed in 10% buffered formalin, the morphology of the cells/tissue will invariably be less well preserved, and there is a good likelihood that the signal will migrate either in the cell to some other compartment (e.g., nuclear to cytoplasm) or will simply migrate out of the cell into the overlying solution, as seen in Figure 4-2. Thus, I do not recommend using any alcohol- or acetone-based fixative for in situ hybridization, or, as we will see, immunohistochemistry. For similar reasons, I do not recommend using frozen, unfixed tissue for either in situ hybridization or immunohistochemistry.

Before we discuss the possibility of "rescuing" the signal in tissues fixed in either denaturing fixatives, or no fixative at all, let's briefly remind ourselves of the consequences of adding certain chemicals to formalin for in situ hybridization.

ADDING SOME ADDITIONAL INGREDIENTS TO FORMALIN

Sometimes various chemicals are added to formalin. This is done because of the belief that such chemicals can better preserve the natural appearance of the cell, which is, after all, the major goal for the surgical pathologist who has to review the biopsy. The most common of these chemicals are Bouin's solution and Zenker's solution. The bottom line is that they contain chemical additives (picric acid or a heavy metal such as mercury or zinc). These fixatives are typically used in certain biopsies such as from the bone marrow and testes. When you think about it, the formalin cross-linking does such a good job of preserving the cell's shape that it is not clear what added advantage either picric acid or a heavy metal may provide. Heavy metal presents toxicity issues in addition to the well-documented nasal irritation and carcinogenic potential of formalin. But there is another simple yet dramatic point. Picric acid and heavy metals are able to degrade DNA and RNA. Within a few hours of fixation, the HPV DNA and RNA can be so degraded that a standard-sized probe of between 40 and 80 nucleotides is no longer able to recognize it! That is, picric acid, for example, can degrade the 8000 base pair HPV DNA into fragments so small that our probes can no longer detect it. This is amazing! Consider how rapidly these agents must attack the structure of DNA and RNA such that they can no longer be detected after a few hours of fixation. As an aside, the ability of picric acid to degrade DNA and RNA may give us some insight into why DNA, RNA, and proteins that are present in formalin-fixed, paraffin-embedded tissues slowly degrade over time. Part of picric acid's ability to degrade DNA and RNA may relate to its oxidizing ability. Indeed, it has been used for many years as a preservative of protein products such as silk, as well as being used as an explosive! Many anti-oxidizing agents have been employed to stop the damage done to DNA and RNA by agents such as picric acid.

UNFIXED CELLS/TISSUE

It is certainly understandable that some investigators have devoted much energy to optimizing in situ hybridization with unfixed, frozen samples. This is the simplest way to avoid the formalin-induced dense latticework of cross-linked macromolecules that our probe and other in situ reagents must transverse to achieve successful in situ hybridization. Also, we can be assured that the DNA, RNA, and proteins in cryostat sections that are unfixed will be pristine and basically identical to the natural state with no breakage in their phosphodiester bonds. We can also avoid the protease digestion and/or antigen retrieval steps in such samples. However, I have not been able to get as strong a signal on a consistent basis with cryostat sections when compared to formalin-fixed, paraffin-embedded tissues. Further, the morphology, as already indicated, suffers with cryostat sections where so-called ice crystals are

commonly seen. Also, I have not been able to get good co-expression data with cryostat sections, in the sense that the initial in situ hybridization or immunohistochemical reaction has not been optimal and the tissue morphology usually suffers after two separate in situ-based reactions.

A reasonable theory to explain the relative superiority of formalin-fixed, paraffin-embedded tissues to frozen, unfixed tissue may center around the three-dimensional cross-linked network created by the formalin fixation. If we accept the hypothesis that the cross-linking three-dimensional "cage" is essential to the generation and confinement of the signal with in situ hybridization, then it follows that frozen, unfixed tissues will not perform as well.

We can have the "best of both worlds." If you have access to cryostat sections, then fixing the unstained silane-coated slide in formalin for a few hours will generate the three-dimensional macromolecule cross-linked cage, yet the DNA and RNA will be pristine and undamaged. Of course, in most cases, frozen, unfixed tissue is not available for a given case. Still, if you have access to cryostat sections, then I strongly recommend that you fix some in 10% buffered formalin after sectioning. You will get tissue sections that truly are optimal for in situ hybridization and immunohistochemistry.

Now let's get back to the theoretical aspect of in situ hybridization. We have established that formalin fixation is superior to the denaturing fixatives and cryostat sections for in situ hybridization. The reasons are two-fold:

1. Astro ngsign al
2. Ab etter-definedsign al

Our working hypothesis is that these results reflect the fact that formalin can generate the three-dimensional macromolecule cross-linked cage around any given DNA or RNA target and the other fixatives (or lack of fixative) cannot. If true, could we create the three-dimensional macromolecule cross-linked network in ethanol/acetone-fixed/cryostat-unfixed sections? Let's try to do this. We already established in Chapter 3 that ISHPerfect can regenerate the three-dimensional protein cross-linked cage in old formalin-fixed, paraffin-embedded tissues. Let's treat tissues fixed in ethanol, formalin, or cryostat sections; incubate them in ISHPerfect; and then do the in situ hybridization. We will continue to use microRNA in situ hybridization as our model system, since the miRNAs we are using (miR-31 and -328) are produced in abundance in the tissue we are using for these experiments, which is lymph node. Before reviewing these data, please remind yourself how acetone and ethanol fixatives worked for these microRNAs in lymph nodes after in situ hybridization by looking at Figures 3–18 and 3–19. No signal was seen under these conditions. The signal was evident for the cryostat sections, but was poor and more diffuse relative to formalin-fixed tissues. Now look at Figure 4-3. Note these key points:

1. A signal is evident in the acetone-fixed tissue after ISHPerfect treatment (panel A), but only if protease digestion is used prior to in situ hybridization.
2. No pretreatment for the acetone- (panel B) and ethanol-fixed tissue treated with ISHPerfect eliminated the signal. This is consistent with the hypothesis that ISHPerfect is "building" a three-dimensional protein

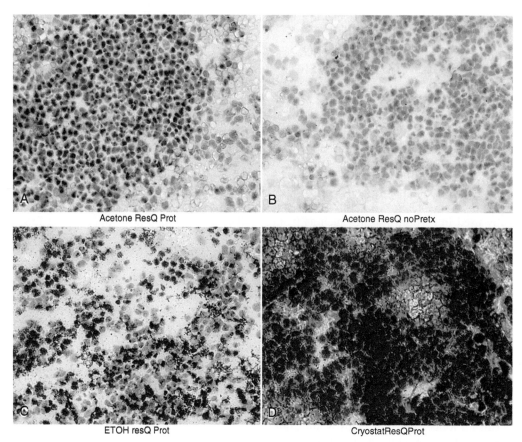

Figure 4-3 Response of denaturing and cryostat sections to ISHPerfect in their performance with microRNA in situ hybridization. Tissues fixed in acetone or ethanol showed basically no signal when analyzed for microRNAs by in situ hybridization (see Figures 3-18 and 3-19). However, after pretreatment with ISHPerfect, the acetone-fixed tissue showed a moderate signal in many of the cells with protease pretreatment (panel A), but not if no pretreatment was done (panel B). Similar results were noted with the ethanol-fixed tissue; a well-defined good signal was seen in many of the cells after pretreatment with ISHPerfect and protease digestion (panel C). An even more dramatic increase in the microRNA in situ hybridization signal was seen after pretreatment with ISHPerfect and protease digestion with the cryostat sections (panel D). These data suggest that denaturing fixatives, by denaturing the proteins around microRNAs, may reduce the ability to completely regenerate the three-dimensional protein cross-linked cage by ISHPerfect.

cross-linked cage around the microRNA that needs protease digestion for maximal access to the target miRNAs

3. A strong signal is also evident in the ethanol-fixed tissue after ISHPerfect treatment (panel C). However, like the acetone-fixed tissue "rescued" by this treatment, it is not quite as robust as the formalin-fixed tissue.

4. The cryostat section, on the other hand, after treatment with ISHPerfect, yields a signal for these miRNAs after protease digestion as strong as that seen with formalin. These data certainly suggest that the denaturing fixatives cause some "damage" to our prospective cross-linked protein cage that the ISHPerfect cannot fix.

5. The signals for the ethanol, acetone, and cryostat sections are all now very sharply defined after the ISHPerfect treatment, as compared to before when they were both weak and diffuse (see Figures 3-18 and 3–19), which probably represented much diffusion out of the cell, as seen in Figure 4-2.

So, to summarize, we have accumulated more data in support of the hypothesis that the three-dimensional

macromolecule cross-linked network is essential for *optimal* in situ hybridization. And by optimal, I mean both the *intensity* of the signal and the sharp *subcellular localization* of the signal. A graphic representation of these new data is provided in Figures 4-4 and 4-5. Of course, I am not suggesting that we use cryostat-, ethanol-, or acetone-fixed tissues after the "rejuvenation" step. Rather, we are using the denaturing fixative/cryostat sections treated with ISHPerfect as an additional system to better understand the biochemical processes occurring inside the cell when we do in situ hybridization or immunohistochemistry. In this way, we can better understand the system and, thus, how to manipulate it to our advantage.

A QUICK REVIEW OF THE CONSEQUENCES OF FORMALIN FIXATION IN PERFORMING IN SITU HYBRIDIZATION

Before we leave the topic of cell fixation, I do not want to give the impression that formalin fixation does not create some problems for successful immunohistochemistry

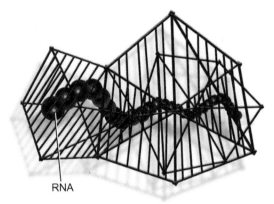

RNA

Formalin Induced Cross-linked Cage

DNA Target, Denaturing Fixative

Figure4-4 Schematic diagram of the differential effects of formalin fixation versus denaturing fixation on the proteins that surround DNA and RNA targets. Formalin fixation, via its ability to cross-link the macromolecules of the cell, will create a three-dimensional cage around the DNA and RNA targets. This will produce a complex matrix that includes variable pore sizes, plus ionic and hydrogen bond potential that can be manipulated to produce an intense signal with minimal background and little to no diffusion. In comparison, the denaturing fixatives do not create such a three-dimensional protein cross-linked network that cannot either maximize the signal or confine it to the part of the cell where the target is located.

DNA Target, Denaturing Fixative

DNA Target after ISH/IHCPerfect

Figure4-5 Schematic diagram of the effect of ISHPerfect on tissues fixed with denaturing fixatives. The figure shows an artist's conception of the effect of ISHPerfect on the biochemical microenvironment around the DNA or RNA target of interest. The ISHPerfect regime creates a relatively poor three-dimensional protein cross-linked cage that yields better results than the denaturing fixatives can do by themselves, but not as well as the formalin-fixed cage (Figure 4-4).

the signal and minimize poor morphology or signal loss. I also give you some food for thought by asking a question that we address later in this chapter. That question is: does the three-dimensional protein cage that surrounds each epitope, RNA, and DNA sequence show much or little variability in a given formalin-fixed, paraffin-embedded tissue?

It is my best guess that an understanding of the answer to this question will give us much insight into the marked variability of the pretreatment and other conditions with in situ hybridization. We get back to this question shortly.

STEP TWO: PUTTING THE FIXED CELLS OR TISSUE ON A GLASS SLIDE

FORMALIN-FIXED, PARAFFIN-EMBEDDED TISSUES

Most likely you have seen the acronym FFPE. This, of course, stands for formalin-fixed, paraffin-embedded tissues. It makes sense that this term is used because most likely more than 99% of the samples you will test by in situ hybridization or immunohistochemistry will be formalin-fixed, paraffin-embedded tissues.

The next step in the in situ hybridization process will be to have a histotechnologist cut the tissue in the paraffin block onto glass slides. This task may sound routine, but it is not! You need a well-trained histotechnologist. The job is to place the tissue block in a small plastic cassette that keeps the tissue in place while the histotechnologist rotates a super-sharp blade capable of cutting the tissue in 4-micron ribbons. More specifically, the histotechnologist keeps cutting the wax that contains the tissue which creates a "single-file" arrangement of wax/tissue that is placed in a water bath. This is commonly referred to as a

or in situ hybridization or that it is clearly better than any other type of fixative/nonfixative. It is the preferable fixative, but it certainly creates some problems. The three-dimensional latticework induced by formalin can make it very difficult to generate a signal, especially if the tissues were fixed for a very long time. The main problem, however, relates to the marked variability in the pretreatment conditions required for optimal in situ hybridization or immunohistochemistry in a given tissue. The worse part of this problem is that there is no way, to my knowledge, to predict which specific pretreatment (or lack thereof) will be optimal for a given tissue. This side effect of formalin fixation will necessitate a well-thought-out game plan from case to case if we are to get useful data in our in situ hybridization and immunohistochemistry experiments. We discuss this point in great detail when we go over specific protocols for immunohistochemistry and in situ hybridization. Suffice it to say for now that, since no two tissues are exactly the same size or contain the same exact cells, and have been exposed to different times of formalin fixation, the three-dimensional latticework of the cells from tissue to tissue will differ. The side effect of this is that we will need to adjust the pretreatment conditions precisely to find the "key" that unlocks this three-dimensional latticework in such a way that we optimize "tissue ribbon." The histotechnologist then takes a glass slide, puts it in the water, goes underneath the tissue/wax, and then moves the slide up to capture the wax/tissue. The slide is then allowed to dry at an angle to remove most of the water, and then the drying process is concluded at 60°C in an oven, typically for 30–60 minutes.

The skill of the histotechnologist is a key point for several reasons. If some water or a bubble gets under the tissue, or if the tissue folds while it is being placed on the glass slide, then the likelihood that the tissue will fall off the slide during in situ hybridization is much increased. Also, as we have and will continue to discuss, it is very important to get *three* sections per slide and to place subsequent sections on serially labeled slides. In my experience, a good histotechnologist can do each easily, but not so if the person is either not well trained or not good at his or her job. I have tremendous respect for the work of a good histotechnologist, both for my research and diagnostic work; a well-prepared tissue section on a glass slide is so important for good results.

THE IMPORTANCE OF PRETREATED SLIDES: SILANE-COATED SLIDES

When I started performing in situ hybridization in 1988, we realized quickly that untreated glass slides do not work well, for the simple reason that the tissue section tends to fall off during the in situ hybridization process. We tried many different pretreatment conditions to stop this problem. Pretreatment conditions designed to keep the tissue on the slide during the entire in situ hybridization process included glues, clam extracts, and polylysine. The latter certainly improved the rate at which the tissue kept on the glass slide during the in situ hybridization process. However, sections still fell off from time to time. It took a chemical commonly used to treat glass in industry to affect the charge of the glass to make in situ hybridization and immunohistochemistry more efficient by basically eliminating the chance that the tissue would fall off the section during the process. This chemical is silane or, more accurately, organosilane.

The key feature of organosilanes is that they consist of carbon and silicon bonds, which means that they are organically like carbon, in that the bound silicon is tetravalent and tetrahedral. Organosilanes are widely used in industry because when they coat glass, they have two primary effects: (1) to clean the glass and, subsequently, repel water and dirt; and (2) to affect the surface charge of the glass and other properties such that the glass can bind much more avidly to a wide range of chemicals and substances, including paraffin-embedded tissue sections or cell suspensions placed on the glass.

It is very unusual, in my experience, for tissue sections or cells to fall off the silane-coated slide when doing in situ hybridization or immunohistochemistry. The most common reason nowadays for tissue to fall off a slide during in situ hybridization, based on many emails and correspondences I have received, is that the researcher's assistant inadvertently did not use silane-coated slides but thought he or she did. This is not to say that all tissues will remain adhered to a silane-coated slide during in situ hybridization. If the paraffin block has a lot of blood or poorly fixed adipose tissue, for example, then the section may detach from the slide during in situ hybridization. However, in most instances, and certainly when using well-fixed standard surgical biopsies, tissue on silane-coated slides tends to adhere so well that you would need a scalpel blade to remove it! Although there are simple protocols to silane-coat regular glass microscopic slides, my recommendation is to purchase these slides from one of the many companies that sells them, typically under names such as *plus* slides, *Superfrost* slides, *charged* slides, and so on.

There is one problem that I see from time to time when using silane-coated slides. Sometimes the organosilane works so well that the slide becomes too hydrophobic. When that happens, when you add an aqueous component to the slide during the in situ hybridization or immunohistochemistry reaction, such as the primary antibody in the latter, the solution beads up and refuses to stay over the section. This problem, in my experience, is found mostly with tissue microarray (TMA) slides, and only if you are using an automated platform. The one time I see the problem with non-TMA samples on an automated platform is when I do co-expression analyses, but only with the second reaction, not the first.

Recall that automated platforms use a reagent whose function is to prevent the primary antibody (or probe for in situ hybridization) from drying out over the tissue when performing an experimental run. Such solutions function as "liquid coverslips" and must possess oil-like properties that are lighter than water so that they can cover the aqueous reaction and not mix with it. It is not clear to me why some TMA slides are so hydrophobic that the aqueous solution is "forced" above the oil phase. The solution to this problem, however, is straightforward. You should not use the automated platform and, instead, do the in situ hybridization or immunohistochemistry reaction manually. As we discuss shortly, the simplest way to do manual in situ hybridization or immunohistochemistry is to use *polypropylene* coverslips. They can be cut to size, will withstand the high temperatures of the immunohistochemical or in situ hybridization reaction, and will prevent drying of the in situ hybridization reagents.

THE IMPORTANCE OF PLACING MULTIPLE SERIAL SECTIONS PER SILANE-COATED GLASS SLIDE

It is time to stress this simple but important point again. Even if the lab could fix tissues for the exact same time in the exact same volume of formalin, tissue size will invariably vary, as will the cellular composition of any given tissue. Further, I theorize that individual epitopes, DNA, and RNA targets may also vary a great deal in a given tissue in their optimal pretreatment regime. The end result is that tissues will show marked variation one to the next in terms of the optimal pretreatment conditions needed for optimal in situ hybridization. The practical point for us doing in situ hybridization is that we *always* want three sections per slide. This way, we can vary important variables such as protease digestion time on a given slide and, thus, not only increase our chances of success but also do so using far fewer slides and reagents. Now,

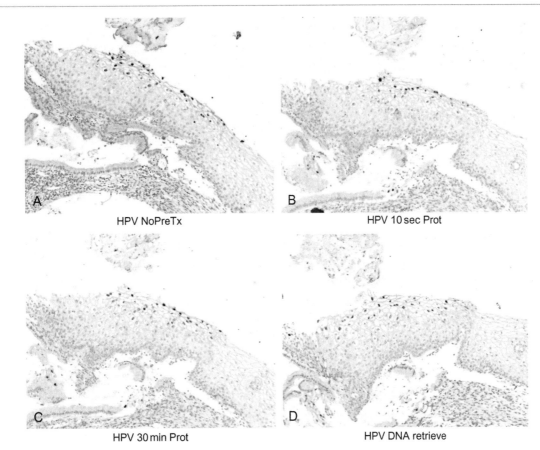

A — HPV NoPreTx

B — HPV 10 sec Prot

C — HPV 30 min Prot

D — HPV DNA retrieve

Figure 4-6 The value of placing multiple tissue sections on one silane-coated glass slide. The tissue was a small (8 mm) cervical biopsy that showed CIN 1. Hybrid capture analysis of the corresponding Pap smear showed that the tissue contained a high-risk HPV type. By placing multiple sections on a given glass slide, we needed only two slides to show that a strong signal with the HPV 16/18 probe cocktail (Enzo Life Sciences) occurred with no pretreatment (panel A), 10 seconds of protease digestion (panel B), 30 minutes of protease digestion (panel C), or DNA retrieval (95°C for 30 minutes in an aqueous solution) (panel D).

of course, formalin-fixed tissues in paraffin blocks vary in size from a few millimeters to about 20 mm. Thus, for 20 mm tissues, you can, at best, get 1½ to 2 sections per slide, whereas for small surgical biopsies (often around 8–10 mm) you can get 3–4 sections per slide. Still, by using polypropylene coverslips cut to size, you can do 3–4 different reactions on the same slide, which, again, will much increase the odds that you will use the conditions that are optimal for successful in situ hybridization in that particular tissue. Figure 4-6 shows an example in which the HPV in situ hybridization results were equivalent whether we did no pretreatment, protease digestion, or antigen retrieval. This situation is *unusual*. Figure 4-7 shows the more common example in which, if we did in situ hybridization for human papillomavirus (HPV) on this biopsy and used only protease digestion, then the data would have greatly underestimated the amount of HPV DNA that was actually present in this lower genital tract biopsy. I can think of no more important practical, operational point than this one for successful in situ hybridization (and immunohistochemistry): you need to vary the key variables, including the pretreatment regime and probe concentration for any given tissue/target, to increase the likelihood that you will be using the

conditions that are optimal *for that specific tissue*. And, again, the optimizing data shown in Figure 4-7 were obtained by using just two glass slides because multiple sections were placed per slide. Of course, if only one section was placed per slide, then at least four slides would have been tested to generate the data in Figure 4-7.

REMOVING THE PARAFFIN WAX FROM THE TISSUE SAMPLE

At this stage we have a series of unstained slides provided to us by our histotechnologist. The slides are organosilane-coated, and each slide contains from 2 to 5 levels of the tissue (depending on the size of the biopsy) that are each 4 microns away from the subsequent sample. Typically, I ask the histotechnologist to prepare 5 to 10 unstained slides per block, again depending on the size of the tissue. Another simple but important point is to have the histotechnologist label each successive slide as 1, 2, 3, and so on. The reason this labeling is important is based on the fact that most histotechnologists will cut the paraffin-embedded tissue in sections that are 3–5 microns thick. Most cells are far larger than 4 microns; red blood cells are usually about 8 microns, and larger cells such as

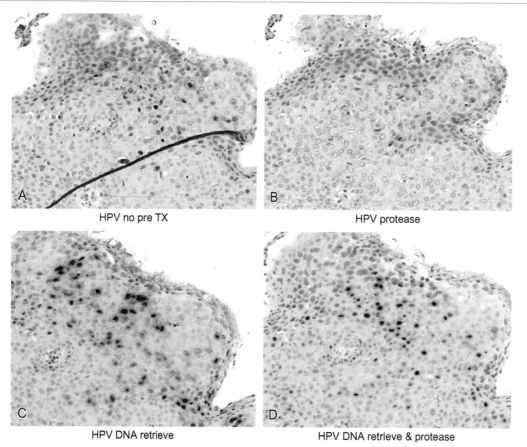

HPV no pre TX

HPV protease

HPV DNA retrieve

HPV DNA retrieve & protease

Figure4-7 The value of placing multiple tissue sections on one silane-coated glass slide. The tissue was a small (10 mm) vulvar biopsy that showed a condyloma. Over 95% of these lesions contain HPVs 6/11. By placing multiple sections on a given glass slide, we needed only two slides to show that a weak signal with the HPV 6/11 probe cocktail (Enzo Life Sciences) occurred with no pretreatment (panel A) or 4 minutes of protease digestion (panel B). However, DNA retrieval (95°C for 30 minutes in an aqueous solution) yielded a strong signal without (panel C) or with (panel D) 10 seconds of protease digestion. Also note that the sections are serial sections, so we examined the same cells. This is valuable because the copy number of HPV can vary a lot in different parts of the lesion.

squamous cells and motor neurons can be 30+ microns in size. Thus, the consecutive sections often contain the same cells. That is to say, slide 2 and slide 3 will contain many of the exact same cells cut at different parts of the cell. We call such sections *serial sections*. The analysis of serial sections by in situ hybridization or immunohisto-chemistry for two different DNA/RNA or proteins often gives us more information than if we had examined, for example, the first and last slides that came from the block. The reason is that we are examining the same groups of cells, as is evident in Figure 4-8. Note that as we progress from serial section A to B to C and finally to D, you can follow the progression of the exact same cells that are with each section separated by 4 microns. You can also see with these serial sections how reproducible the MCM_2 immunohistochemical results are from section to section. Look at Figures 4-6 and 4-7 again; these figures represent the serial section analyses of HPV DNA in situ hybridization under different experimental conditions. Analyzing serial sections this way is the pathologic equivalent of doing a CT or MRI scan of a part of the body.

To remove the paraffin, you place the slides in a fresh solution of xylene. Since xylene requires special disposal

conditions, some laboratories prefer to use other solutions such as Sub-X. Although these solutions will remove paraffin, you need to increase the incubation time to at least 20 minutes. When you are using fresh xylene, 3 minutes of incubation is sufficient to remove all the paraffin wax. To remove the xylene, incubate the tissues in a fresh solution of 100% ethanol. By "fresh solution," I mean one that has not processed more than 75 slides for in situ hybridization or immunohistochemistry, which, for my lab, is typically one week's work. Let me add that this discussion presupposes that you are doing the in situ hybridization or immunohistochemistry manually (by hand). If you are using an automated platform, then typically the machine will remove the paraffin wax from the tissue as part of its function. If you are doing the process manually, after incubating the slide in 100% ethanol, you only need to air dry the slide, which I do in a 37°C incubator.

What can go wrong at this step whereby we remove the wax? One problem is that the xylene may be contaminated with water. You can easily recognize this because a glass slide exposed to fresh xylene is truly as "clear as glass," whereas the slide exposed to water-contaminated

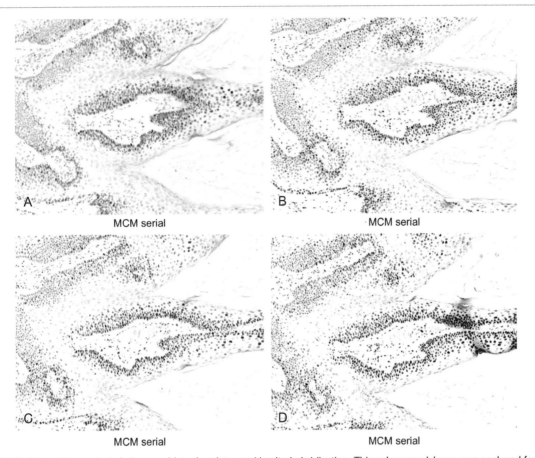

A   MCM serial

B   MCM serial

C   MCM serial

D   MCM serial

Figure4-8   Serial section analysis in immunohistochemistry and in situ hybridization. This vulvar condyloma was analyzed for the protein MCM$_2$ by immunohistochemistry. This protein is a marker of the proliferative index of the cells and, thus, is expressed mostly toward the basal epithelia. By examining serial sections, we can use certain landmarks to find the exact same group of cells in each section. In this case, the biopsy showed a distinctive "fin-shaped" area of the dermis in the center of the biopsy. Note that the cells show the exact same pattern of signal in the serial sections (panels A to D), which reflects the fact that the sections are 4 microns apart and the squamous cells range from about 10 to 30 microns in size.

xylene will be opaque. If you look closely at the slide with contaminated xylene, you will see the very minute droplets of water that are giving the slide its opaqueness. This problem, which is much more common in the summer months owing to the higher ambient humidity, is easily solved by realizing that it is *not* xylene's fault! Xylene per se cannot absorb water. The fault lies with the 100% ethanol in which the slides are washed before the xylene. As anyone who has used high-ethanol-containing fuel probably knows, you need to use a "water absorber;" otherwise, the ethanol will absorb enough water to eventually damage some components of the engine, such as those that contain neoprene. It is the same with mounting slides. If you are experiencing water droplets in the xylene, the first thing to try is to change the 100% ethanol. If that does not work (and it won't if the humidity in the lab is too high), then the simple "trick of the trade" is to rinse the slides with 100% ethanol and then air dry by placing the slide on a hot plate set to either 45°C or 60°C. When the slide dries completely, put it in fresh xylene, and the water droplets will no longer be a problem. Representative data from using this simple remedy areillu strated in Figure4-9 .

Another problem that can arise at this stage is that the investigator may decide to speed up the drying of the slide by placing the slide on a hot plate after the incubation in 100% ethanol and prior to immunohistochemistry or in situ hybridization. Let's think about how this might ruin the in situ hybridization or immunohistochemistry experiment. The paraffin wax is, of course, solid at room temperature. It has to be heated to about 65°C for it to become liquid. While the temperature of the slide is being raised from room temperature to 65°C, the paraffin wax is absorbing most of the heat and, in this manner, protecting the DNA/RNA/proteins in the tissue from damage that could include major conformational change, nicks from the breakage of phosphodiester/peptide bonds, and proteolysis that could ultimately lead to the loss of the signal due to, for example, loss of the exposed epitope (for immunohistochemistry) or diffusion of the target/probe signal (for either immunohistochemistry or in situ hybridization). I have received correspondence from quite a few investigators who have experienced problems with immunohistochemistry or in situ hybridization because they had exposed to the slide to high temperatures (usually 60°C or 95°C) in an attempt to either

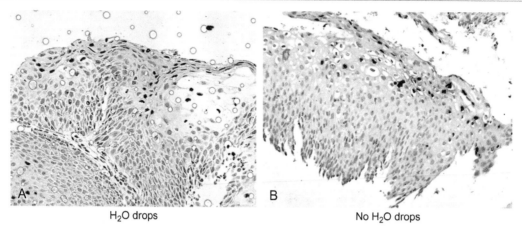

A	B
H₂O drops	No H₂O drops

Figure4-9   Water-contaminated xylene and in situ hybridization. This CIN 1 lesion showed a strong signal after HPV in situ hybridization. However, the ethanol was contaminated with water that was transferred to the xylene. This caused small water droplets in the image (panel A). When fresh 100% ethanol was used or the slide was dried prior to placing it in the xylene, the water droplets were no longerp resent(panelB) .

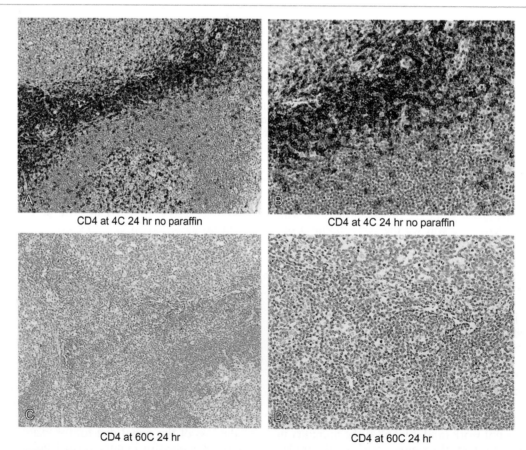

CD4 at 4C 24 hr no paraffin	CD4 at 4C 24 hr no paraffin
CD4 at 60C 24 hr	CD4 at 60C 24 hr

Figure4-10   Effect of preheating the slide to 60°C for 24 hours after removal of the paraffin and prior to immunohistochemistry. This tonsil was deparaffinized and then exposed to either 4°C or 60°C for 24 hours prior to immunohistochemistry for CD4. The CD4 signal is intense in the tissue sections exposed to 4°C for 1 day (panel A and, at higher magnification, panel B). However, the signal was completely lost if the serial section was exposed to dry heat (60°C for 24 hours) prior to the immunohistochemical reaction (panel C and, at higher magnification, panel D). Consider this question: if the three-dimensional macromolecule cross-linked cage hypothesis is correct, and the dry heat damaged this "cage," could we resuscitate it by treating the tissue with IHCPerfect?

help remove the paraffin wax or increase the rate at which the slide dries after deparaffinization. I can especially understand the former situation. I have seen TMAs with very thick paraffin on the slide that was added to protect the tissue in storage. It is a simple and useful trick to expose the slide to 60°C–95°C, which melts the wax, tilt the slide while hot, and then have most of the wax drain off the slide. Since treating the slides that contain the tissue sections with high temperatures is relatively common, let's discuss what problems may ensue with our in situ hybridization results.

THE IMPORTANCE OF NOT EXPOSING THE SLIDES (EITHER PRIOR TO OR, ESPECIALLY, AFTER REMOVAL OF THE PARAFFIN WAX) TO DRY HEAT

Figures 4-10 through 4-13 show data that underscore why it is important not to expose the tissue sections that are on the unstained silane-coated slide directly to dry heat prior to in situ hybridization or immunohistochemistry. Here, the tissue is tonsil, and the target is either CD4 or CD45. These proteins were chosen because many cells in the tonsil will express them, *and* the positive cells will show an intense signal. Thus, any diminution of the signal would be obvious. I examined many serial section slides so that I would be examining the same cells under different experimental conditions. In a typical experiment, I analyzed two sets of slides. In the first set of two slides, the paraffin was not removed. These slides were placed at either 4°C or 60°C, respectively. The 60°C was "dry heat." That is, I used a hot plate set to 60°C and placed the slides directly onto the surface of the hot plate. The other two slides had their paraffin wax removed with sequential washes in xylene and 100% ethanol, were air dried, and then placed at either 4°C or 60°C, respectively. The slides were incubated for 24 hours at the respective temperature and then immunohistochemistry was done for CD4. Figure 4-10 shows that placing the deparaffinized tissue on the glass slide at 60°C for 24 hours eliminated the CD4 signal. The fact that the high temperature was the cause of the lost signal is clear when we compare these data to serial sections where the deparaffinized tissue was kept at 4°C for 24 hours.

It may not be too surprising that exposing the deparaffinized tissues at 60°C for 24 hours prior to immunohistochemical eliminated the signal. But how about 2 hours at 60°C? These data are presented in Figure 4-11. Here,

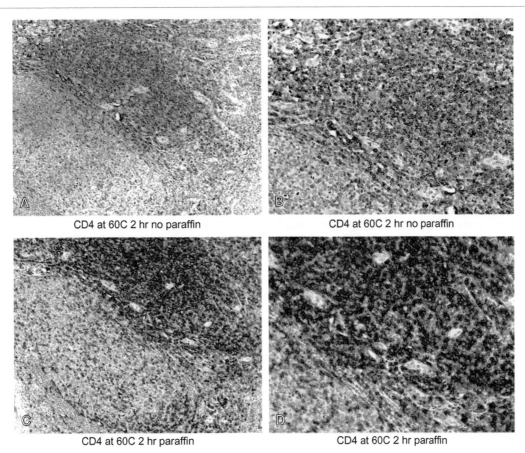

CD4 at 60C 2 hr no paraffin

CD4 at 60C 2 hr no paraffin

CD4 at 60C 2 hr paraffin

CD4 at 60C 2 hr paraffin

Figure4-11 Effect of preheating the slide to 60°C for 2 hours prior to immunohistochemistry. These experiments were similar to those shown in Figure 4-10, but the time of exposure to 60°C was much shorter, and the effect of leaving the paraffin on the slide was compared to removing the paraffin prior to immunohistochemistry. Note that if the paraffin was removed during the dry heat, even 2 hours at 60°C caused a marked decrease in signal (panels A and B). The paraffin protected the tissue from this deleterious effect of the dry heat (panels C and D); the target is again CD4, and the tissue is the serial sections of the tonsil shown in Figure 4-10.

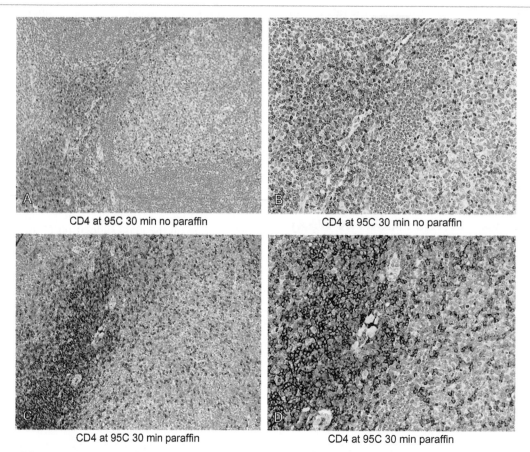

CD4 at 95C 30 min no paraffin   CD4 at 95C 30 min no paraffin

CD4 at 95C 30 min paraffin   CD4 at 95C 30 min paraffin

Figure 4-12   Effect of preheating the slide to 95°C for 30 minutes prior to immunohistochemistry. These experiments were similar to those shown in Figures 4-10 and 4-11, but the temperature was now 95°C and the dry heat exposure time was reduced to 30 minutes. The effect of leaving the paraffin on the slide was again compared to removing the paraffin prior to immunohistochemistry. Note that if the paraffin was removed during the dry heat, even ½ hour at 95°C caused a marked decrease in signal (panels A and B). The paraffin protected the tissue from this deleterious effect of the dry heat (panels C and D); the target is again CD4, and the tissue is the serial sections of the tonsil shown in Figure 4-10.

we also addressed the question: did the paraffin wax protect the tissue from the deleterious effects of the heat? Here is a summary of the data presented in Figure 4-11:

1. There is a clear-cut reduction in the CD4 signal after exposure of the deparaffinized tissue to 60°C for 2 hours prior to immunohistochemistry.
2. This diminution of the signal, though still evident, is *less* if the tissue is not deparaffinized during the exposure of the slide to the dry heat. In other words, the paraffin protects, at least partly, the tissue from the deleterious effects of the high temperature. You may recall that paraffin block also protected the in situ hybridization and immunohistochemistry signals from degradation in aged blocks. Is there any link between these two sets of data? In a short while, we will see that there is, but first some more data.

In the next series of experiments, I used 95°C for 30 minutes and examined the effects exposure of the tissue to this temperature/time pretreatment regime had on the CD4 immunohistochemistry signal. These data are presented in Figure 4-12. It is evident that just 30 minutes of exposure

of deparaffinized tissue to 95°C greatly reduced the immunohistochemistry signal. The same results were evident with HPV in situ hybridization (data not shown). It is also clear from Figure 4-12 that exposing the tissue to 95°C for 30 minutes *without* removing the paraffin protected the tissue somewhat from the marked reduction of the signal.

What is the basis for this reduction in signal that comes from exposure of the deparaffinized tissue to dry heat? If you look closely at the images in Figures 4-10 through 4-12, it is clear that the morphology of the deparaffinized tissue after exposure to the 60°C or 95°C heat was excellent. Thus, the explanation is not that the cells were so damaged by the brief exposure to heat that they lost their integrity. No, something more subtle, but still dramatic, occurred. There are two possibilities.

1. The epitope underwent such conformational change that it was no longer available to the antibody during immunohistochemistry; or
2. The epitope was not affected by the dry heat, but the "three-dimensional cross-linked" protein skeleton induced by formalin fixation was so affected by the dry heat that this caused the marked diminution of signal.

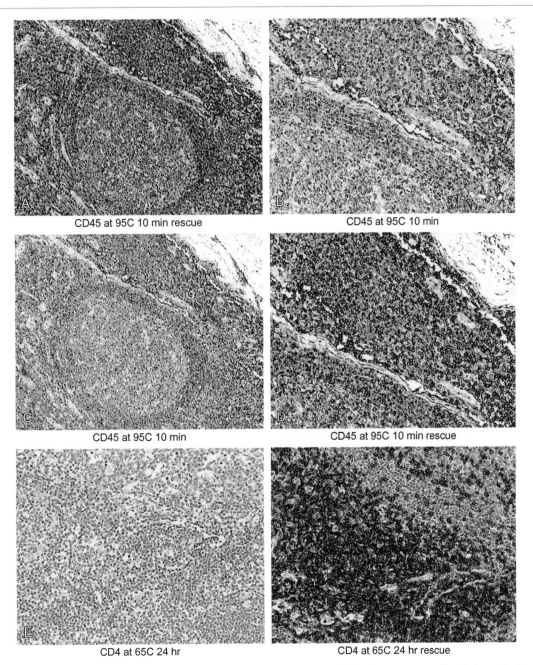

CD45 at 95C 10 min rescue

CD45 at 95C 10 min

CD45 at 95C 10 min

CD45 at 95C 10 min rescue

CD4 at 65C 24 hr

CD4 at 65C 24 hr rescue

Figure4-13   Rescuing of the immunohistochemistry signal that was diminished with dry heat exposure. Panels A and D show that treatment of the tissue with the IHCPerfect reagents rescued the CD45 signal that was much reduced after exposure of the tissue to the dry heat of 95°C at 10 minutes (panels B and C). The CD4 signal, as shown in Figure 4-10, was lost if the tissue was exposed to 65°C for 24 hours after removal of the paraffin wax (panel E). This signal was regenerated if, after the dry heat exposure for 24 hours, the tissue was treated with IHCPerfect prior to immunohistochemistry (panel F). These data give additional credence to the importance of the three-dimensional protein cross-linked cage surrounding CD4 and CD45 to the final immunohistochemical result.

Let's think about this as it relates to a fundamental point of in situ hybridization and immunohistochemistry. If the explanation was possibility #1, then we should be able to get a signal if we simply employ steps that are very good at exposing hidden antigens, such as increased antigen retrieval time and/or increased protease digestion time. I have tried such experiments many times to no avail; I can never recover the signal after too much dry heat exposure. Plus, I have had the same exact experience with DNA and

RNA targets, such that even 1 hour of exposure of the tissue on the glass slide to 60°C *after* removing the paraffin will basically eliminate the signal. DNA/RNA targets are not like epitopes where conformational changes could explain the loss of exposure of the target sequence. So we are left with the intriguing conclusion that formalin fixation is creating a three-dimensional network that is serving the same function as an "ion exchange and hydrogen bond resin." Though, to be complete (and as discussed at length

95

in Chapter 3), we need to also include the hydrophobic and hydrophilic environments created by the formalin-induced three-dimensional macromolecule cross-linked network, as well as the pore size. The dry heat, which certainly could damage proteins, must be so perturbing this network that the target/probe and epitope/primary antibody complex simply can no longer generate an optimal signal. This could reflect a variety of factors, such as reduced "docking" of the primary antibody/probe and other reagents to the epitope/nucleic acid target, and increased diffusion of the detection reagents from the site of the epitope/nucleic acid target, to name just two possibilities.

These CD4 data highlight another simple yet important point: namely, in order to become as good as possible with immunohistochemistry and in situ hybridization, you need to become proficient in anatomic pathology. I do not mean that you need to be able to diagnose complicated tumors and such. What I mean is that you need to have a solid foundation in the basics of the morphology and pattern recognition of cells and tissues. We are going to accomplish this in Chapter 6. There, you will see that it really is not too difficult with a little training to differentiate cancer from benign tissue, to differentiate epithelial cells from fibroblasts and from blood vessels. And you will see that just by being able to differentiate the nucleus, the nuclear membrane, the cytoplasm, and the cell membrane, you can go a long way in determining whether or not a signal has been optimized, as well as determining the significance of a the histologic distribution of a given target. In the case of CD4, the observation that the signal was cell-membrane-based, intense, and in the area of the tonsil where the T-cells (and not B-cells or macrophages typically are found) will assure you that your CD4 immunohistochemistry protocol is working correctly and that the 2 hours of dry heat after removing the paraffin wax from the tissue did indeed nearly eliminate the signal.

If possibility #2 above is correct, then we should be able to regenerate the signal by "rejuvenating" the three-dimensional macromolecule cross-linked network as we discussed at length in Chapter 3. Thus, I took the tissues that were exposed to 95°C for 10 minutes and 60°C for 24 hours (after the paraffin wax was removed from the tissue) and did the immunohistochemistry again after I treated the tissues with the proprietary reagents that regenerate the three-dimensional macromolecule cross-linked network. These data are presented in Figure 4-13. First, note that even 10 minutes of exposure of the slide to 95°C (after removal of the wax) was sufficient to reduce the CD45 signal. However, subsequent exposure of the tissue to the three-dimensional macromolecule cross-linked network regenerating agents (which I will refer to as IHCPerfect) rescued the signal. Resurrecting the signal by repairing the three-dimensional macromolecule cross-linked network was even more dramatic for the tissues exposed to 65°C for 24 hours (Figure 4-13).

Thus, to summarize, do *not* expose the tissue on the glass slides to dry heat, especially if the wax was already removed from the tissue; otherwise, your signal will be much reduced or lost. Also, we have more evidence that the three-dimensional macromolecule cross-linked network induced by formalin fixation is critical in understanding how in situ hybridization and

immunohistochemistry work and in rejuvenating the signal in aged blocks, or in tissues exposed to dry heat.

THE PREPARATION OF FORMALIN-FIXED CELLS FOR IN SITU HYBRIDIZATION

I use a simple protocol when dealing with cells that have been fixed in buffered formalin. You will recall that the cells were fixed by simply removing the growth media and adding the 10% neutral-buffered formalin. Allow the formalin to fix the cells for at least 4 hours, although overnight fixation will certainly be fine. This way, when you start to remove the cells, they will be protected by their strong three-dimensional cross-linked cytoskeleton and their morphology will be excellent. Remove them first, however gently, and then fix them, and then you end up with cells that are very much distorted.

After removing the cells, put them into a conical tube (either 15 ml or 50 ml, depending on the number of cells that were grown). Then spin down at 5,000 rpm for 5–10 minutes, decant the fluid, and then add from 1 to 5 ml of RNase- and DNase-free water to the pellet, depending on its size. Then suspend the cells suspended in the sterile water by robust vortexing. Next, place 10–20 μl aliquots of the cell suspension on the glass slide and put the slide in a 37°C incubator until the water evaporates. As with tissue, put at least *three* such dots per silane-coated slide. Also, as with deparaffinized tissue, do *not* expose the slide to dry heat (60°C or above); otherwise, you will most likely lose the signal.

STORAGE OF SLIDES AND TISSUE BLOCKS WITH IN SITU HYBRIDIZATION

Investigators often ask me how long and under what conditions they should retain their paraffin-embedded blocks and unstained tissue sections that were placed on the positively charged silane-coated slides. When I was trained more than 25 years ago, I was taught that both the paraffin-embedded blocks and the unstained slides that contained the formalin-fixed tissues that came from those blocks would last for many decades if stored at room temperature. This makes sense. Formalin is such a powerful fixation agent that it is hard to imagine the DNA/RNA or protein changing over time in such materials. Indeed, I have been invited to over a dozen large pathology labs in my career to examine the facility, and in each case, the paraffin blocks and the stained/unstained slides were always stored at room temperature. The slides were either stored in light-tight boxes or in long trays that fitted into light-tight filing cabinets.

One advantage to having been in the field for more than 25 years is that I have accumulated hundreds of paraffin blocks and the resultant unstained plus slides. Specifically, I have thousands of unstained plus slides that contain tissue sections obtained from 1987 through 2012. Thus, it is an easy matter for me to determine if the age of the paraffin block or the age of the slide prepared from a given paraffin block has any effect on the signal with in situ hybridization.

As we discussed in Chapter 3, the age of the unstained slide has a very strong effect on the signal with in situ hybridization! The samples I used to test were tissue sections that were prepared on unstained plus slides in 1996–1997 using tissues proven to be HPV positive by in situ hybridization. I retained the HPV positive in situ hybridization result that I achieved in 1996–1997. I still had unstained plus slides also prepared in 1996–1997 that were serial sections from the sample I tested more than 15 years ago. By comparing the strength of the in situ hybridization signal in the serial sections that I tested in 1996–1997 and then again in 2012, I could easily determine if the HPV DNA present on the unstained slide had lost any of its "potency." As we discussed in Chapter 3, there was indeed a strong decrease in the HPV in situ hybridization signal in the aged slides/blocks. The same phenomenon was evident for RNA in situ hybridization and immunohistochemistry. To summarize these data:

1. There is approximately a 70% *decrease* in the strength of the in situ hybridization and immunohistochemistry signal in tissue sections that were put on glass slides more than 10 years ago when compared to the signal seen when the slides were initially prepared.
2. There is approximately a 40%–50% *decrease* in the strength of the in situ hybridization and immunohistochemistry signal in tissue sections if recent recuts were made of aged paraffin blocks when compared to the signal seen when the slides were initially prepared.
3. The signal could be regenerated by treating the slides with biochemical agents that regenerated the three-dimensional macromolecule cross-linked network.

So, clearly, the simple act of storing unstained slides or paraffin-embedded blocks at room temperature will result in a decreased signal over time. Although the data generated here was with blocks that were 10–21 years old, I have seen the diminution in signal in blocks that are only a few years old, although the amount of reduction is usually less marked. However, formalin-fixed cell preps also show the reduction in signal, and usually, it is marked after only a few years of storage. The bottom line is clear: the age of the paraffin block you are using for research is a key variable that must be taken into account when interpreting the data. If you are comparing data from blocks that are a few years old versus blocks that are many years old, the results will be skewed and comparisons will not be valid, unless you regenerate the signal, as discussed in detail in the preceding chapter.

STEP THREE: PRETREATMENT OF THE TISSUE FOR SUCCESSFUL IN SITU HYBRIDIZATION

At this stage, we have obtained our formalin-fixed tissues or cell preparations and placed them on silane-coated glass slides. Our histotechnologist has placed multiple serial sections per slide and has labeled each slide in sequential order. We have kept the unstained slides at room temperature in a light-tight box and now are ready for the next step.

Up until now in this chapter, we have been combining general ideas that apply to immunohistochemistry and in situ hybridization. Here is where we diverge from that

schema. The reason is that the pretreatment of tissues with in situ hybridization is different enough from immunohistochemistry to warrant separate coverage. Please don't get me wrong; there is still a lot in common between the two methods relative to pretreatment of the samples. But let's consider a clear-cut difference between immunohistochemistry and in situ hybridization. With immunohistochemistry, we are trying to detect the epitope of a protein that, by definition, will be part of the three-dimensional protein cross-linked network that invariably is created when the tissue or cells are exposed to formalin. And, of course, since it is a protein, it will be very sensitive to any pretreatment that may affect protein-protein cross-links, such as protease digestion or antigen retrieval, which we refer to as "wet heat" now and then. The reason we use the term "wet heat" for antigen retrieval, besides being more descriptive of the actual process, is that we want to, with as much emphasis as possible, differentiate it from exposing the deparaffinized tissue to dry heat. The latter, as we have discussed, invariably makes proteins/DNA/RNA either less detectable or undetectable with in situ hybridization or immunohistochemistry. Although "wet heat" can also decrease the signal with immunohistochemistry or in situ hybridization, in many cases it will greatly facilitate the signal.

DNA and RNA targets can be cross-linked to proteins by formalin and, as such, are a part of the three-dimensional molecular network generated by formalin fixation. However, they are not going to be directly affected by protease digestion or antigen retrieval. Rather, the *exposure* of the DNA/RNA targets may be affected by pretreatments such as protease digestion and antigen retrieval. Further, the *diffusibility* of the signal generated by DNA or RNA targets may also be affected by pretreatments that affect the protein cross-linked network. However, the *structure* or *conformation* of the DNA or RNA target should not be affected by protease digestion or antigen retrieval. With this in mind, let's look at the various possible ways we can pretreat tissues to improve our results with in situ hybridization.

NO PRETREATMENT

When I was trained to do in situ hybridization in the late 1980s, I was told that we had to do protease digestion in order to get a signal with in situ hybridization. This made sense to me. Since formalin fixation creates such a complex three-dimensional macromolecule network, and our target for in situ hybridization is usually nuclear (back then we all were doing in situ hybridization exclusively for HPV DNA), it seemed logical that the large 80–120 base pair probes and other reagents could not possibly get into the nucleus until we punched some serious holes into the three-dimensional labyrinth of the fixed cells by doing protease digestion.

Despite how logical the preceding description sounds, when you examine Figures 4-6 and 4-7, it is clear that protease digestion will not result in a better signal for HPV DNA by in situ hybridization for all tissues! In Figure 4-6, the HPV in situ hybridization signal was equally strong whether or not protease digestion or antigen retrieval or no pretreatment was done. In Figure 4-7, the HPV in situ hybridization signal clearly was

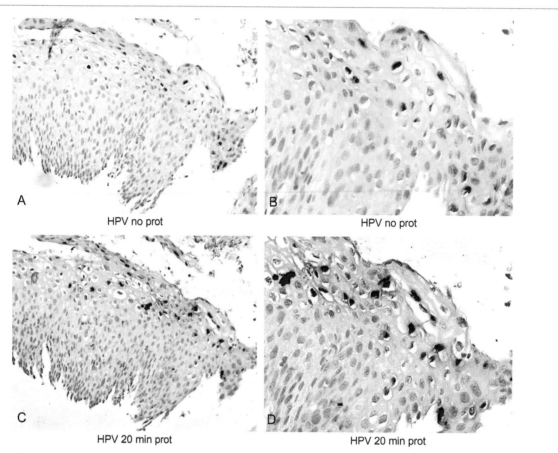

Figure4-14   Need for protease digestion to maximize the signal with HPV in situ hybridization. This CIN 1 biopsy contained high copy numbers of HPV DNA. Note that the signal is moderately strong if no pretreatment was done (panel A and, at higher magnification, panel B). If protease digestion for 20 minutes preceded the in situ hybridization, then the signal intensity and the percentage of dysplastic cells that had detectable HPV DNA increased (panels C and, at higher magnification, panel D).

reduced by protease digestion prior to the denaturation/hybridization.

Of course, there are tissues in which you *must* predigest with a protease in order to maximize the signal with in situ hybridization. An example of such is seen in Figure 4-14. The tissues in Figures 4-6, 4-7, and 4-14 look similar; they are each surgical biopsies done by a gynecologist who immediately fixed the tissue in formalin. How can they be acting so differently with respect to something as basic as pretreatment in protease?

The answer is simple: the tissues may look the same, but it is a certainty that they were not exposed to formalin for the same amount of time. One may have been obtained Friday afternoon and stayed in formalin for 3+ days. The other may have been done Monday evening, brought immediately to pathology lab, and then exposed to only 4 hours of formalin fixation before the paraffin embedding started. And, to remind you, there is no way we can ever determine which of the two scenarios may be correct.

When you think about it, it is amazing that no pretreatment at all can be optimal in quite a few cases for in situ hybridization. You would think that the three-dimensional macromolecule cross-linked network created by formalin fixation would create such a barrier to probe/

reagent entry to the nucleus (or even the cytoplasm) that unless you punched many holes into it, there would be no way to have success with in situ hybridization. The fact that formalin-fixed, paraffin-embedded tissues that appear well fixed in the sense that the morphology is beautifully preserved during in situ hybridization can give excellent in situ hybridization results with no pretreatment suggests to me that pore size may indeed play a role with this process. The tissues that do not need protease digestion for optimal in situ hybridization may have been fixed in such a way that the pore size is already optimal for allowing the probe/other reagents to access and stay with the target.

PROTEASE DIGESTION

Whatever theoretical construct may be correct to explain why some tissues need protease predigestion for optimal DNA or RNA in situ hybridization, and others do not, the bottom line is that we cannot predict this accurately. Thus, I recommend that you consider the following regime when testing a formalin-fixed, paraffin-embedded tissues or cell preparation for the first time: treat at least one section on the slide to short-term protease, treat the other to long-term protease, and leave the last section untreated.

I prefer to use proteinase K when doing protease digestion with in situ hybridization (or immunohistochemistry). Proteinase K offers several advantages:

1. It is a broad spectrum protease.
2. The proteinase K solution is very stable and can be used for many months when stored at 4°C.
3. Its activity can be easily modulated. You only have to increase the temperature of the reaction and/or the salt concentration of the solution to increase the activity of proteinase K.

I would like to stress that there is no reason to prefer proteinase K over the other proteases that are commercially available. Indeed, I used pepsin/pepsinogen for many years before I decided to make proteinase K my "default" protease. Pepsin offers multiple advantages:

1. It is also a broad spectrum protease.
2. It tends to have less specific activity than proteinase K and, thus, tends not to cause overdigestion.
3. It can be easily inactivated by simply raising the pH.

A disadvantage of pepsin is that you need to either make the solution right at the time it will be used or need to freeze it and thaw it immediately before use. In comparison, the proteinase K solutions can be kept at 4°C for many months and still maintain their activity.

I have already stressed the point that we cannot predict from the hematoxylin and eosin stain, or any other characteristics of the tissue, which cases will require no pretreatment, protease digestion, or antigen retrieval. We have already seen examples of cases in which predigestion with protease was essential to obtain the optimal intense signal, which is, of course, the primary operational goal of our work. However, in the cases that need protease digestion prior to probe/target denaturation/hybridization, what is the optimal time for protease digestion?

TIME OF PROTEINASE K DIGESTION

I ran a series of experiments using HPV positive lower genital tract biopsies. I chose them because they already were documented to contain relatively high copy numbers of HPV. I varied the protease digestion from 10 seconds to 2 hours and recorded the relationship of the time of protease digestion to the strength of the hybridization signal. The data are presented in Figure 4-15.

Note that these data tell us some important information about in situ hybridization:

1. Although the formalin-fixed, paraffin-embedded tissues each look equivalent under the microscope, they show markedly variable results when comparing the strength of the in situ hybridization signal to the amount of time of protease digestion.
2. Some CIN 1/condylomatous lesions show a *stronger* signal with increased time of protease digestion.
3. Some CIN 1/condylomatous lesions show a *weaker* signal with increased time of protease digestion.
4. Some CIN 1/condylomatous lesions show *no change* in the signal with increased time of protease digestion.

Representative photomicrographs of the data are presented in Figure 4-16.

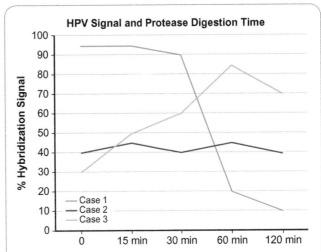

Figure 4-15 Graphic representation of the relationship of protease digestion time to the strength of the hybridization signal for HPV in situ hybridization. The HPV positive tissues were exposed to various protease digestion times ranging from 15 minutes to 2 hours and then tested for HPV DNA by in situ hybridization. As is evident, histologically equivalent tissues from the same site showed marked variability in the response to the time of protease digestion and the strength of the in situ hybridization signal.

I also want to stress that these data are very representative of what you will experience in your lab on a daily basis when you do in situ hybridization. It is understood that tissues will vary widely in the "broad picture" of pretreatments to optimize in situ hybridization: *no pretreatment*, *protease* digestion, or DNA/RNA *retrieval*. But it is more complicated than that. Within the last two categories are important variables that can affect the final in situ hybridization signal. As we see in Figures 4-15 and 4-16, one such variable is the length of *time* we predigest the tissue in the protease. As we will explore after two more paragraphs, the *temperature* at which the protease digestion is done also can affect the optimization of the in situ hybridization signal. It is very important to me in this book that we don't fall into the habit of doing protocols without always trying to figure out the reason for any specific step. Although such "cookbooks" are useful, they don't allow us to delve into the biochemical bases of the different steps of the protocol. So let's stop for a second and see whether we can make some sense of the data that relate to pretreatment and optimizing in situ hybridization.

It is logical to assume that longer incubation times in proteinase K result in larger pore sizes and a lesser "hydrogen and ionic bond potential" of the cell. If the proteinase K cuts through many peptide bonds on a given protein, it makes it more likely that this segment of protein may be able to dissociate from the remainder of the peptide, which, in turn, should increase pore size. Concomitantly, the hydrogen and ionic bond potential of the R side chains on the amino acids that are part of the protein that has been destroyed by the proteinase digestion will likewise be lost. This should reduce the hydrogen and ionic bond potential of this region of the

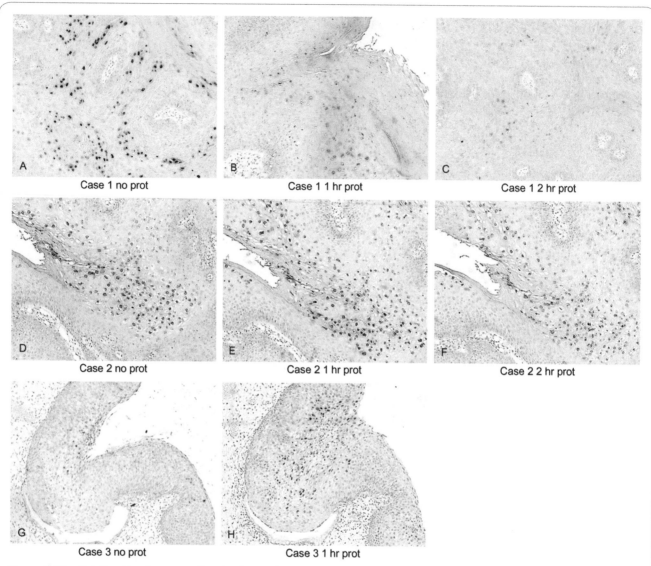

Figure4-16 Relationship of protease digestion time to the strength of the hybridization signal for HPV in situ hybridization. This figure shows representative data from the graph of Figure 4-15. Note that in case 1 the serial section shows a strong signal with no pretreatment (panel A), which is much decreased after 1 hour (panel B) or 2 hours (panel C) of protease digestion. The serial sections of case 2 show equivalent results for no pretreatment (panel D), 1 hour (panel E), and 2 hours (panel F) of protease digestion. Case 3 shows an increase in the signal when no pretreatment (panel G) was compared to 1 hour of protease digestion (panel H) with a reduction of the signal in the serial section if the protease digestion was increased to 2 hours (data not shown).

three-dimensional protein cross-linked "cage." Thus, the data regarding protease digestion and optimizing in situ hybridization may be telling us that in some tissues the formalin fixation may have created a relatively "weak" pore size/ionic and hydrogen bond barrier. Such tissues will do fine with in situ hybridization with no pretreatment (as seen in Figure 4-16). If we give them protease digestion or expose them to DNA retrieval, this relatively weak three-dimensional protein/protein cross-linked network is basically destroyed, and the in situ hybridization signal is either reduced or lost (Figure 4-16, case 1). Of course, we are presupposing that an intact and strong three-dimensional protein/protein cross-linked network is critical to sin situ hybridization. I maintain that our DNA and RNA "rescue" experiments, in which we take old blocks

and rejuvenate the signal, strongly support that the latter statement is true. Similarly, we can predict that some tissues will, after formalin fixation, have such a strong three-dimensional protein/protein cross-linked cage around the DNA and RNA that it limits the entry of our in situ hybridization reagents and/or access to the DNA or RNA probe. Such tissues need robust protease digestion and/or DNA retrieval to optimize the in situ hybridization result. Indeed, as we have seen, some such tissues need *both* protease digestion and DNA retrieval for optimal in situ hybridization. These two theoretical possibilities are presented in Figures 4-17, 4-18, and 4-19. Assuming that these hypotheses are correct, it follows that there will be tissues that are intermediate, where protease digestion and/or antigen retrieval will have relatively little effect on

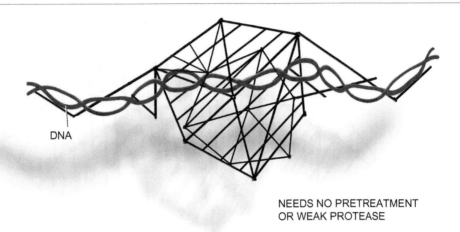

NEEDS NO PRETREATMENT
OR WEAK PROTEASE

Figure4-17   Graphic representation of the importance of the three-dimensional cross-linked protein cage to the signal with DNA in situ hybridization: small cage. In this graphic depiction, the small cage allows for an optimal signal for HPV in situ hybridization with no pretreatment or "weak" pretreatment with the protease.

the signal. Of course, we have already seen examples of this in this book. Realizing that any such theory will need to be a bit oversimplistic, let's, for the sake of discussion, refer to these three scenarios as "strongly fixed tissues," "intermediately fixed tissues," and "weakly fixed tissues." Let me stress again that these designations reflect a complex interplay between formalin fixation time, temperature of fixation, volume and shape of tissue, cellular components of the tissue, and so on, *and* that we cannot predict ahead of time which tissue will fall into what category.

Let's now look at how the *temperature* of the protease digestion can affect the in situ hybridization signal.

THE EFFECT OF TEMPERATURE OF THE PROTEASE DIGESTION TIME ON THE IN SITU HYBRIDIZATION SIGNAL

We have previously discussed the fact that you can increase the specific activity of proteinase K by increasing the temperature of the reaction to 60°C. Another way to increase the activity of proteinase K is to raise the concentration of salt in the solution.

For those who extract DNA or RNA from formalin-fixed, paraffin-embedded tissues, perhaps the most important trick of the trade is to realize that you must absolutely maximize the protease digestion. Otherwise, the RNA and DNA molecules won't be accessible to the probe because of persistent protein-DNA and protein-RNA cross-links, or because many of the nucleic acid molecules will be lost at the purification step. Get rid of the protein-DNA and protein-RNA cross-links by maximizing the protease digestion, and then you will precipitate out relatively intact nucleic acids assuming, as we have discussed, that the formalin-fixed, paraffin-embedded tissues are not too many years old.

How about in situ hybridization: can augmenting the proteinase K specific activity enhance our signal? I predict that you know the answer; it is at times *yes* and at times *no*!

Figure 4-20 shows examples (for HPV and microRNA-125b) where increasing the temperature from room

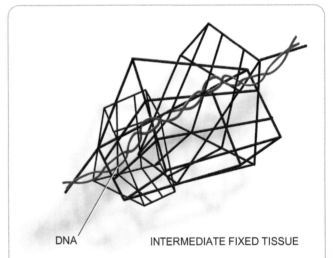

DNA   INTERMEDIATE FIXED TISSUE

Figure4-18   Graphic representation of the importance of the three-dimensional cross-linked protein cage to the signal with DNA in situ hybridization: intermediate cage. In this graphic depiction, the intermediate cage allows for an optimal signal for HPV in situ hybridization with "intermediate" pretreatment conditions.

temperature to 60°C augmented the DNA and RNA signals, respectively. In most cases, increasing the temperature further to 95°C reduced the signals, as is evident from the photos for the microRNA. Thus, if you have evidence to suggest that you are working with strongly fixed tissues, then increasing the temperature of the protease to 60°C may help. However, I simply have seen too many examples where increasing the temperature to 60°C or 95°C damaged the cellular morphology and/or reduced the signal (Figure 4-20) that I do not recommend using 60°C as the temperature for the protease digestion with in situ hybridization unless you are certain that the tissue was fixed for a very long time (weeks to months) in 10% neutral-buffered formalin (i.e., a strongly fixed tissue).

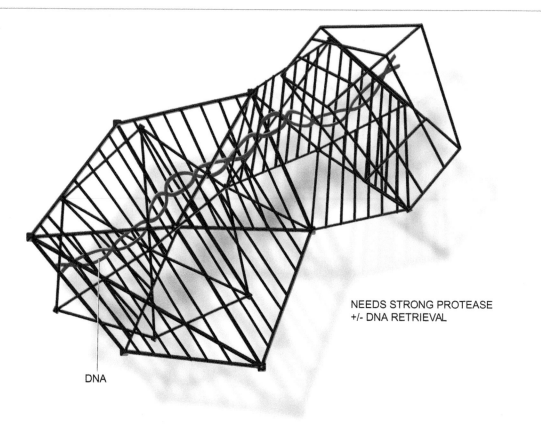

NEEDS STRONG PROTEASE
+/- DNA RETRIEVAL

DNA

Figure4-19 Graphic representation of the importance of the three-dimensional cross-linked protein cage to the signal with DNA in situ hybridization: large cage. In this graphic depiction, the large cage allows for an optimal signal for HPV in situ hybridization only after rigorous protease digestion at times with DNA retrieval.

Figure 4-21 shows an example of a tissue from a person with a disease called focal epithelial hyperplasia, which is also known as Heck's disease. It was sent to me as part of a study by my friend and colleague Dr. Cynthia Magro, a superior dermatopathologist. These lesions often contain HPV type 13. As you can see, the tissue was negative for HPV 13 when the protease digestion step was done at 60°C. However, when the protease digestion was done at room temperature, a strong signal for HPV 13 was seen. This is thus an example of a tissue that I refer to as an "intermediate fixed tissue." By the way, note another beautiful example of how we can measure the specificity of the reaction by using our knowledge of surgical pathology. Note that some cells in the lesion show large halos. These are called koilocytes, so named more than 50 years ago by the great cytopathologist Dr. Leopold Koss. These halos are formed when HPV proliferates in a cell and basically forces water to accumulate in the cell, pushing the organelles to the side. Thus, they are a good marker of high copy HPV DNA. This is evident in the figure, as the signal for HPV 13 is basically limited to the koilocytes. Figure 4-21 also shows an example of what I refer to as a "weakly fixed tissue." Note that the signal was augmented after 10 seconds of protease digestion, but that longer protease digestion much reduced the signal. The fact that some tissues show an improvement in the signal after only 10 seconds of protease digestion suggests that this is not a pore size issue, but rather the

ionic and hydrogen bond potential of the three-dimensional protein cross-linked cage that surrounds the DNA or RNA molecule.

DNA AND RNA RETRIEVAL

Of course, the title of this section could also be "antigen retrieval." Purists would correctly say that when we expose formalin-fixed, paraffin-embedded tissues to 95°C in solution with in situ hybridization, we are not exposing antigens, so we should not use the term "antigen retrieval." People who prefer to simplify concepts would indicate that the term "antigen retrieval" refers to a *process* in the field of in situ-based molecular pathology whereby formalin-fixed, paraffin-embedded tissues are exposed to aqueous solutions heated to 95°C in order to affect the three-dimensional macromolecule cross-linked cage and, thus, expose the target of interest. Clearly, either idea is equally valid. Since this chapter is dedicated to in situ hybridization, I refer to the process for the rest of this chapter as "DNA and RNA retrieval."

From our model of the three-dimensional structure of formalin-fixed, paraffin-embedded tissues, it is no surprise that aqueous solutions at high temperatures could be helpful to us. The three-dimensional macromolecule cross-linked network has as its backbone protein-protein cross-links. These protein-protein cross-links will certainly undergo conformational changes in response to

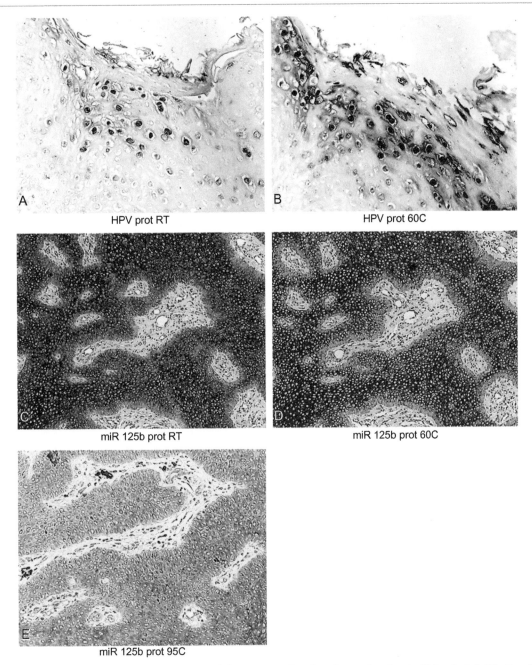

A — HPV prot RT

B — HPV prot 60C

C — miR 125b prot RT

D — miR 125b prot 60C

E — miR 125b prot 95C

Figure4-20   The effect of increasing the temperature of the protease digestion step on the signal with in situ hybridization. The serial sections in panels A and B show that if the protease digestion step was done at 60°C (panel B), then the HPV in situ hybridization signal was much stronger as compared to doing the protease digestion step at room temperature (panel A). Similarly, the in situ signal for microRNA 125b was increased with a protease digestion temperature of 60°C (panel D) versus room temperature (panel C). However, increasing the protease digestion temperature to 95°C reduced the microRNA signal (panel E).

the high aqueous temperatures. The two predicted consequences of these conformational changes are:

1. The altering of the structure of the three-dimensional macromolecule cross-linked cage that surrounds the DNA or RNA target such that pore size and the ionic/hydrogen bond potential as well as hydrophobic/hydrophilic microenvironments inside the individual cells will change.

2. The direct exposure of DNA and RNA sequences due to the alterations of the complex protein cage that surrounds it.

We have already seen multiple examples in which DNA and RNA retrieval has augmented the signal with in situ hybridization, such as Figure 4-7. Figure 4-22 shows another example of this for a vulvar intraepithelial neoplasia that contains HPV 16 DNA. Note that the HPV

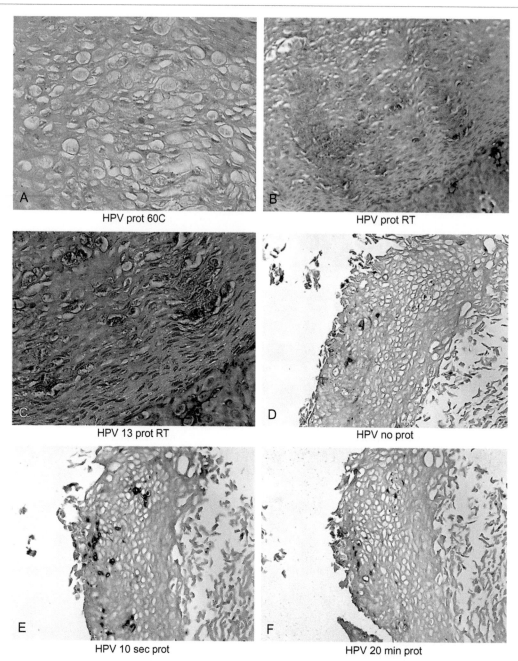

Figure4-21   The effect of increasing the temperature of the protease digestion step on the signal with in situ hybridization. Panels A–C are serial sections from a case of Heck's disease, which is due to infection by HPV 13. Note that no signal was evident if the tissue was incubated in the protease solution at 60°C (Panel A), but was present if the protease digestion was done at room temperature (panel B). Panel C shows how the viral DNA localized to the koilocytes of the lesion. Panels D–F show the serial sections of another "weakly fixed tissue" where the HPV in situ hybridization signal was enhanced by very short protease digestion time (panel E versus D), but reduced with longer protease digestion times (panel F).

DNA signal is weak if either protease digestion or DNA retrieval was done. In order to see a marked enhancement of the HPV signal with in situ hybridization, you must do protease digestion and DNA retrieval. Thus, this tissue is a good example of a "strongly fixed tissue." Also, note that the signal and tissue morphology are well preserved when both DNA retrieval and protease digestion are done on the tissue. Figure 4-23 shows an example in which the in situ hybridization signal is enhanced, this

time for RNA (microRNA-200) after RNA retrieval. Note the clear-cut signal in the tissue pretreated with RNA retrieval and the lack of any signal for the serial section (and, thus, same cells) pretreated with protease. There are other points in Figure 4-23, however, that are worth stressing:

1. The standard negative control is shown in Panel A. This can be either omission of the probe or use of a

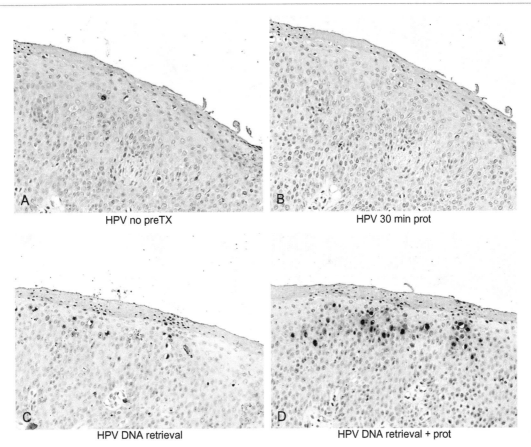

HPV no preTX

HPV 30 min prot

HPV DNA retrieval

HPV DNA retrieval + prot

Figure4-22   An example of a "strongly fixed tissue" and the need for DNA retrieval and protease digestion to maximize the signal for HPV DNA. A very weak signal was seen in this VIN lesion for HPV in situ hybridization if no pretreatment or protease digestion was done (panels A and B, respectively). Although the signal increased with DNA retrieval (95°C for 30 minutes, panel C), the most intense signal required DNA retrieval with protease digestion (panel D).

scrambled probe. We pay a great deal of attention on the negative controls for in situ hybridization in Chapter6.
2. Panels B and C show another key feature that helps us in determining the specificity of the microRNA in situ hybridization reaction—specifically, the histopathologic interpretation of the signal. Note in panel B how the signal localized to one part of the spinal cord. This part is called the ventral horn. This is the part of the spinal cord where the large motor neurons are located. Also note that the adjacent part of the spinal cord was negative for the microRNA. This part of the spinal cord carries the fascicles from the cortex of the brain and, thus, is mostly axons with supporting cells and few neurons. The key point is that a specific histologic distribution of a signal is good presumptive evidence of a specific reaction. Of course, this must always be viewed with the negative controls, but still the importance of a solid foundation in histopathology cannot be understated for those doing in situ hybridization or immunohistochemistry. Indeed, that is the reason why we spend one entire chapter on the topic.

Though I realize we have not yet discussed the histopathology of normal tissues, there is one other point from Figure 4-23 that increases our confidence in the specificity of the reaction. It is evident in panel C. Note that the positive cells are very large and have comet-like tails (arrow). In the spinal cord (and central brain), this is diagnostic of neurons. Localization of a DNA or RNA target to one cell type is a useful indicator of specificity, since background, typically, will affect different cell types.

The bottom line with DNA or RNA retrieval is that tissues that we are referring to as strongly fixed tissues will require either protease digestion and/or DNA/RNA retrieval to maximize the in situ hybridization signal. An obvious question is: what percentage of formalin-fixed, paraffin-embedded tissues are "strongly fixed tissues" versus "weakly fixed tissues." Or, in other words, what percentage of formalin-fixed, paraffin-embedded tissues will yield maximum signal with no pretreatment versus protease digestion (at times with DNA or RNA retrieval) versus DNA/RNA retrieval alone? Of course, these numbers won't be too useful if coming from one laboratory because the fixation characteristics of the tissues from that lab may be quite different from the laboratory

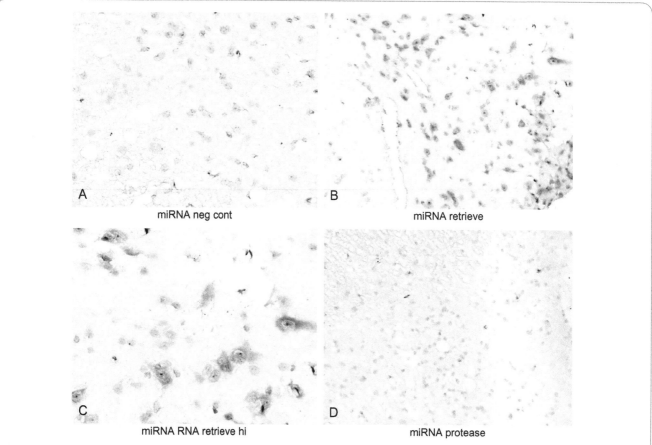

Figure 4-23   The importance of different pretreatment conditions for the in situ detection of microRNAs. This section of brain showed no signal with the scrambled probe (panel A) and a strong signal after RNA retrieval (panel B and, at higher magnification, panel C). Note that some of the microRNA positive cells have the cytologic features of neurons, and that the signal was present in their cytoplasm. No signal was seen if protease digestion was used (panel D).

you work in. But I can give you my experiences for HPV DNA in situ hybridization, where I have tested thousands of cases over time by in situ hybridization. In my experience, about 30% of HPV+formalin-fixed, paraffin-embedded tissues give a strong signal with *no* pretreatment, another 40% require intermediate protease digestion or DNA retrieval, and the remaining 30% require either strong protease digestion and/or DNA retrieval with protease digestion. I am assuming that the first group corresponds to the weakly fixed tissues; the second group, the intermediately fixed tissues; and the last group, to the strongly fixed tissues.

We discuss protocols for DNA and RNA in situ hybridization in Chapter 7. Suffice it to say for now that the many observations and experimental data we have generated for weakly fixed tissues, intermediately fixed tissues, and strongly fixed tissues will lead us to a *key* part of our in situ hybridization protocol. *When you are testing new formalin-fixed, paraffin-embedded tissues for DNA or RNA, I recommend using two slides with multiple sections per slide and doing the following pretreatment:*

No pretreatment.
Variable protease digestion (5 minutes to 60 minutes)
DNA or RNA retrieval (30 minutes at 95°C) with and without protease digestion.

This simple protocol will cost you only two slides of work and will cover more than 95% of the optimal conditions for DNA/RNA in situ hybridization. The protocol is the logical conclusion of all the data we have examined and our theoretical constructs for in situ hybridization seen in Figures 4-17 through 4-19.

Before we go to the next step (the probe), there is one more question that we should address now relative to pretreatment of formalin-fixed, paraffin-embedded tissues for in situ hybridization. The question is simple:

For a given formalin-fixed, paraffin-embedded tissue, will the pretreatment regimes differ for RNA, DNA, and protein targets?

Again, I am confident that you know the answer based in part on our theoretical discussions. The answer is *yes* (at least in most cases).

I examined a series of HPV positive formalin-fixed, paraffin-embedded tissues for HPV DNA, microRNA (miR-26, abundant in squamous epithelia), and several proteins including cytokeratin AE 1/3, which, again, should be present in squamous cells. I optimized serial sections for the HPV DNA, microRNA-26, and AE 1/3 plus some other proteins. Representative data are presented in Figures 4-24 and 4-25. Note that for HPV DNA the optimal condition for this tissue was DNA retrieval.

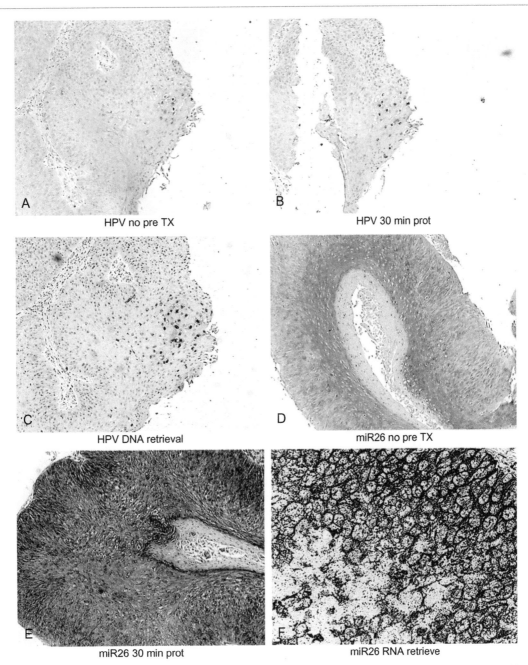

Figure4-24 Different targets in the same tissue required different pretreatment regimes: Part A. Panels A–C show the HPV in situ hybridization signal was optimal after DNA retrieval in this tissue. However, when a different region of the same tissue was analyzed for microRNA-26, then protease digestion yielded the strongest signal (panel F). Only a weak signal was seen with no pretreatment (panel E), and RNA retrieval caused a good signal, but there was some loss of tissue morphology (panel G).

For the protein AE 1/3, the optimal condition was protease *plus* antigen retrieval. However, for microRNA-26a, the optimal condition was intermediate protease digestion. For the RNA target, RNA retrieval led to a strong signal but an associated loss in tissue morphology and, as evident in Figures 4-24 and 4-25, the signal was much stronger for the intermediate protease digestion.

The data from Figures 4-24 and 4-25 are fascinating. I should add that these data are very representative of

the data you get when comparing the optimal conditions for the detection of a DNA, RNA, and protein molecule in a given tissue. The data suggest that the terms "strongly fixed tissues," "intermediately fixed tissues," and "weakly fixed tissues" may be overly broad. It may be more accurate to refer to "strongly fixed epitopes/DNA/RNA," "intermediately fixed epitopes/DNA/RNA," and "weakly fixed epitopes/DNA/RNA" in a *given* tissue. This hypothesis is illustrated in

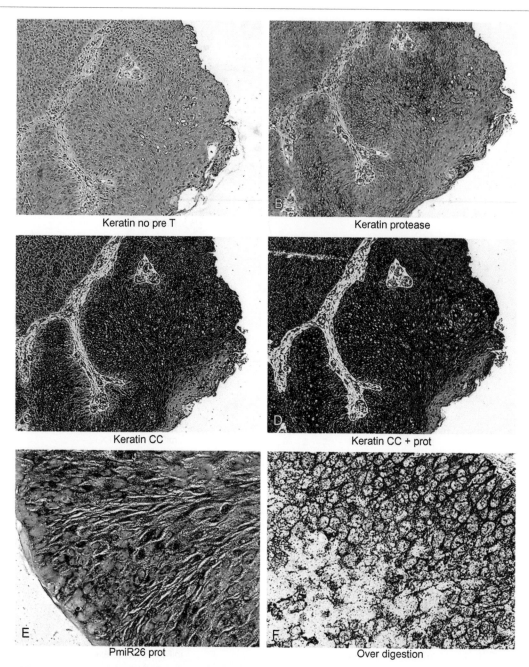

Keratin no pre T

Keratin protease

Keratin CC

Keratin CC + prot

PmiR26 prot

Over digestion

Figure 4-25 Different targets in the same tissue required different pretreatment regimes: Part B. This tissue is from the same biopsy shown in Figure 4-24. Note that the immunohistochemical signal for cytokeratin is strongest with antigen retrieval and protease digestion (panels A–D, with antigen retrieval and protease digestion being panel D). Panel E shows at high magnification that the optimal condition for microRNA-26 detection in this tissue was protease digestion, and also shows at high magnification the overdigestion noted in Figure 4-24. The fact that different targets from the same tissue can show such variable optimal pretreatment conditions despite having the same fixation time suggests that each epitope may be surrounded by a biochemically unique three-dimensional macromolecule cross-linked cage.

Figure 4-26. As seen, this theory is based on the idea that a given epitope, RNA, or DNA target will have a unique cross-linked protein "cage" around it. The unique cage will simply reflect the density and biochemical/structural characteristics of the proteins that surround the epitope/DNA/RNA prior to fixation. The putative variation in the biochemical make-up of the "cage" relative to each potential target may explain, in part, the marked variation in the pretreatment regimes needed for optimal in situ hybridization and immunohistochemistry.

STEP FOUR: CHOOSING A PROBE FOR IN SITU HYBRIDIZATION

The next step with in situ hybridization to discuss is the probe.

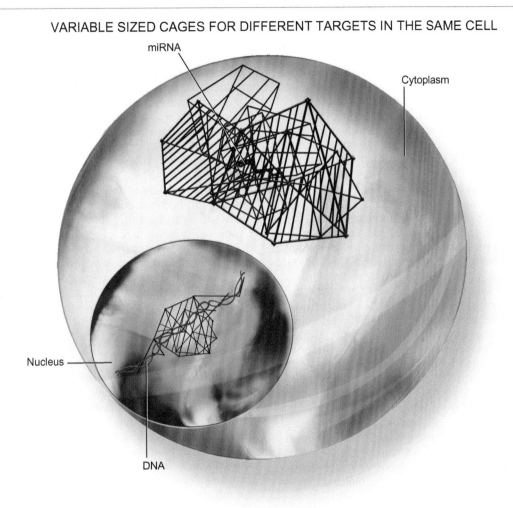

VARIABLE SIZED CAGES FOR DIFFERENT TARGETS IN THE SAME CELL

miRNA

Cytoplasm

Nucleus

DNA

Figure4-26 Graphic representation that variably sized cages may surround different RNA, DNA, and epitope targets in a given tissue. This figure represents the artist's representation that one way to explain the marked differences in optimal pretreatment conditions for different targets in the same cell is to hypothesize that they are surrounded by variably sized "cages." These "cages" would represent the different protein density that surrounds any DNA, RNA, or protein epitope in the living state after variable cross-linking with formalin.

PROBE TYPE (GENOMIC PROBE VERSUS OLIGOPROBE VERSUS LNA PROBE)

In the 1990s and the first decade of the 2000s, most people who did in situ hybridization made their own probes. This necessitated cloning the DNA or RNA (cDNA) sequence of interest, growing it in a plasmid so as to get many micrograms of the cloned DNA, purifying it, then labeling the DNA with a reporter nucleotide, and purifying it again. It was very important to clone fragments that were at least 100 nucleotides in size. Indeed, technicians typically tried to clone fragments of at least 1000 nucleotides in size. The reason was that these large DNA or cDNA sequences were optimal for labeling probes with either one of the two standard methods: nick translation or random primers. Nick translation works much the same way the primer independent signal works with in situ PCR: DNase digestion creates small "nicks" via breakage of some phosphodiester bonds (also created by the high temperatures of embedding with formalin-fixed,

paraffin-embedded tissues) that a DNA polymerase will "see" and "repair." The DNA polymerase does the repair by using its own endonuclease activity to "cut away" one of the DNA strands at the nick, and then uses the remaining single strand as a template to make the new DNA. Random primer labeling of a probe works in a process at least tangentially related to PCR. A large array of "primers," each approximately 6 base pairs long, is added to denatured DNA. The sheer numbers of primers allows for some hybridization to the target, which then induces the polymerase to make double-stranded DNA using the target sequence as the template. With each process (random primers and nick translation), you add a reporter nucleotide to the reagents (usually digoxigenin dUTP or biotin- dUTP), which then serves as the label in the newly synthesized probe. On average, about 1 labeled nucleotide is included with every 20 nucleotides that are added to the growing DNA molecule.

Although I may be making this process sound like ancient history, nowadays it is unusual to clone the target

109

sequence in a plasmid or equivalent vehicle, grow it in culture, purify, and label the DNA for in situ hybridization. Rather, probes are usually made by a commercial source already labeled and ready to go.

Another issue at the early stages of in situ hybridization about 20 years ago was that oligoprobes were very difficult to use, because they rarely showed a strong signal but readily showed background. This is easily predictable from the melting curve dynamics evident with a small oligoprobe (about 20 nucleotides) and its target. Indeed, a major reason in situ PCR was invented and used a lot in the 1990s and early 2000s was that technicians had to amplify the DNA or RNA target in order to have a good chance of getting a strong signal with in situ hybridization when using a 20 nucleotide oligoprobe. Again, it all makes perfect sense today. If you are using a probe that can only hold one or a few reporter nucleotides that have a relatively low melting temperature, then the odds are stacked against you when doing in situ hybridization.

This problem was solved with the LNA modification. As we have discussed, the LNA modification nicely demonstrates how important the "breathability" of macromolecules is to successful in situ hybridization. Indeed, LNA probes contain either one or two reporter nucleotides. By putting in a bridge between oxygen 2 and carbon 4, and dramatically reducing the ability of the hybridized DNA sequence that has the LNA-modified nucleotides to separate from the target by basically not allowing the ribose sugar to "pivot" around the base, you increase the melting curve by about 6°C for every LNA-modified nucleotide! Since most LNA-modified 20 nucleotide oligoprobes have 6 LNA-modified nucleotides, this translates to an increase in the melting temperature (Tm) of about 36°C!

The major drawback of the LNA-modified labeled probes is their price. We discuss ways to deal with this in Chapter 7 by simply using the lowest possible concentration of the LNA probe when doing in situ hybridization, and by minimizing the amount of probe cocktail that we use.

Of course, if you have access to large amounts of cloned target sequences that are 1000 base pairs in size or larger, labeling such target sequences via a kit is very simple. In that case, these so-called genomic probes or full-length probes do work very well for in situ hybridization, as well as any LNA probe. The one recommendation I make is *not* to use a small (e.g., 20 nucleotides) oligoprobe that is not LNA modified, because such probes strongly stack the deck against you to get some good in situ hybridization data.

LABELING THE PROBE: CHOOSING A REPORTER NUCLEOTIDE

The three most commonly used reporter nucleotides are *biotin, digoxigenin,* and *fluorescein.* Of course, biotin necessitates the use of a streptavidin conjugate. The avidity of the binding between avidin and biotin is among the strongest bonds in all of molecular biology; this is how streptavidin got its name. Digoxigenin requires the use of conjugated antidigoxigenin antibody. Fluorescein tags can be viewed directly if your laboratory has an expensive dark field microscope. In my experience, I get a much

better signal-to-background ratio with fluorescein probes by using a conjugated antifluorescein antibody, which leads to a color-based signal evident with any microscope.

The main advantage of biotin-labeled probes is that streptavidin-reporter enzyme complexes are readily available for in situ hybridization. Also, the streptavidin conjugates are very stable and do not have to be made fresh. The main disadvantage is that some tissues, such as liver, may contain endogenous biotin. In my experience, this is usually not a large problem.

The main advantage of digoxigenin-labeled probes is that digoxigenin is a plant-related chemical and is not found in mammalian tissues. The main disadvantage is that the antibody solution should be made fresh prior to the in situ hybridization reaction.

The main advantage of fluorescein-labeled probes is that you have two separate ways to see the signal: directly with dark field microscopy or indirectly by using an antifluorescein-reporter enzyme complex. The main disadvantages are that, when using a dark field microscope to view the data, you may encounter background problems and the signal does not last very long.

THE PROBE COCKTAIL

I have tried pretty much all the probe diluents that are commercially available. They all work well. They contain various mixtures of salt, dextran sulfate, and, at times, formamide. The dextran sulfate works as a "probe displacer." That is, it effectively raises the concentration of the probe in the aqueous phase due to its ability not to absorb water (or the probe suspended in the water). The dextran sulfate also aids in preventing the probe from drying out during the denaturation and hybridization steps. The formamide (when present) and salt, of course, are added to directly impact the melting curve of the probe/target hybridized complex, as we discussed in Chapter 2.

My recommendation is to purchase an in situ probe diluent and use it exclusively. In my work, I have gotten consistent good results with the probe diluents from Enzo Life Sciences and Exiqon.

AGE OF THE PROBE

Like many pathologists, I tend to keep "things" for many years. This does not just include slides. In the refrigerator, I have stored probes (mostly HPV) that are between 10 and 20 years old. I do this to allow me to easily address these questions:

1. Do DNA probes tend to degrade over time?
2. If they do degrade, what is the cut-off point in years when we can expect that the HPV probe is no longer active?

I should note that the probes are stored at 4°C. They include biotin-, digoxigenin-, and fluorescein-labeled probes. They are all "genomic probes" that were made via nick translation or random primers from the 8000 base pair full HPV genome.

As we discussed at length in the preceding chapter, we know that the HPV DNA in formalin-fixed, paraffin-embedded tissues degrades over time. It degrades

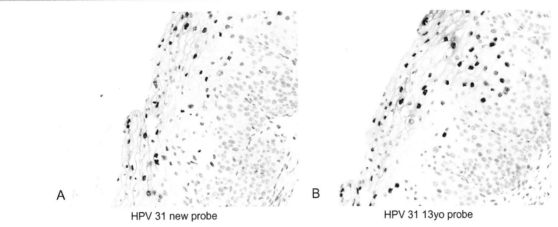

A   HPV 31 new probe   B   HPV 31 13yo probe

Figure 4-27 The effect of the age of the probe on the signal with in situ hybridization. Panel A shows a strong in situ hybridization signal for HPV 31 in this CIN 1 lesion. The probe was purchased right before the in situ hybridization was done. A serial section was analyzed with an HPV 31 probe that was bought 13 years ago. The signal with this old probe is still excellent (panel B).

to the point that if the block is 10+ years old, then the HPV DNA has degraded from its original size of 8000 base pairs to fragments of around 100 base pairs in size. Thus, I was expecting the HPV probes to do the same. However, look at the data in Figure 4-27. I compared the intensity of the HPV DNA in situ hybridization signal for a CIN 1 lesion using probes that were either 13 years old or about 1 year old. As is evident, there is no difference in the signal! This is the consistent result I obtained when comparing old probes (5 years old or more) to probes that were made within the past year.

I must admit to some frustration when I am using a commercial probe on an automated platform and I get the error message "the probe has expired." This is for a probe that is usually 6 months old! I have used such probes with a manual in situ hybridization protocol and consistently got excellent results.

Of course, the data shown in Figure 4-27 simply show that the HPV probe is still able to generate a strong signal with in situ hybridization. It does not prove that the HPV DNA probe did not degrade over time. Still, if the probe had degenerated to a size of about 30 base pairs, then it would not generate a good signal with in situ hybridization. Thus, even if the probe had partly degraded, the labeled fragments must still be at least 50 base pairs or larger in size.

The final point to make is that the lack of apparent degradation of the HPV probes over time is in sharp contrast to the marked degradation of the HPV DNA in the formalin-fixed, paraffin-embedded tissues. This again raises the possibility that the DNA nicks due to the high temperatures that the tissue is exposed to during the paraffin embedding may precipitate the long-term, continued slow degradation of the macromolecule over time.

PROBE CONCENTRATION

Of all the potential issues with the probe when doing in situ hybridization, by far the most important is the probe *concentration*. The probe has two binding choices when

we perform in situ hybridization: it can bind to the target sequence, or it can bind nonspecifically to other macromolecules. Both binding possibilities will be positively correlated to the concentration of the probe. Of course, other variables such as protease digestion, and the degree of formalin fixation will also be factors that can influence both signal and background. I cannot understate the importance of probe concentration to background. I consider it the "first suspect" when either our laboratory has a background issue in an experiment or if another laboratory comes to us for help with a background problem.

Since probe concentration is such a key variable for signal and background, I perform many experiments in which I plot signal as a function of probe concentration. In such experiments I typically exploit the fact that the target I am looking for is specific for a certain cell type. Hence, a colorimetric signal in those cells is presumed to be signal. Figure 4-28 shows some representative data from such experiments using HPV DNA as the assay system. Note that there is a strong relationship between probe concentration and hybridization signal intensity *relative* to the pretreatment regime. This certainly makes sense because the optimal pretreatment regime—in this case, protease digestion—should allow better diffusion of the probe and, thus, augment the signal for the lower concentrations of the probe.

In summary, probe concentration is a critical variable for both signal and background with in situ hybridization. Clearly, our goal is to find the lowest concentration that will yield the strongest signal, which, thus, would minimize the risk of background. Different tissues, even if they look histologically very similar, may require different probe concentrations for an optimal signal. Also, for a given tissue, the optimal probe concentration may change depending on the pretreatment regime (no pretreatment versus protease digestion versus DNA or RNA retrieval). Hence, as we discuss in detail in Chapter 7, I strongly recommend that you use two different and disparate probe concentrations when testing a new tissue with in situ hybridization for either DNA or RNA targets.

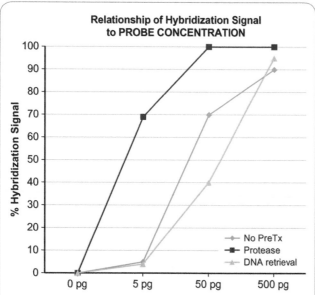

Figure4-28 Relationship of the probe concentration and the pretreatment conditions to the signal intensity with in situ hybridization. In these experiments, different concentrations of the HPV DNA probe were used and the hybridization signal was determined for no pretreatment (blue line), protease digestion (red line), and DNA retrieval (green line). As is evident, the protease digestion was associated with a stronger intensity of the signal for a given concentration of the probe, which is consistent with the hypothesis that the increased pore size facilitated the entry of the probe to the target.

STEP FIVE: DENATURATION AND HYBRIDIZATION OF THE PROBE

It was typical in the 1990s that colorimetric in situ hybridization protocols lasted for two days or more. The primary reason was that the probe and target DNA or, less commonly, RNA sequences were separately denatured and only after a long "pre-incubation" series of washes. Today it is uncommon to find an in situ hybridization protocol that lasts more than 1 day, and many of the automated platforms can be finished in a few hours. One major reason for the much more rapid in situ hybridization protocols is that it is now routine to co-denature the probe and target nucleic acid sequence.

CO-DENATURATION OF PROBE AND DNA TARGET/POLYPROPYLENE COVERSLIPS

I much prefer polypropylene coverslips when doing in situ hybridization. Polypropylene is a clear plastic that usually comes in the form of a double-sided bag. It can be autoclaved. By using sterile gloves when cutting the plastic, you can use the inner part of the coverslip for RNA in situ hybridization and, of course, be assured that it has no fingerprints (and, thus, no exogenous RNase) that could ruin the reaction.

There is also a practical, economic-driven reason to use polypropylene coverslips for in situ hybridization. If you do in situ hybridization using an automated platform, the protocol will require a large amount of probe cocktail

no matter how large the tissue. Typically, this volume is 100 μl. However, by cutting the polypropylene coverslips to size (just slightly larger than the tissue), you can use much less volume of probe cocktail. Indeed, for biopsies that are really small, say 3 mm, I have successfully performed in situ hybridization on them using as little as 1 μl of probe cocktail! In most of the in situ hybridization experimentsId o,5–10 μl of probe cocktail is sufficient.

We use a denaturation temperature of 95°C–100°C when doing in situ hybridization for DNA targets. At these temperatures, the probe cocktail tends to form bubbles under the coverslip. If these are not removed, then "dead zones" will occur in the in situ hybridization reaction, which inevitably leads to background. These small bubbles can easily be removed by gently pushing them with a toothpick.

HYBRIDIZATION AND DIFFUSION TIME

After we have co-denatured the DNA or RNA probe and target, the labeled single-stranded probe can then begin to diffuse through the cell. We have theorized that the labeled probe will encounter many forces on the way that can impede its movement. These forces include hydrogen bonds and ionic bonds with specific R side chains in the cross-linked proteins, hydrophilic and hydrophobic pockets in the three-dimensional protein/protein cross-linked cage, and the pore size of this "cage" that surrounds the nucleic acid target sequence. If we recognize that some of these forces may actually assist the in situ hybridization reaction by allowing the probe and target sequence to more likely "dock" with each other, these forces clearly have the potential to slow down the diffusion of the probe into the hundreds of thousands of cells present in the typical formalin-fixed, paraffin-embedded tissue.

By plotting hybridization time versus the intensity of the in situ hybridization signal, we can get a good idea of the relative diffusion time of the probe to the target. Since we have predicted that probe diffusion is likely related to pore size, as well as other previously listed variables that all can be influenced by pretreatment regimes such as protease digestion and DNA retrieval, let's look at the "probe diffusion data" in relationship not just to hybridization time, but also the pretreatment regime. These data arep resentedin Figures4-29th rough4-31 .

Note how dramatically different the data can be relative to the specific tissue. Figure 4-29 shows the hybridization kinetics for a weakly fixed tissue, which I am defining as a tissue in which protease digestion, no pretreatment, and DNA retrieval gave comparable results for in situ hybridization. Note that the diffusion kinetics are nearly identical whether the HPV positive tissue was not pretreated, protease digested, or subject to DNA retrieval (these data not shown). This suggests that the pore size (and other variables) are already near optimal after fixation, and that neither protease digestion nor heating the tissue in solution at 95°C for 30 minutes was sufficient to have a negative effect on the in situ hybridization reaction. This is the reason I refer to the tissue as a "weakly" fixed tissue.

Compare these data to the kinetics of an HPV positive tissue that requires protease digestion for an optimal

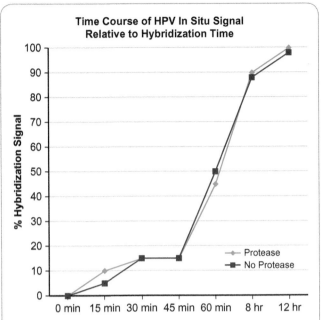

Figure 4-29 Relationship of the pretreatment conditions to the signal intensity with in situ hybridization relative to hybridization time: weakly fixed tissues. In these experiments, the HPV DNA hybridization signal intensity was equivalent with increasing hybridization times whether there was no pretreatment (red line) or protease digestion (blue line). This suggests that pore size was already sufficient for optimal diffusioni nt hisb iopsy.

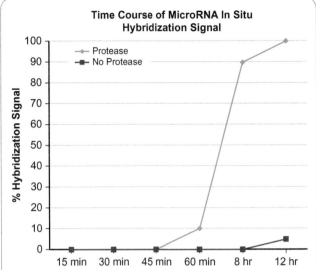

Figure 4-31 Relationship of the pretreatment conditions to the signal intensity with microRNA in situ hybridization relative to the hybridization time: strongly fixed tissues. In these experiments, the microRNA hybridization signal intensity was strongest as the hybridization time was increased when protease digestion (blue line) was used, whereas basically no signal was seen if protease digestion was omitted (red line). This suggests that pore size was much too small for optimal diffusion in this biopsy and needed strong protease digestion to maximize.

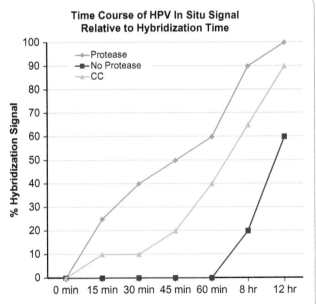

Figure 4-30 Relationship of the pretreatment conditions to the signal intensity with in situ hybridization relative to hybridization times: intermediate fixed tissues. In these experiments, the HPV DNA hybridization signal intensity was strongest as the hybridization time was increased when protease digestion (blue line) was used and intermediate for DNA retrieval (CC, green line). This suggests that pore size was too small for optimal diffusion in this biopsy and needed either protease digestion or DNA retrieval to maximize.

signal. These data are presented in Figure 4-30. Note here that increased hybridization time for the tissue section not pretreated did lead to an increased signal, albeit nowhere near 100%. Also note that DNA retrieval enhanced the signal, but not quite to the level of protease digestion. This lets us theorize that this formalin-fixed, paraffin-embedded tissue is a bit more strongly fixed than the tissue illustrated in Figure 4-29. Finally, look at Figure 4-31. This is for microRNA in situ hybridization. Here, basically no signal is ever evident if there was no pretreatment, even after 12 hours of hybridization! This represents a "strongly fixed tissue." These data suggest that the pore size for microRNA targets may be much smaller than the pore size for HPV DNA targets in the same tissue. Certainly, I have not seen a microRNA that can be detected optimally in situ with no pretreatment.

Another simple but important point comes from analyzing all of the preceding "kinetics of hybridization data" results. That point is that for optimal pretreatment, we note that 50% of the signal is evident after about 1 hour of hybridization, and that 95%+ of the signal requires about 10 hours of hybridization. So, yes, if you want to do very short hybridization times of 30 minutes or so, it will work *assuming every other variable is optimized.* However, so that you are able to maximize the signal, the hybridization time should go overnight. This is why my protocol calls for overnight hybridization.

Now let's compare the HPV DNA in situ hybridization kinetics data to that for microRNA in situ hybridization. These data, presented in Figure 4-32, are for a tissue that required protease digestion to generate the maximum signal and used the same strong protease digestion times.

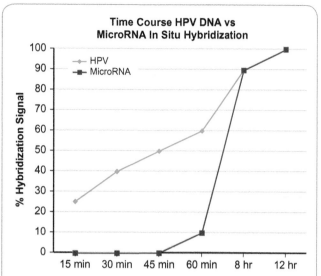

Figure4-32 Comparison of the hybridization time course for HPV DNA versus microRNA in the same cells. The signal for the HPV DNA appears more rapidly than the signal for microRNA analyzed in the same tissue, although after 8 hours the two signals became equally intense. This suggests that there are more barriers to diffusion for the microRNA probe than the HPV probe.

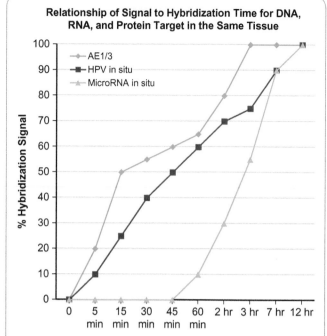

Figure4-33 Comparison of the hybridization time course for HPV DNA versus microRNA and cytokeratin in the same cells. The signal for the HPV DNA and cytokeratin appeared more rapidly than the signal for microRNA analyzed in the same tissue, although over time the three signals became equally intense. This suggests that there were more barriers to diffusion for the microRNA probe than the other targets, even though the microRNA and keratin were in the same cellular compartment.

Note that the diffusion time for the microRNA is longer in the same cells when compared to the diffusion time for the HPV DNA target. This may reflect the fact that microRNAs tend to be tightly bound to a variety of proteins that are part of the RISC (RNA-induced silencing complex).

Although I was not surprised there was a difference in the hybridization kinetics for the RNA and DNA targets, I was surprised that the difference was so dramatic. More to the point, why would the cytoplasm-based microRNA have a hybridization kinetic curve that was "much further to the right" than the nucleus-based HPV DNA? Obviously, there is less distance to traverse for the probe to get to the cytoplasm versus the nucleus, so why is the microRNA signal so much slower to develop?

I analyzed more tissues for these two targets, and the results were equivalent. More such data are presented in Figure 4-33, but here I added the hybridization kinetics of a cytoplasmic protein AE 1/3, since it was in the same cellular compartment as the microRNA.

Clearly, HPV DNA and cytokeratin show nearly identical hybridization kinetics, whereas the time course for the signal for the microRNA is much slower. Why?

I think the answer lies with the three-dimensional protein/protein cross-linked network. All we need to postulate is that the protein cage around the microRNA is much more "dense" than the cages around the cytokeratin and HPV DNA, and these data make sense.

There is another way to look at these interesting, if not a bit puzzling, data. In the living cell, which of the two situations is more likely?

1. The different RNA and DNA molecules in the cell are surrounded by basically the same density of proteins, which show equivalent ionic and hydrogen bondp otential.
2. The different RNA and DNA molecules in the cell are surrounded by a markedly variable density of proteins, which show much disparity in their ionic and hydrogen bond potential.

If we accept possibility #2 as the more likely hypothesis, then the data we have just seen (such as in Figures 4-32 and 4-33) not only make sense, but are predictable. Also, it would explain why tissues differ so much in the optimal conditions to detect a given DNA, RNA, or protein, since the "cross-linked" cage around the specific target probably will differ from tissue to tissue. This point clearly is important to our understanding of why the relationship between pretreatment and various targets is so variable.

IMPORTANCE OF THE COPY NUMBER OF THE TARGET TO THE SIGNAL INTENSITY WITH IN SITU HYBRIDIZATION

The last point to discuss relative to the hybridization kinetics is the copy number. As with any diffusion-based experiment, the higher the amount of the target, then the quicker you will reach 100% of the hybridization signal and the stronger that signal will be. This is easy to see any time you do an experiment that involves HPV in situ hybridization. As we discussed in the preceding chapter, you will see some cells with intense signal, whereas others right next door show weak to no signal. The reason is that HPV DNA copy number varies considerably in a given lesion. It tends to be much greater in the koilocytes that reach the surface, which, clearly, will facilitate the spread of the virus via sexual contact.

The practical point of the positive correlation between copy number and the rapidity and strength of the hybridization signal is that high copy events can be seen after only 30 minutes or so of hybridization *but low copy events cannot be reliably visualized* after such a short hybridization time. This is another reason I prefer overnight hybridization times.

STEP SIX: THE STRINGENT WASH AND IN SITU HYBRIDIZATION

A review of the key elements of the melting curve reminds us that the temperature, formamide concentration, and salt concentration of the stringent wash will have an effect on our signal. This is certainly true. However, it may surprise you to hear that not one of these variables is as important as another component of our stringent wash!

ROLE OF SALT CONCENTRATION AND FORMAMIDE IN THE STRINGENT WASH

As indicated before, I do not use formamide in the stringent wash. It is a toxic compound that has carcinogenic properties. Although it can help reduce background, there are other ways (such as the probe concentration!) that let us adequately control background. As far as the salt concentration, I prefer to keep it constant at 15 microM.

IMPORTANCE OF THE TEMPERATURE TO THE STRINGENT WASH

Temperature is an important variable with the stringent wash. The standard is to consider room temperature as a low stringent temperature, 45°C as an intermediate temperature, and 60°C–70°C as a high stringent temperature. This is certainly a reasonable approach. However, it has been my experience that when keeping the salt concentration constant and not using formamide, varying the stringent wash from high to low stringency has relatively little effect if you are using a "full-length" probe (of 50–100 base pairs) or an LNA probe. Such varying stringencies do matter if you are using an oligoprobe with no LNA modification, as these probes are very sensitive to relatively slight changes in the stringent wash. However, since I strongly recommend that you do *not* use such probes, let me end by saying that my recommendation is to use a 45°C wash for the in situ hybridization reactions, realizing that lower or higher temperatures will probably yield similar results if you follow the other key points of the protocols outlined in the later chapters.

IMPORTANCE OF BOVINE SERUM ALBUMIN

In my experience, the most important chemical of the stringent wash is bovine serum albumin (BSA). BSA's function with in situ hybridization,

immunohistochemistry and PCR is to coat the glass and plastic surfaces, which, in turn, minimizes the adsorption of the probe/antibody/Taq polymerase, respectively, to the glass or plastic. The BSA also coats the surface of the formalin-fixed, paraffin-embedded tissues and, thus, minimizes the nonspecific absorption of the detection reagents on the tissue. The reason I say "coats the tissues" is that we routinely incubate the tissues in the stringent wash solution that contains BSA for 10 minutes. Ten minutes is way too short a period of time for the BSA to diffuse into the tissue to any degree, but is sufficient for it to coat the surface of the tissue as well as the glass and the plastic.

The results are impressive if you omit BSA from the stringent wash. They are shown in Figure 4-34. Although we have discussed this point before, I think it bears repeating. Note the horrific background in Figure 4-34 if the BSA is omitted from the wash! This is why I say that BSA is the most important chemical in the stringent wash for in situ hybridization (as well as for immunohistochemistry)!

POSSIBILITY OF LOSS OF SIGNAL WITH STRINGENT WASH: BACTERIAL CONTAMINATION

I have had several investigators come to me for help with background issues in which altering the probe (or primary antibody) concentration made no difference. I was able to visit one of the laboratories. It became clear that the stringent wash solution, which contained 2% BSA, was being stored at room temperature for extended periods of time. Of course, BSA is an excellent growth medium for bacteria. A review of a drop of the solution under the microscope revealed many bacteria. Keeping all the reagents the same and preparing a new solution of the stringent wash solution revealed a robust signal. Thus, the bacteria in the stringent wash were simply inactivating the alkaline phosphatase conjugate, with the end result that the signal was lost. Realizing this is a rare event, it does remind us that we need to take precautions against bacterial overgrowth when using solutions that contain BSA.

STEP SEVEN: THE DETECTION STEP OF IN SITU HYBRIDIZATION SIGNAL

At this stage, we have formalin-fixed, paraffin-embedded tissues in which some cells have the target of interest. A labeled probe is now attached to the target. That label is typically biotin, digoxigenin, or fluorescein. The next step is basically identical for immunohistochemistry and in situ hybridization. We need to add a chemical that will attach to the reporter label. This chemical also needs to be conjugated to the enzyme that will create the precipitate that will allow us to see the target under the microscope.

Let's now examine the time kinetics of perhaps the most common conjugate: streptavidin-alkaline phosphatase. You may recall that the affinity of streptavidin and biotin is among the strongest in molecular pathology. Thus, when streptavidin "finds" the biotin on the probe, it immediately and strongly binds to it. So, the key

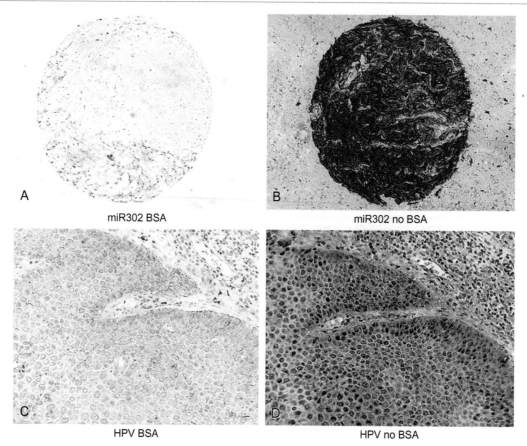

A   miR302 BSA

B   miR302 no BSA

C   HPV BSA

D   HPV no BSA

Figure4-34   Importance of BSA to the background with in situ hybridization. In these experiments analyzing microRNA-302 and HPV 6 by in situ hybridization, little to no signal was evident and was expected based on the histologic features of the tissues if BSA was included in the stringent wash (panels A and C, respectively). However, if all other conditions were identical but BSA omitted, then very high background ensued (panels B and D, respectively).

to the time kinetics of streptavidin-biotin binding will be the rate of diffusion of the conjugate to the target/probe complex.

I did a series of experiments examining the time course of streptavidin-alkaline phosphatase binding to biotin-labeled HPV probes in CIN/condylomatous lesions. If our theory of the three-dimensional macromolecule cross-linked cage is correct, then we will note different diffusion times for the streptavidin-alkaline phosphatase conjugate when comparing various tissues. Representative data are presentedin Figure4-35 .

Clearly, there is a disparity in the rate of diffusion of the streptavidin-alkaline phosphatase conjugate in these three representative tissues. This disparity parallels the variability in the rate of diffusion of basically every step of in situ hybridization (or immunohistochemistry), as we saw in some of the previous figures. I find these data very interesting, as they underscore that formalin-fixed, paraffin-embedded tissues are both variable and dynamic. The corollary is that, by manipulating a few variables, we can dramatically increase the chances of both successful in situ hybridization or immunohistochemistry, as well as successful co-expression analysis.

It is also clear from Figure 4-35 that 30–45 minutes is sufficient to generate maximum diffusion of the

streptavidin-alkaline phosphatase conjugate in cases 1 and 3. However, some tissues, such as those in case 2, may require much more time for the streptavidin-alkaline phosphatase conjugate to reach near 100% access to the biotin. The latter likely represents less than optimal pretreatment of the tissues that can occur with the "strongly fixed tissues," and underscores again the importance of doing various disparate pretreatments on each new biopsy we test to increase the chances that one of these pretreatments will allow maximum diffusion of our reagents into the fixed cells.

BIOTIN-BASED

In about 45% of my in situ hybridization experiments, I use a biotin-based system. In another 45%, I use digoxigenin-tagged probes. In the remainder of my in situ hybridization experiments, I use fluorescein-labeled probes. This is probably quite typical of many in situ hybridization laboratories. Certainly, the different reporter nucleotides give equivalent performances with in situ hybridization, both in terms of signal and background.

Biotin was the first reporter nucleotide that was used wide-scale with in situ hybridization. Initially, its main

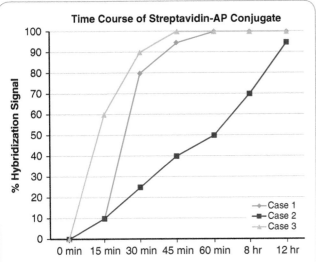

Figure4-35   The time course of streptavidin/alkaline phosphatase diffusion can vary with different tissues. In these experiments, the same pretreatment regime was used. The incubation time in the streptavidin-alkaline phosphatase conjugate was altered to see how these different time points affected the strength of the hybridization signal. As is evident, it appears that the streptavidin-alkaline phosphatase conjugate diffused much more slowly within the cells of case 2, whereas the other two cases showed a maximum signal for the HPV DNA after about 45 minutes incubation.

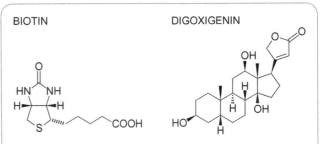

Figure4-36   The structures of digoxigenin and biotin. These are each relatively small molecules that have little ability to enter into hydrogen or ionic bonds and, thus, should diffuse readily in formalin-fixed, paraffin-embedded tissues

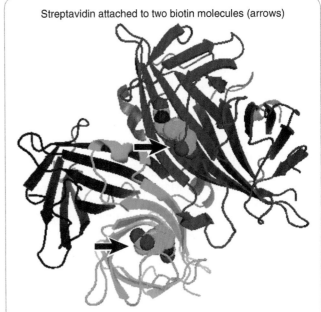

Figure4-37   The structure of streptavidin. Streptavidin is much larger than biotin or digoxigenin. However, and perhaps more importantly, since it is a protein, it will be able to interplay with the protein cage surrounding any given epitope/DNA/RNA target such that ionic and hydrogen bond potential, hydrophobic/philic pockets, and pore size may all influence its diffusibility in the formalin-fixed, paraffin-embedded tissues.

competitor was the radioactive isotopes that were attached to a nucleotide. However, as early as the early 1990s, laboratories were reporting that a biotin-based in situ hybridization system could perform as well as one that used ^{3}H or [35]S. There have been many advances with the biotin-based in situ hybridization system over the past 25 years. Thus, in my experience, I do not use radioactive isotopes any more for in situ hybridization. This is not to say that radioactive isotopes cannot be useful for in situ hybridization. For one, they give excellent subcellular localization although biotin-labeled probes, especially if the signal has been optimized, are pretty good in this regard. Radioactive probes also offer a logical system to use with the colorimetric system for co-expression analyses. I do not discuss radioactive isotopes elsewhere in this book. If you want more information, I strongly recommend the December 2010 issue of *Methods*, which is devoted to in situ hybridization. There are several excellent review articles about radioactive in situ hybridization in that issue.

As previously indicated, you can readily find streptavidin-alkaline phosphatase and streptavidin-peroxidase conjugates from various commercial laboratories. They come ready to use and, when stored at 4°C, will remain active for, in many cases, years.

DIGOXIGENIN-BASED

Since endogenous biotin can be present in certain tissues, such as the liver, there was a strong effort several decades ago to find an alternative to biotin-based in situ hybridization and immunohistochemistry. This search led to a reporter nucleotide that is found only in the plant

kingdom and, thus, could not per se cause background when dealing with formalin-fixed, paraffin-embedded mammalian tissues. This reporter nucleotide is digoxigenin. It is a steroid that is produced by the same genus of plants (*Digitalis*) from which the heart medicine digitalin is made. Indeed, the term "genin," at the end of digoxigenin, refers to the fact that digoxigenin is basically digitalis without the sugar part of the molecule.

The structures of biotin and digoxigenin are presented in Figure 4-36. My purpose of showing these structures is to remind you that these haptens are very small molecules, especially when considered in the context of the 8000 base pair HPV DNA genome or the 100 base pair biotin-labeled HPV DNA probe. They have little ionic or hydrogen bond potential. So they should readily diffuse in formalin-fixed, paraffin-embedded tissues. To further underline this point,

Figure 4-37 shows the structure of streptavidin. Inside the streptavidin are two biotin molecules (arrows), which, again, gives us an idea of why we can use these chemicals and get such good results with in situ hybridization and immunohistochemistry. When we look at Figure 4-35, it is clear that the streptavidin-alkaline phosphatase conjugate (obviously much bigger than the streptavidin molecule alone) is able to readily diffuse into cells such that 30 minutes is sufficient in most tissues to get a maximum signal. Recall that the diffusion rate of small molecules in water is $10^{-6}\,cm^2/sec$. This translates to a diffusion rate of 10 microns in 16 minutes. Ten microns is the approximate size of many of the cells in formalin-fixed, paraffin-embedded tissues that we are examining with in situ hybridization. This would suggest that the pores are "wide open" and not hindering diffusion of our reagents during in situ hybridization and immunohistochemistry under optimal conditions. These calculations suggest to me that pore size is not the only explanation as far as the diffusion and retention of our reagents with in situ hybridization. That is, other factors such as ionic and hydrogen bonds, hydrophobic and hydrophilic areas, and breathability of the macromolecules we are working with may be as important or, in some cases, even more important than pore size.

Since digoxigenin is strongly immunogenic, it is a simple matter to generate an antibody against it. Digoxigenin can be covalently added to proteins or nucleic acids, which makes it very useful in in situ hybridization or immunohistochemistry. Many commercial laboratories make the antibody directed against digoxigenin that is conjugate either to peroxidase or alkaline phosphatase.

FLUORESCEIN-BASED

Fluorescein is the last of the commonly used reporter tags. Its structure is shown in Figure 4-38. Again, note the small size of this organic compound relative to many of the other molecules we routinely use when doing in situ hybridization or immunohistochemistry. It has an absorbance maximum at 494 nm and an emission maximum at 521 nm, which, of course, is responsible for the green-colored emission that we can see when using dark field microscopy. The fact that it produces various colors in solution is why we come into contact with the chemical in places other than the in situ hybridization or immunohistochemistry lab. It is a common food additive (D&C yellow #7 and #8).

As previously discussed, I, on occasion, will examine the signal due to a fluorescein-tagged nucleotide with a dark field camera. However, I prefer to use a color-based system when using a fluorescein tag for two reasons:

1. The fluorescein tag fades after a week or so, and the slide needs to be stored at 4°C; in comparison, the color-based precipitate will not fade, and the slide can be stored at room temperature.
2. The morphology is much better appreciated with a color-based in situ hybridization system than with a dark field image.

Thus, when I use a fluorescein-tagged nucleotide in the probe, I follow up the reaction with an antibody directed against fluorescein that is conjugated with either alkaline phosphatase or peroxidase.

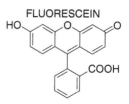

Figure 4-38 The structure of fluorescein. Like biotin and digoxigenin, fluorescein is a relatively small molecule that has little ability to enter into hydrogen or ionic bonds and, thus, should diffuse readily in formalin-fixed, paraffin-embedded tissues

DAB (3, 3′ - diaminobenzidine)

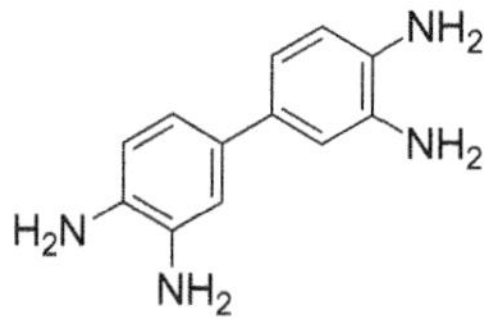

Figure 4-39 The structure of DAB (3, 3'- diaminobenzidine). Like biotin and digoxigenin, DAB is a relatively small molecule that has little ability to enter into ionic bonds, although it can form hydrogen bonds with our other key chemicals during in situhyb ridizationa ndi mmunohistochemistry.

REPORTER ENZYME

In situ hybridization and immunohistochemistry each commonly use one of two reporter enzymes: alkaline phosphatase or peroxidase. If you do a serial section analysis for a given DNA, RNA, or protein and use alkaline phosphatase in one reaction and peroxidase in the other, the results will be equivalent. The chromogen for peroxidase is DAB. DAB is short for 3,3'-diaminobenzidine. It is a known carcinogen and, thus, you need to wear gloves while using it. As seen in Figure 4-39, it is a small molecule. Peroxidase, in the presence of hydrogen peroxide, will react with DAB to form an insoluble brown precipitate. I have been impressed how permanent the precipitate is, in the sense that any attempt I have made to remove a DAB signal (prolonged acetone washing, acid alcohol treatment, boiling for 30 minutes in a stringent wash) did not recover the precipitate.

There are two common chromogens that are used with alkaline phosphatase. They are Fast Red and NBT/BCIP.

Nitro blue tetrazolium, or NBT, is a relatively large molecule (although still small compared to other reagents we are using with in situ hybridization). It is used in pathology for reasons other than in situ hybridization. Specifically, since it can serve as an oxidant, NBT has been used as a diagnostic test for phagocyte function. There are several diseases, including chronic granulomatous disease, in which the macrophages and neutrophils are unable to make the free radicals needed to kill the bacteria. Thus, the bacteria will grow in the cell, and the NBT test will be negative. As evident, the actual alkaline phosphatase substrate is BCIP. Alkaline phosphatase hydrolyses

NBT (nitro blue tetrazolium)

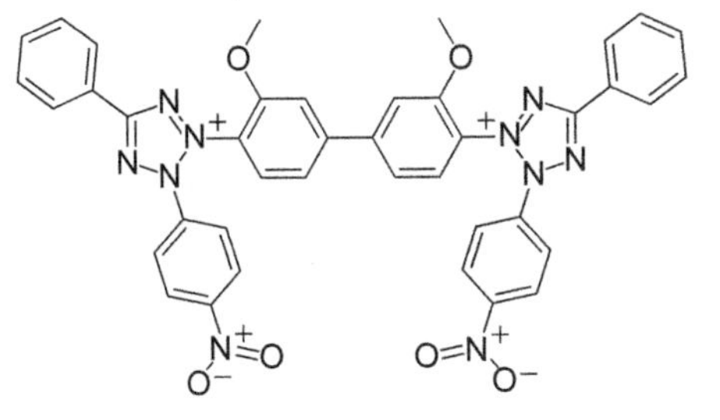

BCIP (5-bromo-4-chloro-3-indolyl phosphate)

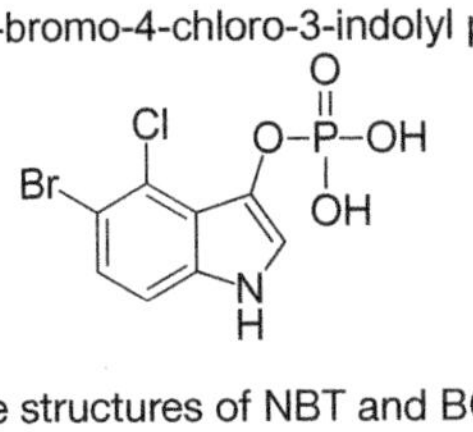

Figure 4-40 The structures of NBT and BCIP (nitro blue tetrazolium and 5-bromo-4-chloro-3-indolyl phosphate). Although NBT is relatively large, these compounds still are small enough as to most likely readily diffuse in the aqueous solution of formalin-fixed, paraffin-embedded tissues.

5-bromo-4-chloro-3-indolyl phosphate to 5-bromo-4-chloro-3-indole and inorganic phosphate. The 5-bromo-4-chloro-3-indole, in turn, is oxidized by the NBT to form an insoluble dark blue diformazan precipitate after reduction. As with the DAB precipitate, I have not been able to find any treatment that will remove the NBT/BCIP precipitate and still preserve the cell morphology. Of course, this is a good thing, as we do not want the in situ hybridization precipitate to fade over time. The structures of NBT and BCIP are presented in Figure 4-40.

There are other chemicals that can be used with alkaline phosphatase. The most common is called Fast Red. This chromogen used to be soluble in alcohol, requiring mounting in an aqueous media. Now, you can get Fast Red precipitates that are alcohol and xylene insoluble. Another relatively common chromogen is AEC (3-Amino-9-EthylCarbazole), which can be used with peroxidase. Yet other chromogens are available; which one(s) your laboratory will use is mostly a matter of choice, since they each give equivalent signals.

Since we will spend a lot of time on co-expression (including simultaneous detection of three targets), I prefer to use just three chromogens: NBT/BCIP, DAB, and Fast Red. This way, I can become very familiar with their nuances. Let's discuss various nuances and issues with these three chromogens. Again, this is not to say that other chromogens are not available. It is that, for this book, we present data just for NBT/BCIP, DAB, and Fast Red.

BACKGROUND ISSUES WITH THE DIFFERENT CHROMOGENS
DAB

The best known background issue with regards to the chromogen is with DAB. It has long been known that if you do not add an H_2O_2 blocking step prior to adding the primary antibody, background may well ensue. It has long been assumed that the reason for this background was endogenous peroxidase in the tissue.

Although we discussed DAB-related background in the preceding chapter, since this is an important point, I want to discuss the topic again. We did a series of experiments to try to better understand the basis for the background with DAB/peroxidase systems for in situ hybridization or immunohistochemistry. We examined placentas because they are rich in the cell type that commonly shows background with DAB/peroxidase: red blood cells. As is evident from Figure 4-41, when we test placenta with no primary antibody and no pretreatment, the red blood cells may show a dark brown color indicative of background. The fact that we see this background with no primary antibody tells us that it likely reflects some endogenous factor in the tissue. If this factor was endogenous peroxidase, then it should be easy to inactivate it with strong treatment with a protease or with incubation in formalin prior to the immunohistochemical reaction. However, as we can see in Figure 4-41, the background signal in the red blood cells was still present after the tissue was treated with a strong protease digestion followed by formalin post-fixation prior to immunohistochemistry without a primary antibody. This suggests that there was some chemical in red blood cells, not peroxidase, that was reacting with the exogenous peroxidase to produce the background. If we incubated the tissue in H_2O_2 prior to immunohistochemistry, the background was gone (panel D). Another important point evident in this figure is seen when these same experiments were done with a primary antibody—in this case, cytokeratin AE 1/3. This protein was present in high copy number in the trophoblasts. Note that with H_2O_2 pretreatment, we see a strong signal for the cytokeratin and no background. If the H_2O_2 pretreatment was omitted, the background was evident and there was a concomitant decrease in the signal. Reduced signal with high background is a common observation and may reflect, in part, the fact that the key reagents such as DAB are being used up rapidly in the background reaction, which, in turn, reduces their effective concentration for the signal.

Whatever the explanation, these data show that the use of H_2O_2 incubation prior to the primary antibody is a key step in the immunohistochemistry or in situ hybridization protocol using DAB/peroxidase. A trick of the trade that I use is to include 3% H_2O_2 in the probe cocktail/primary antibody and subsequent reagents to make certain that all background has been eliminated.

Another problem with DAB/peroxidase systems has nothing to do with background. Rather, it has to do with the fact that there are some pigments that are relatively common in cells that are brown. Specifically, both melanin and hemosiderin stain brown. Melanin will be evident as small brown granules that are found in benign melanocytes and in melanoma cells. Any skin biopsy will contain melanocytes; the darker colored the skin, the more melanin-producing cells you will see. A very common benign tumor of skin is a nevus, in which many benign melanocytes form the lesion. Hemosiderin is a much coarser brown pigment that comes from the breakdown of blood. Hence, it is common wherever a prior biopsy had been

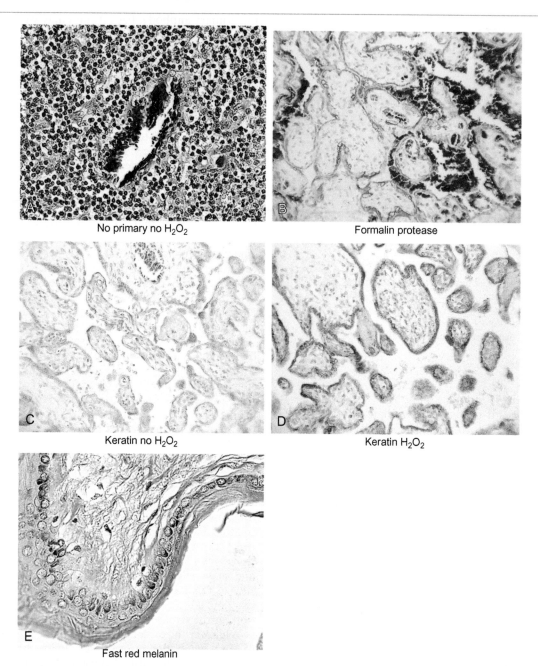

No primary no H₂O₂

Formalin protease

Keratin no H₂O₂

Keratin H₂O₂

Fast red melanin

Figure4-41   DAB background. DAB can cause background staining evident when no primary antibody or probe is used. Red blood cells are a common source of background (panel A). Protease digestion or post formalin fixation will not eliminate this DAB background (panel B), suggesting that it is not due to endogenous peroxidase activity. The background will also reduce the signal (panel C); these trophoblasts should be strongly positive for cytokeratin. The simplest way to eliminate this background is to use H_2O_2 incubation either prior to in situ hybridization/immunohistochemistry or, as I prefer, at each step of the process (panel D). Finally, pigments such as melanin stain darkly brown and preclude the use of DAB as a chromogen; rather, we can use Fast Red for the signal for keratin (panel E).

done, or in tumors, since they may show areas of hemorrhage. Hemosiderin is usually found in macrophages because they function as the primary phagocyte of the tissues. They are common in the lung in patients with chronic heart failure. The reason is that people with congestive heart failure get small hemorrhages in the capillaries of the terminal air sacs that release the red blood cell pigment. Indeed, the macrophages in the alveoli that contain hemosiderin are often called "cardiac macrophages."

If you are doing in situ hybridization or immunohistochemistry using tissues that contain melanin positive cells or hemosiderin, then DAB is to be avoided. Rather, you can use either NBT/BCIP or Fast Red/AEC (Figure 4-41, panel E).

DAB sometimes (thankfully rarely) will form a precipitate that persists through the coverslipping step of the in situ hybridization reaction. This precipitate is easy to recognize because the brown color is usually large

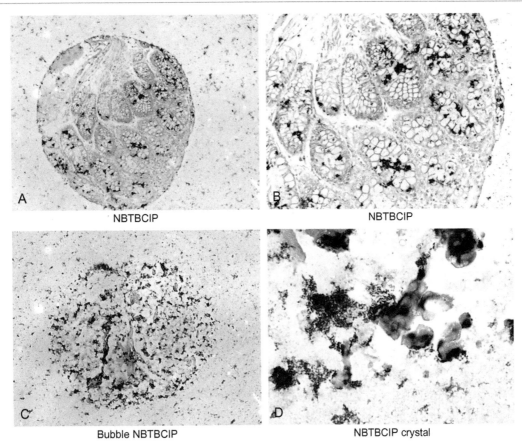

A — NBTBCIP
B — NBTBCIP
C — Bubble NBTBCIP
D — NBTBCIP crystal

Figure4-42   NBT/BCIP background. Panels A and B show the crystallization of NBT/BCIP over the tissue of this TMA (tissue microarray) core biopsy. This could have been prevented by agitating the slide as soon as the precipitate became noticeable. Panels C and D show what may be called a "bubble" effect. This is where, after the probe was denatured with in situ hybridization, a small bubble appeared under the plastic coverslip. Such bubbles cause the probe to dry and stick to the tissue, serving as the nidus of this NBT/BCIP precipitation.

and extracellular. This can usually be removed by rinsing the slides for a minute in acetone prior to doing the counterstain.

NBT/BCIP

Since NBT/BCIP signals are based in alkaline phosphatase, it was assumed that tissues that possess a lot of alkaline phosphatase (such as the placenta, small bowel, colon, and kidney) are more prone to background when using NBT/BCIP. Certainly, these tissues are prone to background when using NBT/BCIP. This type of background is easy to recognize, as it stains just the very surface (called the brush border) of the epithelial cells in the bowel and kidney tubular cells as well as the trophoblasts. There is a drug that can inhibit the activity of alkaline phosphatase that has been used to reduce this background. It is called levamisole and has been used for many years as an antiworm agent in animals; it is also commonly added to cocaine. However, the forms of alkaline phosphatase in the bowel and trophoblasts are not susceptible to levamisole inhibition. More importantly, in my experience, this type of background can be eliminated simply by adjusting the pretreatment regimes and the probe concentration.

Crystallization of NBT/BCIP, in my experience, is more common a problem than for DAB. There are two keys ways to reduce the NBT/BCIP crystals when doing in situ hybridization:

1. Add the NBT/BCIP to the high pH solution needed for maximum alkaline phosphatase activity only after the solution is brought to 37°C. Certainly, do not add the NBT/BCIP to cold pH9.5 alkaline phosphatase detection solution.
2. Use the lowest possible concentration of NBT/BCIP.

I ran a series of experiments in which I purposely generated NBT/BCIP crystals. I did this by purposely using too high a concentration of the NBT/BCIP (350 μg/ml of NBT and 175 μg/ml of BCIP). At the same time, I used too high a concentration of the probe (1 μg/ml). A concentration of probe this high increased the chances that the probe would nonspecifically bind to the glass and tissue surface and, thus, serve as a nidus of extracellular NBT/BCIP precipitation. Representative data are provided in Figure 4-42. Note the large, coarse nature of the crystals. Note in the very low magnification that the nonspecific precipitate of NBT/BCIP is much more likely to occur over tissue than over glass. Importantly, see how the nonspecific precipitate is not present in the nucleus, cytoplasm, or extracellular space, but rather seems to be located a little bit above the plane of the tissue. This is a good indicator of background. Finally, note the rounded appearance of the NBT/BCIP crystals in panel C. The reason this pattern occurred is that I intentionally let a bubble form and stay over the tissue after the denaturation step. The NBT/BCIP in this microbubble precipitated after the solution dried out.

By using the lowest possible concentration of NBT/BCIP (in my experience 200–250 µg/ml of NBT and 100–125 µg/ml of BCIP) and adding it to the high pH solution at 37°C, and by using the correct probe concentration and pretreatment conditions, you should not have much of a problem with NBT/BCIP crystals. They can, like DAB precipitates that occur on top of the tissue or glass, usually be removed by a brief dip in acetone. Still, the trick of the trade is to very carefully monitor any NBT/BCIP reaction over the first few minutes. If you see crystals forming, then agitation of the slide (moving it up and down in the chromogen) should eliminate the crystals.

FAST RED

Fast Red typically produces background in red blood cells and in the lining of trophoblasts. Of course, the red blood cell background is more typical of peroxidase systems and the trophoblast background more common of alkaline phosphatase systems. This is another reason why I do not think the basis of background is residual endogenous peroxidase or alkaline phosphatase activity. Of course, the other reason is that we cannot eliminate background with strong protease digestion and/or post-formalin fixation, which certainly should be able to inactivate any enzyme. Whatever the explanation, any such background with Fast Red can usually be eliminated with a reduction of the probe concentration or adjustment of the pretreatment regime.

STEP EIGHT: THE COUNTERSTAIN FOR IN SITU HYBRIDIZATION SIGNAL

The final point to discuss in this chapter is the counterstain. For Fast Red reactions, you can choose between no counterstain and hematoxylin. For DAB reactions, you can choose between hematoxylin, Nuclear Fast Red, and no counterstain. For NBT/BCIP reactions, the choice is between no counterstain and Nuclear Fast Red.

NO COUNTERSTAIN

The value of doing no counterstain is two-fold:

1. In general, you can see the signal more readily.
2. Weak nuclear-based signals that might be masked by a counterstain may now be appreciated.

Of course, the negative part of not using a counterstain is that it makes it difficult to see the morphology of the positive and negative cells in general, and the overall tissue in particular. A trick of the trade when using no counterstain is to allow the colorimetric reaction to proceed a bit longer than usual, because this will tend to give

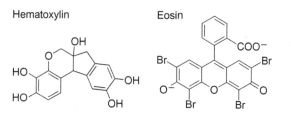

Figure 4-43 The structures of hematoxylin and eosin. The structures of hematoxylin (which binds avidly and rapidly to nucleic acids) and eosin (which binds avidly and rapidly to proteins) is presented. The signal from each stain, unlike the precipitates from DAB and NBT/BCIP, are easily removed from the cells and will not interfere with in situ hybridization or immunohistochemistry.

the entire tissue a weak color that allows you to better appreciate the histologic detail.

In my experience, I routinely omit the counterstain when working with a nuclear-based target. I then examine the data under the microscope "wet" before the counterstain/coverslip step. I then decide whether to use a very weak hematoxylin or Nuclear Fast Red counterstain (a few dips) or not.

HEMATOXYLIN-BASED COUNTERSTAIN

Hematoxylin is a small organic compound extracted from a tree that forms strong complexes with certain metal ions, such as iron. Its structure is shown in Figure 4-43. These complexes strongly bind to DNA and RNA, which indicates that it has some carcinogenic properties. When you use a full-strength solution of hematoxylin, the avidity for nucleic acids is so strong that even 10 seconds of exposure will stain the nuclei a light blue. To get the maximum blue color in the nuclei, you need to incubate the tissue in the hematoxylin for about 2 minutes.

Of course, you cannot use the blue color of the hematoxylin with NBT/BCIP, since the precipitate is the same color blue. You need to use Nuclear Fast Red as the nuclear counterstain if using NBT/BCIP as the chromogen. Since hematoxylin basically stains only the nuclei, it can be used at maximum intensity if the signal is cytoplasmic based.

NUCLEAR FAST RED COUNTERSTAIN

Nuclear Fast Red, as it name implies, stains nuclei pink/red and does not stain the cytoplasm. It can be used with either DAB/NBT and BCIP signals but not, of course, with Fast Red chromogen-stained tissue. If the NBT/BCIP or DAB signal is cytoplasmic, I like to use the maximum intensity of Nuclear Fast Red colorization (about 5 minutes with the full-strength solution) in order to get the most complete image of the tissue as possible. If the signal is nuclear based, then I prefer to obtain a much weaker nuclear counterstain so as not to obscure the signal.

EOSIN AND CYTOPLASMIC COUNTERSTAINING

Clearly, hematoxylin and Nuclear Fast Red stain, for all intents and purposes, just the nucleus. What if we have a nuclear signal and we want to stain the cytoplasm? In this case, use eosin.

Interestingly, the base compound of eosin is fluorescein (Figure 4-43). Specifically, it results from the action of bromine on fluorescein. It will stain any protein an intense pink. The more concentrated the protein, the stronger the pink signal. Hence, red blood cells, muscle fibers, and keratin (as three examples) show an intense pink cytoplasmic stain with eosin. Cells with less concentrated protein in their cytoplasm (for example, noncommitted lymphocytes) will show either a weak pink stain of the cytoplasm or no stain at all.

Considering that many of the reagents with in situ hybridization required an optimal diffusion time of 30–60 minutes, it is amazing to consider that 10–20 seconds of exposure to eosin can produce a strong cytoplasmic

pink color in the cells that have a high protein content in their cytoplasm. On the other hand, the rapid staining of cytoplasmic proteins by eosin does help us understand, to some degree, how a chemical such as proteinase K, which is not very large but, of course, larger than eosin, is able to have clear-cut effects on the in situ hybridization results after only 10–20 seconds of incubation.

The suggested readings [1–361] are provided to give a relatively broad overview of the utility of in situ hybridization in both the diagnostic and research fields. Not surprisingly perhaps, there is a focus on the detection of viral nucleic acids. Plus, to continue to emphasize the overlap between in situ hybridization and immunohistochemistry, many of the articles describe both in situ hybridization and immunohistochemistry analysis.

SUGGESTED READINGS

References on in situ hybridization

[1] Akao I, Sato Y, Mukai K,e tal. Detection of Epstein-Barr virus DNA in formalin-fixed paraffin-embedded tissue of nasopharyngeal carcinoma using polymerase chain reaction andin situ h ybridization. Laryngoscope 1991;101:279–83.

[2] Aksamit A, Mourrain P, Sever J, Major E. Progressive multifocal leukoencephalopathy: investigation of three cases using in situ hybridization with JC virus biotinylated DNA probe. AnnNe urol 1985;18:490–6.

[3] Angerer LM, Angerer RC. Localizationo f mRNAsb y in si tuh ybridization. MethodsCe ll Biol 1991;35:37–71.

[4] Arai E, Chow K, Li C, Tokunaga M, Katayama I. Differentiationb etweenc utaneousf ormo f adult T cell leukemia/lymphoma and cutaneous T cell lymphoma by in situ hybridization using a humanT c ellle ukemiavir us-1DNAp robe. Am JP athol 1994;144:15–20.

[5] Arrieta JJ, Rodriguez-Inigo E, Ortiz-Movilla N, etal. Insitu d etectiono fh epatitisCvir usRNA insalivaryglan ds. AmJ P athol 2001;158:259–64.

[6] Asif M, Khadim M, Mushtaq S, et al. Determination of her-2/neu by chromogenic in situ hybridization on borderline (2+) immunohistochemistry cases in carcinoma breast. Asian Pac J Cancer Prev 2011;12:211–4.

[7] Atwood K, Henderson A, Kacian D, Eicher E. On the feasibility of mapping low-multiplicity genesb y in situ h ybridization. BirthDe fects OrigArtic S er 1975;11:59–61.

[8] Avissar N, Ornt DB, Yagil Y,e tal. Human kidney proximal tubules are the main source of plasmaglu tathionep eroxidase. AmJ P hysiol 1994;266:C367–75.

[9] Avvedimento VE, DiL auro R, Monticelli A,e tal. Mappingo fh umanth yroglobulin gene on the long arm of chromosome 8b yin situ h ybridization. HumGe net 1985;71:163–6.

[10] Awaya H, Takeshima Y, Furonaka O, Kohno N, Inai K. Geneamp lificationan d protein expression of EGFR and HER2 by chromogenic in situ hybridization and immunohistochemistry in atypical adenomatous hyperplasisan dad enocarcinomao fth elu ng. J ClinP athol 2005;58:1076–80.

[11] Babic A, Loftin IR, Stanislaw S, et al. The impact of pre-analytical processing on staining quality for H&E, dual hapten, dual color in situ hybridization and fluorescent in situ hybridizationa ssays. Methods 2010;52:287–300.

[12] Baek IJ, Seo DS, Yon JM, et al. Tissue expression and cellular localization of phospholipid hydroperoxide glutathione peroxidase(PHGPx)mRNAin male mic e. JM olHis tol 2007;38:237–44.

[13] Baek IJ, Yon JM, Lee SR, et al. Differential expression of gastrointestinal glutathione peroxidase (GI-GPx) gene during mouse organogenesis. AnatHis tolE mbryol 2011;40:210–8.

[14] Bagasra O, Harris T. In situ PCR protocols. MethodsM olB iol 2006;334:61–78.

[15] Baldino Jr. F, Ruth JL, Davis LG. Nonradioactive detection of vasopressin mRNA within s ituh ybridizationh istochemistry. Exp Neurol 1989;104:200–7.

[16] Ballester M, Galindo-Cardiel I, Gallardo C, etal. Intranucleard etectiono fAf ricans wine fever virus DNA in several cell types from formalin-fixed and paraffin-embedded tissues usingan ewin s ituh ybridizationp rotocol. JVir olM ethods 2010;168:38–43.

[17] Baloglu G, Haholu A, Kucukodaci Z, et al. The effects of tissue fixation alternatives on DNA content:as tudyo nn ormalc olont issue. Appl ImmunohistochemM olM orphol 2008;16:485–92.

[18] Basyuk E, Bertrand E, Journot L. Alkaline fixation drastically improves the signal of in situ hybridization. NucleicAc idsRe s 2000;28:E46.

[19] Bauman JG, Wiegant J, Borst P, van Duijn P. A new method for fluorescence microscopical localization of specific DNA sequences by in situ hybridization of fluorochromelabelled RNA. ExpCe llRe s 1980;128:485–90.

[20] Becker JL, Miller F, Nuovo GJ, et al. Epstein-Barr virus infection of renal proximal tubule cells: possible role in chronic interstitial nephritis. JClin I nvest 1999;104:1673–81.

[21] Beckmann AM, Myerson D, Daling JR, etal. Detectionan dlo calizationo fh uman papillomavirus DNA in human genital condylomas by in situ hybridization withb iotinylatedp robes. JM edVir ol 1985;16:265–73.

[22] Berge-Lefranc JL, Carouzou G, De Micco C, Fragu P, Lissitzky S. Quantificationo f thyroglobulin messenger RNA by in situ hybridization in differentiated thyroid cancers. Difference between well-differentiated and moderatelyd ifferentiatedh istologict ypes. Cancer 1985;56:345–50.

[23] Berge-Lefranc JL, Cartouzou G, Mattei MG, etal. Localizationo ft het hyroglobulinge neb y ins ituh ybridizationt oh umanc hromosomes. HumGe net 1985;69:28–31.

[24] Berges BK, Wolfe JH, Fraser NW. Stable levels of long-term transgene expression driven by the latency-associated transcript promoter in ah erpess implexvir ust ypelve ctor. MolT her 2005;12:1111–9.

[25] Bhatt B, Sahinoglu T, Stevens C. Nonisotopic in situ hybridization. Gene mapping and cytogenetics. Methods Mol Biol 1998;80:405–17.

[26] Bilous M, Morey A, Armes J, Cummings M, Francis G. Chromogenicin s ituh ybridisation testing for HER2 gene amplification in breast cancer produces highly reproducible results concordant with fluorescence in situ hybridisationan dimmu nohistochemistry. Pathology 2006;38:120–4.

[27] Bloch B, Popovici T, Le Guellec D, et al. In situ hybridization histochemistry for the analysis of gene expression in the endocrine and central nervous system:t issues:a 3- yeare xperience. J NeurosciRe s 1986;16:183–200.

[28] Bloch B, Popovici T, Levin M, Tuil D, Kahn A. Transferringe nee xpression visualized in oligodendrocytes of the rat brain by using in situ hybridization and immunohistochemistry. ProcNat lAc adS ci US A 1985;82:6706–10.

[29] Blough RI, Smolarek TA, Ulbright TM, Heerema NA. Bicolorfl uorescence in situ hybridization on nuclei from formalin-fixed, paraffin-embedded testicular germ cell tumors: comparison with standard metaphasean alysis. CancerGe netCyt ogenet 1997;94:79–84.

[30] Blum HE, Haase AT, Harris JD, Walker D, Vyas GN. Asymmetricr eplicationo fh epatitis B virus DNA in human liver: demonstration of cytoplasmic minus-strand DNA by blot analysesan din s ituh ybridization. Virology 1984;139:87–96.

[31] Bonds LA, Barnes P, Foucar K, Sever CE. Acetic acid-zinc-formalin: a safe alternative to B-5 fixative. Am J Clin Pathol 2005;124:205–11.

[32] Borzi RM, Piacentini A, Monaco MC, et al. A fluorescent in situ hybridization method in flow cytometry to detect HIV-1 specific RNA. J Immunol Methods 1996;193:167–76.

[33] Botella MA, Quesada MA, Kononowicz AK, et al. Characterization and in situ localization of a salt-induced tomato peroxidase mRNA. Plant Mol Biol 1994;25:105–14.

[34] Brandwein M, Nuovo G, Ramer M, Orlowski W, Miller L. Epstein-Barr virus reactivation in hairy leukoplakia. Mod Pathol 1996;9:298–303.

[35] Brannagan 3rd TH, Nuovo GJ, Hays AP, Latov N. Human immunodeficiency virus infection of dorsal root ganglion neurons detected by polymerase chain reaction in situ hybridization. Ann Neurol 1997;42:368–72.

[36] Breborowicz J, Tamaoki T. Detection of messenger RNAs of alpha-fetoprotein and albumin in a human hepatoma cell line by in situ hybridization. Cancer Res 1985;45:1730–6.

[37] Brickley SG, Aller MI, Sandu C, et al. TASK-3 two-pore domain potassium channels enable sustained high-frequency firing in cerebellar granule neurons. J Neurosci 2007;27:9329–40.

[38] Broitman-Maduro G, Maduro MF. In situ hybridization of embryos with antisense RNA probes. Methods Cell Biol 2011;106:253–70.

[39] Brousset P, Butet V, Chittal S, Selves J, Delsol G. Comparison of in situ hybridization using different nonisotopic probes for detection of Epstein-Barr virus in nasopharyngeal carcinoma and immunohistochemical correlation with anti-latent membrane protein antibody. Lab Invest 1992;67:457–64.

[40] Brown WT, Devine EA, Nolin SL, Houck Jr. GE, Jenkins EC. Localization of chromosome 21 probes by in situ hybridization. Ann N Y Acad Sci 1985;450:69–83.

[41] Burke AP, Anderson D, Benson W, et al. Localization of human immunodeficiency virus 1 RNA in thymic tissues from asymptomatic drug addicts. Arch Pathol Lab Med 1995;119:36–41.

[42] Bzorek Sr. M, Petersen BL, Hansen L. Simultaneous phenotyping and genotyping (FICTION-methodology) on paraffin sections and cytologic specimens: a comparison of 2 different protocols. Appl Immunohistochem Mol Morphol 2008;16:279–86.

[43] Calmels TP, Mazurais D. In situ hybridization: a technique to study localization of cardiac gene expression. Methods Mol Biol 2007;366:159–80.

[44] Cannizzaro LA, Aronson MM, Emanuel BS. In situ hybridization and translocation breakpoint mapping. II. Two unusual t(21;22) translocations. Cytogenet Cell Genet 1985;39:173–8.

[45] Carr EL, Eales K, Soddell J, Seviour RJ. Improved permeabilization protocols for fluorescence in situ hybridization (FISH) of mycolic-acid-containing bacteria found in foams. J Microbiol Methods 2005;61:47–54.

[46] Chang CC, Huang TY, Shih CL, et al. Whole-mount identification of gene transcripts in aphids: protocols and evaluation of probe accessibility. Arch Insect Biochem Physiol 2008;68:186–96.

[47] Chiou SH, Chow KC, Yang CH, Chiang SF, Lin CH. Discovery of Epstein-Barr virus (EBV)-encoded RNA signal and EBV nuclear antigen leader protein DNA sequence in pet dogs. J Gen Virol 2005;86:899–905.

[48] Cho EY, Choi YL, Han JJ, Kim KM, Oh YL. Expression and amplification of Her2, EGFR and cyclin D1 in breast cancer: immunohistochemistry and chromogenic in situ hybridization. Pathol Int 2008;58:17–25.

[49] Cho EY, Han JJ, Choi YL, Kim KM, Oh YL. Comparison of Her-2, EGFR and cyclin D1 in primary breast cancer and paired metastatic lymph nodes: an immunohistochemical and chromogenic in situ hybridization study. J Korean Med Sci 2008;23:1053–61.

[50] Cho I, Sugimoto M, Mita S, et al. In vivo proviral burden and viral RNA expression in T cell subsets of patients with human T lymphotropic virus type-1-associated myelopathy/tropical spastic paraparesis. Am J Trop Med Hyg 1995;53:412–8.

[51] Cho WS, Chae C. PCR detection of Actinobacillus pleuropneumoniaeapxIV gene in formalin-fixed, paraffin-embedded lung tissues and comparison with in situ hybridization. Lett Appl Microbiol 2003;37:56–60.

[52] Choi YJ. In situ hybridization using a biotinylated DNA probe on formalin-fixed liver biopsies with hepatitis B virus infections: in situ hybridization superior to immunochemistry. Mod Pathol 1990;3:343–7.

[53] Christensen BB, Sternberg C, Andersen JB, et al. Molecular tools for study of biofilm physiology. Methods Enzymol 1999;310:20–42.

[54] Chu WS, Liang Q, Tang Y, et al. Ultrasound-accelerated tissue fixation/processing achieves superior morphology and macromolecule integrity with storage stability. J Histochem Cytochem 2006;54:503–13.

[55] Cinque P, Brytting M, Vago L, et al. Epstein-Barr virus DNA in cerebrospinal fluid from patients with AIDS-related primary lymphoma of the central nervous system. Lancet 1993;342:398–401.

[56] Cmarko D, Koberna K. Electron microscopy in situ hybridization: tracking of DNA and RNA sequences at high resolution. Methods Mol Biol 2007:213–28.

[57] Cote BD, Uhlenbeck OC, Steffensen DM. Quantitation of in situ hybridization of ribosomal ribonucleic acids to human diploid cells. Chromosoma 1980;80:349–67.

[58] Cui YL, Holt AG, Lomax CA, Altschuler RA. Deafness associated changes in two-pore domain potassium channels in the rat inferior colliculus. Neuroscience 2007;149:421–33.

[59] Dandachi N, Dietze O, Hauser-Kronberger C. Evaluation of the clinical significance of HER2 amplification by chromogenic in situ hybridization in patients with primary breast cancer. Anticancer Res 2004;24:2401–6.

[60] Dedes KJ, Lopez-Garcia MA, Geyer FC, et al. Cortactin gene amplification and expression in breast cancer: a chromogenic in situ hybridization and immunohistochemical study. Breast Cancer Res Treat 2010;124:653–66.

[61] Delfour C, Roger P, Bret C, et al. RCL2, a new fixative, preserves morphology and nucleic acid integrity in paraffin-embedded breast carcinoma and microdissected breast tumor cells. J Mol Diagn 2006;8:157–69.

[62] Delord B, Poveda JD, Astier-Gin T, Gerbaud S, Fleury HJ. Detection of the bunyavirus Germiston in VERO and Aedes albopictus C6/36 cells by in situ hybridization using cDNA and asymmetric RNA probes. J Virol Methods 1989;24:253–64.

[63] Den Herder J, Lievens S, Rombauts S, Holsters M, Goormachtig S. A symbiotic plant peroxidase involved in bacterial invasion of the tropical legume Sesbania rostrata. Plant Physiol 2007;144:717–27.

[64] Derradji H, Bekaert S, Van Oostveldt P, Baatout S. Comparison of different protocols for telomere length estimation by combination of quantitative fluorescence in situ hybridization (Q-FISH) and flow cytometry in human cancer cell lines. Anticancer Res 2005;25:1039–50.

[65] Devine EA, Nolin SL, Houck GE, Jenkins EC, Brown WT. Chromosomal localization of several families of repetitive sequences by in situ hybridization. AmJ Hu mGe net 1985;37:114–23.

[66] Do HM, Hong JK, Jung HW,e tal. Expression of peroxidase-like genes, H2O2 production, and peroxidase activity during the hypersensitive response to Xanthomonas campestris pv. vesicatoria in Capsicum annuum. MolP lantM icrobeI nteract 2003;16:196–205.

[67] Dobler T, Springauf A, Tovornik S,e tal. TRESK two-pore-domain K+ channels constitute a significant component of background potassium currents in murine dorsal root ganglion neurones. JP hysiol 2007;585:867–79.

[68] Doring F, Scholz H, Kuhnlein RP, Karschin A, Wischmeyer E. NovelDr osophilat wo-pore domain K channels: rescue of channel function by heteromeric assembly. EurJ Ne urosci 2006;24:2264–74.

[69] Downs-Kelley E, Shehata BM, Lopez-Terrada D, et al. Theu tilityo fF OXO1fl uorescence in situ hybridization (FISH) in formalin-fixed paraffin-embedded specimens in the diagnosis of alveolar rhabdomyosarcoma. DiagnM ol Pathol 2009;18:138–43.

[70] Dryja TP, Morton CC. Mappingo fs even polymorphic loci on human chromosome 13 by in situ hybridization. HumGe net 1985;71:192–5.

[71] Dutlih B, Bebear C, Taylor-Robinson D, Grimont PA. Detectiono fCh lamydia trachomatis by in situ hybridization with sulphonated total DNA. AnnI nstP asteur Microbiol 1988;139:115–27.

[72] Dvorakova M, Dhir R, Bastacky SI, et al. Renalo ncocytoma:ac omparative clinicopathologic study and fluorescent in-situ hybridization analysis of 73 cases with long-term follow-up. DiagnP athol 2010;5:32–40.

[73] Eberspaecher U, Becker A, Bringmann P, vand erM erwe L, Donner P. Immunohistochemical localization of zona pellucida proteins ZPA, ZPB and ZPC in human, cynomolgus monkey and mouse ovaries. CellT issueRe s 2001;303:277–87.

[74] Edwards AA, Szluinska M, Lloyd DC. Reconstruction of doses from ionizing radiation using fluorescence in situ hybridization techniques. BrJ Rad iol 2007;80:S63–7.

[75] Eichele G, Diez-Roux G. High-throughput analysis of gene expression on tissue sections by in situ hybridization. Methods 2011;53:417–23.

[76] Emanuel BS. Chromosomal in situ hybridization and the molecular cytogenetics of cancer. Surv Synth Pathol Res 1985;4:269–81.

[77] Enomoto H, Inoue S, Matsuhisa A,e tal. Development of a new in situ hybridization

method for the detection of global bacterial DNA to provide early evidence of a bacterial infection in spontaneous bacterial peritonitis. J Hepatol 2012;56:85–94.

[78] Euscher E, Davis J, Holzman I, Nuovo GJ. Coxsackie virus infection of the placenta associated with neurodevelopmental delays in the newborn. Obstet Gynecol 2001;98:1019–26.

[79] Falconi M, Teti G, Zago M, et al. Effect of fixative on chromatin structure and DNA detection. Microsc Res Tech 2007;70:599–606.

[80] Falla N, Van V, Bierkens J, et al. Characterization of a 5-fluorouracil-enriched osteoprogenitor population of the murine bone marrow. Blood 1993;82:3580–91.

[81] Fallert BA, Reinhart T. Improved detection of simian immunodeficiency virus RNA by in situ hybridization in fixed tissue sections: combined effects of temperatures for tissue fixation and probe hybridization. J Virol Methods 2002;99:23–32.

[82] Ferrer B, Bermudo R, Thomson T, et al. Paraffin-embedded cell line microarray (PECLIMA): development and validation of a high-throughput method for antigen profiling of cell lines. Pathobiology 2005;72:225–32.

[83] Fiche JB, Buhot A, Calemczuk R, Livache T. Temperature effects on DNA chip experiments from surface plasmon resonance imaging: isotherms and melting curves. Biophys J 2007;92:935–46.

[84] Fidlerova H, Senger G, Kost M, Sanseau P, Sheer D. Two simple procedures for releasing chromatin from routinely fixed cells for fluorescence in situ hybridization. Cytogenet Cell Genet 1994;65:203–5.

[85] Filion G, Laflamme C, Turgeon N, Ho J, Duchaine C. Permeabilization and hybridization protocols for rapid detection of Bacillus spores using fluorescence in situ hybridization. J Microbiol Methods 2009;77:29–36.

[86] Fink SR, Belongie KJ, Paternoster SF, et al. Validation of a new three-color fluorescence in situ hybridization (FISH) method to detect CHIC2 deletion, FIP1L1/PDGFRA fusion and PDGFRA translocations. Leuk Res 2009;33:843–6.

[87] Finkle D, Quan ZR, Asghari V, et al. HER2-targeted therapy reduces incidence and progression of midlife mammary tumors in female murine mammary tumor virus huHER2-transgenic mice. Clin Cancer Res 2004;10:2499–511.

[88] Fletcher JA. DNA in situ hybridization as an adjunct in tumor diagnosis. Am J Clin Pathol 1999;112:S11–8.

[89] Forghani B, Dupuis KW, Schmidt NJ. Rapid detection of herpes simplex virus DNA in human brain tissue by in situ hybridization. J Clin Microbiol 1985;22:656–8.

[90] Forkert PG. CYP2E1 is preferentially expressed in Clara cells of murine lung: localization by in situ hybridization and immunohistochemical methods. Am J Respir Cell Mol Biol 1995;12:589–96.

[91] Forkert PG, Jackson AC, Parkinson A, Chen S. Diminished expression of CYP1A1 in urethane-induced lung tumors in strain A/J mice: analysis by in situ hybridization and immunohistochemical methods. Am J Respir Cell Mol Biol 1996;14:444–53.

[92] Fried J, Ludwig W, Psenner R, Schleifer KH. Improvement of ciliate identification and quantification: a new protocol for fluorescence in situ hybridization (FISH) in combination

with silver stain techniques. Syst Appl Microbiol 2002;25:555–71.

[93] Fritzsche FR, Pianca S, Gaspert A, et al. Silver-enhanced in situ hybridization for detection of polyomavirus DNA in patients with BK virus nephropathy. Diagn Mol Pathol 2011;20:105–10.

[94] Gabusi E, Lattes C, Fiorentino M, D'Errico A, Grigioni WF. Expression of Epstein-Barr virus-encoded RNA and biological markers in Italian nasopharyngeal carcinomas. J Exp Clin Cancer Res 2001;20:371–6.

[95] Gaulier A, Fourcade C, Szekeres G, Pulik M. Bone marrow one step fixation-decalcification in Lowy FMA solution: an immunohistological and in situ hybridization study. Pathol Res Pract 1994;190:1149–61.

[96] Gelmetti D, Grieco V, Rossi C, Capucci L, Lavazza A. Detection of rabbit haemorrhagic disease virus (RHDV) by in situ hybridisation with a digoxigenin labelled RNA probe. J Virol Methods 1998;72:219–26.

[97] Gendron FP, Mongrain S, Laprise P, et al. The CDX2 transcription factor regulates furin expression during intestinal epithelial cell differentiation. Am J Physiol Gastrointest Liver Physiol 2006;290:G310–8.

[98] Gerami P, Beilfuss B, Haghighat Z, et al. Fluorescence in situ hybridization as an ancillary method for the distinction of desmoplastic melanomas from sclerosing melanocytic nevi. J Cutan Pathol 2011;38:329–34.

[99] Gerami P, Zembowicz A. Update on fluorescence in situ hybridization in melanoma: state of the art. Arch Pathol Lab Med 2011;135:830–7.

[100] Gijzen M, Miller SS, Bowman LA, et al. Localization of peroxidase mRNAs in soybean seeds by in situ hybridization. Plant Mol Biol 1999;41:57–63.

[101] Goldmann T, Dromann D, Marzouki M, et al. Tissue microarrays from HOPE-fixed specimens allow for enhanced high throughput molecular analyses in paraffin-embedded material. Pathol Res Pract 2005;201:599–602.

[102] Gomes SA, Nascimento JP, Siqueira MM, et al. In situ hybridization with biotinylated DNA probes: a rapid diagnostic test for adenovirus upper respiratory infection. J Virol Methods 1985;12:105–10.

[103] Gressens P, Langston C, Martin JR. In situ PCR localization of herpes simplex virus DNA sequences in disseminated neonatal herpes encephalitis. J Neuropathol Exp Neurol 1994;53:469–82.

[104] Groben R, Medlin L. In situ hybridization of phytoplankton using fluorescently labeled rRNA probes. Methods Enzymol 2005;395:299–310.

[105] Gupta J, Gendelman HE, Naghashfar Z, et al. Specific identification of human papillomavirus type in cervical smears and paraffin sections by in situ hybridization with radioactive probes: a preliminary communication. Int J Gynecol Pathol 1985;4:211–8.

[106] Guterman KS, Hair LS, Morgello S. Epstein-Barr virus and AIDS-related primary central nervous system lymphoma. Viral detection by immunohistochemistry, RNA in situ hybridization, and polymerase chain reaction. Clin Neuropathol 1996;15:79–86.

[107] Gutstein HB. In situ hybridization in neural tissues. Methods Mol Med 2003;84:95–105.

[108] Ha SK, Choi C, Chae C. Development of an optimized protocol for the detection of

classical swine fever virus in formalin-fixed, paraffin-embedded tissues by seminested reverse transcription-polymerase chain reaction and comparison with in situ hybridization. Res Vet Sci 2004;77:163–9.

[109] Hada M, Zhang Y, Feiveson A, Cucinotta FA, Wu H. Association of inter- and intrachromosomal exchanges with the distribution of low- and high-LET radiation-induced breaks in chromosomes. Radiat Res 2011;176:25–37.

[110] Hagiwara T, Hattori J, Kaneda T. PNA-in situ hybridization method for detection of HIV-1 DNA in virus-infected cells and subsequent detection of cellular and viral proteins. Methods Mol Biol 2006;326:139–49.

[111] Hair LS, Nuovo G, Powers JM, et al. Progressive multifocal leukoencephalopathy in patients with human immunodeficiency virus. Hum Pathol 1992;23:663–7.

[112] Hamberg A, Ringler S, Krakowka S. A novel method for the detection of porcine circovirus type 2 replicative double stranded viral DNA and nonreplicative single stranded viral DNA in tissue sections. JVe tDiagn I nvest 2007;19:135–41.

[113] Hara T, Ooi A, Kobayashi M, et al. Amplification of c-myc, K-sam, and c-met in gastric cancers: detection by fluorescence in situ hybridization. Lab Invest 1998;78:1143–53.

[114] Hargrave SL, Jung JC, Fini ME, et al. Possible role of the vitamin E solubilizer in topical diclofenac on matrix metalloproteinase expression in corneal melting: an analysis of postoperative keratolysis. Ophthalmology 2002;109:343–50.

[115] Harper SJ, Bailey E, McKeen CM, et al. A comparative study of digoxigenin, 2, 4-dinitrophenyl, and alkaline phosphatase as deoxyoligonucleotide labels in non-radioisotopic in situ hybridization. JClin Pathol 1997;50:686–90.

[116] Hayashi Y, Watanabe J, Nakata K, Fukayama M, Ikeda H. An oveld iagnostic method of Pneumocystis carinii. In situ hybridization of ribosomal ribonucleic acid with biotinylated oligonucleotide probes. Lab Invest 1990;63:576–80.

[117] Heath CV, Copeland CS, Amberg DC, et al. Nuclear pore complex clustering and nuclear accumulation of poly(A)+ RNA associated with mutation of the Saccharomyces cerevisiae RAT2/NUP120 gene. JCe llB iol 1995;131:1677–97.

[118] Helmy MF, Ahmed HM, Saber MA. Detection of hepatitis B virus DNA by in situ tissues hybridization in corneal of HBsAG positive donors. JE gyptP ublicHe althAs soc 1995;70:243–55.

[119] Hendzel MJ, Bazett-Jones DP. Fixation-dependent organization of core histones following DNA fluorescent in situ hybridization. Chromosoma 1997;106:114–23.

[120] Henke RT, Eun Kim S, Maitra A, Paik S, Wellstein A. Expressiona nalysiso fmRNA in formalin-fixed, paraffin-embedded archival tissues by mRNA in situ hybridization. Methods 2006;38:253–62.

[121] Herbergs J, Speel EJ, Ramaekers FC, et al. Combination of lamin immunocytochemistry and in situ hybridization for the analysis of chromosome copy numbers in tumor cell areas with high nuclear density. Cytometry 1996;23:1–7.

[122] Herrington CS, Graham AK, Flannery DM, Burns J, McGee JO. Discriminationo fc losely

homologous HPV types by nonisotopic in situ hybridization: definition and derivation of tissue melting temperatures. Histochem J 1990;22:545–54.

[123] Heyland A, Price DA, Bodnarova-Buganova M, Moroz LL. Thyroid hormone metabolism and peroxidase function in two non-chordate animals. J Exp Zool B Mol Dev Evol 2006;306:551–66.

[124] Hirokawa J, Watakabe A, Ohsawa S, Yamamori T. Analysis of area-specific expression patterns of RORbeta, ER81 and Nurr1 mRNAs in rat neocortex by double in situ hybridization and cortical box method. PLoS One 2008;3:e3266.

[125] Hirose T, Sugiyama S. A simple DNA characterization method using fiber-fluorescence in situ hybridization performed without DNA fragmentation. Photochem Photobiol 2011;87:470–3.

[126] Hiruta J, Mazet F, Ogasawara M. Restricted expression of NADPH oxidase/peroxidase gene (Duox) in zone VII of the ascidian endostyle. Cell Tissue Res 2006;326:835–41.

[127] Hoefler H, Childers H, Montminy MR, et al. In situ hybridization methods for the detection of somatostatin mRNA in tissue sections using antisense RNA probes. Histochem J 1986;18:597–604.

[128] Hoffmeister-Ullerich SA, Herrmann D, Kielholz J, Schweizer M, Schaller HC. Isolation of a putative peroxidase, a target for factors controlling foot-formation in the coelenterate hydra. Eur J Biochem 2002;269:4597–606.

[129] Hokari R, Kurihara C, Nagata N, et al. Increased expression of lipocalin-type-prostaglandin D synthase in ulcerative colitis and exacerbating role in murine colitis. Am J Physiol Gastrointest Liver Physiol 2011;300:G401–8.

[130] Hopman AH, Kamps MA, Smedts F, et al. HPV in situ hybridization: impact of different protocols on the detection of integrated HPV. Int J Cancer 2005;115:419–28.

[131] Horns T, Jeske H. Localization of abutilon mosaic virus (AbMV) DNA within leaf tissue by in situ hybridization. Virology 1991;181:580–8.

[132] Ichinose H, Miyazaki M, Koji T, et al. Detection of cytokine mRNA-expressing cells in peripheral blood of patients with IgA nephropathy using non-radioactive in situ hybridization. Clin Exp Immunol 1996;103:125–32.

[133] Inagaki H, Nonaka M, Nagaya S, et al. Monoclonality in gastric lymphoma detected in formalin-fixed paraffin-embedded endoscopic biopsy specimens using immunohistochemistry, in situ hybridization, and polymerase chain reaction. Diagn Mol Pathol 1995;4:32–8.

[134] Ishii T, Omura M, Mombaerts P. Protocols for two- and three-color fluorescent RNA in situ hybridization of the main and accessory olfactory epithelia in mouse. J Neurocytol 2004;33:657–69.

[135] Iwasaki M, Jikko A, Le AX. Age-dependent effects of hedgehog protein on chondrocytes. J Bone Joint Surg Br 1999;81:1076–82.

[136] Jarvela S, Helin H, Haapasalo H, et al. Amplification of the epidermal growth factor receptor in astrocytic tumours by chromogenic in situ hybridization: association with clinicopathological features and patient survival. Neuropathol Appl Neurobiol 2006;32:441–50.

[137] Jilbert A. In situ hybridization protocols for detection of viral DNA using radioactive and nonradioactive DNA probes. Methods Mol Biol 2000;123:177–93.

[138] Jilbert AR, Burrell CJ, Gowans EJ, Rowland R. Histological aspects of in situ hybridization. Detection of poly (A) nucleotide sequences in mouse liver sections as a model system. Histochemistry 1986;85:505–14.

[139] Jiwa M, Steenbergen RD, Zwaan FE, et al. Three sensitive methods for the detection of cytomegalovirus in lung tissue of patients with interstitial pneumonitis. Am J Clin Pathol 1990;93:491–4.

[140] Jubb AM, Pham TQ, Frantz GD, Peale Jr. FV, Hillan KJ. Quantitative in situ hybridization of tissue microarrays. Methods Mol Biol 2006;326:255–64.

[141] Jung KY, Baek IJ, Yon JM, et al. Developmental expression of plasma glutathione peroxidase during mouse organogenesis. J Mol Histol 2011;42:545–56.

[142] Junying J, Herrmann K, Davies G, et al. Absence of Epstein-Barr virus DNA in the tumor cells of European hepatocellular carcinoma. Virology 2003;306:236–43.

[143] Kang JK, Sul D, Kang JK, et al. Effects of lead exposure on the expression of phospholipid hydroperoxidase glutathione peroxidase mRNA in the rat brain. Toxicol Sci 2004;82:228–36.

[144] Kapranos N, Kontogeorgos G, Frangia K, Kokka E. Effect of fixation on interphase cytogenetic analysis by direct fluorescence in situ hybridization on cell imprints. Biotech Histochem 1997;72:148–51.

[145] Kashiwagi S, Kumasaka T, Bunsei N, et al. Detection of Epstein-Barr virus-encoded small RNA-expressed myofibroblasts and IgG4-producing plasma cells in sclerosing angiomatoid nodular transformation of the spleen. Virchows Arch 2008;453:275–82.

[146] Keherly MJ, Papaconstantinou J, McCombs JL. A sequential staining technique for karyotypic analysis of interspecific somatic cell hybrids. Exp Cell Res 1993;205:361–4.

[147] Khalidi HS, Lones MA, Zhou Y, Weiss LM, Medeiros LJ. Detection of Epstein-Barr virus in the L & H cells of nodular lymphocyte predominance Hodgkin's disease: report of a case documented by immunohistochemical, in situ hybridization, and polymerase chain reaction methods. Am J Clin Pathol 1997;108:687–92.

[148] Khalil SH. Book review: PCR in situ hybridization: protocols and applications. Ann Saudi Med 1997;17:491.

[149] Kim HJ, Yoo TW, Park DI, et al. Gene amplification and protein overexpression of HER-2/neu in human extrahepatic cholangiocarcinoma as detected by chromogenic in situ hybridization and immunohistochemistry: its prognostic implication in node-positive patients. Ann Oncol 2007;18:892–7.

[150] Kim J, Chae C. Optimized protocols for the detection of porcine circovirus 2 DNA from formalin-fixed paraffin-embedded tissues using nested polymerase chain reaction and comparison of nested PCR with in situ hybridization. J Virol Methods 2001;92:105–11.

[151] Kim O, Yi SJ. Optimization of in situ hybridization protocols for detection of feline herpesvirus 1. J Vet Med Sci 2003;65:1031–2.

[152] Knabe C, Berger G, Gildenhaar R, Klar F, Zreiqat H. The modulation of osteogenesis in vitro by calcium titanium phosphate coatings. Biomaterials 2004;25:4911–9.

[153] Knoll JH, Lichter P, Bakdounes K, Eltoum IE. In situ hybridization and detection using nonisotopic probes. Curr Protoc Mol Biol 2007 [Chapter 14, Unit 14.7].

[154] Kotoula V, Cheva A, Barbanis S, Papadimitriou CS, Karkavelas G. hTERT immunopositivity patterns in the normal brain and in astrocytic tumors. Acta Neuropathol 2006;111:569–78.

[155] Kudo M, Matsuo Y, Nakasendo A, et al. Potential clinical benefit of the in situ hybridization method for the diagnosis of sepsis. J Infect Chemother 2009;15:23–6.

[156] Kuhns M, McNamara A, Mason A, Campbell C, Perrillo R. Serum and liver hepatitis B virus DNA in chronic hepatitis B after sustained loss of surface antigen. Gastroenterology 1992;103:1649–56.

[157] Kumamoto H, Sasano H, Taniguchi T, et al. Chromogenic in situ hybridization analysis of HER-2/neu status in breast carcinoma: application in screening of patients for trastuzumab (Herceptin) therapy. Pathol Int 2001;51:579–84.

[158] Kunikowska G, Jenner P. Alterations in m-RNA expression for Cu,Zn-superoxide dismutase and glutathione peroxidase in the basal ganglia of MPTP-treated marmosets and patients with Parkinson's disease. Brain Res 2003;968:206–18.

[159] Kupper H, Seib LO, Sivaguru M, Hoekenga OA, Kochian LV. A method for cellular localization of gene expression via quantitative in situ hybridization in plants. Plant J 2007;50:159–75.

[160] Lalevee N, Monier B, Senatore S, Perrin L, Semeriva M. Control of cardiac rhythm by ORK1, a Drosophila two-pore domain potassium channel. Curr B iol 2006;16:1502–8.

[161] Lambros MB, Simpson PT, Jones C, et al. Unlocking pathology archives for molecular genetic studies: a reliable method to generate probes for chromogenic and fluorescent in situ hybridization. Lab Invest 2006;86:398–408.

[162] Landegent JE, Jansen in de Wal N, van Ommen GJ, et al. Chromosomal localization of a unique gene by non-autoradiographic in situ hybridization. Nature 1985;317:175–7.

[163] Lane JE, Olivares-Villagomez D, Vnencak-Jones CL, McCurley TL, Carter CE. Detection of Trypanosoma cruzi with the polymerase chain reaction and in situ hybridization in infected murine cardiac tissue. Am J Trop Med Hyg 1997;56:588–95.

[164] Larsson LI, Hougaard DM. Optimization of non-radioactive in situ hybridization: image analysis of varying pretreatment, hybridization and probe labelling conditions. Histochemistry 1990;93:347–54.

[165] Lau YF. Detection of Y-specific repeat sequences in normal and variant human chromosomes using in situ hybridization with biotinylated probes. Cytogenet Cell Genet 1985;39:184–7.

[166] Lauter G, Söll I, Hauptmann G. Two-color fluorescent in situ hybridization in the embryonic zebrafish brain using differential detection systems. BMC Dev Biol 2011;11:43.

[167] Lebo RV, Su Y. Positional cloning and multicolor in situ hybridization. Principles and protocols. Methods Mol Biol 1994;33:409–38.

[168] Lecuyer E, Parthasarathy N, Krause HM. Fluorescent in situ hybridization protocols in Drosophila embryos and tissues. Methods Mol Biol 2008;420:289–302.

[169] Lee YM, Jippo T, Kim DK, et al. Alteration of protease expression phenotype of mouse peritoneal mast cells by changing the microenvironment as demonstrated by in situ hybridization histochemistry. Am J Pathol 1998;153:931–6.

[170] Leenman EE, Panzer-Grümayer RE, Fischer S, et al. Rapid determination of Epstein-Barr virus latent or lytic infection in single human cells using in situ hybridization. Mod Pathol 2004;17:1564–72.

[171] Lei KI, Chan LY, Chan WY, Johnson PJ, Lo YM. Quantitative analysis of circulating cell-free Epstein-Barr virus (EBV) DNA levels in patients with EBV-associated lymphoid malignancies. Br J Haematol 2000;111:239–46.

[172] Lewis FA, Andrew A, Cross D, Quirke P. Comparison of in situ hybridization and in situ PCR for the localization of human papilloma virus in formalin-fixed paraffin embedded tissue (Meeting Abstract). J Pathol 1995;176(Suppl): 1995.

[173] Lewis SA, Cowan NJ. Temporal expression of mouse glial fibrillary acidic protein mRNA studied by a rapid in situ hybridization procedure. J Neurochem 1985;45:913–9.

[174] Li JY, Gaillard F, Moreau A, et al. Detection of translocation t(11;14) (q13;q32) in mantle cell lymphoma by fluorescence in situ hybridization. Am J Pathol 1999;154:1449–52.

[175] Li Vigni R, Bianchi UA, Carosi G, et al. Successful application of indirect in-situ polymerase chain reaction to tissues fixed in Bouin's solution. Histopathology 1999;35:134–43.

[176] Lichter P, Fischer K, Joos S, et al. Efficacy of current molecular cytogenetic protocols for the diagnosis of chromosome aberrations in tumor specimens. Cytokines Mol Ther 1996;2:163–9.

[177] Lie ES, Heyden A, Johannesen MK, Boysen M, Brandtzaeg P. Detection of human papillomavirus in routinely processed biopsy specimens from laryngeal papillomas: evaluation of reproducibility of polymerase chain reaction and DNA in situ hybridization procedures. Acta Otolaryngol 1996;116:627–32.

[178] Lin CC, Draper PN, De Braekeleer M. High-resolution chromosomal localization of the beta-gene of the human beta-globin gene complex by in situ hybridization. Cytogenet Cell Genet 1985;39:269–74.

[179] Li-Ning TE, Ronchetti R, Torres-Cabala C, Merino MJ. Role of chromogenic in situ hybridization (CISH) in the evaluation of HER2 status in breast carcinoma: comparison with immunohistochemistry and FISH. Int J Surg Pathol 2005;13:343–51.

[180] Liu CQ, Shan L, Balesar R, et al. A quantitative in situ hybridization protocol for formalin-fixed paraffin-embedded archival post-mortem human brain tissue. Methods 2010;52:359–66.

[181] Liu H, Jiravanichpaisal P, Soderhall I, Cerenius L, Soderhall K. Antilipopolysaccharide factor interferes with white spot syndrome virus replication in vitro and in vivo in the crayfish Pacifastacus leniusculus. J Virol 2006;80:10365–10371.

[182] Lloveras E, Sole F, Florensa L, et al. Contribution of cytogenetics and in situ hybridization to the study of monoclonal gammopathies of undetermined significance. Cancer Genet Cytogenet 2002;132:25–9.

[183] Lossi L, Gambino G, Salio C, Merighi A. Direct in situ RT-PCR. Methods Mol Biol 2011;789:111–26.

[184] Lucassen PJ, Goudsmit E, Pool CW, et al. In situ hybridization for vasopressin mRNA in the human supraoptic and paraventricular nucleus; quantitative aspects of formalin-fixed paraffin-embedded tissue sections as compared to cryostat sections. J Neurosci Methods 1996;64:133.

[185] Lupu F, Alves A, Anderson K, Doye V, Lacy E. Nuclear pore composition regulates neural stem/progenitor cell differentiation in the mouse embryo. Dev Cell 2008;14:831–42.

[186] Macrae VE, Davey MG, McTeir L, et al. Inhibition of PHOSPHO1 activity results in impaired skeletal mineralization during limb development of the chick. Bone 2010;46:1146–55.

[187] Magenis RE, Donlon TA, Tomar DR. Localization of the beta-globin gene to 11p15 by in situ hybridization: utilization of chromosome 11 rearrangements. Hum Genet 1985;69:300–3.

[188] Manuelidis L. Individual interphase chromosome domains revealed by in situ hybridization. Hum Genet 1985;71:288–93.

[189] Mariette X, Gozlan J, Clerc D, Bisson M, Morinet F. Detection of Epstein-Barr virus DNA by in situ hybridization and polymerase chain reaction in salivary gland biopsy specimens from patients with Sjogren's syndrome. Am J Med 1991;90:286–94.

[190] Marquez A, Wu R, Zhao J, Shi Z. Evaluation of epidermal growth factor receptor (EGFR) by chromogenic in situ hybridization (CISH) and immunohistochemistry (IHC) in archival gliomas using bright-field microscopy. Diagn Mol Pathol 2004;13:1–8.

[191] Martin J. In situ hybridization to pre-fixed polytene chromosomes. Trends Genet 1990;6:238.

[192] Martin-DeLeon PA, Wolf SF, Persico G, et al. Localization of glucose-6-phosphate dehydrogenase in mouse and man by in situ hybridization: evidence for a single locus and transposition of homologous X-linked genes. Cytogenet Cell Genet 1985;39:87–92.

[193] Martins SA, Prazeres DM, Fonseca LP, Monteiro GA. Quantitation of non-amplified genomic DNA by bead-based hybridization and template mediated extension coupled to alkaline phosphatase signal amplification. Biotechnol Lett 2010;32:229–34.

[194] Maser RL, Magenheimer BS, Calvet JP. Mouse plasma glutathione peroxidase. cDNA sequence analysis and renal proximal tubular expression and secretion. J Biol Chem 1994;269:27066–27073.

[195] Mattei MG, Baeteman MA, Heilig R, et al. Localization by in situ hybridization of the coagulation factor IX gene and of two polymorphic DNA probes with respect to the fragile X site. Hum Genet 1985;69:327–31.

[196] Mattei MG, Philip N, Passage E, et al. DNA probe localization at 18p113 band by in situ hybridization and identification of a small supernumerary chromosome. Hum Genet 1985;69:268–71.

[197] Mayr D, Heim S, Weyrauch K, et al. Chromogenic in situ hybridization for Her-2/neu-oncogene in breast cancer: comparison of a new dual-colour chromogenic in situ hybridization with immunohistochemistry and fluorescence in situ hybridization. Histopathology 2009;55:716–23.

[198] Mazzotti G, Cocco L, Manzoli FA. High resolution detection of human metaphase chromosomes. Ital J Anat Embryol 1995;100(Suppl. 1):31–8.

[199] McAllister HA, Rock DL. Comparative usefulness of tissue fixatives for in situ viral nucleic acid hybridization. J Histochem Cytochem 1985;33:1026–32.

[200] McCarthy UM, Urquhart KL, Bricknell IR. An improved in situ hybridization method for the detection of fish pathogens. J Fish Dis 2008;31:669–77.

[201] Mehraein Y, Lennerz C, Ehlhardt S, et al. Latent Epstein-Barr virus (EBV) infection and cytomegalovirus (CMV) infection in synovial tissue of autoimmune chronic arthritis determined by RNA- and DNA-in situ hybridization. ModP athol 2004;17:781–9.

[202] Mehta A, Maggioncalda J, Bagasra O, et al. In situ DNA PCR and RNA hybridization detection of herpes simplex virus sequences in trigeminal ganglia of latently infected mice. Virology 1995;206:633–40.

[203] Mino-Kenudson M, Mark EJ. Reflex testing for epidermal growth factor receptor mutation and anaplastic lymphoma kinase fluorescence in situ hybridization in non-small cell lung cancer. Arch Pathol Lab Med 2011;135:655–64.

[204] Moatamed NA, Nanjangud G, Pucci R,e tal. Effecto fis chemict ime,fi xation time, and fixative type on HER2/neu immunohistochemical and fluorescence in situ hybridization results in breast cancer. AmJ Clin Pathol 2011;136:754–61.

[205] Moench TR, Gendelman HE, Clements JE, Narayan O, Griffin DE. Efficiency of in situ hybridization as a function of probe size and fixationt echnique. JVir olM ethods 1985;11:119–30.

[206] Mogensen J, Pedersen S, Hindkjaer J, Kolvraa S, Bolund L. Primedin s itu(PRINS)lab eling of RNA. MethodsM olB iol 1994;33:265–75.

[207] Montone KT, Litzky LA, Feldman MD, et al. Ins ituh ybridizationf orCo ccidioides immitis 5.8s ribosomal RNA sequences in formalin-fixed, paraffin-embedded pulmonary specimens using a locked nucleic acid probe: a rapid means for identification in tissue sections. DiagnM olP athol 2010;19:99–104.

[208] Moon IS, Cho SJ, Jin I, Walikonis R. A simple method for combined fluorescence in situ hybridization and immunochemistry. Mol Cells 2007;24:76–82.

[209] Morasso MI. Detection of gene expression in embryonic tissues and stratified epidermis by in situ hybridization. MethodsM olB iol 2010;585:253–60.

[210] Moreno SG, Laux G, Brielmeier M, Bornkamm GW, Conrad M. Testis-specific expression of the nuclear form of phospholipid hydroperoxide glutathione peroxidase (PHGPx). BiolCh em 2003;384:635–43.

[211] Morii E, Jippo T, Tsujimura T, et al. Abnormal expression of mouse mast cell protease 5 gene in cultured mast cells derived from mutant mi/mi mice. Blood 1997;90:3057–66.

[212] Morrison C. Fluorescent in situ hybridization and array comparative genomic hybridization: complementary techniques for genomic evaluation. Arch Pathol Lab Med 2006;130:967–74.

[213] Motohashi T, Tabara H, Kohara Y. Protocols for large scale in situ hybridization on C. elegans larvae. WormBook 2006:1–8.

[214] Motoi T, Kumagai A, Tsuji K, Imamura T, Fukusato T. Diagnostic utility of dual-color break-apart chromogenic in situ hybridization for the detection of rearranged SS18 in formalin-fixed, paraffin-embedded synovial sarcoma. Hum Pathol 2010;41:1397–404.

[215] Munke M, de Martinville B, Lieber E, Francke U. Minute chromosomes replacing the Y chromosome carry Y-specific sequences by restriction fragment analysis and in situ hybridization. Am J Med Genet 1985;22:361–74.

[216] Munne S, Weier HU, Stein J, Grifo J, Cohen J. A fast and efficient method for simultaneous X and Y in situ hybridization of human blastomeres. J Assist Reprod Genet 1993;10:82–90.

[217] Murakami T, Hagiwara T, Yamamoto K, et al. A novel method for detecting HIV-1 by non-radioactive in situ hybridization: application of a peptide nucleic acid probe and catalysed signal amplification. J Pathol 2011;194:130–5.

[218] Musilova P, Kubickova S, Rubes J. Assignment of porcine cyclin-dependent kinase 4 (CDK4) and oncogene c-mos (MOS) by nonradioactive nonfluorescence in situ hybridization. Anim Genet 2002;33:145–8.

[219] Nakamura S, Tourtellotte WW, Shapshak P, Darvish M. The detection of papovavirus nucleotide sequences in cortical neurons of a patient with progressive multifocal leucoencephalopathy (PML) by in situ hybridization. Rinsho Shinkeigaku 1985;25:1148–51.

[220] Narayanswami S, Dvorkin N, Hamkalo BA. Nucleic acid sequence localization by electron microscopic in situ hybridization. Methods Cell Biol 1991;35:109–32.

[221] Navarro B, Daros JA, Flores R. Reverse transcription polymerase chain reaction protocols for cloning small circular RNAs. J Virol Methods 1998;73:1–9.

[222] Neely LA, Patel S, Garver J, et al. A single-molecule method for the quantitation of microRNA gene expression. Nat Methods 2006;3:41–6.

[223] Negro F, Berninger M, Chiaberge E, et al. Detection of HBV-DNA by in situ hybridization using a biotin-labeled probe. J Med Virol 1985;15:373–82.

[224] Ntoulia M, Kaklamanis L, Valavanis C, et al. Her-2 DNA quantification of paraffin-embedded breast carcinomas with LightCycler real-time PCR in comparison to immunohistochemistry and chromogenic in situ hybridization. Clin Biochem 2006;39:942–6.

[225] Nuovo G. The utility of immunohistochemistry and in situ hybridization in placental pathology. Arch Pathol Lab Med 2006;130:979–83.

[226] Nuovo GJ, Richart RM. Buffered formalin is the superior fixative for the detection of HPV DNA by in situ hybridization. Am J Pathol 1989;134:837–42.

[227] Oberman F, Yisraeli JK. Two protocols for nonradioactive in situ hybridization to Xenopus oocytes. Trends Genet 1995;11:83–4.

[228] Olsson M, Ekblom M, Fecker L, Kurkinen M, Ekblom P. cDNA cloning and embryonic expression of mouse nuclear pore membrane glycoprotein 210 mRNA. Kidney Int 1999;56:827–38.

[229] Ooi A, Kobayashi M, Mai M, Nakanishi I. Amplification of c-erbB-2 in gastric cancer: detection in formalin-fixed, paraffin-embedded tissue by fluorescence in situ hybridization. Lab Invest 1998;78:345–51.

[230] Ozer A, Khaoustov VI, Mearns M, et al. Effect of hepatocyte proliferation and cellular DNA synthesis on hepatitis B virus replication. Gastroenterology 1996;110:1519–28.

[231] Paes MV, Lenzi HL, Nogueira AC, et al. Hepatic damage associated with dengue-2 virus replication in liver cells of BALB/c mice. Lab Invest 2009;89:1140–51.

[232] Palop JJ, Roberson ED, Cobos I. Step-by-step in situ hybridization method for localizing gene expression changes in the brain. Methods Mol Biol 2011;670:207–30.

[233] Pauletti G, Godolphin W, Press MF, Slamon DJ. Detection and quantitation of HER-2/neu gene amplification in human breast cancer archival material using fluorescence in situ hybridization. Oncogene 1996;13:63–72.

[234] Pazouki S, Hume R, Burchell A. A rapid combined immunocytochemical and fluorescence in situ hybridisation method for the identification of human fetal nucleated red blood cells. Acta Histochem 1996;98:29–37.

[235] Pearson BJ, Eisenhoffer GT, Gurley KA, et al. Formaldehyde-based whole-mount in situ hybridization method for planarians. Dev Dyn 2009;238:443–50.

[236] Penault-Llorca F, Bilous M, Dowsett M, et al. Emerging technologies for assessing HER2 amplification. Am J Clin Pathol 2009;132:539–48.

[237] Perez-de-Luque A, Gonzalez-Verdejo CI, Lozano MD, et al. Protein cross-linking, peroxidase and beta-1,3-endoglucanase involved in resistance of pea against Orobanche crenata. J Exp Bot 2006;57:1461–9.

[238] Petersen BL, Sorensen MC, Pedersen S, Rasmussen M. Fluorescence in situ hybridization on formalin-fixed and paraffin-embedded tissue: optimizing the method. Appl Immunohistochem Mol Morphol 2004;12:259–65.

[239] Pineau I, Barrette B, Vallieres N, Lacroix S. A novel method for multiple labeling combining in situ hybridization with immunofluorescence. J Histochem Cytochem 2006;54:1303–13.

[240] Pirnik Z, Kiss A. Detection of oxytocin mRNA in hypertonic saline Fos-activated PVN neurons: comparison of chromogens in dual immunocytochemical and in situ hybridization procedure. Endocr Regul 2002;36:23–39.

[241] Pollock AM, Toner M, McMenamin M, Walker J, Timon CI. Absence of Epstein-Barr virus encoded RNA and latent membrane protein (LMP1) in salivary gland neoplasms. J Laryngol Otol 1999;113:906–8.

[242] Popescu NC, Amsbaugh SC, Swan DC, DiPaolo JA. Induction of chromosome banding by trypsin/EDTA for gene mapping by in situ hybridization. Cytogenet Cell Genet 1985;39:73–4.

[243] Pouchkina NN, Stanchev BS, McQueen-Mason SJ. From EST sequence to spider silk spinning: identification and molecular characterisation of Nephila senegalensis major ampullate gland peroxidase NsPox. Insect Biochem Mol Biol 2003;33:229–38.

[244] Press MF, Slamon DJ, Flom KJ, et al. Evaluation of HER-2/neu gene amplification and overexpression: comparison of frequently used assay methods in a molecularly characterized cohort of breast cancer specimens. J Clin Oncol 2002;20:3095–105.

[245] Price PM, Hirschhorn K. In situ hybridization for gene mapping. Birth Defects Orig Artic Ser 1975;11:225–31.

[246] Price PM, Hirschhorn K. In situ hybridization for gene mapping. Cytogenet Cell Genet 1975;14.

[247] Price PM, Hirschhorn K. In situ hybridization of chromosome loci. Fed Proc 1975;34:2227–32.

[248] Pringle JH, Primrose L, Kind CN, Talbot IC, Lauder I. In situ hybridization demonstration of poly-adenylated RNA sequences in formalin-fixed paraffin sections using a biotinylated oligonucleotide poly d(T) probe. J Pathol 1989;158:279–86.

[249] Prodon F, Hanawa K, Nishida H. Actin microfilaments guide the polarized transport of nuclear pore complexes and the cytoplasmic dispersal of Vasa mRNA during GVBD in the ascidian Halocynthia roretzi. DevB iol 2009;330:377–88.

[250] Qian X, Bauer RA, Xu HS, Lloyd RV. In situ hybridization detection of calcitonin mRNA in routinely fixed, paraffin-embedded tissue sections: a comparison of different types of probes combined with tyramide signal amplification. ApplI mmunohistochemM ol Morphol 2001;9:61–9.

[251] Qian X, Guerrero RB, Plummer TB, Alves VF, Lloyd RV. Detectiono fh epatitisCvir us RNA in formalin-fixed paraffin-embedded sections with digoxigenin-labeled cRNA probes. DiagnM olP athol 2004;13:9–14.

[252] Quirk S, Chou WG, Polakowska R, et al. More precise localization of the human factor IX gene by in situ hybridization. Cytogenet Cell Genet 1985;39:121–4.

[253] Rajan S, Wischmeyer E, Karschin C, et al. THIK-1 and THIK-2, a novel subfamily of tandem pore domain K+ channels. JB iol Chem 2001;276:7302–11.

[254] Ramesh KH, Macera MJ, Verma RS. Rapid denaturation improves chromosome morphology and permits multiple hybridizations during fluorescence in situ hybridization. BiotechHis tochem 1997;72:141–3.

[255] Rasheed MS, Selth LA, Koltunow AM, Randles JW, Rezaian MA. Single-stranded DNA of Tomato leaf curl virus accumulates in the cytoplasm of phloem cells. Virology 2006;348:120–32.

[256] Recordati C, Radaelli E, Simpson KW, Scanziani E. As impleme thodf ort he production of bacterial controls for immunohistochemistry and fluorescent in situ hybridization. JM olHis tol 2008;39:459–62.

[257] Reisinger J, Rumpler S, Lion T, Ambros PF. Visualizationo fe pisomalan d integrated Epstein-Barr virus DNA by fiber fluorescencein s ituh ybridization. IntJ Can cer 2006;118:1603–8.

[258] Riethdorf S, Hoegel B, John B, et al. Prospective multi-centre study to validate chromogenic in situ hybridization for the assessment of HER2 gene amplification in specimens from adjuvant and metastatic breast cancer patients. JCan cerRe sC linOn col 2011;137:261–9.

[259] Robertson CS, Martin BA, Atkinson M. Varicella-zoster virus DNA in the oesophageal myenteric plexus in achalasia. Gut 1993;34:299–302.

[260] Robertson KL, Vora GJ. Locked nucleic acid flow cytometry-fluorescence in situ hybridization (LNA flow-FISH): a method for bacterial small RNA detection. J Vis Exp 2012:e3655.

[261] Rodriguez-Esteban C, Capdevila J, Kawakami Y, Izpisua Belmonte JC. Wnt signaling and PKA control nodal expression and left-right determination in the chick embryo. Development 2001;128:3189–95.

[262] Rodriguez-Inigo E, Casqueiro M, Bartolome J, et al. Detection of TT virus DNA in liver biopsies by in situ hybridization. Am J Pathol 2000;156:1227–34.

[263] Rojo G, Garcia-Beato R, Vinuela E, Salas ML, Salas J. Replication of African swine fever virus DNA in infected cells. Virology 1999;257:524–36.

[264] Romero A, Martín M, Cheang MC, et al. Assessment of Topoisomerase II Î± status in breast cancer by quantitative PCR, gene expression microarrays, immunohistochemistry, and fluorescence in situ hybridization. Am J Pathol 2011;178:1453–60.

[265] Rooker SM, Liu B, Helms JA. Role of Wnt signaling in the biology of the periodontium. Dev Dyn 2010;239:140–7.

[266] Ross JS, Fletcher JA, Bloom KJ, et al. HER-2/neu testing in breast cancer. Am J Clin Pathol 2003;120(Suppl):S53–71.

[267] Ross JS, Yang F, Kallakury BV, et al. HER-2/neu oncogene amplification by fluorescence in situ hybridization in epithelial tumors of the ovary. Am J Clin Pathol 1999;111:311–6.

[268] Roth J, Saremaslani P, Warhol MJ, Heitz PU. Improved accuracy in diagnostic immunohistochemistry, lectin histochemistry and in situ hybridization using a gold-labeled horseradish peroxidase antibody and silver intensification. Lab Invest 1992;67:263–9.

[269] Rummelt V, Wenkel H, Rummelt C, et al. Detection of varicella zoster virus DNA and viral antigen in the late stage of bilateral acute retinal necrosis syndrome. Arch Ophthalmol 1992;110:1132–6.

[270] Rummukainen JK, Salminen T, Lundin J, Joensuu H, Isola JJ. Amplification of c-myc oncogene by chromogenic and fluorescence in situ hybridization in archival breast cancer tissue array samples. Lab Invest 2001;81:1545–51.

[271] Russom A, Haasl S, Brookes AJ, Andersson H, Stemme G. Rapid melting curve analysis on monolayered beads for high-throughput genotyping of single-nucleotide polymorphisms. Anal Chem 2006;78:2220–5.

[272] Ryan JL, Morgan DR, Dominguez RL, et al. High levels of Epstein-Barr virus DNA in latently infected gastric adenocarcinoma. Lab Invest 2009;89:80–90.

[273] Rychkova N, Stahl S, Gaetzner S, Felbor U. Non-heparan sulfate-binding interactions of endostatin/collagen XVIII in murine development. Dev Dyn 2005;232:399–407.

[274] Sakuma K, Uozaki H, Chong JM, et al. Cancer risk to the gastric corpus in Japanese, its correlation with interleukin-1beta gene polymorphism (+3953*T) and Epstein-Barr virus infection. Int J Cancer 2005;115:93–7.

[275] Sanchez-Velasco P, Ocejo-Vinyals JG, Flores R, et al. Simultaneous multiorgan presence of human herpesvirus 8 and restricted lymphotropism of Epstein-Barr virus DNA sequences in a human immunodeficiency virus-negative immunodeficient infant. J Infect Dis 2001;183:338–42.

[276] Sato Y, Demura T, Yamawaki K, et al. Isolation and characterization of a novel peroxidase gene ZPO-C whose expression and function are closely associated with lignification during tracheary element differentiation. Plant Cell Physiol 2006;47:493–503.

[277] Sawitzke AL, Chapman SC, Bleyl SB, Schoenwolf GC. Improvements in histological quality and signal retention following in situ hybridization in early chick embryos using plastic resin and recolorization. Biotech Histochem 2005;80:35–41.

[278] Saxe DF, Takahashi N, Hood L, Simon MI. Localization of the human myelin basic protein gene (MBP) to region 18q22----qter by in situ hybridization. Cytogenet Cell Genet 1985;39:246–9.

[279] Schimak MP, Toenshoff ER, Bright M. Simultaneous 16S and 18S rRNA fluorescence in situ hybridization (FISH) or LR White sectioned demonstrated in Vestimentifera (Siboglinidae) tubeworms. Acta Histochem 2012;114:122–30.

[280] Schneider M, Vogt Weisenhorn DM, Seiler A, et al. Embryonic expression profile of phospholipid hydroperoxide glutathione peroxidase. Gene Expr Patterns 2006;6:489–94.

[281] Selvarajan S, Bay BH, Choo A, et al. Effect of fixation period on HER2/neu gene amplification detected by fluorescence in situ hybridization in invasive breast carcinoma. J Histochem Cytochem 2002;50:1693–6.

[282] Shao JY, Zhang Y, Li YH, et al. Comparison of Epstein-Barr virus DNA level in plasma, peripheral blood cell and tumor tissue in nasopharyngeal carcinoma. Anticancer Res 2004;24:4059–66.

[283] Shapshak P, Tourtellotte WW, Nakamura S, et al. Subacute sclerosing panencephalitis: measles virus matrix protein nucleic acid sequences detected by in situ hybridization. Neurology 1985;35:1605–9.

[284] Sher D, Fishman Y, Melamed-Book N, Zhang M, Zlotkin E. Osmotically driven prey disintegration in the gastrovascular cavity of the green hydra by a pore-forming protein. FASEB J 2008;22:207–14.

[285] Shibata Y, Fujita S, Takahashi H, Yamaguchi A, Koji T. Assessment of decalcifying protocols for detection of specific RNA by non-radioactive in situ hybridization in calcified tissues. Histochem Cell Biol 2000;113:153–9.

[286] Simone G, Mangia A, Malfettone A, et al. Chromogenic in situ hybridization to detect EGFR gene copy number in cell blocks from fine-needle aspirates of non small cell lung carcinomas and lung metastases from colo-rectal cancer. J Exp Clin Cancer Res 2010;29:125.

[287] Simonetti E, Veronico P, Melillo MT, et al. Analysis of class III peroxidase genes expressed in roots of resistant and susceptible wheat lines infected by Heterodera avenae. Mol Plant Microbe Interact 2009;22:1081–92.

[288] Smith MD, Ahern M, Coleman M. The use of combined immunohistochemical labeling and in situ hybridization to colocalize mRNA and protein in tissue sections. Methods Mol Biol 2006;326:235–45.

[289] Smith SJ, Kotecha S, Towers N, Latinkic BV, Mohun TJ. XPOX2-peroxidase expression and the XLURP-1 promoter reveal the site of embryonic myeloid cell development in Xenopus. Mech Dev 2002;117:173–86.

[290] Sollberg S, Peltonen J, Uitto J. Combined use of in situ hybridization and unlabeled antibody peroxidase anti-peroxidase methods: simultaneous detection of type I procollagen mRNAs and factor VIII-related antigen epitopes in keloid tissue. Lab Invest 1991;64:125–9.

[291] Solstad T, Stenvik J, Jorgensen TO. mRNA expression patterns of the BPI/LBP molecule in the Atlantic cod (Gadus morhua L.). Fish Shellfish Immunol 2007;23:260–71.

[292] Spacek M, Hubacek P, Markova J, et al. Plasma EBV-DNA monitoring in Epstein-Barr virus-positive Hodgkin lymphoma patients. APMIS 2011;119:10–16.

[293] Stylianopoulou E, Lykidis D, Ypsilantis P, et al. Ar apidan dh ighlys ensitiveme thod of non radioactive colorimetric in situ hybridization for the detection of mRNA on tissue sections. PLoSOn e 2012;7:e33898.

[294] Su MT, Golden K, Bodmer R. X-gal staining of Drosophila embryos compatible with antibody staining or in situ hybridization. Biotechniques 1998;24:918–20. 922

[295] Sukarova-Angelovska E, Piperkova K, Sredovska A, Ilieva G, Kocova M. Implementation of fluorescent in situ hybridization (FISH) as a method for detecting microdeletion syndromes—our first experiences. Prilozi 2007;28:87–98.

[296] Suwanwalaikorn S, Van Auken M, Kang MI,e tal. Sites electivityo fo steoblastge ne expression response to thyroid hormone localized by in situ hybridization. AmJ Physiol 1997;272:E212–7.

[297] Suwiwat S, Pradutkanchana J, Ishida T, Mitarnun W. Quantitativean alysiso fc ell-free Epstein-Barr virus DNA in the plasma of patients with peripheral T-cell and NK-cell lymphomas and peripheral T-cell proliferative diseases. JClin Vir ol 2007;40:277–83.

[298] Suzuki SO, Iwaki T. Non-isotopic in situ hybridization of CD44 transcript in formalin-fixed paraffin-embedded sections. Brain research. BrainRe sP rotoc 1999;4:29–35.

[299] Swain GP, Jacobs AJ, Frei E, Selzer ME. A method for in situ hybridization in wholemounted lamprey brain: neurofilament expression in larvae and adults. ExpNe urol 1994;126:256–69.

[300] Taiambas E, Alexpopoulou D, Lambropoulou S,e tal. Targetingt opoisomeraseI Iain endometrial adenocarcinoma: a combined chromogenic in situ hybridization and immunohistochemistry study based on tissue microarrays. IntJ Gyn ecolCan cer 2006;16:1424–31.

[301] Tajbakhsh S, Samarbaf-Zadeh AR, Moosavian M. Comparisono ffl uorescent in situ hybridization and histological method for the diagnosis of Helicobacter pylori in gastric biopsy samples. MedS ciM onit 2008;14:BR183–BR187.

[302] Takada S, Yoshino T, Taniwaki M, et al. Involvement of the chromosomal translocation t(11;18) in some mucosa-associated lymphoid tissue lymphomas and diffuse large B-cell lymphomas of the ocular adnexa: evidence from multiplex reverse transcriptase-polymerase chain reaction and fluorescence in situ hybridization on using formalin-fixed, paraffin-embeddeds pecimens. ModP athol 2003;16:445–52.

[303] Talley EM, Solorzano G, Lei Q, Kim D, Bayliss DA. CNS distribution of members of the two-pore-domain (KCNK) potassium channel family. J Neurosci 2001;21:7491–505.

[304] Tamura K, Fukao Y, Iwamoto M, Haraguchi T, Hara-Nishimura I. Identification and characterization of nuclear pore complex components in Arabidopsis thaliana. Plant Cell 2010;22:4084–97.

[305] Tan LH, Do E, Chong SM, Koay ES. Detection of ALK gene rearrangements in formalin-fixed, paraffin-embedded tissue using a fluorescence in situ hybridization (FISH) probe: a search for optimum conditions of tissue archiving and preparation for FISH. Mol Diagn 2003;7:27–33.

[306] Tanaka H, Matsuhisa A, Nagamori S, et al. Characteristics and significance of albumin-positive hepatocytes in analbuminemic rats. Eur J Cell Biol 1990;53:267–74.

[307] Tang KF, Redman RM, Pantoja CR, et al. Identification of an iridovirus in Acetes erythraeus (Sergestidae) and the development of in situ hybridization and PCR method for its detection. J Invertebr Pathol 2007;96:255–60.

[308] Tang L, Tanaka Y, Enomoto N, Marumo F, Sato C. Detection of hepatitis C virus RNA in hepatocellular carcinoma by in situ hybridization. Cancer 1995;76:2211–6.

[309] Tanner M, Gancberg D, Di Leo A, et al. Chromogenic in situ hybridization: a practical alternative for fluorescence in situ hybridization to detect HER-2/neu oncogene amplification in archival breast cancer samples. Am J Pathol 2000;157:1467–72.

[310] Teo CG, Ringler S, Krakowka S. JC virus genomes in progressive multifocal leukoencephalopathy: detection using a sensitive non-radioisotopic in situ hybridization method. J Pathol 1989;157:135–40.

[311] Terenghi G, Polak JM. Preparation of tissue sections and slides for mRNA hybridization. Methods Mol Biol 1994;28:187–91.

[312] Thakurta AG, Gopal G, Yoon JH, Saha T, Dhar R. Conserved nuclear export sequences in Schizosaccharomyces pombe Mex67 and human TAP function in mRNA export by direct nuclear pore interactions. J Biol Chem 2004;279:17434–17442.

[313] Tholouli E, Hoyland JA, Di Vizio D, et al. Imaging of multiple mRNA targets using quantum dot based in situ hybridization and spectral deconvolution in clinical biopsies. Biochem Biophys Res Commun 2006;348:628–36.

[314] Tolivia J, Navarro A, del Valle E, et al. Application of Photoshop and Scion image analysis to quantification of signals in histochemistry, immunocytochemistry and hybridocytochemistry. Anal Quant Cytol Histol 2006;28:43–53.

[315] Todorovic-Rakovic N, Jovanovic D, Neskovic-Konstantinovic Z, Nikolic-Vukosavljevic D. Prognostic value of HER2 gene amplification detected by chromogenic in situ hybridization (CISH) in metastatic breast cancer. Exp Mol Pathol 2007;82:262–8.

[316] Tournier I, Bernuau D, Poliard A, Schoevaert D, Feldmann G. Detection of albumin mRNAs in rat liver by in situ hybridization: usefulness of paraffin embedding and comparison of various fixation procedures. J Histochem Cytochem 1987;35:453–9.

[317] Trapp BD, Small JA, Pulley M, Khoury G, Scangos GA. Dysmyelination in transgenic mice containing JC virus early region. Ann Neurol 1988;23:38–48.

[318] Trask B, van den Engh G, Landegent J, in de Wal NJ, van der Ploeg M. Detection of DNA sequences in nuclei in suspension by in situ hybridization and dual beam flow cytometry. Science 1985;230:1401–3.

[319] Treilleux I, Mallein-Gerin F, Le Guellec D, et al. [Detection of mRNA by in situ hybridization in tissues fixed in Bouin's fixative and embedded in paraffin] Detection des ARNm par hybridation in situ sur des tissus fixes dans le liquide de Bouin et inclus en paraffine. Ann Pathol 1991;11:47–53.

[320] Tse C, Brault D, Gligorov J, et al. Evaluation of the quantitative analytical methods real-time PCR for Her-2 gene quantification and ELISA of serum HER-2 protein and comparison with fluorescence in situ hybridization and immunohistochemistry for determining HER-2 status in breast cancer patients. Clin Chem 2005;51:1093–101.

[321] Tsiambas E, Karameris A, Lazaris AC, et al. EGFR alterations in pancreatic ductal adenocarcinoma: a chromogenic in situ hybridization analysis based on tissue microarrays. Hepatogastroenterology 2006;53:452–7.

[322] Tsiambas E, Karameris A, Tiniakos DG, Karakitsos P. Evaluation of topoisomerase IIa expression in pancreatic ductal adenocarcinoma: a pilot study using chromogenic in situ hybridization and immunohistochemistry on tissue microarrays. Pancreatology 2007;7:45–52.

[323] Tyrrell L, Elias J, Longley J. Detection of specific mRNAs in routinely processed dermatopathology specimens. Am J Dermatopathol 1995;17:476–83.

[324] Uehara F, Ohba N, Nakashima Y, et al. A fixative suitable for in situ hybridization histochemistry. J Histochem Cytochem 1993;41:947–53.

[325] Uehara H, Kunitomi Y, Ikai A, Osada T. mRNA detection of individual cells with the single cell nanoprobe method compared with in situ hybridization. J Nanobiotechnology 2007;5:7.

[326] Umland O, Ulmer AJ, Vollmer E, Goldmann T. HOPE fixation of cytospin preparations of human cells for in situ hybridization and immunocytochemistry. J Histochem Cytochem 2003;51:977–80.

[327] Urieli-Shoval S, Meek RL, Hanson RH, et al. Preservation of RNA for in situ hybridization: Carnoy's versus formaldehyde fixation. J Histochem Cytochem 1992;40:1879–85.

[328] Valencia A, Cervera J, Such E, et al. A new reliable fluorescence in situ hybridization method for identifying multiple specific cytogenetic abnormalities in acute myeloid leukemia. Leuk Lymphoma 2010;51:680–5.

[329] van de Rijke FM, Vrolijk H, Sloos W, Tanke HJ, Raap AK. Sample preparation and in situ hybridization techniques for automated molecular cytogenetic analysis of white blood cells. Cytometry 1996;24:151–7.

[330] Van Patten SM, Donaldson LF, McGuinness MP, et al. Specific testicular cellular localization and hormonal regulation of the PKIalpha and PKIbeta isoforms of the inhibitor protein of the cAMP-dependent protein kinase. J Biol Chem 1997;272:20021–20029.

[331] Varghese R, Gagliardi AD, Bialek PE, et al. Overexpression of human stanniocalcin affects growth and reproduction in transgenic mice. Endocrinology 2002;143:868–76.

[332] Verma S, Frambach GE, Seilstad KH, et al. Epstein-Barr virus-associated B-cell lymphoma in the setting of iatrogenic immune dysregulation presenting initially in the skin. J Cutan Pathol 2005;32:474–83.

[333] Vocaturo A, Novelli F, Benevolo M, et al. Chromogenic in situ hybridization to detect HE-2/neu gene amplification in histological and ThinPrep-processed breast cancer fine-needle aspirates: a sensitive and practical method in the trastuzumab era. Oncologist 2006;11:878–86.

[334] Vollmer E, Galle J, Lang DS, et al. The HOPE technique opens up a multitude of new possibilities in pathology. Rom J Morphol Embryol 2006;47:15–19.

[335] Wahl GM, Vitto L, Padgett RA, Stark GR. Single-copy and amplified CAD genes in Syrian hamster chromosomes localized by a highly sensitive method for in situ hybridization. MolCe llB iol 1982;2:308–19.

[336] Wakamatsu N, King DJ, Seal BS, Brown CC. Detectiono fNe wcastled iseasevir us RNA by reverse transcription-polymerase chain reaction using formalin-fixed, paraffin embedded tissue and comparison with immunohistochemistry and in situ hybridization. JVe tDiagn I nvest 2007;19:396–400.

[337] Wang J, Kawde AN, Jan MR. Carbon-nanotube-modified electrodes for amplified enzyme-based electrical detection of DNA hybridization. BiosensB ioelectron 2004;20:995–1000.

[338] Wang J, Zhou HY, Salih E, et al. Site-specific in vivo calcification and osteogenesis stimulated by bone sialoprotein. CalcifT issue Int 2006;79:179–89.

[339] Wang NS, Wu ZL, Zhang YE, Liao LT. Existence and significance of hepatitis B virus DNA in kidneys of IgA nephropathy. WorldJ Gastroenterol 2005;11:712–6.

[340] Warner CK, Whitfield SG, Fekadu M, Ho H. Proceduresf orr eproducibled etectiono f rabies virus antigen mRNA and genome in situ in formalin-fixedt issues. JVir olM ethods 1997;67:5–12.

[341] Warren KC, Coyne KJ, Waite JH, Cary SC. Use of methacrylate de-embedding protocols for in situ hybridization on semithin plastic sections with multiple detection strategies. JHis tochemCyt ochem 1998;46:149–55.

[342] Weiss LM, Chen YY. Effects of different fixatives on detection of nucleic acids from paraffin-embedded tissues by in situ hybridization using oligonucleotide probes. JHis tochemCyt ochem 1991;39:1237–42.

[343] Wente SR, Blobel G. A temperature-sensitive NUP116 null mutant forms a nuclear envelope seal over the yeast nuclear pore complex thereby blocking nucleocytoplasmic traffic. JCe llB iol 1993;123:275–84.

[344] Wilkens L, Werner M, Nolte M, et al. Influence of formalin fixation on the detection of cytomegalovirus by polymerase chain reaction in immunocompromised patients and correlation to in situ hybridization, immunohistochemistry, and serological data. DiagnM olP athol 1994;3:156–62.

[345] Willmore-Payne C, Metzger K, Layfield LJ. Effects of fixative and fixation protocols on assessment of Her-2/neu oncogene

amplification status by fluorescence in situ hybridization. ApplI mmunohistochemM ol Morphol 2007;15:84–7.

[346] Wixom CR, Albers EA, Weidner N. Her2 amplification: correlation of chromogenic in situ hybridization with immunohistochemistry andfl uorescencein s ituh ybridization. Appl Immunohistochem Mol Morphol 2004;12:248–51.

[347] Wood NB, Sheikholeslami M, Pool M, Coon JS. PCRp roductiono fad igoxigenin-labeled probe for the detection of human cytomegalovirusin t issues ections. DiagnM ol Pathol 1994;3:200–8.

[348] Wright CA, Haffajee Z, van Iddekinge B, Cooper K. Detectiono fh erpess implexvir us DNA in spontaneous abortions from HIV-positive women using non-isotopic in situ hybridization. JP athol 1995;176:399–402.

[349] Wu LT, Chu KH. Characterization of an ovary-specific glutathione peroxidase from the shrimp Metapenaeus ensis and its role inc rustaceanr eproduction. CompB iochem PhysiolB B iochemM olB iol 2010;155:26–33.

[350] Xu X, Pan Y, Wang X. Alterations in the expression of lipid and mechano-gated two-pore domain potassium channel genes in rat brainfo llowingc hronicc erebralis chemia. BrainRe sM olB rainRe s 2004;120:205–9.

[351] Xue SA, Labrecque LG, Lu QL, et al. Promiscuous expression of Epstein-Barr virus genes in Burkitt's lymphoma from the centralAf ricanc ountryM alawi. IntJ Can cer 2002;99:635–43.

[352] Yagyuu T, Ikeda E, Ohgushi H, et al. Hard tissue-forming potential of stem/progenitor cells in human dental follicle and dental papilla. ArchOr alB iol 2010;55:68–76.

[353] Yamamoto T, Nakamura Y, Kishimoto K, etal. Epstein-Barrvir us(EBV)-infectedc ells were frequently but dispersely detected in T-cell lymphomas of various types by in situ hybridization with an RNA probe specific toE BV-specificn uclearan tigen1 . VirusRe s 1999;65:43–55.

[354] Yan F, Wu X, Crawford M, et al. The search for an optimal DNA, RNA, and protein detection by in situ hybridization, immunohistochemistry, and solution-based methods. Methods 2010;52:281–6.

[355] Yang W, Maqsodi B, Ma Y, et al. Direct quantification of gene expression in homogenates of formalin-fixed, paraffin-embeddedt issues. Biotechniques 2006;40:481–6.

[356] Yang YQ, Li XT, Rabie AB, Fu MK, Zhang D. Humanp eriodontalligame ntc ellse xpress osteoblastic phenotypes under intermittent forcelo adingin vit ro. FrontB iosci 2006;11:776–81.

[357] Yilmaz O, Demiray E. Clinical role and importance of fluorescence in situ hybridization method in diagnosis of H pylori infection and determination of clarithromycin resistancein Hp ylorie radicationt herapy. WorldJ Gas troenterol 2007;13:671–5.

[358] Yong VC, Ong KW, Sidik SM, Rosli R, Chong PP. Amo difiedin s ituRT -PCR method for localizing fungal-specific gene expression in Candida-infected mice renal cells. JM icrobiolM ethods 2009;79:242–5.

[359] Yoshimura S, Suemizu H, Taniguchi Y, et al. The human plasma glutathione peroxidase-encoding gene: organization, sequence and localizationt oc hromosome5q 32. Gene 1994;145:293–7.

[360] Zhang DT, Hu LH, Yang YZ. Detection of three common G6PD gene mutations in Chinesein dividualsb yp robeme ltingc urves. ClinB iochem 2005;38:390–4.

[361] Zhang S, Troyer DL, Kapil S, et al. Detection of proviral DNA of bovine immunodeficiency virus in bovine tissues by polymerase chain reaction(PCR)an dP CRin s ituh ybridization. Virology 1997;236:249–57.

The Basics of Immunohistochemistry

5

INTRODUCTION

The previous chapter focused on the basics of in situ hybridization, with a strong focus on the time kinetics and biochemistry of the processes. I intentionally included a lot of information about immunohistochemistry. I did so to underline one of the main foundations of this textbook: specifically, in situ hybridization and immunohistochemistry have so much in common that it may be more logical to unite them as the in situ-based molecular pathology tests.

Still, the fact that the epitope of proteins is a direct part of the three-dimensional protein cross-linked network of formalin-fixed, paraffin-embedded tissues whereas DNA and RNA targets are only surrounded by this cage does result in some important differences. These differences are the focus of this chapter.

STEP ONE: FIXING THE TISSUE OR CELLS

DIFFERENT FIXATIVES AND IMMUNOHISTOCHEMISTRY

In the early days of immunohistochemistry, which can be defined as the time before antigen retrieval, there were many epitopes that were basically undetectable with immunohistochemistry when using formalin-fixed, paraffin-embedded tissues. This naturally led to investigators trying unfixed cells/tissues and other noncross-linking fixatives, with variable success. To this date, companies that sell antibodies for research and/or diagnostic work often make a distinction between "antibody validated in frozen tissues" and "antibody validated in formalin-fixed, paraffin-embedded tissues."

As we discussed in previous chapters, it certainly makes a great deal of sense to use either frozen, unfixed tissues or tissues fixed in denaturing agents for immunohistochemistry. It seems logical that this should greatly reduce the chance of a false-negative reaction with immunohistochemistry due to "hidden epitopes." It is certainly understandable that some investigators have devoted a great deal of energy to optimizing in situ hybridization with unfixed, frozen samples. This is the simplest way to avoid the formalin-induced dense latticework of cross-linked

macromolecules that our probe and other in situ reagents must transverse to achieve successful in situ hybridization. Also, we can be assured that the DNA, RNA, and proteins in cryostat sections that are unfixed will be pristine and basically identical to the natural state. We can also avoid the protease digestion and/or antigen retrieval steps in such samples.

However, we can see two potential problems by using frozen, fixed tissues or alcohol/acetone-fixed tissues for immunohistochemistry. First, the lack of a strong three-dimensional protein cross-linked cytoskeleton may render the tissue more susceptible to unwanted morphologic changes. At worst, the morphology is destroyed; in other cases, the crisp morphologic detail we regularly see with formalin-fixed, paraffin-embedded tissues is likely not going to be evident. However, the second reason may be the more important. The complex cross-linked network of proteins may well play a key role in successful immunohistochemistry in facilitating both the movement and the "docking" of the key immunohistochemistry reagents to the epitope of interest. As we saw in previous chapters, this three-dimensional protein cross-linked cage may indeed be the essential variable for maximizing the immunohistochemical signal and, just as importantly, in limiting its diffusion from the subcellular component from which it arises. We will see shortly that the three-dimensional protein cross-linked cage may indeed be the essential variable in limiting background as well. Let's also discuss a bit how this idea of a rigid "organic-based cage" is used in solid-state organic chemistry, because this may not be too far removed from what is occurring every time we perform immunohistochemistry inside the fixed cells.

Ion exchange resins are commonly used in organic chemistry and also in many household applications. This type of resin consists of an insoluble matrix made from an organic substrate that can polymerize. The resultant material has a complex three-dimensional structure of pores, on the surface of which are sites that can trap a wide variety of chemicals such as magnesium/calcium (water softener), organic chemicals, and other poisons such as heavy metals (water purifier) and unwanted chemicals in the body such as cholesterol (this ion exchange resin is called cholestyramine). The trapping of ions (or the ionic change of a large chemical) takes place with simultaneous releasing of other ions; thus, the

133

DOI: http://dx.doi.org/10.1016/B978-0-12-415944-0.00005-X

process is called *ion exchange*. In industry, it is easy to generate different ion exchange resins based on pore size and charge for different purposes. Of course, these ion exchange resins are not based solely on the ionic charge of the molecules that are being exposed to the resin. Other forces such as hydrogen bonding, hydrophobicity, antibodies that recognize specific amino acids, or probes that recognize specific DNA sequences and such can be incorporated into the resin. You can easily see the analogy with the three-dimensional protein cross-linked network of a formalin-fixed cell.

We have explored a lot of data that suggest formalin-fixed, paraffin-embedded tissues contain the equivalent of a complex organic "ion exchange and hydrogen bonding resin." If we realize that any such model and term needs, at this stage, to be oversimplistic, the different forces that may facilitate the "trapping" and movement of our key reagents with immunohistochemistry include:

1. Ion exchange (amino acid R side chains with + or − charge)
2. Hydrogen bonding (amino acid R side chains with hydrogen bonding potential)
3. Hydrophobic/hydrophilic pockets (amino acid R side chains that favor such environments)
4. "Breathability" of proteins (probably influenced primarily by the physical constraints of the actual cross-linking)

Since frozen, unfixed tissues do not possess this complex three-dimensional macromolecule cross-linked network, it follows that they may not be able in most cases to outperform formalin-fixed, paraffin-embedded tissues for immunohistochemistry using a given antigen. Of course, in previous chapters we saw data with frozen, unfixed tissues that supported this statement. It was only when we ourselves created this "three-dimensional protein cross-linked cage" on frozen (cryostat) tissue that we were able to get it to perform as well as formalin-fixed, paraffin-embedded tissues for microRNA in situ hybridization and immunohistochemistry.

At this stage, I have tested thousands of primary antibodies by immunohistochemistry in which the antibody was listed as "suitable for frozen unfixed tissue, not validated in formalin-fixed, paraffin-embedded tissues." In my experience, about 85% of such antibodies can yield an excellent signal with immunohistochemistry. The key is simply to vary the predigestion conditions and the primary antibody concentration, as we discuss at length in Chapter 8 on protocol. This certainly makes sense based on the data we discussed with regards to ethanol, acetone, unfixed, and formalin-fixed, paraffin-embedded tissues in previous chapters. The theory regarding the three-dimensional protein cross-linked network induced by formalin fixation would predict that any immunohistochemical reaction, when optimized, should be superior in formalin-fixed, paraffin-embedded tissues when compared to unfixed/denatured fixed tissues. This is not to say that determining the optimal conditions in the latter is trivial. It takes a series of experiments and a sharp eye for the histopathologic distribution of the signal. Still, if you do the side-by-side experiments, I think

you will see that formalin-fixed, paraffin-embedded tissues will, in at least 90% of cases, yield a signal with any antibody with immunohistochemistry and, in all cases, when optimized, yield better data than using tissues that are unfixed or fixed with acetone/ethanol/equivalent denaturing fixative.

Remember that we can have the best of both worlds. If you have access to cryostat sections, then fixing the unstained silane-coated slide in formalin for a few hours will generate the three-dimensional macromolecule cross-linked cage, yet the DNA and RNA will be pristine and undamaged. Of course, in most cases, frozen, unfixed tissue is not available for a given case. Still, if you have access to cryostat sections, I strongly recommend that you fix some in 10% buffered formalin after sectioning. You will get tissue sections that truly are optimal for in situ hybridization and immunohistochemistry.

The literature on frozen sections for immunohistochemistry is quite variable. Many papers indicate that formalin-fixed, paraffin-embedded tissue sections perform overall just as well as cryostat sections. Some papers have concluded that formalin-fixed, paraffin-embedded tissues are optimal for nuclear antigens, but that cryostat sections may be preferable for cell membrane antigens. Again, I recommend you try these comparisons for yourself and make your own decision, remembering to use the necessary conditions to ensure that you have optimized each detection assay.

BOUIN'S FIXATIVE AND IMMUNOHISTOCHEMISTRY

The data on Bouins's fixative and in situ hybridization are very consistent, as discussed in the preceding chapter. The picric acid in Bouin's solution rapidly degrades DNA and RNA and, thus, a few hours of fixation with Bouin's solution is enough to markedly weaken the in situ hybridization signal. The data on Bouin's solution and immunohistochemical are less clear. Some groups describe that this fixative will not interfere with immunohistochemistry, whereas other groups note that prolonged fixation in Bouin's solution (as short as a few hours) may be sufficient to weaken the signal with immunohistochemistry.

The data on other noncross-linking fixatives and immunohistochemistry are likewise somewhat variable. In general, most papers indicate no large advantage for immunohistochemistry when using tissues fixed with the denaturing fixatives compared to formalin-fixed, paraffin-embedded tissues.

My recommendation is to always start with formalin-fixed, paraffin-embedded tissues when doing immunohistochemistry. If you have procured fresh tissue, then fix part of it in formalin. If you use the optimization protocol for formalin-fixed, paraffin-embedded tissues and immunohistochemistry that we discuss in Chapter 7, I am confident that you will get a robust signal with excellent morphology for at least 95% of the antigens present in "physiologically active" copy numbers. I recommend resorting to cryostat sections only if all your efforts with immunohistochemistry and formalin-fixed, paraffin-embedded tissues fail. However, before you go to cryostat sections in this scenario, I recommend testing the unfixed tissue for the protein by Western blot. A negative result with Western blot would reveal what probably is the most common reason you cannot detect an antigen by immunohistochemistry: the epitope is either absent or present in very low copy numbers in the tissue.

A Quick Review of the Consequences of Formalin Fixation in Performing Immunohistochemistry

This is a good time in the book to restate something that is very important to me: I don't want this book to be a "cookbook." Although the protocols are certainly a key part of any book like this, it is important that you understand the theoretical foundations used to derive the protocols.

As you know, I have proposed that different proteins, RNA, and DNA targets show such marked variability in the pretreatment conditions needed to optimize the accessibility of the target to the probe/antibody, even in a given tissue, because each target likely has a biochemically distinct "cage" that varies greatly from target to target. I hope the theory is useful as you troubleshoot your own laboratory's in situ hybridization and immunohistochemistry data. But let's forget the theory for a moment and look at some simple but useful data.

I reviewed the optimizing protocols I generated for several hundred antibodies for immunohistochemistry and compared these data for my optimizing protocols for RNA (microRNA) and DNA (HPV) in situ hybridization. Of course, each optimization was done with formalin-fixed, paraffin-embedded tissues. In each case, I compared *no* pretreatment to *protease* digestion to *antigen retrieval.* Here is the breakdown for optimal immunohistochemistry:

Optimal Pretreatment Conditions for Immunohistochemistry

1. No pretreatment	Optimal for about 5% of targets
2. Protease digestion	Optimal for about 15% of targets
3. Antigen retrieval	Optimal for about 80% of targets

It was very rare to find an antigen that gave excellent results with all three possible basic pretreatment regimes: nothing, protease, or antigen retrieval. It was also relatively rare to see antigen retrieval and protease predigestion give equivalent strong results for immunohistochemistry; one of the two pretreatment regimes usually was clearly better.

In comparison, here are the data I generated after optimizing in situ hybridization on hundreds of tissues:

Optimal Pretreatment Conditions for In Situ Hybridization

1. No pretreatment	Optimal for about 10% of targets
2. Protease digestion	Optimal for about 15% of targets
3. DNA/RNA retrieval	Optimal for about 5% of targets
4. Protease digestion or DNA/RNA retrieval	Optimal for about 50% of targets
5. All three	Optimal for about 20% of targets

Again, to clarify the preceding table, category 4 refers to in situ hybridization where *either* protease digestion *or* DNA/RNA retrieval each gave excellent and equivalent results, whereas category 5 refers to situations in which I got superior in situ hybridization results whether *no* pretreatment, protease digestion, or DNA/RNA retrieval was done.

Why is it that in the vast majority of cases when doing immunohistochemistry, antigen retrieval is required for optimal signal, whereas for in situ hybridization you are much more likely to get optimal/excellent results whether you do protease digestion and/or DNA/RNA retrieval alone? Indeed, it is not unusual when doing immunohistochemistry in which antigen retrieval is optimal to see the protease digestion step either eliminate or weaken the signal. Why does this happen? Can we fit these observations into the model of the three-dimensional protein cross-linked cage that we have discussed at length in this book?

A fundamental difference between protein targets and RNA/DNA targets is that the former are actually a component of the three-dimensional protein/protein cross-linked network, whereas the latter are "innocent bystanders" that are linked to the cage but probably don't play a major role in its "construction" or structure.

Let's make two suppositions:

1. Since DNA and RNA targets are encased within the three-dimensional protein/protein cross-linked network, whereas protein epitopes are part of this cage, pore size will be a more important issue for in situ hybridization than immunohistochemistry.
2. Protease digestion is more efficient at enlarging pores in the three-dimensional protein/protein cross-linked network than antigen retrieval.

If these two suppositions are true, this would explain in part the disparity in the optimal pretreatment conditions with immunohistochemistry versus in situ hybridization. That is, since the protein epitope is part of the three-dimensional protein cross-linked cage, pore size should not be a major issue, but, rather, the conformation of the epitope in the cage should be the key. Antigen retrieval would, of course, directly affect the conformation of the epitope. DNA and, especially, microRNA targets, if encased in a dense cross-linked cage, would require strong protease digestion with or without DNA/RNA retrieval. This certainly would also explain the observation that microRNA in situ hybridization usually requires very strong protease, even if the protein epitopes (or HPV DNA target) in the same tissue do not. RNA retrieval by itself usually does not work for miRNA in situ hybridization. Of course, we need to keep doing experiments to test these suppositions; more such data will be presented later in the book. For now, I want to make you aware of these disparities so we can approach both our protocols and troubleshooting thinking process based on these theoretical constructs (see Figures 5-1 and 5-2). Figure 5-1 shows an artist's conception of the relationship of the epitope to the three-dimensional protein cross-linked network. Panel A shows the most common scenario: since the epitope is part of the cage and slightly hidden due to the formalin fixation, antigen retrieval will nicely expose the antigen and optimize the signal. Panel B is less common though certainly not unusual; the epitope is now on the "edge"

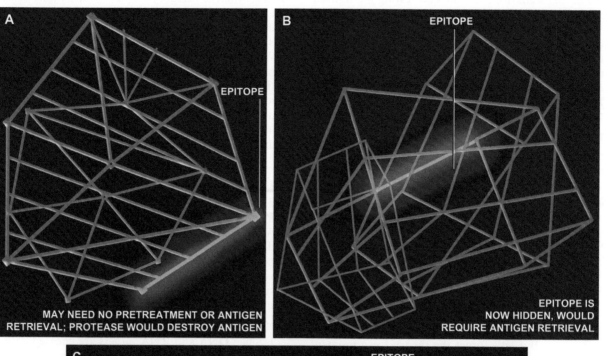

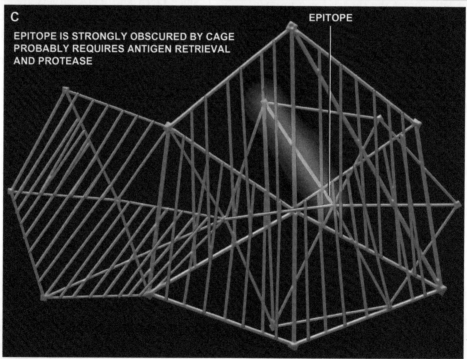

Figure5-1   Artist's conception of why antigen retrieval is usually optimal for immunohistochemistry. Panel B shows the artist's conception of the epitope being part of the three-dimensional protein cross-linked cage and partly obscured by it. Antigen retrieval will be sufficient in most cases to optimize the immunohistochemistry signal. Compare this to panel A, where the epitope is on the "edge" of the cage; here, either no pretreatment or weak antigen retrieval should suffice to optimize the signal, and protease digestion should destroy the signal. Panel C shows an epitope deeply hidden in the cage and surrounded by a dense complex protein latticework. Such epitopes usually need strong protease digestion and, at times, antigen retrieval, in order to optimize the signal.

of the cage and, though antigen retrieval or no pretreatment can give us a good signal, protease digestion will usually destroy the signal. Panel C is relatively rare—a well-hidden epitope that needs antigen retrieval and protease to expose it. Panel C for immunohistochemistry is equivalent to the common scenario for microRNAs seen in Figure 5-2. The miRNA is deeply encased in the three-dimensional protein cross-linked cage and needs some serious protease digestion for the in situ hybridization reagents to access it.

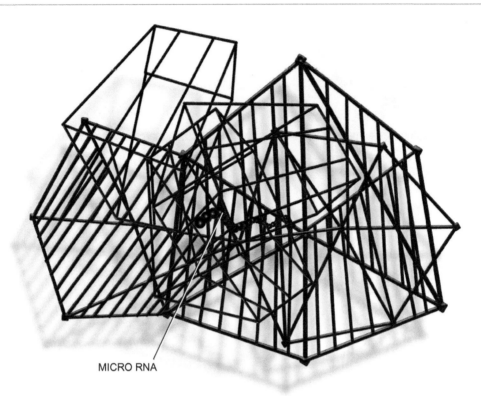

MICRO RNA

CAGE MORE DIFFICULT TO PENETRATE
1) NEEDS STRONG PRETREATMEMT
2) DIFFUSION IS SLOWER

Figure 5-2   Artist's conception of why strong protease is usually optimal for microRNA in situ hybridization. This figure shows the artist's conception of the microRNA being deeply buried within the three-dimensional dense protein cross-linked cage and, of course, distinct from it. This state may reflect the formalin-fixed sequela of the usual state of microRNAs in the living cell. Such targets usually require strong protease digestion; I have not seen a microRNA that could be optimized after in situ hybridization with no pretreatment in any type of tissue including, of course, formalin-fixed, paraffin-embedded tissues.

STEP TWO: PUTTING THE FIXED CELLS OR TISSUE ON A GLASS SLIDE

Of course, putting fixed cells or tissue on a glass slide is identical to the step in situ hybridization. I want to stress that when doing immunohistochemistry we *always* want three sections per slide. This way, we can vary important variables such as protease digestion time on a given slide and, thus, not only increase our chances of success, but also do so using far fewer slides and reagents. As a reminder, formalin-fixed tissues in paraffin blocks vary in size from a few millimeters to about 20 mm. Thus, for 20 mm tissues you can, at best, get 1½ to 2 sections per slide, whereas for small surgical biopsies (often around 8–10 mm) you can get 3–4 sections per slide. Still, by using polypropylene coverslips cut to size, you can do 3–5 different reactions on the same slide, which, again, will much increase the odds that you will use the conditions that are optimal for successful immunohistochemistry. Figure 5-3 shows an example for the optimization of the immunohistochemistry detection of the tumor suppressor gene LATS2. Note the complete absence of a signal with no pretreatment or protease digestion (panels A and B).

A strong signal is seen with antigen retrieval, whereas antigen retrieval plus protease digestion still give a strong signal but background is now evident. Also note that the sections are serial sections, so we are looking at the same groups of cells. Figure 5-4 shows the exact same area at subsequent serial sections, but now tested for CD45 and smooth muscle actin. Like LATS2, CD45 and smooth muscle actin proteins should be present in the cytoplasm of the positive cells (although LATS2 can migrate to the nucleus, but this is not important for this discussion). I included this illustration for three reasons:

1. Even when you have the *same cells* in the same tissue and examine epitopes from the *same* cellular compartment, you can see different results for different pretreatment regimes. For example, note that you can see a signal for both CD45 and smooth muscle actin with either no pretreatment (weak) or protease digestion alone (moderate), whereas no signal is seen for these two pretreatment regimes for LATS2 (Figure 5-3). Antigen retrieval and antigen retrieval plus protease digestion each show intense signals for CD45 and SMA with little background. These types of data lend credence to the theory that each epitope

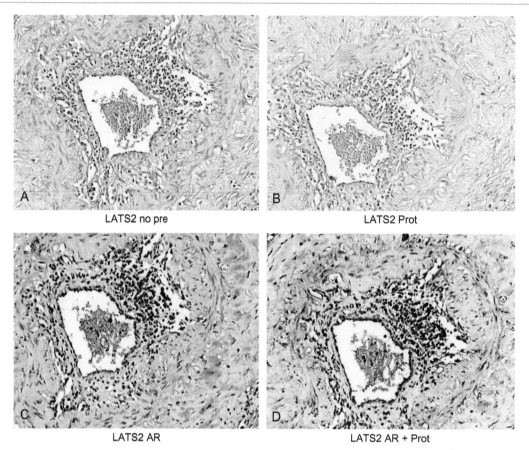

LATS2 no pre

LATS2 Prot

LATS2 AR

LATS2 AR + Prot

Figure 5-3 Optimization of the immunohistochemistry for LATS2 in the lung. The data are typical of many immunohistochemical optimizing experiments and would be consistent with the model shown in Figure 5-1, panel A. No signal is seen with no pretreatment (panel A) or protease digestion (panel B). An intense signal is seen with antigen retrieval (panel C) and with antigen retrieval plus protease (panel D), but the latter is associated with background.

will have its own unique three-dimensional protein/protein cross-linked network with its unique ionic/hydrogen bond potential, hydrophobic/hydrophilic pockets, breathability, and pore size.

2. When you place three sections per slide (this was from a biopsy of a lung for cancer), all the experiments shown in Figures 5-3 and 5-4 could be done on just four slides.

3. The simplest form of co-expression analyses is seen in Figure 5-4. Whenever you have two targets that are seen in different cellular compartments (for example, nucleus or cytoplasm) and/or are present in different cell populations, you can add both antibodies at the same time when doing the immunohistochemistry and get nice co-expression data. In this case, CD45 is found in the lymphocytes, whereas smooth muscle actin is found in the smooth muscle cells that line the larger blood vessels and bronchioles.

REMOVING THE PARAFFIN WAX FROM THE TISSUE SAMPLE

Removing the paraffin wax from the tissue sample is identical to the procedure for in situ hybridization. The key "trick of the trade" is to realize that the xylene solution will be "exhausted" over time. That is to say, after you deparaffinize enough tissues through a given xylene solution, it will lose its ability to completely remove all the paraffin from the slide. This will be easy to recognize. After you remove the paraffin, hold the slide up to the light. If you see any streaking, then the xylene is indeed exhausted. The problem is easy to solve; simply use fresh xylene. In my experience, a typical Copeland jar of xylene is able to deparaffinize about 200 slides before it becomes "exhausted."

THE IMPORTANCE OF NOT EXPOSING THE SLIDES (EITHER PRIOR TO OR, ESPECIALLY, AFTER REMOVAL OF THE PARAFFIN WAX) TO DRY HEAT

Though we discussed this topic in Chapter 2, it is worth briefly discussing again. I have had enough investigators contact me about signal reduction only to realize that they were exposing the tissues to dry heat (from 60°C up to 95°C) for a variety of reasons, such as inactivating the protease or drying the slides in between steps, that it is worth repeating. Figure 5-5 shows an example of a lymph node exposed to 95°C for 30 minutes (A and B) as well as 10 minutes (C and D) with a diminution of the signal

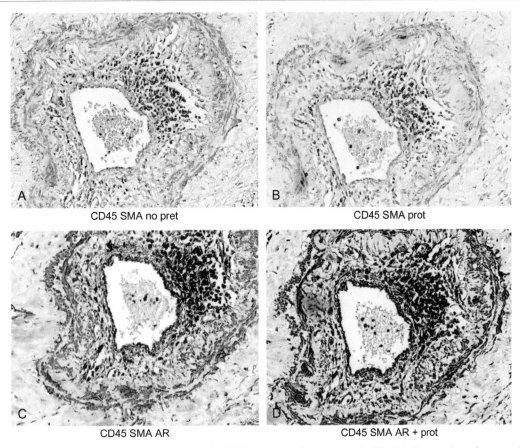

CD45 SMA no pret

CD45 SMA prot

CD45 SMA AR

CD45 SMA AR + prot

Figure5-4   Optimization of the immunohistochemistry for CD45 and smooth muscle actin (SMA) in the serial sections used to optimize LATS2 in the lung. Serial sections similar to those tested in Figure 5-3 for LATS2 were analyzed for CD45 and SMA. Note a signal is evident for each epitope with no pretreatment (panel A) and with protease digestion (panel B). The signal is stronger for each with antigen retrieval (panel C) and with antigen retrieval plus protease (panel D).

in each case. Note that the reduction of the CD45 signal is much more dramatic if the tissue is exposed to the dry heat *after* removing the paraffin (panel B) as compared to using dry heat before the paraffin wax is removed from the slide (panel A). I realize that we can "rescue" this lost signal by using "regenerating agents" that can recreate the three-dimensional protein/protein cross-linked network (panel C versus D), but, of course, we would rather avoid having to do that.

THE PREPARATION OF FORMALIN-FIXED CELLS FOR IMMUNOHISTOCHEMISTRY

I recommend the same exact protocol for cell suspensions and immunohistochemistry that I proposed in Chapter 4 for in situ hybridization. The key is to fix the cells while they are still in the growth-culture medium dish/flask before removing them. This way, you will get excellent morphology.

STORAGE OF SLIDES AND TISSUE BLOCKS WITH IMMUNOHISTOCHEMISTRY

The topic of storing slides and tissue blocks is so important for immunohistochemistry that I would like to

revisit some of the data we discussed previously, but with a different focus. Let's look at the data regarding the degeneration of the HPV DNA for in situ hybridization signal over time (Table 3-1, Chapter 3) and compare this to the equivalent data for the degeneration of the immunohistochemistry signal for either p16 or cytokeratin AE 1/3 over time (Table 3-9). Since HPV DNA is nuclear based, I purposely tested two antigens, one nuclear based (p16) and one cytoplasmic based (cytokeratin AE 1/3), to see if the two proteins showed much variance in the degeneration of the signal over time. They did not, but rather showed nearly identical degeneration of the signal over time.

Note that the degree of degeneration of the HPV DNA signal (58.2%) is comparable to the degradation of the immunohistochemistry signal over time, which is a bit higher at 82.0%. Figure 5-6 shows two examples of lymph nodes that were obtained in 1990 and 1992, respectively, showing a much weaker signal for CD3 than a lymph node biopsied in 2011. As you know, CD3 is a T-cell marker that you would expect to be present in high copy number in the T-cells that concentrate in the interfollicular zone of the lymph node, which is where the photographs were taken.

So, to repeat, storing unstained slides or paraffin-embedded blocks at room temperature will result in

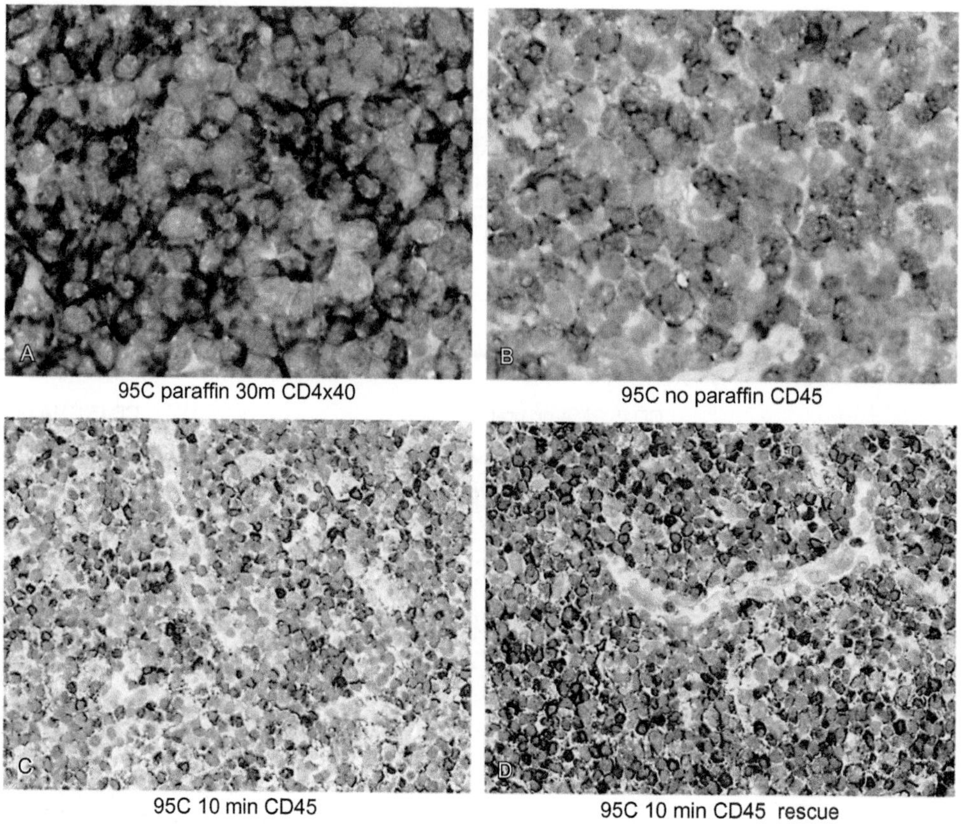

Figure5-5 Effecto fd ryhe at and removal of paraffin on the immunohistochemistry signal. The signal for CD45 is still evident, although a bit reduced, if the tissue was exposed to 95°C for 30 minutes prior to immunohistochemistry, but only if the paraffin was not removed (panel A). Removal of the paraffin prior to exposure to the same amount of dry heat led to a marked reduction in the signal (panel B). The reduced signal with exposure to dry heat (95°C for 10 minutes) after removal of paraffin (panel C) was reversed by the subsequent treatment of the tissue with ICHPerfect (panelD) .

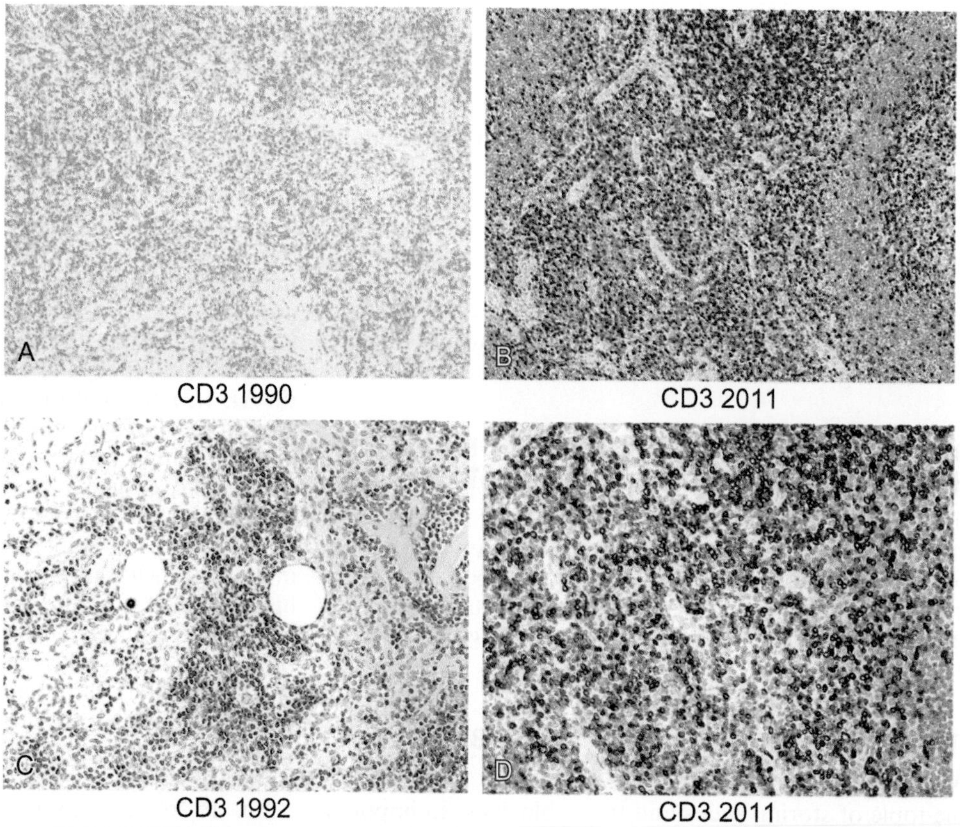

95C paraffin 30m CD4x40

95C no paraffin CD45

95C 10 min CD45

95C 10 min CD45 rescue

Figure5-6 Effecto ft he age of the formalin-fixed, paraffin-embedded tissue on the immunohistochemistry signal. The signal for CD3 was not evident in this lymph node tissue that was fixed and embedded in paraffin in 1990 (panel A), and was reduced in intensity in the lymph node tissue that was embedded in 1992 (panel C). In comparison, note the strong signal for CD3 in the lymph node embedded in 2011 (panel B and, at higher magnification,i np anelD) .

CD3 1990

CD3 2011

CD3 1992

CD3 2011

a decreased immunohistochemistry signal over time. Although the data generated previously were with blocks that were 10–21 years old, I have seen the diminution in signal in blocks that are only a few years old, although the amount of reduction is usually less marked. However, formalin-fixed cell preps also show the reduction in signal, and usually, it is marked after only a few years of storage. Let's look at an example of this for the anti-cancer virus reovirus (see Figure 5-7). These tumor cells (a melanoma cell line) were exposed to reovirus. It was demonstrated that the virus was able to kill some of the cancer cells by, in part, reducing the expression of miR-let-7d, which, in turn, upregulated the active form of caspase-3 that ultimately led to the apoptotic death of the cancer cells. Note in Figure 5-7 that the reovirus signal was much diminished after only 2 years of storage (panel A shows the original signal, and panel B shows the loss of the signal after 2 years of storage). Also note that we were able to regenerate the signal by reinforcing the three-dimensional protein/protein cross-linked cage (panel C, with panel D showing a negative control, which is the cancer cell line not exposed to the virus). Unless you employ this "antigen rescue" process, it is important to remember that the age of the paraffin block you are using for research is a key variable that must be taken into account when interpreting the data, especially if comparing a set of data from relatively new blocks versus relatively old formalin-fixed, paraffin-embedded tissues.

THE USE OF SLIDES ALREADY STAINED WITH HEMATOXYLIN AND EOSIN

It happens now and then that all you have available for immunohistochemistry are slides already stained in hematoxylin and eosin. These slides rarely (in my experience, never!) are silane-coated slides. The reasons that only these already-stained slides are available include that the area of interest (such as a metastasis) is only present on these sections and not the deeper serial sections, the block is no longer available, and that the tissue was so small that there is no tissue left in the block.

It is not difficult to remove the blue and pink colors of a slide stained in hematoxylin and eosin:

1. Put the slide in xylene and wait for the coverslip to detach from the slide. This may take a week or longer if the slide was coverslipped many years ago.
2. Put the slide in another container of xylene for 30 minutes to remove the Permount.
3. Wash the slide in a solution of 70% alcohol to which has been added a 1:20 dilution of 1N HCl (so-called acidalc ohol).
4. Examine the slide every 5 minutes. You will see the color leaving the tissue. When the tissue is clear, wash the slide in running water for 10 minutes.

The problem with such sections is that since the slides are not coated with silane, the tissue will tend to fall off during the in situ hybridization (or

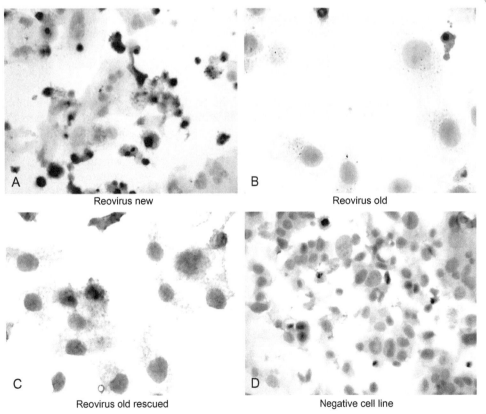

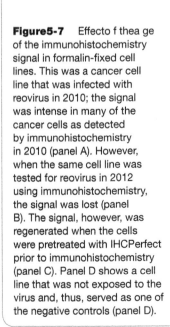

Figure5-7 Effecto f thea ge of the immunohistochemistry signal in formalin-fixed cell lines. This was a cancer cell line that was infected with reovirus in 2010; the signal was intense in many of the cancer cells as detected by immunohistochemistry in 2010 (panel A). However, when the same cell line was tested for reovirus in 2012 using immunohistochemistry, the signal was lost (panel B). The signal, however, was regenerated when the cells were pretreated with IHCPerfect prior to immunohistochemistry (panel C). Panel D shows a cell line that was not exposed to the virus and, thus, served as one of the negative controls (panel D).

Reovirus new

Reovirus old

Reovirus old rescued

Negative cell line

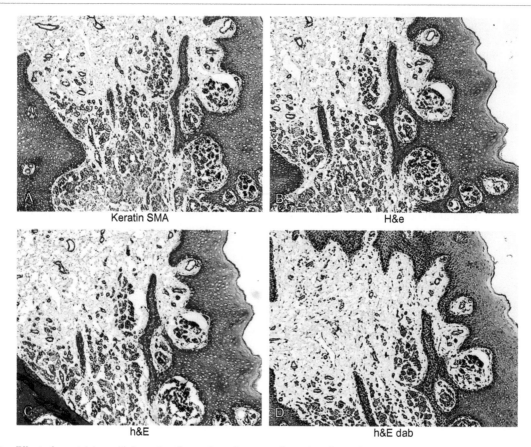

Figure5-8   Effect of prestaining with hematoxylin, eosin, or hematoxylin and eosin on the signal with immunohistochemistry. This tissue, which was a benign nevus with strong melanin staining, was unstained (panel A), stained with hematoxylin (panel B), stained with eosin (panel C), or stained with both hematoxylin plus eosin (panel D) prior to immunohistochemistry for keratin and smooth muscle actin. No counterstain was used. Note the intense signal for keratin in the squamous epithelia and the SMA signal in the small vessels. Also note that the signal is identical for panels A, B, and C; plus there is no residual hematoxylin or eosin staining. In panel D, the signal is equivalent with DAB when compared to the Fast Red chromogen. However, the melanocytes in the lower center of the photo (panel D) show a false-positive brown color due to the melanin pigment; nevus cells would be keratin and SMA negative.

immunohistochemistry) reaction. A "trick of the trade" is to put the dried slides on a 45°C plate for 30 minutes, which will improve adherence. But, of course, do not use higher temperatures because such dry heat will reduce or eliminate the signal.

Before leaving this section, let's answer another question: would either eosin or hematoxylin interfere with the immunohistochemistry staining process? To address this question, we stained multiple tissues either with hematoxylin, eosin, both, or neither; then we used immunohistochemistry using antigen retrieval. The tissues were simultaneously analyzed for two epitopes: smooth muscle actin (SMA) and keratin AE1/3. Since these two targets are found in mutually exclusive cell types (smooth muscle cells and epithelial cells, respectively), it will be easy to differentiate the two signals. Representative data are shown in Figure 5-8. Note three things:

1. Neither eosin nor hematoxylin staining interferes with the immunohistochemistry signal.
2. Although both the hematoxylin and eosin strongly stained the tissue prior to immunohistochemistry, the blue and pink colors, respectively, are washed

out during the immunohistochemistry process. Of course, this means that you do not really have to do the acid alcohol rinse to remove the hematoxylin and eosin on already-stained slides. It also means that when doing co-expression analyses, you must do the counterstain (if you are using such a step) after the second reaction, and not the first reaction.
3. The lesion is a nevus and, thus, many of the cells are positive for the brown pigment melanin. As is evident from Figure 5-8, panel D, the melanocytes with the ample melanin pigment would incorrectly be called positive for keratin or smooth muscle actin. We know this must be a false-positive result because of two observations: (1) prior data from many groups shows that melanocytes do not express keratin AE1/3 or smooth muscle actin; (2) when Fast Red is used as the chromogen in the serial sections, the melanocytes are negative.

Figure 5-9 shows the equivalent data at higher magnification. Another reason I included these six images is to stress to you the value of serial section analyses. When you look at panels A through F, note that there is a wishbone-shaped

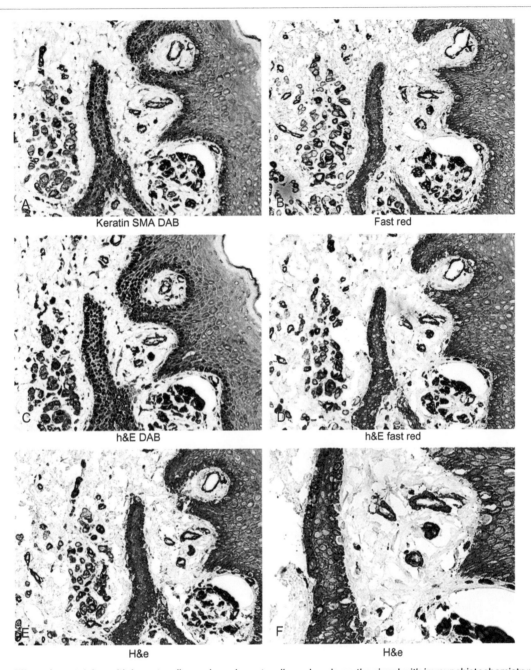

Keratin SMA DAB

Fast red

h&E DAB

h&E fast red

H&e

H&e

Figure5-9   Effect of prestaining with hematoxylin, eosin, or hematoxylin and eosin on the signal with immunohistochemistry: higher magnification. These panels show the same data as Figure 5-8 at higher magnification. Note that when you use histologic landmarks, such as the island of squamous cells that is in the center of the photographs, it is straightforward to show the same exact cells at successive sections. These images also show the lack of residual hematoxylin and eosin staining in these sections after immunohistochemistry, as well as the false-negative result for the melanocytes (left side and lower right side of each panel) in the DAB-stained sections (panels A and C due to their endogenous melanin pigment).

fragment of epithelial tissue in the center. We refer to such parts of the tissue as "serial section landmarks" because it allows you to easily find the same group of cells in each serial section. The trick of the trade is to find a relatively large landmark, such as a large island of epithelia or a blood vessel, which allows you to easily find the same group of cells in the next section. This all underscores the importance of labeling each serial section slide so you can be sure that you are indeed examining the actual serial sections.

STEP THREE: PRETREATMENT OF THE TISSUE FOR SUCCESSFUL IMMUNOHISTOCHEMISTRY

Let's revisit some data we discussed earlier in this chapter. It relates to the data from many optimizing experiments with new antibodies where we compared no pretreatment to protease digestion to antigen retrieval.

Here is the breakdown for the compilation for optimal immunohistochemistry in these many cases:

Nop retreatment	Optimal for about 5% of targets
Proteased igestion	Optimal for about 15% of targets
Antigenr etrieval	Optimal for about 80% of targets

Clearly, the discussion on pretreatment regimes for immunohistochemistry is quite a bit simpler than for in situ hybridization. The reason is that in about 80% of the cases of optimizing an antibody that is new to your laboratory, antigen retrieval will be the winner. The rare times that either no pretreatment or protease digestion gives better results than for antigen retrieval with immunohistochemistry, it is typically for an epitope that is present in the cytoplasm. I have never seen a nuclear-based protein target give its optimal signal with either no pretreatment or protease digestion; all nuclear targets, in my experience, required antigen retrieval for optimal detection.

PROTEASE DIGESTION

I only use proteinase K when doing immunohistochemistry. Certainly, you can use pronase or pepsin or any other protease if you prefer. Many laboratories, ours included, use two (or more) different concentrations of proteinase K. These are sometimes referred to as the "weak" and the "strong" proteinase K digestion solutions. Given the ability of proteinase K to disrupt the three-dimensional protein/protein cross-linked network as well as the epitope itself, you will not be surprised to hear that you may get a very different signal with immunohistochemistry depending on the strength of the proteinase K solution. This is illustrated in Figure 5-10. Note that the signal for cytokeratin is much more dependent on the strength of the protease digestion solution than that for CD45. Each proteinase K solution (weak and strong) was applied for 4 minutes.

TIME OF PROTEINASE K DIGESTION AND TIME OF THE ANTIGEN RETRIEVAL

Let's pause here for a moment and recall the "bigger picture" that applies for both immunohistochemistry and in situ hybridization. In each case, we are dealing with a three-dimensional protein/protein cross-linked cage that is intimately involved with our target. In this chapter, of course, our target is an epitope that is part of the three-dimensional

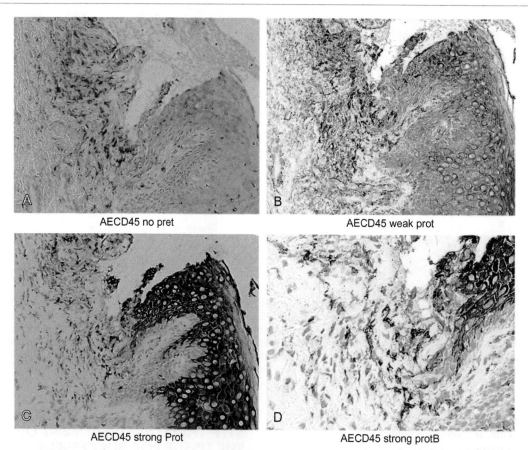

AECD45 no pret

AECD45 weak prot

AECD45 strong Prot

AECD45 strong protB

Figure5-10   Effect of the strength of the protease on the immunohistochemical signal. This inflamed skin biopsy was simultaneously tested for CD45 and cytokeratin AE 1/3. Note that with no pretreatment the dermal signal for CD45 is evident but weak, whereas the keratin signal is barely discernible (panel A). Digestion in a weak protease solution increased each signal in the serial section (panel B). However, if the protease digestion solution has much stronger activity due to a higher concentration of proteinase K, then the cytokeratin signal is now intense and optimized (panel C). Panel D shows the distinct CD45 and cytokeratin signals at higher magnification at the epidermal/dermalj unction.

protein/protein cross-linked network. When we do immunohistochemistry, the kinetics of the diffusion of the reagents is important, as we have already seen, so that the primary antibody concentration will be a factor for both signal and background. Also, the *type* of pretreatment regime (protease, antigen retrieval) and the *length of time* of the protease or antigen retrieval will affect the signal directly as a consequence of their effect on the three-dimensional protein/protein cross-linked network and the epitope.

It follows from the preceding discussion that it would be helpful if we plotted the strength of the immunohistochemistry signal relative to the *primary antibody concentration*, the *time duration* of the protease digestion, and the *time duration* of the antigen retrieval. In effect, we are repeating the equivalent experiments that we did for DNA and RNA targets and in situ hybridization. I did a series of experiments in which this was done for the following antigens: nucleolin; neuron-specific enolase (NSE); GFAP; and interleukin-6 (IL-6). I chose these antigens for the following reasons:

1. Nucleolin—Itisan uclear-basede pitope.
2. NSE—It is a cytoplasmic-based epitope that is present in very large cells (neurons).
3. GFAP—It is a cytoplasmic-based epitope present in the very long and fine dendritic processes of intermediate-sized cells (astrocytes).

4. IL-6—It is a cell-membrane-based epitope present in small cells (lymphocytes).

The data from these experiments are provided in Figures 5-11 through 5-14. First, let's look at the data in Figure 5-11 for NSE. Note that at the lower concentration of 1:600 a signal is only seen with antigen retrieval. I indicated previously that this was a very common scenario with immunohistochemistry. But, as is evident from this figure, the situation is a bit more complicated. A signal is evident with either the "strong protease digestion (30 minutes of protease digestion)" or "weak protease digestion (4 minutes of protease digestion)" if you use a higher concentration of NSE (1:400 dilution). Still, the protease digestion never is able to generate as strong a signal, even with the higher concentration of NSE, as the antigen retrieval. This suggests that the protease and antigen retrieval are affecting different variables with the immunohistochemistry for NSE, with the antigen retrieval's mechanism being more efficient for producing a strong signal. My hypothesis is that pore size is a minor variable with immunohistochemistry for NSE, in the sense that protease digestion, which probably affects pore size more than antigen retrieval, can increase the signal with higher concentrations of NSE. However, it is logical to assume that exposure of the antigen via

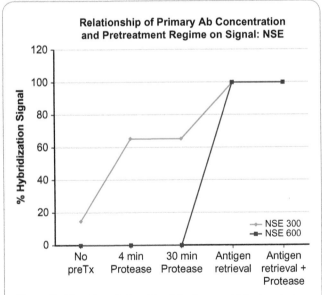

Figure 5-11 The relationship of the immunohistochemical signal for neuron-specific enolase (NSE) in neurons to the pretreatment regime and the concentration of the primary antibody. In these experiments, serial sections of brain were tested for NSE by immunohistochemistry using a dilution of the primary antibody of either 1:300 (blue line) or 1:600 (red line). As is evident, neither concentration of the primary antibody generated a signal if there was no pretreatment. An equivalent signal was seen with either short- or long-term protease digestion, but only at the higher concentration of the NSE. These are not the expected data if pore size was the key variable for entry of the primary antibody to the epitope. Antigen retrieval generated intense and equivalent signals at each concentration of the primary antibody.

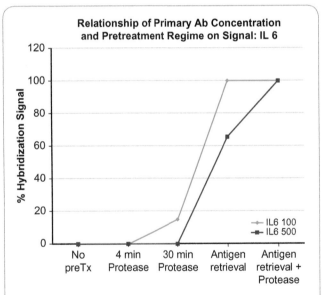

Figure 5-12 The relationship of the immunohistochemical signal for IL-6 in lymphocytes and macrophages in the tonsil to the pretreatment regime and the concentration of the primary antibody. In these experiments, serial sections of tonsil were tested for IL-6 by immunohistochemistry using a dilution of the primary antibody of either 1:100 (blue line) or 1:500 (red line). As is evident, neither concentration of the primary antibody generated a signal if there was no pretreatment. A weak signal was seen with the protease digestion, but only with the higher concentration of the primary antibody and after strong protease digestion. Antigen retrieval again generated the strongest signal, although only at the higher concentration of the primary antibody. The lower concentration of the primary antibody could also generate the maximum signal, but it required both antigen retrieval and protease digestion.

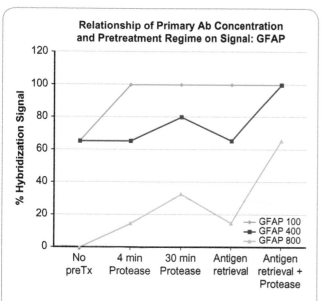

Figure5-13 The relationship of the immunohistochemical signal for GFAP in astrocytes in the brain to the pretreatment regime and the concentration of the primary antibody. In these experiments, serial sections of brain were tested for GFAP by immunohistochemistry using a dilution of the primary antibody of either 1:100 (blue line), 1:400 (red line), or 1:800 (green). As is evident, the two higher concentrations produced a signal with no pretreatment and strong signals with protease digestion that were similar to the signal seen with antigen retrieval. However, the lowest concentration produced only weak signals for each of these pretreatments (protease digestion or antigen retrieval), which is certainly suggestive of a "diffusion" type process in which antibody concentration is strongly related to the signal. Interestingly, note that antigen retrieval and protease not only yield the strongest signals, but also give a remarkably strong signal for the very low concentration of the primary antibody.

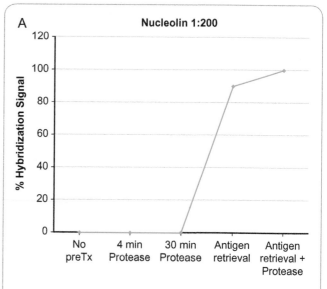

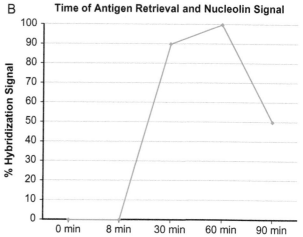

Figure5-14 The relationship of the immunohistochemical signal for nucleolin in the dysplastic cells of CIN lesions to the pretreatment regime and the concentration of the primary antibody. As evident in Panel A of Figure 5-14, nucleolin, at a concentration of 1:200, showed a very common pattern of optimization with immunohistochemistry. There was no signal if the pretreatment step was omitted or if protease digestion was done for either 4 minutes or 30 minutes. However, an intense signal was seen with antigen retrieval for 30 minutes and with protease digestion, but only if it was done after the antigen retrieval. Panel B then addresses this question: does the hybridization signal vary with the time of antigen retrieval? As is evident, no signal was seen with 8 minutes of antigen retrieval (95°C in an EDTA solution). The signal was strong at 30 minutes, maximized at 60 minutes of antigen retrieval, and then decreased after 90 minutes of antigen retrieval. This is a typical pattern for many epitopes that show optimal signal with antigen retrieval although, in many cases, the signal is either strongest at 30 minutes of antigen retrieval or equivalent at 30 and 60 minuteso fa ntigenr etrieval.

conformational change, best done by antigen retrieval, is the key variable with immunohistochemistry for NSE.

Now let's compare these data to that for IL-6 (Figure 5-12). Note that the IL-6 data give us the same general information about immunohistochemistry as does NSE. Specifically, the data indicate that protease digestion and antigen retrieval affect the results with immunohistochemistry via two separate pathways. The protease digestion pathway is the least efficient in the sense that you see a weaker signal compared to immunohistochemistry and only at the higher concentrations of the primary antibody. One possible interpretation is that "punching holes" into the three-dimensional protein/protein cross-linked network may allow the primary antibody better access to the epitope, but not nearly as efficiently as if you change the conformation of the cross-linked protein cage via antigen retrieval. This makes sense for cytoplasmic-based epitopes, especially if the epitope is on the outer part of the three-dimensional protein cross-linked cage that includes the IL-6 protein. What about the theory that the protease digestion is also partly destroying the epitope and, thus, is not as efficient as antigen retrieval in generating a signal? This is unlikely when you examine Figures 5-11 and 5-12. Note that if we add antigen retrieval and protease digestion, we

still see a strong signal. However, I should add that, in my experience, combining antigen retrieval and protease digestion often leads to either background or overdigestion issues with immunohistochemistry (see Figure 5-1).

Now let's compare the preceding data to those for GFAP, which is presented in Figure 5-13. GFAP is unusual in the sense that it is present not in the relatively large volume of the cytoplasm of the cell, like IL-6 or NSE, but rather in the many fine arborizing processes of astrocytes. Note that here is an antibody that can show good signals either with no pretreatment or protease digestion as well as antigen retrieval, but only at higher concentrations of the primary antibody. Indeed, at a concentration of 1:100 protease digestion and antigen retrieval, each yields very strong signals, with no pretreatment not far behind. Also note that at the lower concentrations of the primary antibody, protease digestion actually gives better signals than antigen retrieval. This raises the possibility that, for GFAP, exposure of the antigen and pore size may each come into play. I must add that I am surprised in this regard with the signal with no pretreatment at the higher concentrations. But this raises another simple yet key difference between immunohistochemistry and in situ hybridization: with in situ hybridization, we are looking for one unique DNA or RNA sequence. For immunohistochemistry, we are looking at a signal that represents *multiple* epitopes and, if the primary antibody is polyclonal, then multiple primary antibodies that will show variable affinity for the many different epitopes present in the immunizing protein. It is not difficult to envision, for GFAP, some epitopes that are represented graphically in Figure 5-1, panel A, others as represented in Figure 5-1, panel B, and yet others as represented in Figure 5-1, panel C. This simple point could certainly explain Figure 5-13, as it would represent the sum of epitope/primary antibody hybrids, some of which could be accessed by no pretreatment and others that would require antigen retrieval and protease digestion. Interestingly, note that antigen retrieval and protease not only yield the strongest signals, but also give a remarkably strong signal for the very low concentration of the primary antibody. This leads to another trick of the trade that was especially important when primary antibodies were very expensive and hard to come by. If you want to use the lowest possible concentration of the primary antibody, then in many cases antigen retrieval plus protease digestion will achieve this goal. You may encounter two problems, though, with this scenario:

1. Overdigestion
2. Background, especially if the primary antibody concentration is too high

Now that we have analyzed the data for these cytoplasmic-based epitopes, let's look at the nucleolin data presented in Figure 5-14. As evident in panel A, this nuclear-based epitope behaves much like NSE and IL-6, in the sense that antigen retrieval clearly gives us a stronger signal than protease digestion. This raises the possibility that nuclear versus cytoplasmic is not the key variable. Rather, it is the biochemical features of the three-dimensional protein/protein cross-linked network cage surrounding an epitope and the specific location of the epitope in the cage that are the key variables for understanding how to optimize the immunohistochemistry signal. In other words, the difference in distance

between the cytoplasm and nucleus is probably not sufficient to have an effect on the immunohistochemistry results. This certainly is consistent with the observation that the diffusion time of our key reagents with immunohistochemistry and in situ hybridization tends to be very short. Finally, note in panel B that increased antigen retrieval times can enhance the signal but, in my experience, are often associated with background and/or overdigestion. It certainly seems that 30 minutes at 95°C is sufficient to alter the conformation of most proteins that are part of the three-dimensional protein/protein cross-linked network. I have not encountered, in my own experience, any epitope that required 90 minutes of antigen retrieval to get an optimal signal.

ANTIGEN RETRIEVAL: THE SOLUTIONS

Most antigen retrieval protocols suggest either citrate or EDTA-based buffers, although many others have been proposed. Out of curiosity, I did a series of experiments comparing the two antigen retrieval solutions provided by Ventana Medical Systems (CC1 and CC2) with a solution of 1XSSC and double distilled water. I have consistently noted that the CC1 and CC2 solutions (EDTA and citrate buffer) give equivalent performances. This outcome was not too surprising. I was very surprised, however, when I saw the data with 1XSSC and water, which is provided in Figure 5-15. Let me stress that these results are representative; I did many such experiments and got these types of data time after time. Note the weaker signal for the 1XSSC solution. I assumed that the relatively high salt concentration may have stabilized the components of the three-dimensional protein/protein cross-linked network, such that the conformational changes were much less compared to CC1. Hence, the weaker signal. But when distilled water was used for antigen retrieval, the signal was dramatic; the signal was much enhanced. This outcome certainly suggests that double distilled water may be more efficient at causing conformational changes in the three-dimensional protein/protein cross-linked network when compared to solutions that contain electrolytes.

We discuss protocols for immunohistochemistry in Chapter 7. Still, let's summarize the data and discussions presented to this point and consider a logical protocol for optimizing any new antibody for immunohistochemistry. Certainly, we want to make sure that we use protease, antigen retrieval, and antigen retrieval plus protease. I still like to test a section with no pretreatment. My thinking is that if I see a signal with no pretreatment, then it is more likely that the tissue will be "protease dependent" and not "antigen retrieval-dependent." When testing new formalin-fixed, paraffin-embedded tissues with a new primary antibody, I recommend using two slides with multiple sections per slide and doing the following:

Nop retreatment
Proteased igestion(30min utes)
Antigen retrieval (30 minutes at 95°C) with and without protease digestion

This simple protocol will cost you only two slides of work and will cover more than 95% of the optimal conditions for immunohistochemistry.

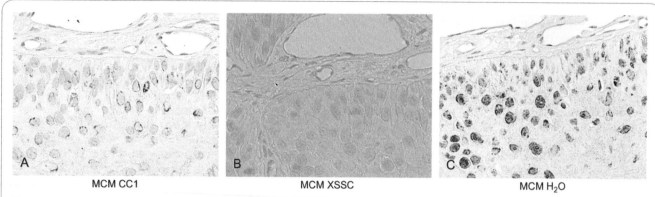

MCM CC1   MCM XSSC   MCM H₂O

Figure5-15   The effect of different antigen retrieval solutions on the signal with immunohistochemistry. Serial sections of this high-grade squamous intraepithelial lesion (SIL) were analyzed for the nuclear protein nucleolin. As we saw in Figure 5-14, nucleolin requires antigen retrieval for an optimal signal. These three serial section slides were tested for nucleolin at a 1:200 dilution after antigen retrieval in the cell conditioning #1 solution (Ventana Medical System, panel A), a solution of 1 XSSC (panel B), and double distilled water (panel C). As is evident, a moderate signal was seen if the antigen retrieval solution was the cell conditioning 1 (CC1) solution, no signal was seen with the 1XSSC solution, and an intense signal was seen if the antigen retrieval solution was simply double distilled water. These data suggest that electrolytes in the antigen retrieval solution may stabilize the three-dimensional protein cross-linked cage associated with the epitope and, thus, make it less likely that the epitope will be exposed by the antigen retrieval process.

STEP FOUR: CHOOSING A PRIMARY ANTIBODY FOR IMMUNOHISTOCHEMISTRY

MONOCLONAL VERSUS POLYCLONAL

Of course, any protein will have multiple areas that can serve as the binding site of antibodies that are directed against it. Each area (epitope) will show different affinities for the specific antibody directed against it. Assuming that a given protein will show much variation as you proceed down its sequence with respect to its conformation and its specific R side chains, it is reasonable to speculate that each different epitope of a given protein may be associated with unique three-dimensional protein/protein cross-linked "cages."

A fundamental issue with immunohistochemistry is whether to use a monoclonal or polyclonal antibody. Each has its advantages and disadvantages.

Monoclonal antibodies, which are typically abbreviated by the companies that sell them as mAB, are made by one clone of immune cells and, thus, will bind to one specific epitope of a given protein. Multiple myeloma is a disease in which the malignant cells are able to make antibodies. So monoclonal antibodies are typically made by fusing myeloma cells with benign cells from a given animal that have been immunized with the epitope of interest. The fused cells (called *hybridoma* cells) are placed in a growth medium that only allows them to proliferate, and they will continue to proliferate as immortal cells, thanks to the myeloma cells' ability in this regard. The separate malignant cells and the benign unfused mononuclear cells simply do not grow. The monoclonal antibodies are then purified, which, interestingly, often involves ion exchange chromatography. This step is important in producing a "true" monoclonal product, since the process of growing and purifying the monoclonal antibodies will generate some biochemical changes in a small population of these antibodies which, in turn, will likely have different affinities for the epitope.

Polyclonal antibodies are sometimes referred to as *antisera*. The reason is that they represent the many different antibodies a B-cell population in an intact animal will secrete when presented with a specific antigen. In general, the larger and more diverse the antigenic molecule, the greater numbers of different antibodies that are generated that will, in turn, react with different epitopes on the antigenic molecule. Small polypeptides and non-protein antigens must be linked to larger proteins (such as bovine serum albumin) to increase immunogenicity. Generally, the larger the immunogenic protein, the better. Since these antibodies are typically purified from the sera of animals exposed to the antigen, the term *antisera* is very descriptive of the process. To maximize the amount of polyclonal antibodies generated in an animal after antigenic exposure, the laboratory will use immune-stimulatory agents called *adjuvants*.

Many (if not most) mammals and some nonmammals (such as chickens) can be used for generating polyclonal antibodies. Since chickens are more distant evolutionarily from mammals, they tend to produce stronger affinity antibodies when challenged with an antigen that is typical to a mammal. Indeed, humans have been used to isolate polyclonal antibodies whenever they are exposed to an infectious disease that generates high titers of protective antibodies (such as hepatitis A). The antisera can help protect people against the disease if they have been recently exposed to the pathogen. Still, most polyclonal antibodies are made in rabbits and mice.

The primary advantage of monoclonal antibodies is that you are analyzing only one epitope. This logically would increase specificity. However, polyclonal antibodies offer a more "diversified portfolio" of antigenic determinants that may increase the sensitivity of the reaction. In my experience, I see little difference between immunohistochemistry performed with monoclonal versus polyclonal antibodies and, true to my Vermont roots, I tend to go with the less expensive product.

PROBE TYPE (RABBIT VERSUS MOUSE VERSUS OTHER ANIMAL)

The key point regarding probe type here is two-fold:

1. Make sure to check the source of the primary antibody. Most companies abbreviate primary antibodies made in mice as ms, in rabbits as rb, and in goats asgt.
2. Just as importantly, make sure that the detection kit for immunohistochemistry that you are using is able to detect the animal that was the source of the primaryan tibody.

For example, the Ventana Universal Ultraview system detects only primary antibodies made in mice or rabbits. So if you use a primary antibody made in a goat, the Ventana Ultraview system will give a negative result. This is a relatively common reason for a false-negative result with immunohistochemistry. You should always expect this when you use a high concentration of antibody with antigen retrieval and see no signal at all in a tissue that should contain the antigen of interest. This is a very easy problem to solve. The trick of the trade is to add the primary antibody, wait 20 minutes, and then add a rabbit- (or mouse-) based secondary antibody that will detect any antibody generated in a goat (or whatever other animal may have been used to make the primary antibody). By using this trick of the trade, you convert the system to a mouse- or rabbit-based system, which the system can now process to a signal.

LABELING THE ANTIBODY

In immunohistochemistry, the secondary antibody is usually conjugated to a chemical that will serve as the reporter system. Many such chemicals are available. The same biotin/streptavidin system we commonly use with in situ hybridization is still the most popular reporter system for immunohistochemistry. I much prefer this to labeling the secondary antibody directly with the reporter enzyme (peroxidase or alkaline phosphatase).

AGE OF THE PRIMARY ANTIBODY

As I indicated in the preceding chapter, I, like many pathologists, tend to keep "things" for many years. This does not just include slides. I have stored in the refrigerator primary antibodies that are between 10 and 20 years old. I have done this to allow me to easily address these questions:

1. Do primary antibodies tend to degrade over time?
2. If they do degrade, what is the cutoff point in years when we can expect that the primary antibody is no longerac tive?

I should note that the primary antibodies were stored at 4°C and not at −20°C. They were mostly polyclonal antibodies, given their age.

As we discussed at length in the preceding chapter, we know that the HPV DNA in formalin-fixed, paraffin-embedded tissues degrades over time. It degrades to the point that if the block is 10+ years old, then the HPV DNA has degraded from its original size of 8000 base pairs to fragments of around 100 base pairs in size. I reasoned that the epitope in the formalin-fixed, paraffin-embedded tissues likewise degrades over time. However, I did appreciate that perhaps the three-dimensional protein/protein cross-linked network may have kept the epitope in its original conformation. I also assumed that the primary antibody probably degraded over time. Thus, I was expecting that the old primary antibodies would either not work or would work poorly when compared to antibodies for the same epitope that I recently had purchased. This hypothesis, to my surprise, was wrong. For example, look at the data in Figure 5-16. I compared the intensity of a primary antibody against melanin A that was 13 years old versus one that was a few months old. As is evident, there is no difference in the signal! Figure 5-17 shows the results for a primary antibody against chromogranin that was 14 years old. Note the intense signal in the cells in the base of the crypts of the rectum (these are the same cells that form carcinoid tumors of the rectum).

I must again admit to some frustration when I am using a commercial primary antibody on an automated platform and I get the error message "the antibody has expired" when the antibody is 6 months old. This is not to say that primary antibodies cannot lose some potency over time. I have seen this happen, but it is unusual. More common in my experience is noting different signal intensities when comparing the same primary antibodies from different commercial sources. An example of this is provided in Figure 5-18. Note the much weaker signal for TdT in the serial section marked A (from Company A) when compared to the equivalent primary antibody from another company (Company B, panel B). This makes sense given that different companies will most likely be using different antigenic determinants and adjuvants, and we would expect different affinities for their primary antibodies. Still, in my own experience, I have tested probably more than 100 sets of antibodies from different companies. In most cases, the results are equivalent. Finally, note yet again that the antibody from Company B works as well if it is brand new versus being 8 years old (panel C).

The final point to make is that the lack of apparent degradation of the primary antibody over time is in sharp contrast to the degradation of the three-dimensional protein/protein cross-linked network in the formalin-fixed, paraffin-embedded tissues. This again raises the possibility that "nicks in the three-dimensional protein/protein cross-linked network" due to the high temperatures that the tissue is exposed to during the paraffin embedding may precipitate the long-term, continued slow degradation of the macromolecule over time.

STORAGE OF THE PRIMARY ANTIBODY

As I indicated previously, the aged primary antibodies that still worked well for immunohistochemistry that were between 10 and 20 years old had been stored for that time at 4°C. Since we are often advised to store the primary antibody at −20°C, and to not freeze and thaw it too many

Figure5-16 Thee ffect of the age of the primary antibody on the signal with immunohistochemistry. Serial sections of this lymph node that contained metastatic melanoma were analyzed for the melanoma marker melanin A. Panels A and B show at low and high magnification, respectively, the intense signal for melanin A typical of this tumor; the antibody was purchased a few months prior to the testing. Compare this to the data from a melanin A antibody that was purchased 15 years prior to the testing. The analyses for the old and new antibody were done at the same exact time under identical conditions. Clearly, the 15-year-old antibody performed as well, if not better, than the new antibody.

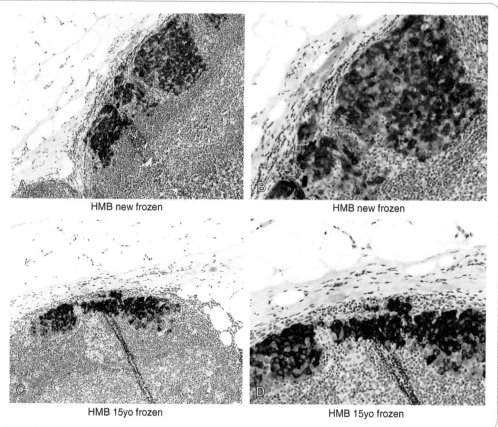

HMB new frozen

HMB new frozen

HMB 15yo frozen

HMB 15yo frozen

Figure 5-17 The effect of the age of the primary antibody on the signal with immunohistochemistry. The tissue is from the colon, and the panels show some of the optimization data for the antigen chromogranin. As with all such experiments, the strength and specificity of the signal are good measures of the fidelity of the assay and the strength of the primary antibody. Note the strong signal in scattered cells in the colonic crypts after antigen retrieval (panels A and B) or antigen retrieval and protease (panel C); no signal was seen with protease digestion alone (panel D). The positive cells are called *argentaffin* cells and are thought to produce chemicals such as serotonin that aid in the movement of the colon wall. The primary antibody in this case was 14 years old and, based on these data, has not lost any appreciable strength over these years while stored at 4°C.

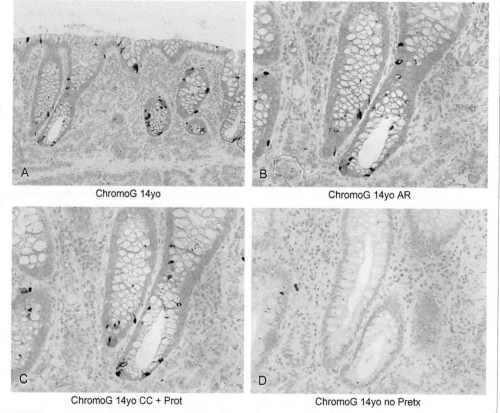

ChromoG 14yo

ChromoG 14yo AR

ChromoG 14yo CC + Prot

ChromoG 14yo no Pretx

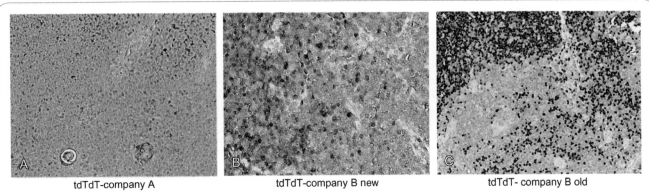

tdTdT-company A | tdTdT-company B new | tdTdT- company B old

Figure5-18 The effect of different commercial sources and age of the primary antibody on the signal with immunohistochemistry. Serial sections of this thymus tissue were analyzed for the protein TdT, which should be present in abundance in this tissue. Note the relatively sparse signal seen in panel A with Company 1's product (panel A). The signal in the serial section was much stronger with Company 2's product (panel B). The primary antibody from Company 2 provided equally strong immunohistochemistry results if the antibody was brand new (panel B) or 8 years old (panel C).

times, I did a series of experiments with serial sections of a melanoma, examining the potency of an antibody against Mel-A that was repeatedly frozen and thawed. These data are shown in Figure 5-19. I saw minimal to no change in the intensity of the signal with multiple freeze/thaw cycles. I prefer to store a large amount of my primary antibody at 4°C and keep a small aliquot at –20°C in case something inadvertent happens to the primary antibody.

ANTIBODY CONCENTRATION

Just as with in situ hybridization, a key variable with immunohistochemistry, both for signal and background, is *concentration*. By "key variable," I mean that it is the first variable I consider adjusting when either the signal is too weak or the background is too strong. The primary antibody has two binding choices when we perform immunohistochemistry: it can bind to the epitope, or it can bind nonspecifically to a variety of macromolecules. Both binding possibilities will be positively correlated to the concentration of the primary antibody. Of course, other variables such as protease digestion, antigen retrieval, and the degree of formalin fixation will also be factors that can influence both signal and background. Still, I always consider the primary antibody concentration to be the "first suspect" when either our laboratory has a background issue in an experiment or if another laboratory comes to us for help with a background problem or a problem of too weak a signal.

Our laboratory (like most, if not all, labs) has done many experiments comparing the intensity of the hybridization signal as well as background to probe concentration. We have seen (as has every other laboratory that has done these experiments) that there is a point, even with optimal pretreatment conditions, where the concentration is too high and background ensues. An example of this is seen in Figure 5-20 for the nuclear neuronal protein Fox3 (also called NeuN). Note that at a concentration of 1:1000 with optimal pretreatment conditions (antigen

retrieval for 30 minutes), an intense signal is seen that is evident only in the neurons (panel A). There is no background in the surrounding neuropil (which consists of the astrocyte and microglial network that surrounds the neurons and their axons/dendrites). Note that in panel B when the concentration is raised to a dilution of 1:200, the neuronal signal is still evident, but with a high degree of staining of the astrocyte/microglial network. The latter must be background, since Fox 3 is not able to bind to such cells in a target-specific manner. Here is another example of where our knowledge of surgical pathology allows us to differentiate signal and background. Now look at the results for the protein Iba1. The optimal pretreatment for this protein is no pretreatment. Note that a strong cell-specific signal is present under these conditions (panel C) in rare cells toward the top of the image; this target should not be present in too many cells in the tonsil. However, when antigen retrieval is done at the same primary antibody concentration, high background is seen in the form of most of the cells showing a brown color (panel D).

Although primary antibody concentration is an important variable in background with immunohistochemistry, it is important to stress that the relationship between primary antibody concentration and background is, in turn, highly dependent on the pretreatment conditions. To illustrate this point, we did a large number of experiments using nucleolin at a dilution of 1:200 to see the relationship between signal and background. The data are highlighted in Figure 5-21. Note the inverse relationship between signal and background. Also note that, in general, background indicates that the problem is with the pretreatment conditions and *not* with the concentration of the primary antibody. Of course, there are exceptions to this rule, as we saw in Figure 5-20. Still, as a general rule, this one works well: if you see background with immunohistochemistry, consider changing the pretreatment conditions as well as the concentration of the primary antibody.

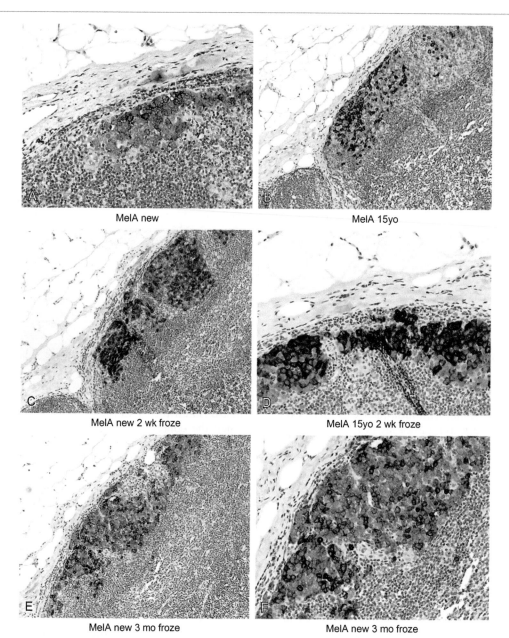

Figure5-19 The effect of repeated freeze/thaw cycles and the age of the primary antibody on the signal with immunohistochemistry. Serial sections of the metastatic melanoma were tested for melanin A (see Figure 5-16), but this time the effect of repeated freeze/thaw cycles was examined. The baseline data are shown in Panel A (new primary antibody, no freeze/thaw). Panel B shows the results with the antibody that was 15 years old. The new antibody was subjected to freeze/thawing for 2 weeks (panel C), as was the 15-year-old antibody (panel D). There was no diminution of signal. Repeated freeze/thawing at 3 months of the new antibody resulted in a good albeit slightlywe akenedsi gnal.

STEP FIVE: THE HYBRIDIZATION STEP

THE RELATIONSHIP OF THE TIME OF HYBRIDIZATION OF THE PRIMARY ANTIBODY TO THE INTENSITY OF THE HYBRIDIZATION SIGNAL
Hybridization and Diffusion Time

When we add the primary antibody to the formalin-fixed, paraffin-embedded tissue, it will start to diffuse through the cell. As with in situ hybridization, we have theorized that the primary antibody (like the labeled probe) will encounter many forces on the way that can impede its movement. These forces include hydrogen bonds and ionic bonds with specific R side chains in the cross-linked proteins, hydrophilic and hydrophobic pockets in the three-dimensional protein/protein cross-linked cage, and the pore size of this "cage" that surrounds the epitope of interest. If we recognize that some of these forces may actually assist the immunohistochemistry reaction by allowing the primary antibody and epitope sequence to

Figure5-20 Thee ffecto f the antibody concentration and the pretreatment conditions on the signal/background ratio. Note the strong nuclear-membrane-based signal present in just neurons for the antigen Fox3 at a 1:1000 dilution (panel A). If the concentration of the primary antibody was raised to 1:200 with the same pretreatment (antigen retrieval), there is high background (panel B). In comparison, panel C shows the results for Iba-1 with no pretreatment in the tonsil. As expected, only scattered cells show the cytoplasmic signal (panel C). If the pretreatment was antigen retrieval, then background was evident, as seen by many cells showing strong colorization. In each case, a good knowledge of the histopathology and the expected distribution of the signal is very helpful in the interpretationo ft hed ata.

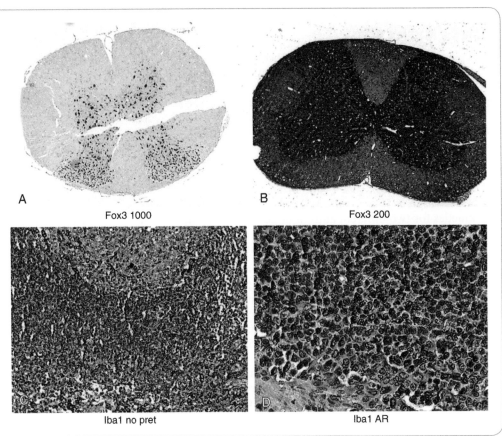

A — Fox3 1000
B — Fox3 200
C — Iba1 no pret
D — Iba1 AR

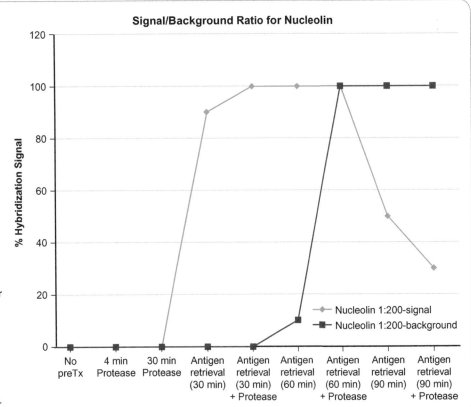

Figure5-21 Ther elationship between signal and background: the effect of pretreatment conditions. Using a concentration of 1:200, we plotted the relationship between signal and background for the nuclear protein nucleolin with a wide variety of pretreatment conditions. Note the inverse relationship between signal and background, and how background is much more likely with prolonged pretreatments such as antigen retrieval greater than 30 minutes with the addition of protease.

153

more likely "dock" with each other, these forces clearly have the potential to slow down the diffusion of the probe into the hundreds of thousands of cells present in the typical formalin-fixed, paraffin-embedded tissue. It should be noted that in the immune system, certain cells (e.g., follicular dendritic cells) have the function of allowing the "docking" of antigens with the primary antibody; it is possible that the equivalent biochemical processes are occurring in the formalin-fixed, paraffin-embedded tissues.

By plotting hybridization time versus the intensity of the immunohistochemistry signal, we can get a good idea of the relative diffusion time of the primary antibody to the target. Figure 5-22, panel A, shows the hybridization kinetics for two different antigen targets (cytokeratin AE1/3 and CD45) in the same formalin-fixed, paraffin-embedded tissues (tonsil). Note three key points:

1. The primary antibody for cytokeratin AE 1/3 diffuses rapidly to its target. Only 15 minutes of diffusion is needed to see 50% of the maximum signal.
2. The primary antibody for CD45 appears to diffuse more slowly through the same tissue, as only 20% of the maximum signal is evident after 15 minutes of incubation.
3. Despite the obvious differences in the initial rate of diffusion, each primary antibody required between 2–3 hours in order to see the maximum signal.

Points 1 and 2 here certainly would be predicted from the model that suggests that each epitope on a different protein probably has a biochemically different three-dimensional protein/protein cross-linked cage associated with it, which, in turn, would affect the diffusion time of the primary antibody. The reason for point 3 is less clear. It suggests that there are two separate phases in the attachment of the primary antibody to its target in immunohistochemistry:

1. An initial diffusion related to factors such as pore size, antibody concentration, accessibility of the epitope.
2. A secondary steady-state phase that is slower than the primary antibody-epitope diffusion kinetics. This, of course, could be the binding of the secondary antibody and/or the activity of the reporter enzyme (peroxidase or alkaline phosphatase), which may change over time as more precipitate accumulates. Recall from Figure 4-33 of Chapter 4 that the streptavidin/alkaline phosphatase conjugate may require 1–2 hours (or more) of diffusion to show its maximum signal, and that the avidity of streptavidin and biotin is much stronger than that between a primary and secondary antibody in immunohistochemistry.

Whatever the explanation for the data represented in Figure 5-22, the bottom line is that 1–2 hours is a logical incubation time for the primary antibody. Under optimal conditions, with an antigen in high copy number, you can see a signal after much shorter hybridization times. However, using, for example, 30 minute incubation time for all primary antibodies lead either to some weak signals or false-negative results in a certain percentage of cases.

Now look at panel B for Figure 5-22. In this experiment, the time kinetics of cytokeratin immunohistochemistry signal were analyzed in the same breast cancer tissue. However, here the comparison was between the carcinoma cells and benign breast epithelial cells. As is well known, the cancer cells have, in general, a smaller copy number of any given epitope typical of the cell type than the corresponding benign cell. Hence, we would predict that the time course of the cytokeratin signal would be slower for the cancer cells than the benign cells, and this is indeed the case.

A

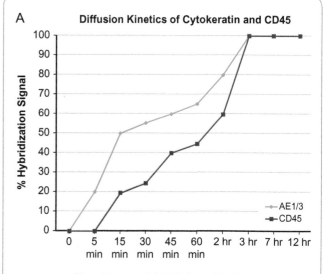

B

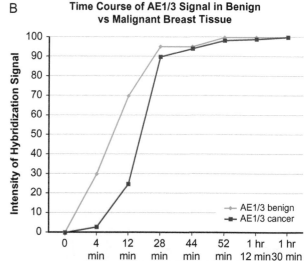

Figure 5-22 The time course of the hybridization of the primary antibody. In these experiments, two different primary antibodies (cytokeratin and CD45) were co-incubated with the tonsil tissue for between 5 minutes and 12 hours, with the intensity of the hybridization signal plotted versus the hybridization time. Note that the signal for the cytokeratin (blue) is generated more quickly than the signal for CD45 (red) in the same tissue. However, after sufficient time, each generates intense signals. This suggests that the three-dimensional protein cross-linked cage for CD45 may be a bit more restrictive to the antibody's movement than the equivalent cagef orc ytokeratin.

Importance of the Copy Number of the Target to the Signal Intensity with Immunohistochemistry

Just as with in situ hybridization, the copy number of the target of interest will clearly influence the signal with immunohistochemistry. This is clearly evident with a protein that closely parallels HPV in CIN lesions: p16. Let's examine Figure 5-23. Note that in this CIN lesion (panel A and, at higher magnification, panel B), the p16 signal varies from intense in the more basal cells to less intense in the cells with the perinuclear halos. This, actually, is the reverse correlation of the HPV DNA copy number as we discussed in Chapter 4, which is weak in the basal cells and very high in the koilocytes in a CIN lesion. As an aside, look at panel C, which is from a CIN lesion from 1997 where the slide was actually cut in 1997. Note the absence of any signal, which is very unusual for a CIN lesion. The 1997 block was available and recuts were made in 2012. The p16 data for this slide are shown in panel D. Note that foci of cells are positive, but that most of the cells are negative. This is not the typical pattern of p16 in a CIN lesion. Rather, the pattern probably represents the pattern of degeneration of the p16 signal in the block over the past 15 years. The data in panels C and D remind us that the paraffin block offers some protection from the degradation of the signal over time, but that this protection is far from complete. However, we can regenerate the p16 signal back to the initial results, as we discussed in detail in the preceding chapter.

STEP SIX: THE DETECTION STEP OF IMMUNOHISTOCHEMISTRY

At this stage, we have formalin-fixed, paraffin-embedded tissues in which some cells have the target of interest. A primary antibody is now attached to the target. The next step is to add the secondary antibody. There are two key features of the secondary antibody:

1. It must recognize any antibody from the animal that was used to make the primary antibody.
2. It is conjugated to some molecule, typically biotin, which will allow us to detect the epitope–primary antibody–secondary antibody complex.

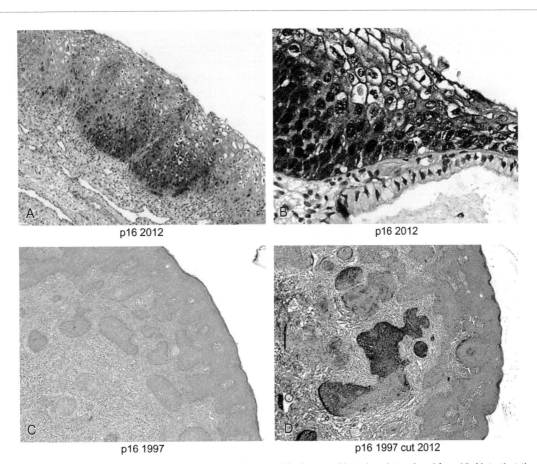

p16 2012

p16 2012

p16 1997

p16 1997 cut 2012

Figure 5-23 The relationship between the epitope copy number and the immunohistochemistry signal for p16. Note that the intensity of the p16 signal varies considerably with the geography of the CIN lesion. The signal is strongest toward the base of the lesion and weaker as we go toward the surface, where the koilocytes are located (panels A and, at higher magnification, B). This is the reverse pattern evident for HPV DNA in such lesions, as we saw in the preceding chapter. Also note that no p16 signal was seen in the CIN lesion tested at the same time as the tissue in panels A and B if the slides were cut in 1997 (panel C), and a variable signal, suggesting differential loss of the detectable epitope, was seen if that same tissue was sectioned in 2012 right before the testing was done (panel D).

Since most companies make primary antibodies in rabbits or mice, the commercially available secondary antibodies typically can detect any antibody derived in a rabbit or mouse. It is easy to find, however, secondary antibody mixtures that will also contain antibodies that will detect any primary antibody from a goat, rat, chicken, and so on. These cocktails are logically called "multilink secondary antibodies." One example is the "Supersensitive Multilink" from Biogenex (catalog #HK340).

The kinetics of binding of the secondary antibody are very similar to that of the primary antibody to the epitope, as seen in Figure 5-22, in the sense that the majority of the reaction is typically done between 1 and 2 hours. Thus, although most manual protocols for immunohistochemistry call for a 20–30-minute incubation with the secondary antibody, I prefer a 1–2 hour incubation, especially when dealing with a target that may have a low copy number.

Let me make brief mention of the washes for the primary and secondary antibody. I use the same wash for immunohistochemistry for each step. The reasoning is that the forces holding together the primary antibody to the epitope and the secondary antibody to the primary antibody are relatively strong. As we showed in Chapter 3, these forces can withstand washes of 60°C. I prefer washes at 37°C and include a low concentration of bovine serum albumin (0.2%) in a solution of PBS. As mentioned before, a trick of the trade I use is to include 3% H_2O_2 in the wash when using peroxidase/DAB as the detection system. The reason I do not focus that much attention on the washes of immunohistochemistry is that, in my opinion, the pretreatment conditions and the primary antibody concentration are the key variables for both signal and background with immunohistochemistry. When these two variables are optimized, there is no need to vary the wash conditions, because background will not be an issue.

The next step is basically identical for immunohistochemistry and in situ hybridization, assuming that our secondary antibody has been conjugated to biotin. Although there are many different reporter systems for immunohistochemistry, perhaps more than for in situ hybridization, biotin is still commonly employed, so I focus on that system. Thus, the next step is to add streptavidin/reporter enzyme conjugate.

Let's discuss again the time kinetics of biotin and streptavidin-alkaline phosphatase. You may recall that the affinity of streptavidin and biotin is among the strongest in molecular pathology. Thus, when streptavidin "finds" the biotin on the secondary antibody, it immediately and strongly binds to it. So, the key to the time kinetics of streptavidin-biotin binding will be the rate of diffusion of the conjugate to the target/primary antibody complex.

In Chapter 4, we discussed these series of experiments in which I examined the time course of streptavidin-alkaline phosphatase binding to biotin complexes. If our theory of the three-dimensional macromolecule cross-linked cage is correct, then we will note different diffusion times for the streptavidin-alkaline phosphatase conjugate when comparing various tissues. Representative data

were presented in Figure 4-33. Let's revisit those data, as they are clearly relevant to immunohistochemistry.

Note that there is a disparity in the rate of diffusion of the streptavidin-alkaline phosphatase conjugate in these three representative tissues. This disparity parallels the variability in the rate of diffusion for different targets in basically every step of immunohistochemistry (and in situ hybridization), as we have seen repeatedly in this book. I find these data very interesting because they underscore that formalin-fixed, paraffin-embedded tissues are both variable and dynamic.

It is also clear from Figure 4-33 that 30–45 minutes is sufficient to generate maximum diffusion of the streptavidin-alkaline phosphatase conjugate. However, some tissues, such as case 2 in Figure 4-33, may require much more time for the streptavidin-alkaline phosphatase conjugate to reach near 100% access to the biotin. The latter likely represents less than optimal pretreatment of the tissues that can occur with the "strongly fixed tissues," and underscores again the importance of doing various disparate pretreatments on each new biopsy you test to increase the chances that one of these pretreatments will allow maximum diffusion of the reagents into the fixed cells. The key point for this protocol is that, to be on the safe side, it is best to incubate the secondary antibody for 1–2 hours.

Before leaving this section, let's discuss one more trick of the trade. It applies to anyone who works with mouse or rabbit formalin-fixed, paraffin-embedded tissues. We discuss mouse here, as this is much more common. When we are using a commercially available kit for immunohistochemistry, the "multilinked" secondary antibody will detect, at a minimum, primary antibodies from either a mouse or rabbit. However, if we are using mouse formalin-fixed, paraffin-embedded tissues, then the secondary antibody may well cause background by "doing its job" and binding to the mouse antibodies in the tissue. As expected, this typically will produce intravascular background, due to the relatively high concentrations of circulating primary mouse antibodies present in the blood.

This problem is easy to solve. You need to use as a primary antibody either rabbit or goat (or basically any animal other than mouse-derived primary antibody). Then, for the secondary antibody, you can use an antibody mix that will recognize only rabbit primary antibodies. This simple way eliminates the anti-mouse secondary antibody background problem. An example of this is seen in Figure 5-24. The tissue is from a mouse injected with cancer cells and reovirus. Note the strong background in panels A and B when a multilink secondary antibody was directed against mouse and rabbit (my primary antibody was rabbit-derived). The background was associated with the blood. This background was so strong that it hid the signal. When I used a secondary antibody that recognized only rabbit primary antibodies (and switched to a DAB system to make the difference more obvious), the background was gone, and the strong perivascular signal for reovirus was evident (panels C and D). This point is important, because this is the way that you demonstrate that reovirus leaves the vasculature and attacks the cancer cells in this location.

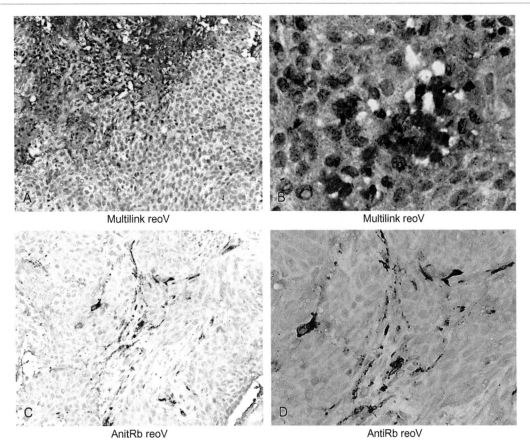

Multilink reoV

Multilink reoV

AnitRb reoV

AntiRb reoV

Figure5-24 Reducing background in mouse tissues by using a primary and secondary antibody not generated in a mouse. The tissue is a cancer that developed in a mouse after injection of the tumor cells that were exposed to the reovirus. The primary antibody was rabbit based. Panels A and B show the results if the secondary antibody was directed against both rabbit and mouse. There is a diffuse signal that shows a strong intravascular component. This is not the expected pattern for reovirus. Hence, this is probably nonspecific staining of the mouse secondary antibody to the many antibodies that are in the vascular system of the animal. In comparison, panels C and D show the results if the secondary antibody was only anti-rabbit. The background is gone, and the viral-specific signal is evident primarily in endothelial cells, which is the expected and logical pattern for this virus after IV injection into the mouse.

REPORTER ENZYMES

The discussion of the reporter enzymes with immunohistochemistry is basically the same as with in situ hybridization, with the exception that it is rare to use NBT/BCIP in the former. Most immunohistochemistry systems use either DAB or Fast Red although, again, many other chromogens are available. Since DAB and Fast Red dominate the market, I focus my attention on them here.

DAB/PEROXIDASE

As we discussed at length in Chapter 4, the best-known background issue with regards to the chromogen is with DAB. It has long been known that if you do not add an H_2O_2 blocking step prior to adding the primary antibody, background may well ensue.

By re-examining Figure 4-39, we can recall one strategy to reduce the peroxidase-associated background. I restate some of the data and conclusions here for simplicity's sake.

When we test placenta with no primary antibody and no pretreatment, the red blood cells may show a dark brown color indicative of background. The fact that we see this background with no primary antibody tells us it probably is due to some endogenous factor. If this factor was endogenous peroxidase, then it should be easy to inactivate it with strong treatment with a protease or with incubation in formalin prior to the immunohistochemical reaction. However, as we can see in Figure 4-39, the background signal in the red blood cells was still present after the tissue was treated with a strong protease digestion, followed by formalin post-fixation prior to immunohistochemistry without a primary antibody. This suggests that there is some chemical in red blood cells, not peroxidase, which is reacting with the exogenous peroxidase to produce the background. If we incubate the tissue in H_2O_2 prior to immunohistochemistry, the background is gone (same figure). Another important point evident in this figure is seen when these same experiments are done with a primary antibody—in this case, cytokeratin AE 1/3. This protein is present in high copy number in the trophoblasts. Note that with H_2O_2 pretreatment, we see a strong signal for the cytokeratin and no background. If the H_2O_2 pretreatment is omitted, the background is evident

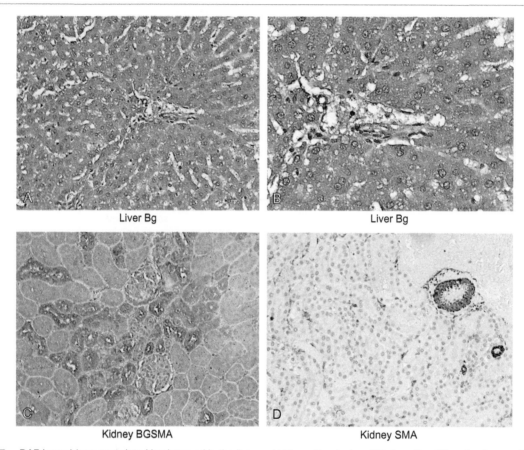

Liver Bg

Liver Bg

Kidney BGSMA

Kidney SMA

Figure5-25   DAB/peroxidase-associated background in the liver and kidney. Panels A and B show the diffuse background staining in the liver using a peroxidase/DAB system. Note that the background is mostly in hepatocytes; the target was CD45, which is not present in these cells. Panel C shows the same tissue in the kidney stained with smooth muscle actin (SMA). Note that the background is mostly in the large tubules of the kidney that surround the glomeruli. These are called the proximal convoluted tubules, and they contain abundant alkaline phosphatase. In panel D the background was not evident, as this is in the collecting duct region of the kidney, which usually shows less background than the cortex (shown in panel C). Note the SMA signal now evident in panel D when the background was eliminated.

and there is a concomitant decrease in the signal. This inverse relationship was also discussed previously in this chapter(Figure5-21).

Whatever the explanation, these data show that the use of H$_2$O$_2$ incubation prior to the primary antibody is a key step in the immunohistochemistry or in situ hybridization protocol using DAB/peroxidase. Also, as a reminder, certain tissues are more prone to background. The liver is probably the most-cited organ as far as endogenous biotin and background with DAB are concerned. An example of this type of background is given in Figure 5-25. The antibody was CD45, so all of the brown color in panels A and B must be background because this is normal liver. As indicated previously, I do not think such background reflects endogenous peroxidase (or alkaline phosphatase for Fast Red). Another reason for this conclusion is presented in Figure 5-25. The kidney is rich in alkaline phosphatase, especially in the proximal convoluted tubular cells. Note how the background in this kidney tested for smooth muscle actin is present in the proximal convoluted tubular cells (panel C), which are easy to recognize because they are much more numerous than the distal convoluted

tubular cells, and also larger. However, to my knowledge, the kidney is not rich in peroxidase. Also note that the background is not present in the same section in the distal collecting ducts (panel D), whereas the actin-based signal is strong. This and similar data suggest to me that background with DAB and peroxidase comes from similar sources—endogenous chemicals unique to certain cell types. The key feature of these cells seems to be their high protein/metabolic content and activity. It is plausible that some of the background may relate to the three-dimensional macromolecule cross-linked cages typical of such cells, which we discuss in more detail later in the book.

Another problem with DAB/peroxidase systems has nothing to do with background. Rather, it has to do with the fact that there are some pigments that are relatively common in cells that are brown. Specifically, both melanin and hemosiderin stain brown. Melanin will be evident as small brown granules that are found in benign melanocytes and in melanoma cells. In Figures 5-8 and 5-9 earlier in this chapter, we showed that any tissue with melanin (or, by extension, hemosiderin) positive cells is best analyzed with Fast Red as the chromogen and not DAB.

FAST RED/ALKALINE PHOSPHATASE

As previously discussed, Fast Red typically produces background in red blood cells and in the lining of trophoblasts. Of course, the red blood cell background was more typical of peroxidase systems and the trophoblast background more common of alkaline phosphatase systems. This is another reason why I do not think the basis of background is residual endogenous peroxidase or alkaline phosphatase activity. Of course, the other reason is that we cannot eliminate background with strong protease digestion and/or post-formalin fixation, which certainly should be able to inactivate any enzyme. Whatever the explanation, any such background with Fast Red can usually be eliminated with a reduction of the probe concentration or adjustment of the pretreatment regime.

SIGNAL/BACKGROUND IN TISSUES FROM OTHER ANIMALS

A very common issue which I am asked for comments on is background when testing tissues that are from animals other than humans. Although mice and rat tissues are the most common, I have been asked for help with background issues using tissues from guinea pigs, horses, pigs, dogs, cats, and other animals.

As you know, specific kits are commercially available to deal with background in mice tissues. Although these kits are useful, in my experience, the causes of background in formalin-fixed, paraffin-embedded tissues other than human are the exact same as with human tissues. The best way to address this issue is to focus on the primary antibody concentration and the pretreatment conditions.

Working with mammalian tissues other than humans also lets us explore a very practical tissue with immunohistochemistry. The question is:

Can we use a primary antibody directed against an epitope in one animal and expect it to work in another animal?

This is an important question for those working with nonhuman tissues, as it may be difficult to find any company that sells primary antibodies directed against dog, cat, cow, horse, or other epitopes.

As I am sure you know, the answer is yes and no! However, it is more yes than no. The key is the degree of homology between the epitope in the "primary animal" for which the primary antibody was raised versus the animal you are interested in testing. Here are the tricks of the trade I use when trying to use a primary antibody made against one animal (usually human or mouse) in the formalin-fixed, paraffin-embedded tissues of another animal:

1. Use a polyclonal antibody. This, by definition, increases your chances, since there will be a much wider diversity of primary antibodies, some of which may have good affinity to the epitope of the animal you are interested in.
2. If possible, choose an epitope that is well conserved across the species.
3. Use a higher concentration of the primary antibody than you did with the "primary animal" tissue.

Figure 5-26 shows some representative data from such experiments. Each tissue (human, dog, pig, chipmunk, and

vole) was tested for CD45 and smooth muscle actin (SMA) at the same time. Such co-expression analyses work because CD45 and SMA will stain completely different cell types (lymphocytes and blood vessel walls, respectively). Note in panel A how the human tissue is strongly positive (tonsil) for CD45 and SMA. The primary antibody was indeed directed against the human CD45 and SMA. Compare these data to those derived from the dog in panels B and C. The tissue is skin with a strong lymphocyte infiltration. Note that the SMA stains the dog blood vessels strongly. Hence, the anti-human SMA antibody shows a lot of homology with the corresponding epitope of the dog. However, no CD45 staining is evident in the lymphocytes. Thus, there is not sufficient homology between the human and dog CD45 epitopes that the primary antibody is directed against. Compare this to panel C (pig spleen) where there are, of course, many CD45+ cells. Note the strong SMA signal and the moderately strong CD45 signal. Thus, the pig CD45 epitopes share better homology with the anti-human CD45 antibodies than does the dog. Finally, panels E and F show that SMA is clearly a very well conserved protein, as the vole and chipmunk CD45 molecules strongly react with the human anti-CD45 antibody. Also note the minimal amount of background in these tissues.

A simple way to determine if a given antibody directed against an epitope in one species may work well in another species is to do some research about the epitope. Did it evolve relatively early or late in the species in question? If you can do sequence analysis, is the mRNA or protein structure well or poorly conserved over different species? Clearly, smooth muscle actin is well conserved. Look at Figure 5-27, where you see that the primary antibody directed against human smooth muscle actin also detects the protein in horse (panel A) and nonmammal species, including a bird (sparrow, panel B) and a goldfish (panel C). On the other hand, CD45 is a relatively late protein in evolution, since it is involved with the T/B-cell immune system. Compare this to the more primitive element of the immune system: the toll-like receptor component. If you examine TLR-7 (the human antigen) expression in other mammals, then the primary antibody will generate a signal. This is evident in panels D and E, where you can see in the spleen of a raccoon a strong signal for SMA and for the protein TLR-7. Cytokeratin AE 1/3 is another conserved protein and, thus, it, as well as SMA, is detectable in the rabbit if primary antibodies made against the human epitopes are used (panel F). Finally, another well-conserved protein is fibrinogen, since blood clotting has been an issue in any animal that has a heart, which obviously includes most animals. Note that the signal for fibrinogen is strong in the goldfish liver (panel C) even though the antibody is directed against the human epitopes.

STEP SEVEN: THE COUNTERSTAIN FOR IMMUNOHISTOCHEMISTRY

The final point to discuss in this chapter is the counterstain. This discussion, again, is basically the same as for in situ hybridization. For Fast Red reactions, you can choose between no counterstain and hematoxylin. For DAB reactions, you can choose between hematoxylin, Nuclear Fast Red, and no counterstain.

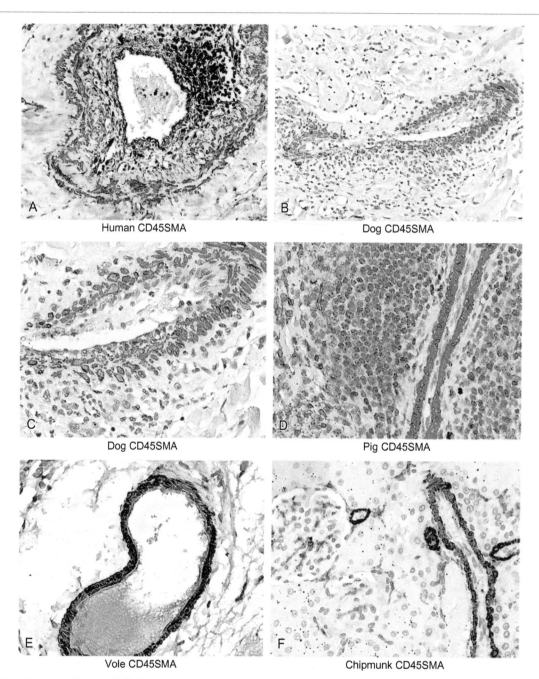

Figure5-26 Detection of antigens in animals other than the species for which the primary antibody was initially directed against: Part A. The tissue in panel A is human, panels B and C are from dog, and panel D is from a pig. Note that each shows smooth muscle and inflammatory cells. The smooth muscle actin (SMA) signal in each animal is intense, whereas the CD45 signal is strong in the human, weak in the pig, and nonexistent in the dog. This result suggests that the SMA antigen is far better conserved against species than the CD45 antigen. Further evidence for the strong conservation of the SMA antigen is seen in panels E and F, where strong signals for such are evident in the vole (panel E) and the chipmunk (panel F).

NO COUNTERSTAIN

The value of doing no counterstain is two-fold:

1. In general, you can see the signal more readily.
2. Weak nuclear-based signals that might be masked by a counterstain may now be appreciated.

Of course, the negative part of not using a counterstain is that it makes it difficult to see the morphology of the positive and negative cells in general, and the overall

tissue in particular. As with in situ hybridization, a trick of the trade for immunohistochemistry when using no counterstain is to allow the colorimetric reaction to proceed a bit longer than usual, because this will tend to give the entire tissue a weak color that allows you to better appreciate the histologic detail.

In my experience, I routinely omit the counterstain when working with a nuclear-based target. I then examine the data under the microscope "wet" before the counterstain/coverslip

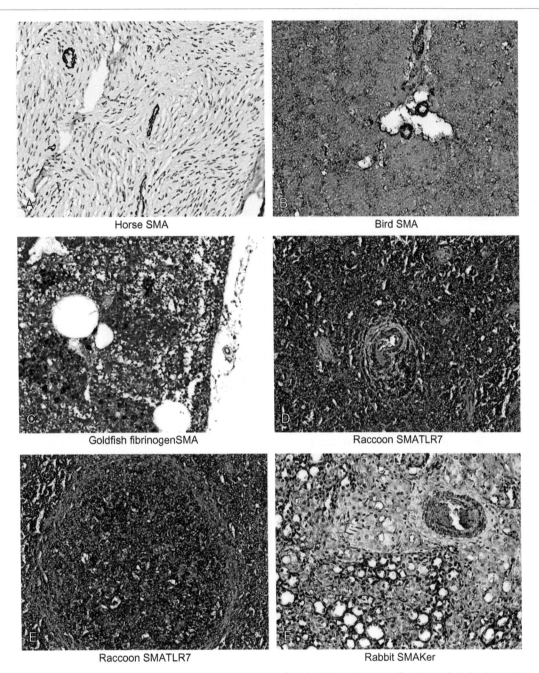

Horse SMA

Bird SMA

Goldfish fibrinogenSMA

Raccoon SMATLR7

Raccoon SMATLR7

Rabbit SMAKer

Figure5-27 Detection of antigens in animals other than the species for which the primary antibody was initially directed against: Part B. The primary antibody directed against human smooth muscle actin also detects the protein in horse (panel A) and nonmammal species including a bird (sparrow, panel B) and a goldfish (panel C). On the other hand, CD45 is a relatively late protein in evolution, since it is involved with the T/B-cell immune system. Compare this to the more primitive element of the immune system: the toll-like receptor component. If you examine TLR-7 (the human antigen) expression in other mammals, then the primary antibody will generate a signal. This is evident in panels D and E, where you can see in the spleen of a raccoon a strong signal for SMA and for the protein TLR-7. Finally, cytokeratin AE 1/3 is another conserved protein and, thus, it, as well as SMA, is detectable in the rabbit if primary antibodies made against the human epitopes are used (panel F).

step. I then decide whether to use a very weak hematoxylin or Nuclear Fast Red counterstain (a few dips) or not.

HEMATOXYLIN-BASED COUNTERSTAIN

Hematoxylin is the standard counterstain for immunohistochemistry when using either DAB or Fast Red as the chromogen. I prefer to do brief dips (a few seconds) of hematoxylin counterstain, since the commercial hematoxylin counterstains tend to be too strong. You want to see the tissue morphology and the nuclei, but you don't want this counterstain dominating the color of the immunohistochemical reaction.

161

NUCLEAR FAST RED COUNTERSTAIN

Nuclear Fast Red, as it name implies, stains nuclei pink/red and does not stain the cytoplasm. It can be used with DAB/NBT signals but not, of course, with Fast Red chromogen-stained tissue. If the DAB signal is cytoplasmic, I like to use the maximum intensity of Nuclear Fast Red colorization (about 5 minutes with the full-strength solution) in order to get the most complete image of the tissue possible. If the signal is nuclear-based, then I prefer to obtain a much weaker nuclear counterstain so as not to obscure the signal.

EOSIN AND CYTOPLASMIC COUNTERSTAINING

Clearly, hematoxylin and Nuclear Fast Red stain, for all intents and purposes, just the nucleus. What if you have a nuclear signal and you want to stain the cytoplasm? In this case, use eosin. As with hematoxylin, I use (with DAB chromogens only, of course) only a few dips in eosin to accentuate a nuclear-based signal with immunohistochemistry. Certainly, under these circumstances, with eosin counterstain, you do not risk masking the nuclear signal as you do if you use hematoxylin.

One last point needs to be made regarding the storage of slides for immunohistochemistry. It revolves around this question: is there any effect upon removing the paraffin from the slide days before you do the immunohistochemical reaction on the immunohistochemistry signal? The answer is *yes*. Note in Figure 5-28 that there is a 70% reduction in the CD3-specific signal in a tonsil if the paraffin was removed from the glass slide about 2 weeks prior to doing the immunohistochemistry reaction. This is reminiscent of the effect we saw previously when we removed the paraffin and then heated the slide to either 60°C or 95°C and, in either case, noted a marked reduction in the immunohistochemistry signal. Indeed, note in the second part of Figure 5-28 that the reduction in the immunohistochemistry signal by removing the paraffin and then waiting 1 week to do the immunohistochemistry reaction is very dependent on the temperature at which the slides were stored. The colder the temperature of storage, the lower was the reduction in the intensity of the immunohistochemistry signal seen. This reminds us that formalin-fixed tissue can be very sensitive to the ambient temperature, and certainly to the removal of paraffin that surrounds each tissue section on a glass slide.

The suggested readings [1–97] are provided to give a relatively broad overview of the utility of immunohistochemistry in both the diagnostic and research fields. Again, there is a focus on the detection of viral nucleic acids. Plus, to continue to emphasize the overlap between in situ hybridization and immunohistochemistry, many of the articles describe both in situ hybridization and immunohistochemistry analysis.

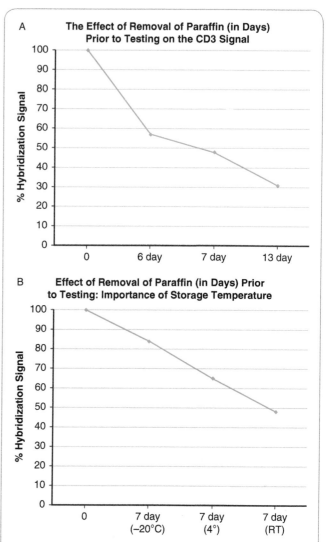

Figure5-28   Importance of the removal of the paraffin wax on the glass slide to the signal with immunohistochemistry. In these experiments, serial sections of a tonsil were placed on glass slides. The paraffin wax was removed and the tissues stored for up to 13 days. Immunohistochemistry reactions were performed during this time period. Note that the intensity of the immunohistochemistry signal was reduced the longer the slides were stored at room temperature without the paraffin wax (left side of image). The bottom part of the image shows that the signal reduction was very dependent on the temperature at which the slides were stored once the paraffin wax was removed. The lower the storage temperature, the lower was the reduction in the intensity of the immunohistochemistry signal.

SUGGESTED READINGS

References on immunohistochemistry

[1] Akutsu S, Miyazaki J. Biochemical and immunohistochemical studies on tropomyosin and glutamate dehydrogenase in the chicken liver. Zoolog Sci 2002;19:275E86.

[2] Aloia AL, Sfanos KS, Isaacs WB, et al. XMRV: a new virus in prostate cancer? Cancer Res 2010;70:10028E33.

[3] Arrieta JJ, Rodriguez-Inigo E, Ortiz-Movilla N, et al. In situ detection of hepatitis C virus RNA in salivary glands. Am J Pathol 2001;158:259E64.

[4] Ashton JC, Appleton I, Darlington CL, Smith PF. Immunohistochemical localization of cerebrovascular cannabinoid CB1 receptor protein. J Cardiovasc Pharmacol 2004;44:517E9.

[5] Asif M, Khadim M, Mushtaq S, et al. Determination of her-2/neu by chromogenic in situ hybridization on borderline (2+) immunohistochemistry cases in carcinoma breast. Asian Pac J Cancer Prev 2011;12:211E4.

[6] Awaya H, Takeshima Y, Furonaka O, Kohno N, Inai K. Gene ampliÞcation and protein expression of EGFR and HER2 by chromogenic in situ hybridization and immunohistochemistry in atypical adenomatous hyperplasia and adenocarcinoma of the lung. J Clin Pathol 2005;58:1076E80.

[7] Baloglu G, Haholu A, Kucukodaci Z, et al. The effects of tissue Þxation alternatives on DNA content: a study on normal colon tissue. Appl Immunohistochem Mol Morphol 2008;16:485E92.

[8] Balyasnikova IV, Metzger R, Franke FE, Danilov SM. Monoclonal antibodies to denatured human ACE (CD 143), broad species speciÞcity, reactivity on paraffn sections, and detection of subtle conformational changes in the C-terminal domain of ACE. Tissue Antigens 2003;61:49E62.

[9] Baroni CD, Pezzella F, Stoppacciaro A, et al. Systemic lymphadenopathy (LAS) in intravenous drug abusers. Histology, immunohistochemistry and electron microscopy: pathogenic correlations. Histopathology 1985;9:1275E93.

[10] Bilous M, Morey A, Armes J, Cummings M, Francis G. Chromogenic in situ hybridisation testing for HER2 gene ampliÞcation in breast cancer produces highly reproducible results concordant with ßuorescence in situ hybridisation and immunohistochemistry. Pathology 2006;38:120E4.

[11] Bosq J, Gatter KC, Micheau C, Mason DY. Role of immunohistochemistry in diagnosis of nasopharyngeal tumors. J Clin Pathol 1985;38:845E8.

[12] Brooks JS, Freeman M, Enterline HT. Malignant ÒtritonÓ tumors. Natural history and immunohistochemistry of nine new cases with literature review. Cancer 1985;55:2543E9.

[13] Bzorek Sr. M, Petersen BL, Hansen L. Simultaneous phenotyping and genotyping (FICTION-methodology) on paraffn sections and cytologic specimens: a comparison of 2 different protocols. Appl Immunohistochem Mol Morphol 2008;16:279E86.

[14] Cho EY, Choi YL, Han JJ, Kim KM, Oh YL. Expression and ampliÞcation of Her2, EGFR and cyclin D1 in breast cancer: immunohistochemistry and chromogenic in situ hybridization. Pathol Int 2008;58:17E25.

[15] Cho EY, Han JJ, Choi YL, Kim KM, Oh YL. Comparison of Her-2, EGFR and cyclin D1 in primary breast cancer and paired metastatic lymph nodes: an immunohistochemical and chromogenic in situ hybridization study. J Korean Med Sci 2008;23:1053E61.

[16] Cserni G. Complete sectioning of axillary sentinel nodes in patients with breast cancer. Analysis of two different step sectioning and immunohistochemistry protocols in 246 patients. J Clin Pathol 2002;55:926E31.

[17] de Armond SJ, Eng LF, Rubinstein LJ. The application of glial Þbrillary acidic (GFA) protein immunohistochemistry in neurooncology. A progress report. Pathol Res Pract 1980;168:374E94.

[18] De Vos R, De Wolf-Peeters C, van den Oord JJ, Desmet V. A recommended procedure for ultrastructural immunohistochemistry on small human tissue samples. J Histochem Cytochem 1985;33:959E64.

[19] Del Valle L, Gordon J, Assimakopoulou M, et al. Detection of JC virus DNA sequences and expression of the viral regulatory protein T-antigen in tumors of the central nervous system. Cancer Res 2001;61:4287E93.

[20] Delfour C, Roger P, Bret C, et al. RCL2, a new Þxative, preserves morphology and nucleic acid integrity in paraffn-embedded breast carcinoma and microdissected breast tumor cells. J Mol Diagn 2006;8:157E69.

[21] Dralle H, Schwarzrock R, Lang W, et al. Comparison of histology and immunohistochemistry with thyroglobulin serum levels and radioiodine uptake in recurrences and metastases of differentiated thyroid carcinomas. Acta Endocrinol 1985;108:504E10.

[22] Ermert L, Hocke AC, Duncker HR, Seeger W, Ermert M. Comparison of different detection methods in quantitative microdensitometry. Am J Pathol 2001;158:407E17.

[23] Fisher CJ, Gillett CE, Vojtesek B, Barnes DM, Millis RR. Problems with p53 immunohistochemical staining: the effect of Þxation and variation in the methods of evaluation. Br J Cancer 1994;69:26E31.

[24] Fowler CB, OÕLeary TJ, Mason JT. Modeling formalin Þxation and histological processing with ribonuclease A: effects of ethanol dehydration on reversal of formaldehyde cross-links. Lab Invest 2008;88:785E91.

[25] Frederiks WM, Mook OR. Metabolic mapping of proteinase activity with emphasis on in situ zymography of gelatinases: review and protocols. J Histochem Cytochem 2004;52:711E22.

[26] Friedman M, Gentile P, Tarectecan A, Bluchs A. Malignant mesothelioma: immunohistochemistry and DNA ploidy analysis as methods to differentiate mesothelioma from benign reactive mesothelial cell proliferation and adenocarcinoma in pleural and peritoneal effusions. Arch Pathol Lab Med 1996;120:959E66.

[27] Gabusi E, Lattes C, Fiorentino M, DÕErrico A, Grigioni WF. Expression of Epstein-Barr virus-encoded RNA and biological markers in Italian nasopharyngeal carcinomas. J Exp Clin Cancer Res 2001;20:371E6.

[28] Grabau D, Ryden L, Ferns̃ M, Ingvar C. Analysis of sentinel node biopsyÑa single-institution experience supporting the use of serial sectioning and immunohistochemistry for detection of micrometastases by comparing four different histopathological laboratory protocols. Histopathology 2011;59:129E38.

[29] Grantzdorffer I, Yumlu S, Gioeva Z, et al. Comparison of different tissue sampling methods for protein extraction from formalin-Þxed and paraffn-embedded tissue specimens. Exp Mol Pathol 2010;88:190E6.

[30] Hayashi Y, Watanabe J, Nakata K, Fukayama M, Ikeda H. A novel diagnostic method of Pneumocystis carinii. In situ hybridization of ribosomal ribonucleic acid with biotinylated oligonucleotide probes. Lab Invest 1990;63:576E80.

[31] Haynes WD, Shertock KL, Skinner JM, Whitehead R. The ultrastructural immunohistochemistry of oncofoetal antigens in large bowel carcinomas. Virchows Arch A Pathol Anat Histopathol 1985;405:263E75.

[32] Hiort O, Kwan PW, DeLellis RA. Immunohistochemistry of estrogen receptor protein in paraffn sections. Effects of enzymatic pretreatment and cobalt chloride intensiÞcation. Am J Clin Pathol 1988;90:559E63.

[33] Hiraku Y, Tabata T, Ma N, et al. Nitrative and oxidative DNA damage in cervical intraepithelial neoplasia associated with human papilloma virus infection. CancerS ci 2007;98:964E72.

[34] Ishitsuka Y, Maniwa F, Koide C, et al. Increased halogenated tyrosine levels are useful markers of human skin ageing, reßecting proteins denatured by past skin inßammation. ClinE xpDe rmatol 2012;37:252E3.

[35] Janssen S, Elema JD, van Rijswijk MH, et al. ClassiÞcation of amyloidosis: immunohistochemistry versus the potassium permanganate method in differentiating AA from AL amyloidosis. ApplP athol 1985;3:29E 38.

[36] Javanmard SH, Moeiny A. Quantitative immunohistochemistry by measuring chromogen signal strength using a C# written program. JRe sM edS ci 2009;14:201E3.

[37] Jiwa M, Steenbergen RD, Zwaan FE, et al. Three sensitive methods for the detection of cytomegalovirus in lung tissue of patients with interstitial pneumonitis. AmJ Clin P athol 1990;93:491E4.

[38] Johansen P, Jensen MK. Enzymecytochemistry and immunohistochemistry in monoclonal gammopathy and reactive plasmacytosis. Acta Pathol Microbiol Scand A 1980;88:377E82.

[39] Jung KY, Baek IJ, Yon JM, et al. Developmental expression of plasma glutathione peroxidase during mouse organogenesis. JM olHis tol 2011;42:545E56.

[40] Kameya T, Tsumuraya M, Adachi I, et al. Ultrastructure, immunohistochemistry and hormone release of pituitary adenomas in relation to prolactin production. VirchowsAr ch A Pathol Anat Histol 1980;387:31E46.

[41] Kashiwagi S, Kumasaka T, Bunsei N, et al. Detection of Epstein-Barr virus-encoded small RNA-expressed myoÞbroblasts and IgG4-producing plasma cells in sclerosing angiomatoid nodular transformation of the spleen. VirchowsAr ch 2008;453:275E82.

[42] Kirbis IS, Maxwell P, Fležar MS, Miller K, Ibrahim M. Externalq ualityc ontrol for immunohistochemistry on cytology samples: a review of UK NEQAS ICC

(cytology module) results. Cytopathology 2011;22:230–7.

[43] Kitahama K, Pearson J, Denoroy L, et al. Adrenergic neurons in human brain demonstrated by immunohistochemistry with antibodies to phenylethanolamine-N-methyltransferase (PNMT): discovery of a new group in the nucleus tractus solitarius. Neurosci Lett 1985;53:303–8.

[44] Kohlberger PD, Obermair A, Sliutz G, et al. Quantitative immunohistochemistry of factor VIII-related antigen in breast carcinoma: a comparison of computer-assisted image analysis with established counting methods. Am J Clin Pathol 1996;105:705–10.

[45] Korsrud FR, Brandtzaeg P. Quantitative immunohistochemistry of immunoglobulin- and J-chain-producing cells in human parotid and submandibular salivary glands. Immunology 1980;39:129–40.

[46] Kotoula V, Cheva A, Barbanis S, Papadimitriou CS, Karkavelas G. hTERT immunopositivity patterns in the normal brain and in astrocytic tumors. Acta Neuropathol 2006;111:569–78.

[47] Kraft O, Sevcik L, Klat J, et al. Detection of sentinel lymph nodes in cervical cancer. A comparison of two protocols. Nucl Med Rev Cent East Eur 2006;9:65–8.

[48] Kumamoto H, Sasano H, Taniguchi T, et al. Chromogenic in situ hybridization analysis of HER-2/neu status in breast carcinoma: application in screening of patients for trastuzumab (Herceptin) therapy. Pathol Int 2001;51:579–84.

[49] Lau JY, Naoumov NV, Alexander GJ, Williams R. Rapid detection of hepatitis B virus DNA in liver tissue by in situ hybridization and its combination with immunohistochemistry for simultaneous detection of HBV antigens. J Clin Pathol 1991;44:905–8.

[50] Leathem A, Atkins N. Fixation and immunohistochemistry of lymphoid tissue. J Clin Pathol 1980;33:1010–2.

[51] Li-Ning TE, Ronchetti R, Torres-Cabala C, Merino MJ. Role of chromogenic in situ hybridization (CISH) in the evaluation of HER2 status in breast carcinoma: comparison with immunohistochemistry and FISH. Int J Surg Pathol 2005;13:343–51.

[52] Lourenco HM, Pereira TP, Fonseca RR, Cardoso PM. HER2/neu detection by immunohistochemistry: optimization of in-house protocols. Appl Immunohistochem Mol Morphol 2009;17:151–7.

[53] Mahad DJ, Ziabreva I, Campbell G, et al. Detection of cytochrome c oxidase activity and mitochondrial proteins in single cells. J Neurosci Methods 2009;184:310–9.

[54] Manavis J, Gilham P, Davies R, Ruszkiewicz A. The immunohistochemical detection of mismatch repair gene proteins (MLH1, MSH2, MSH6, and PMS2): practical aspects in antigen retrieval and biotin blocking protocols. Appl Immunohistochem Mol Morphol 2003;11:73–7.

[55] Marquez A, Wu R, Zhao J, Shi Z. Evaluation of epidermal growth factor receptor (EGFR) by chromogenic in situ hybridization (CISH) and immunohistochemistry (IHC) in archival gliomas using bright-field microscopy. Diagn Mol Pathol 2004;13:1–8.

[56] Matkowskyj KA, Cox R, Jensen RT, Benya RV. Quantitative immunohistochemistry by measuring cumulative signal strength accurately measures receptor number. J Histochem Cytochem 2003;51:205–14.

[57] Mayr D, Heim S, Weyrauch K, et al. Chromogenic in situ hybridization for Her-2/neu-oncogene in breast cancer: comparison of a new dual-colour chromogenic in situ hybridization with immunohistochemistry and fluorescence in situ hybridization. Histopathology 2009;55:716–23.

[58] McWilliam LJ, Harris M. Granular cell angiosarcoma of the skin: histology, electron microscopy and immunohistochemistry of a newly recognized tumor. Histopathology 1985;9:1205–16.

[59] Miles PA, Herrera GA, Mena H, Trujillo I. Cytologic findings in primary malignant carcinoid tumor of the cervix. Including immunohistochemistry and electron microscopy performed on cervical smears. Acta Cytol 1985;29:1003–8.

[60] Mishiro T, Hamamoto S, Furuta K, et al. Quantitative measurement of hepatitis B virus DNA in different areas of hepatic lobules in patients with chronic hepatitis B. J Med Virol 2006;78:37–43.

[61] Miyaji K, Nakagawa Y, Matsumoto K, et al. Overexpression of a DEAD box/RNA helicase protein, rck/p54, in human hepatocytes from patients with hepatitis C virus-related chronic hepatitis and its implication in hepatocellular carcinogenesis. J Viral Hepat 2003;10:241–8.

[62] Morrison HW, Downs CA. Immunological methods for nursing research: from cells to systems. Biol Res Nurs 2011;13:227–34.

[63] Ngwenya LB, Peters A, Rosene DL. Light and electron microscopic immunohistochemical detection of bromodeoxyuridine-labeled cells in the brain: different fixation and processing protocols. J Histochem Cytochem 2005;53:821–32.

[64] Ntoulia M, Kaklamanis L, Valavanis C, et al. Her-2 DNA quantification of paraffin-embedded breast carcinomas with LightCycler real-time PCR in comparison to immunohistochemistry and chromogenic in situ hybridization. Clin Biochem 2006;39:942–6.

[65] Nuovo G. The utility of immunohistochemistry and in situ hybridization in placental pathology. Arch Pathol Lab Med 2006;130:979–83.

[66] Padmanathan A, Yadav M, Gregory AR, Kumar S, Norhanum AW. Human papillomavirus DNA and virus-encoded antigens in cervical carcinoma. Med J Malaysia 1997;52:108–16.

[67] Park CH, Matsuda K, Sunden Y, et al. Persistence of viral RNA segments in the central nervous system of mice after recovery from acute influenza A virus infection. Vet Microbiol 2003;97:259–68.

[68] Paydas S, Ergin M, Tanriverdi K, et al. Detection of hepatitis C virus RNA in paraffin-embedded tissues from patients with non-Hodgkin's lymphoma. Am J Hematol 2004;76:252–7.

[69] Pineau I, Barrette B, Vallieres N, Lacroix S. A novel method for multiple labeling combining in situ hybridization with immunofluorescence. J Histochem Cytochem 2006;54:1303–13.

[70] Privat N, Sazdovitch V, Seilhean D, LaPlanche JL, Hauw JJ. PrP immunohistochemistry: different protocols, including a procedure for long formalin fixation, and a proposed schematic classification for deposits in sporadic Creutzfeldt-Jakob disease. Microsc Res Tech 2000;50:26–31.

[71] Qian X, Bauer RA, Xu HS, Lloyd RV. In situ hybridization detection of calcitonin mRNA

in routinely fixed, paraffin-embedded tissue sections: a comparison of different types of probes combined with tyramide signal amplification. Appl Immunohistochem Mol Morphol 2001;9:61–9.

[72] Qian X, Guerrero RB, Plummer TB, Alves VF, Lloyd RV. Detection of hepatitis C virus RNA in formalin-fixed paraffin-embedded sections with digoxigenin-labeled cRNA probes. Diagn Mol Pathol 2004;13:9–14.

[73] Recordati C, Radaelli E, Simpson KW, Scanziani E. A simple method for the production of bacterial controls for immunohistochemistry and fluorescent in situ hybridization. J Mol Histol 2008;39:459–62.

[74] Rosekrans PC, Meijer CJ, Cornelisse CJ, van der Wal AM, Lindeman J. Use of morphometry and immunohistochemistry of small intestinal biopsy specimens in the diagnosis of food allergy. J Clin Pathol 1980;33:125–30.

[75] Samarawardana P, Singh M, Shroyer KR. Dual stain immunohistochemical localization of p16INK4A and ki-67: a synergistic approach to identify clinically significant cervical mucosal lesions. Appll mmunohistochemM olM orphol 2011;19:514–8.

[76] Skofitsch G, Jacobowitz DM. Distribution of corticotropin releasing factor-like immunoreactivity in the rat brain by immunohistochemistry and radioimmunoassay: comparison and characterization of ovine and rat/human CRF antisera. Peptides 1985;6:319–36.

[77] Spacek M, Hubacek P, Markova J, et al. Plasma EBV-DNA monitoring in Epstein-Barr virus-positive Hodgkin lymphoma patients. APMIS 2011;119:10–16.

[78] Steele KE, Stabler K, VanderZanden L. Cutaneous DNA vaccination against Ebola virus by particle bombardment: histopathology and alteration of CD3-positive dendritic epidermal cells. VetP athol 2001;38:203–15.

[79] Stewart JP, Egan JJ, Ross AJ, et al. The detection of Epstein-Barr virus DNA in lung tissue from patients with idiopathic pulmonary fibrosis. AmJ Re spirCr itCar eM ed 1999;159:1336–41.

[80] Sundquist SJ, Nisenbaum LK. Fast Fos: rapid protocols for single- and double-labeling c-Fos immunohistochemistry in fresh frozen brain sections. JNe urosciM ethods 2005;141:9–20.

[81] Tadokoro C, Yoshimoto Y, Sakata M, et al. Localizationo fh umanp lacental glucose transporter 1 during pregnancy. An immunohistochemical study. HistolHis topathol 1996;11:673–81.

[82] Taiambas E, Alexpopoulou D, Lambropoulou S,e tal. Targetingt opoisomeraseI Iain endometrial adenocarcinoma: a combined chromogenic in situ hybridization and immunohistochemistry study based on tissue microarrays. IntJ Gyn ecolCan cer 2006;16:1424–31.

[83] Taxy JB, Hidvegi DF, Battifora H. Nasopharyngeal carcinoma: antikeratin immunohistochemistry and electron microscopy. AmJ Clin P athol 1985;83:320–5.

[84] Todorovic-Rakovic N, Jovanovic D, Neskovic-Konstantinovic Z, Nikolic-Vukosavljevic D. Prognostic value of HER2 gene amplification detected by chromogenic in situ hybridization (CISH) in metastatic breast cancer. ExpM ol Pathol 2007;82:262–8.

[85] Tse C, Brault D, Gligorov J, et al. Evaluation of the quantitative analytical methods real-time

PCR for Her-2 gene quantification and ELISA of serum HER-2 protein and comparison with fluorescence in situ hybridization and immunohistochemistry for determining HER-2 statusin b reastc ancerp atients. ClinCh em 2005;51:1093–101.

[86] Tsiambas E, Karameris A, Lazaris AC, etal. EGFRalte rationsin p ancreatic ductal adenocarcinoma: a chromogenic in situ hybridization analysis based on tissue microarrays. Hepatogastroenterology 2006;53:452–7.

[87] Tsiambas E, Karameris A, Tiniakos DG, Karakitsos P. Evaluationo fto poisomerase IIa expression in pancreatic ductal adenocarcinoma: a pilot study using chromogenic in situ hybridization and immunohistochemistryo ntissu emic roarrays. Pancreatology 2007;7:45–52.

[88] Tubbs RR, Sheibani K, Weiss RA, et al. Immunohistochemistryo fW arthin'stu mor. Am JClin P athol 1980;74:795–7.

[89] Vollmer E, Galle J, Lang DS, et al. The HOPE technique opens up a multitude of new possibilitiesin p athology. RomJ M orphol Embryol 2006;47:15–19.

[90] Wagner M, Klussmann JP, Dirsch O, et al. Low prevalence of transfusion transmitted virus (TTV)-like DNA sequences in cystadenolymphoma and pleomorphic adenomao ft hes alivaryglan ds. EurAr ch Otorhinolaryngol 2006;263:759–63.

[91] Wakamatsu N, King DJ, Seal BS, Brown CC. Detection of Newcastle disease virus RNA by reverse transcription-polymerase chain reaction using formalin-fixed, paraffin embedded tissue and comparison with immunohistochemistry andin s ituh ybridization. JVe tDiagn I nvest 2007;19:396–400.

[92] Wang NS, Wu ZL, Zhang YE, Liao LT. Existence and significance of hepatitis B virus DNAin k idneyso fI gAn ephropathy. WorldJ Gastroenterol 2005;11:712–6.

[93] Wasserman J, Maddox J, Racz M, Petronic-Rosic V. Updateo n immunohistochemical methods relevant to dermatopathology. ArchP atholL abM ed 2009;133:1053–61.

[94] Wei Q, Javadian A, Lausen N, Fultz PN. Distribution and quantification of human immunodeficiency virus type 1, strain JC499, proviral DNA in tissues from an infected chimpanzee. Virology 2000;276:59–69.

[95] Willmore-Payne C, Metzger K, Layfield LJ. Effects of fixative and fixation protocols on assessment of Her-2/neu oncogene amplification statusb yfl uorescencein s ituh ybridization. Appl Immunohistochem Mol Morphol 2007;15:84–7.

[96] Yoon HJ, Jo BC, Shin WJ, et al. Comparative immunohistochemical study of ameloblastoma andame loblasticc arcinoma. OralS urg Oral Med Oral Pathol Oral Radiol Endod 2011;112:767–76.

[97] Zotter S, Lossnitzer A, Kunze KD, et al. Epithelial markers for paraffin-embedded human tissues. Immunohistochemistry with monoclonal antibodies against milk fat globule antigens. VirchowsAr chA P atholAn at Histopathol 1985;406:237–51.

The Basics of Histologic Interpretations of Tissues

INTRODUCTION

One of the more challenging parts of molecular pathology is that it truly requires two separate and distinct skill sets. For one, you must have a thorough foundation in the basics of molecular biology. An understanding of stringency, homology, the biochemistry of the three-dimensional protein/protein cross-linked network, and so on, is essential for you to become very good at in situ molecular pathology. However, this is not enough. You may have generated the most beautiful immunohistochemistry or in situ hybridization data for a new protein or gene in a formalin-fixed, paraffin-embedded cancer tissue. The pattern may provide enormous insight into the role of that specific protein or RNA into oncogenesis. However, if you cannot interpret that pattern, this information will remain hidden. Obviously, the best way to address this potential situation is to take a course in histopathology. However, that is not practical for a lot of people. Thus, I have written this chapter, in general, for the investigator with no formal training in histopathology. My goal is not to make you capable of diagnosing complex tumors. My goal is to make you able to quickly, confidently, and accurately determine what specific cell type or types contain your target of interest when you perform immunohistochemistry or in situ hybridization. However, if you have formal surgical pathology training, I have included more complex histopathological issues toward the end of the chapter that you may find interesting and challenging!

PART ONE: THE DIFFERENT CELL TYPES IN FORMALIN-FIXED, PARAFFIN-EMBEDDED TISSUES

EPITHELIAL CELLS

Epithelial cells are the most abundant in the body and have many important functions. For example, they cover our skin and line the gastrointestinal tract from top to bottom. They are also the progenitor cell for what is by far the most common type of cancer in humans: the carcinomas. The different types of epithelial cells are easy to recognize under the microscope.

Squamous Cells

A major function of squamous cells is protection. They protect the skin from evaporation and from outside forces that could puncture the skin and, thus, cause bleeding. They produce, in abundance, a protein that is highly characteristic of this cell type: cytokeratin.

Figures 6-1 and 6-2 show the variability that squamous cells can show in formalin-fixed, paraffin-embedded tissues. Let's first focus on what these different images have in common, which are the defining characteristics of squamous cells in general:

1. Theyh aveath ick, *stratified* layer of cells.
2. The mature squamous cells toward the surface are *large* and *flattened*.
3. Squamous cells are linked by bridge-like *tight junctions*.

Figure 6-1, panel A, shows the stratification of these epithelial cells very clearly. As you progress to higher magnification, the large flattened cells at the surface and the tight junctions (arrows, panel D) are clearly evident. Clearly, having such a thick layer of cells in which each cell is holding tightly to its neighboring cells will result in a watertight barrier that can withstand the bumps and bruises of everyday life.

Figure 6-2 also shows squamous epithelia; each of the three features noted previously are evident. However, note that in panel A the squamous cells have large, clear zones. In comparison, panels B–D show, at successive magnifications, that the squamous epithelial cells are lined by a dark pink layer. Before we get into why panel A looks so different from panels B–D, let's remember some basics about hematoxylin and eosin stains that we discussed in previous chapters:

Hematoxylin stains *nucleic acids* blue. *Thus, the blue stain typically corresponds to the nucleus.*
Eosin stains *proteins* pink. *Thus, the pink stain typically corresponds to the cytoplasm.*

Let's examine again Figure 6-2, panel B. Note that it has two dark pink zones (small arrow, surface, and larger arrow, at the base of the epithelia). Also note that the squamous cells between the two arrows have blue nuclei

167

DOI: http://dx.doi.org/10.1016/B978-0-12-415944-0.00006-1

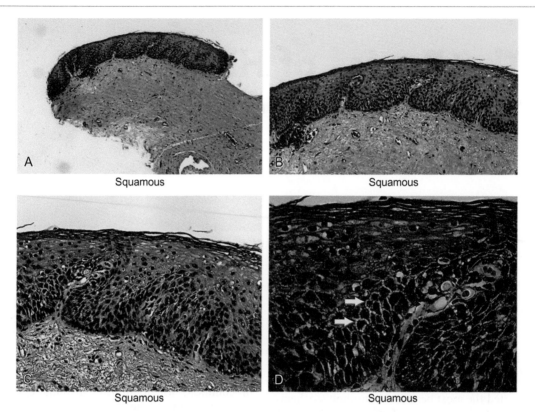

Figure6-1 Epithelial cells—the stratified squamous cells: part A. Panel A shows the stratification of these epithelial cells very clearly. The cells are dark blue and are layered like tiles in about 20 separate rows over the underlying dermis. Note that the dermis is mostly dark pink, owing to the collagen that gives the skin its strength. As you progress to higher magnification, the large flattened cells at the surface and the tight junctions that hold the squamous cells together (arrows, panel D) are evident. Clearly, having such a thick layer of cells in which each cell is holding tightly to its neighboring cells will result in a watertight barrier that can withstand the bumps and bruises of everydayl ife.

and a light blue cytoplasm. So, from what we just noted, we can make the following statements:

1. The material highlighted by the arrows is a densely packedp rotein.
2. Toward the base and middle, the squamous cells show some nucleic acids in their cytoplasm, but toward the surface their cytoplasm contains mostly protein.

What are the densely packed proteins indicated by the arrows? The concentrated protein marked by the small arrow at the surface is *keratin.* Keratin is an excellent water proof agent. There are very common diseases of the skin in which the keratin breaks down, and thus, the person gets small blisters, or "bumps," that tend to ooze fluids due to the loss of this protein barrier. Two common diseases that fit this category are eczema and contact dermatitis (e.g., poison ivy). The other protein (large arrow) is called *collagen.* It is the main protein of the stroma, which we discuss later. Finally, it makes sense that the squamous cells in Figure 6-2, panels B–D, will show RNA (nucleic acids) in their cytoplasm toward the base of the skin, because these cells make the RNAs that are needed for keratin synthesis. Of course, at the surface, the cells don't need RNAs but need the protein for their protection, and thus, the cells at the surface actively make the protein.

Getting back to Figure 6-2, the tissue in panel A is from the cervix, whereas in panels B–D it is from the hands. There is no "need" to protect the cervix with abundant keratin, whereas the skin of the hand is constantly active and, thus, needs much more protection against the elements. Mucosal squamous epithelia are, thus, sometimes referred to as "nonkeratinizing" epithelia, whereas skin epithelia are sometimes called "keratinizing epithelia." I do not like these terms because mucosal squamous epithelia *do* make keratin, and we can use that to our advantage with immunohistochemistry. They simply make much less keratin than the squamous epithelia of skin.

It is amazing how specialized squamous cells can be! Of course, they form the nails of our fingers and toes, as well as our hair. In the animal kingdom, these cells are responsible for hooves, antlers, feathers, scales, and so on! As with most proteins, when they are abundant and concentrated enough, they can form a very strong structure indeed!

Glandular Epithelia

Whereas squamous cells are analogous to the tile on our roof or the armor of a tank, in the sense that one of its main roles is protection, glandular epithelial cells are more involved with direct physiologic functions. This is

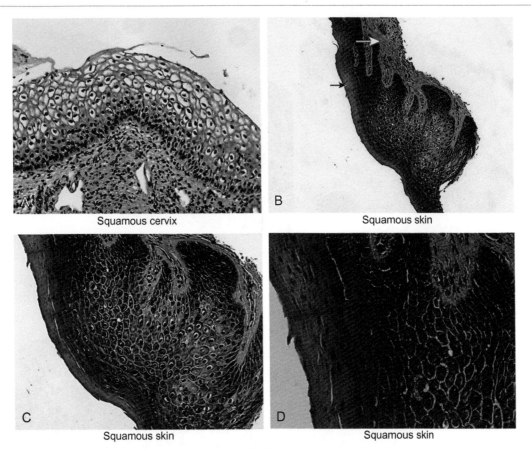

Figure 6-2 Epithelial cells—the stratified squamous cells: Part B. This figure also shows squamous epithelia. However, note that in panel A the squamous cells have large, clear zones. In comparison, panels B–D show, at successive magnifications, that the squamous epithelial cells are lined by a dark pink layer. Panel A represents an example of infection by human papillomavirus (HPV). HPV primarily infects squamous cells and causes the cytoplasm to fill with water. This pushes the organelles and cellular proteins to the side. Since the water has no color, the end result is this clear zone called a *koilocyte*. In comparison, panels B, C, and D show the most typical feature of squamous cells besides their multiple stacked layer pattern of growth. Squamous cells make keratin, and it gives them their strong pink color. If the tissue is under any kind of pressure stress (as occurs routinely on the hands and feet), then the squamous cells will mature into a thick, protective outer coat of keratin. That is why this outer layer is dark pink (panel B, black arrow). The yellow arrow in panel B also marks a thick protein layer, but this is the dermal collagen. Finally, note that in panels C and D the squamous cells at the surface of the skin (left part of image) have small inconspicuous nuclei as compared to the deeper layer of squamous cells. This is called *parakeratosis* and simply refers to the persistence of the nuclei in the squamous cell as it matures into a small "ball of keratin" on the skin's surface.

not to say that squamous epithelia are inert in this regard. But, in general, glandular epithelial cells do not play any protective role but rather have a functional role. The one exception would be that glandular cells can make a substance called *mucin* that can play an important lubricating and protective role in certain organs, such as the stomach.

Glandular cells line our gastrointestinal tract. They also make up the major digestive organs such as the pancreas and gallbladder. They can have an endocrine function evident in the thyroid, which is a modified gland based in glandular epithelia.

Here are the salient features of glandular epithelia and, thus, how we can recognize them in the hematoxylin and eosin:

1. Singlelay ero fc ells
2. Largec olumnarsh apedc ells

Figure 6-3 shows three tissues where glandular epithelia are present: the colon, the cervix, and the

endometrium. Note that at low magnification (panels A and C) the glands consist basically of round to oval tubes. This gives the pattern of many tubes arranged in a well-organized, uniform pattern. Pathologists love to be descriptive about such patterns. One such term they use for this simple uniform pattern of many glands arranged next to each other is a "wallflower" pattern of growth. Note in panel D that the cervix shows the single-cell columnar lining typical of glandular epithelium. However, as noted by the arrow, smaller cells are present at the base of the gland. These are the so-called reserve or stem cells. These are the cells that can make new glandular cells when the older ones undergo the normal process of cellular turnover.

Glandular epithelia, as noted previously, can take on major functional roles. This is clearly evident in the pancreas and thyroid. The pancreas must make most of the digestive enzymes needed to break down the food we eat. One of the many serious problems children with

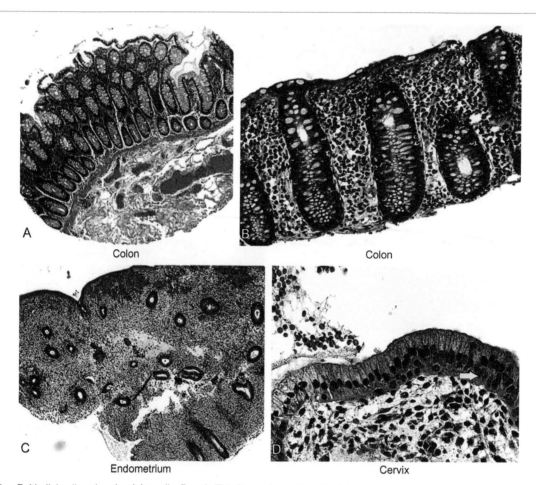

Figure6-3   Epithelial cells—the glandular cells: Part A. This figure shows the glandular epithelia found in the colon (panels A and B), endometrium (panel C), and cervix (panel D). Note that at low magnification (panels A and C) the glands consist basically of round to oval tubes. This gives the pattern of many tubes arranged in a well-organized, uniform pattern. Well-organized glands are always benign. Note that in panel D the cervix shows the single-cell columnar lining typical of glandular epithelium. However, as noted by the arrow, smaller cells are present at the base of the gland. These are the so-called reserve or stem cells. These are the cells that can make new glandular cells when the older ones undergo the normal process of cellular turnover. They are also capable of making squamous cells in a process called *squamous metaplasia*. Finally, note that the glandular cells in panels A, B, and D are tall and show a gray color, whereas the glandular cells in panel C are smaller and don't have this gray material. The gray material is the mucopolysaccharides that these cells make as a necessary lubricant in the colon and cervix. The endometrial cells at this stage (called the *proliferative stage*) do not need to make any secretions; this happens later in the endometrial cycle.

cystic fibrosis face is that their pancreas is destroyed by the disease. Therefore, they cannot properly break down the food they eat for energy and nutrients and, thus, must take digestive enzyme supplements. In the glandular lined organs where the cells have a major secretory function, it is common for the cells to acquire a lot of cytoplasm and form small clusters of cells. These clusters, called *acini,* are depicted in Figure 6-4 for the breast and pancreas. We can think of this as single layer of tall cells where the main duct of the gland made many small buds lined by these cells to maximize the amount of secretory product in a relatively small area.

Glandular epithelia are the progenitor of all the adenocarcinomas. These are among the most common and lethal tumors and include, of course, most colon, breast, prostate, ovarian, uterine, pancreas, and gastric cancers, and many lung cancers.

Since this chapter is being written primarily for the nonpathologist, I am oversimplifying some points. Some benign glands can show stratification. The best example is the endometrium, and it reflects the high mitotic activity the endometrium shows during the proliferative phase. But, in general, the points listed here are accurate and will help you get a base of knowledge to build upon.

Transitional Epithelia

It is easy to remember the features of the last type of epithelia, transitional epithelia. The reason is that it represents a transition from the single-layered, tall glandular cells to the multilayered, flattened squamous cells. Transitional epithelia are only found lining the urinary bladder where these cells routinely face high intraluminal pressures. Thus, a single layer of epithelia would not

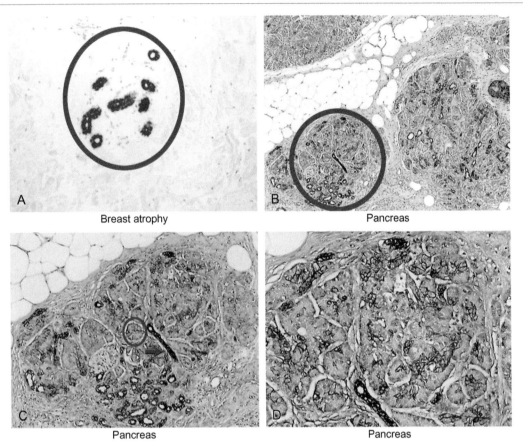

A — Breast atrophy

B — Pancreas

C — Pancreas

D — Pancreas

Figure6-4 Epithelial cells—the glandular cells: Part B. Glandular epithelia are responsible for many diverse functions. This includes the production of many secreted products such as milk (breast, panel A), and the digestive enzymes needed to process the energy from food (the pancreas, panels B–D). Breast and pancreatic epithelia are each patterned as small outpouchings of glandular epithelia (called *acini*) surrounding the small ducts that drain these acini. In the breast these acini atrophy over time. This will produce an atrophic pattern where the small ducts persist but the acini are gone (panel A, highlighted by the keratin immunohistochemical stain). In comparison, look at the pancreas in which atrophy does not occur over time. The glands are arranged in circular lobules (panel B). The lobules, in turn, are composed of ducts (highlighted by the cytokeratin AE1/3 stain as in panel A). The glandular epithelia around the ducts (called the acini; panel C, circle with a red arrow at the duct) do not stain well with AE1/3. The acini cells are responsible for the massive secretions that the pancreas must make daily. As an aside, the duct is responsible for most pancreatic tumors in particular and adenocarcinomas in general. This is why adenocarcinomas are typically positive for AE 1/3.

be sufficient to protect the underlying blood vessels from trauma. Rather, the defining characteristics of transitional epithelia (also known as urothelial epithelia) are as follows:

1. There is a thin, stratified lining of cells
2. The cells at the surface have small caps

The histologic features of transitional epithelia are shown in Figure 6-5. Note that the arrow in panel D highlights the large cap-like cells on the top that form an umbrella-like structure over the other urothelial epithelial cells. Thus, these cells represent the most differentiated form of urothelial cells. Recognizing these umbrella cells is very useful in surgical biopsies. The reason is that malignant cells rarely form the most differentiated form of the epithelia from which they were derived. So the presence of these cap or umbrella cells in a bladder biopsy will reassure the pathologist that the lesion is probably benign. A better example of this rule is cilia. The terminal hair-like cilia on a cell is typical of *glandular* epithelia that need to move something along a corresponding "channel." Good examples of this include the fallopian tube and the lung, which must move the fertilized egg and mucus, respectively. Malignant cells simply do not like to differentiate and, in my experience, there has been no proven example of a malignant tumor that contained cilia. This can be very helpful when interpreting atypical cells in a lung washing.

STROMAL CELLS

It is always easy to find stromal cells after locating the epithelial cells. The stromal cells make up the tissue that is underneath the epithelia. In the skin, the stroma is commonly called the *dermis*. In the gastrointestinal tract, the stroma is often called the *lamina propria* (right under the epithelia) or *submucosa* (deep below the epithelia).

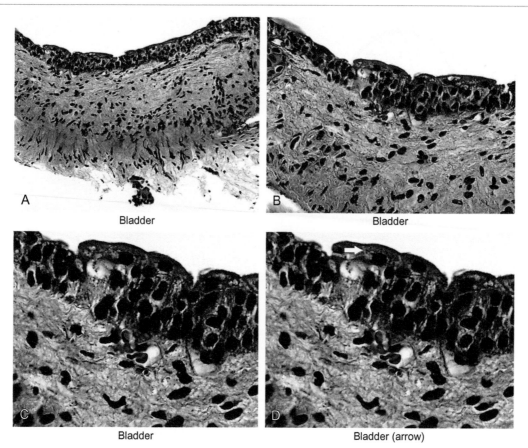

Figure6-5   Epithelial cells—the transitional cell. The bladder is lined by epithelial cells that are somewhere in between the thick layers of squamous cells and the single layer of tall cells of glandular epithelia. Logically, these cells are called transitional cells because they represent a transition between these two disparate epithelial cell types. Note that the bladder tissue is lined by about five rows of transitional cells stacked on each other. No keratin is present or "needed" since the surface of the bladder is not under the kind of pressure that the skin is under. In panel D note the large cap-like cells on the top that form an umbrella-like structure over the other urothelial epithelial cells. This feature of differentiation is any easy way to be sure that the epithelia are benign.

To better understand what cells we will find in the stroma, let's think of its function:

1. Tosu pportth eo verlyinge pithelia
2. To deliver blood and nutrients to the tissues
3. To help remove the waste products from the tissues
4. To help in the immune protection of the tissue

The support function of the stroma is done mostly by one cell: the fibroblast. Its job is to make the main protein of the stroma, collagen, which we discussed regarding Figure 6-1. The delivery and removal of nutrients/waste products is done by small blood vessels that are lined by one simple cell type: the endothelial cell. The small vessels are called the *capillaries* and the *lymphatic channels.* The cells in the stroma that are the "protectors" against foreign invaders are called *macrophages.* Therefore, recognizing the specific cell type in the stroma is not difficult if you remember that, for the most part, all you need to be able to do is to discriminate the *fibroblast* from the *endothelial* cell from the *macrophage.* These points are illustrated next and in Figure 6-6.

Fibroblasts
1. Large, isolated cells with elongate prominent nuclei

2. Associated with extracellular dark pink protein (collagen)

Endothelial cells
1. Large, very flat cells with inconspicuous nuclei
2. Line microvessels (capillaries, which contain red blood cells, or lymphatics, which contain clear fluid)

Macrophages
1. Large cells that tend to form small clusters with round to oval nuclei
2. May contain foreign pigment and/or hemosiderin pigment

In Figure 6-7, panel C, the fibroblast is depicted with a large arrow, the endothelial cell with a small arrow, and a few macrophages are shown inside a small box. Panels D–F are successive magnifications of a cervical biopsy from a woman who recently had a separate cervical biopsy. Therefore, as expected, there was a lot of bleeding. This stimulated the macrophages to proliferate. They are easily seen in panel F, where their abundant large brown pigment is evident in the cytoplasm (arrows). This is the hemosiderin pigment, and is the byproduct of the bleeding that occurred in this area after the initial biopsy.

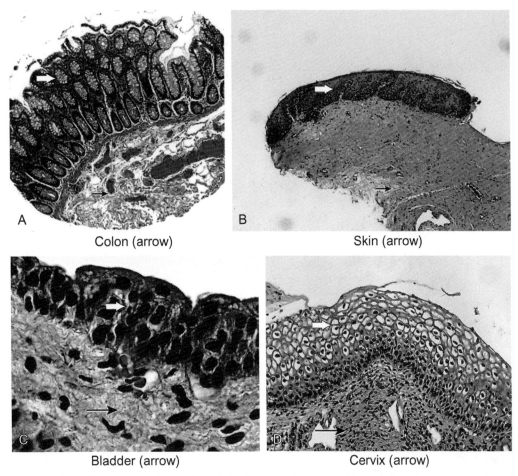

Colon (arrow)

Skin (arrow)

Bladder (arrow)

Cervix (arrow)

Figure6-6   A review of the different types of epithelia and the distinction between epithelia and stroma. This figure is a compilation of the different types of epithelia (glandular, A, squamous B and D, and transitional, C). The large arrow for each figure corresponds to the epithelia and the small arrow to the underlying tissue, which is referred to as the stroma. The stroma is the place where the blood vessels that support the epithelia and the fibroblasts that make the supportive protein collagen are present. The stroma can go by different names; in organs such as the colon, bladder, and cervix, where the organ is hollow, it is often called the submucosa (or, at times, the lamina propria). In the skin the stroma is typically called the dermis.

INFLAMMATORY CELLS

It would be unusual to see inflammatory cells if you biopsied normal tissues. Of course, by definition, surgeons do not routinely biopsy normal-appearing tissues. Therefore, in surgical biopsies you usually encounter inflammatory cells. There are four major types of inflammatory cells that we need to recognize when interpreting immunohistochemistry or in situ hybridization results:

1. Lymphocytes
2. Plasmac ells
3. Eosinophils
4. Neutrophils

Lymphocytes

Lymphocytes are the most common of the cell types we will come across when interpreting immunohistochemistry or in situ hybridization data. Panels A–D in Figure 6-8 show successively increasing magnifications of a cervical biopsy. The majority of cervical biopsies for an atypical Pap smear show lymphocytic infiltration. Here are the features that allow us to identify the lymphocytes under the microscope:

1. Aggregates of small, round cells with dark nuclei and scantc ytoplasm
2. Typically large numbers owing to the fact that some antigenic stimulus is "calling" them to this site

There are many subtypes of lymphocytes. In general, they fall into the category of T-cell, B-cell, natural killer (NK) cell, and memory cell. In each group are many other subcategories. In most cases in which you perform immunohistochemistry or in situ hybridization, the only knowledge you need is the general category. As we discuss in a moment, each category has its own specific antigenic marker that allows us to characterize it accurately.

Plasma Cells

As indicated previously, when you are analyzing a biopsy and see there is inflammation, in most cases the

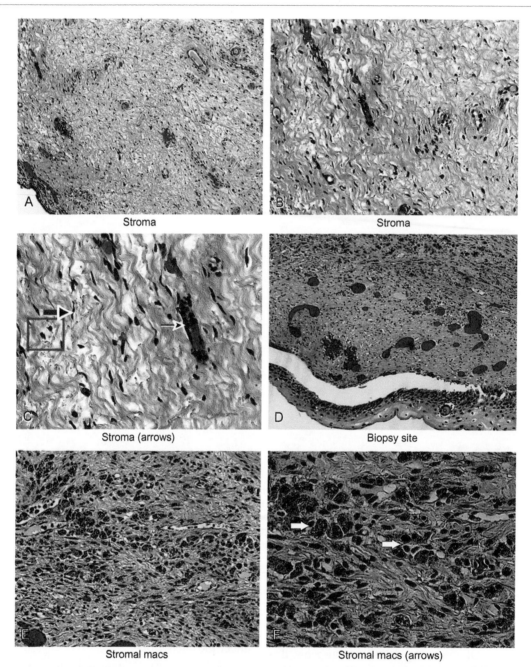

Figure 6-7   The different components of the stroma. This figure shows the different cell types that may be encountered in the stroma. In panel C the fibroblast is depicted with a large arrow, the endothelial cell with a small arrow, and a few macrophages are shown inside a small box. Panels D–F are successive magnifications of a cervical biopsy from a woman who recently had a separate cervical biopsy. The tissue damage from any biopsy will stimulate the macrophages to proliferate. They are easily seen in panel F, where their abundant large brown pigment is evident in the cytoplasm (arrows). This is the hemosiderin pigment, and is the byproduct of the bleeding that occurred in this area after the initial biopsy. Also note in panel D the many small blood vessels that were induced to proliferate in response to the biopsy.

lymphocytes will dominate. They are easily recognized as aggregates of small, round cells with dark blue nuclei and very little cytoplasm. These cells represent the body's attempt to deal with some relatively long-term exposure to a foreign antigen. The inflammation takes many days to start up and usually longer to leave. Hence, it is often referred to as chronic inflammation. Acute inflammation, which we discuss shortly, is much less common and usually reflects either acute trauma or the rapid response to the presence of a bacterial infection.

Once the chronic inflammation is established, the body will try to rid itself of the foreign antigen. One way it will do this is by producing antibodies. Some lymphocytes, called B-cells, will gain the ability to synthesize specific antibodies against the foreign antigen. These cells become metabolic machines, geared to the production of large

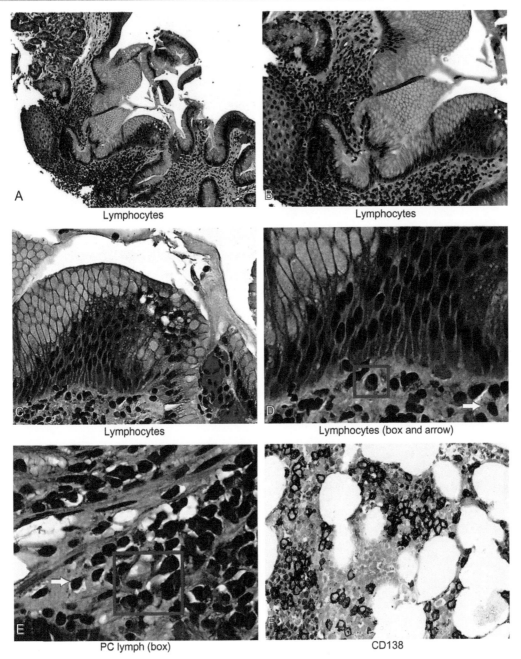

Figure6-8   The inflammatory cells: the lymphocyte and the plasma cell. This figure is from an esophagus with chronic esophagitis. Note the large collection of small blue cells in the stroma under the glandular epithelia in panels A, B, and C. Also note in panel C that there is a collection of stratified squamous cells in the middle of the epithelia. This replacement of the glandular cells by the squamous cells in the cervix is normal and is called squamous metaplasia. Squamous metaplasia is the initiation point of HPV in the cervix. Panels D and E show at higher magnification the plasma cells (box) and the lymphocytes (arrow). The plasma cells show eccentric nuclei, a clear zone around the nucleus, and a bluish-pink cytoplasm. Plasma cells, which are ultimately responsible for making all the antibodies we use in immunohistochemistry, are small protein factories; the bluish-pink cytoplasm reflects the high mRNA and protein content of these cells directed to producing the target-specific antibody. Panel F shows an antigen called CD 138 that is present in most plasma cells. This is from a person who had a malignant tumor of the plasma cells called multiple myeloma. Not surprisingly, many symptoms these patients get relate to the high number of antibodies and proteins that the tumor cells synthesize.

amounts of antibodies. These antibodies are stored in the cytoplasm. Thus, these cells, called *plasma cells,* can be readily distinguished from lymphocytes by their ample cytoplasm. Here are the features that allow us to identify the plasma cells under the microscope:

1. Aggregates of larger, round cells have open nuclei and ample cytoplasm.
2. The chromatin of the nuclei often forms small clusters around the periphery of the nucleus, hence the term "c lock-wheelc hromatin."

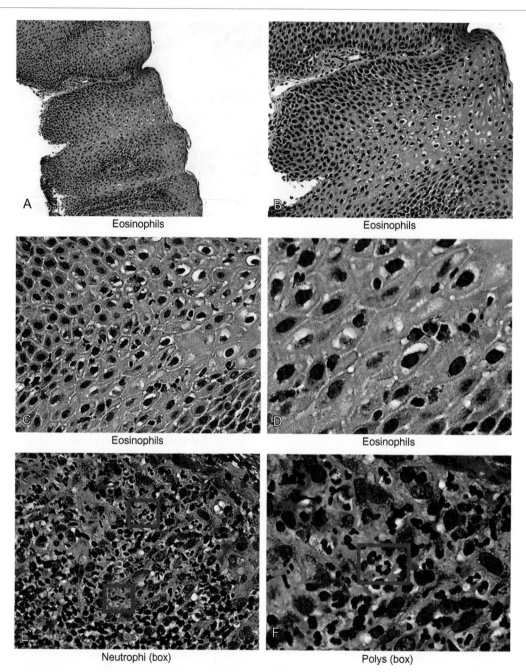

Figure6-9   The inflammatory cells: the eosinophils and neutrophils. Panels A through D show the histology at successively high magnifications; the tissue is the esophagus from a young person who had difficulty swallowing with heartburn that he indicated was related to specific foods. Note that the esophagus is lined by stratified squamous epithelia. Also note that interspersed among the large squamous cells are many smaller inflammatory cells with bilobed nuclei and ample cytoplasm that contains pink granules. The latter gives the eosinophils their bright red color and, of course, their name because they stain strongly with eosin. Panels E and F show the stomach of another person with acute gastritis associated with marked reflux. Note that many small cells are infiltrating the epithelia. They have multilobed nuclei (box, panel F) that give these cells their name; polymorphonuclear leukocytes (polys for short). They are also called neutrophils. Their presence is classic of acute tissue damage and bacterial infection.

Examples of plasma cells are shown in Figure 6-8, where they are enclosed in small boxes. In comparison, the smaller lymphocytes are marked by the arrow.

Eosinophils

Eosinophils are another inflammatory cell type of chronic inflammation. Their presence usually denotes an allergic type response. Hence, they are very commonly seen in diseases such as asthma. Their presence can be diagnostic of specific diseases. For example, there is a disease called eosinophilic esophagitis in which the presence of at least 15 eosinophils per high power field is sufficient to render this diagnosis. Here are the features that allow us to identify the eosinophils under the microscope:

1. Aggregates of intermediate-sized, round cells have open nuclei and ample cytoplasm.
2. The nuclei have two lobes, and the cytoplasm has smallp inkgran ules.

Figure 6-9 presents images from a case of eosinophilic esophagitis. Panels A through D show the histology at successively high magnifications. Note in panel D that the esophagus is lined by stratified squamous epithelia. Also note that interspersed among the large squamous cells are many smaller inflammatory cells with bilobed nuclei and ample cytoplasm that contains pink granules. The latter gives the eosinophils their bright red color and, of course, their name because they stain strongly with eosin.

Neutrophils

Whereas lymphocytes, plasma cells, macrophages, and eosinophils represent the *chronic* reaction of the body to a foreign antigen, neutrophils represent an *acute* reaction to some antigen. Common examples of acute inflammation include exposure to bacterial antigens, and the acute death of cells, such as after radiotherapy, chemotherapy, or ischemia (such as a heart attack). Tissue ulceration is another example where there is acute cell death and tissue damage that results in the infiltration of the tissue by neutrophils.

Figure 6-9 shows some examples of neutrophilic infiltration; the lesion was a stomach ulcer. Here are the features that allow us to identify the neutrophils under the microscope:

1. Aggregates of intermediate-sized, round cells have open nuclei and ample cytoplasm.
2. The nuclei have multiple lobes, and the cytoplasm has small weakly staining granules.

The neutrophils in Figure 6-9 are marked by the small boxes. Note the multilobed nuclei that are the pathogenomic feature of these cells. Sometimes these nuclei look like the hat worn by Mickey Mouse, hence the term "Mickey Mouse nuclei"! As I said, there seems to be no limit to the descriptive ability of pathologists!

TERMS COMMONLY USED TO DESCRIBE FORMALIN-FIXED, PARAFFIN-EMBEDDED TISSUES

There are terms that you will need to become familiar with when interpreting either your own or other people's data using either in situ hybridization or immunohistochemistry. These terms are discussed next.

ATROPHY

Atrophy refers to the thinning of a specific type of tissue. It usually refers to epithelia. Since the epithelia are thinned, by definition, there are far fewer cells per unit area of the biopsy. The most common cause of atrophy is simply aging. Another relatively common cause is the re-epithelialization of a tissue after ulceration or after a biopsy. Another less common cause is chemotherapy effect.

Figure 6-10 shows two different examples of atrophy. Panel A is a normal endometrial biopsy from a 25-year-old woman. Note the abundant glands and stroma. Panel B is the endometrial biopsy for an 81-year-old woman. Note that the stroma is so atrophic that it is barely visible and that the epithelia consist of far fewer cells than is evident in panel A; therefore, the glands are also atrophic. Panel C shows a normal squamous epithelial lining of a cervix in a 32-year-old woman. Compare this to panel D, which is the cervical biopsy of a 30-year-old woman who had a large cervical biopsy taken a few weeks before. The ulceration caused by the biopsy is now re-epithelializing, and the squamous lining is still much thinner than normal. Therefore, it is still atrophic.

An important corollary for atrophic cells and immunohistochemistry plus in situ hybridization is that atrophic cells tend to be metabolically inactive. Therefore, they often do not show the robust protein/DNA/RNA synthesis required for a strong signal with any of our in situ-based molecular pathology methods.

ULCERATION/EROSION

Whereas atrophy is the *thinning* of tissue, and often refers to the epithelial layer, ulceration is the *loss* of the epithelial lining. The lining is usually lost because the epithelial cells have been actively destroyed. If the epithelial lining is much thinned but not completely lost, it is called an *erosion.*

Of course, ulcerations expose the underlying blood vessels in the stroma. Thus, most ulcers readily bleed. The bleeding will be associated with hemosiderin deposition in the tissue. Also, the patient will often show a decreased red blood cell count, referred to as a decreased hematocrit.

Panels A and B of Figure 6-11 show low and high magnifications of the ulceration found in a person with Barrett's esophagus. This disease is marked by the replacement of the normal thick squamous lining with a much thinner glandular lining. Hence, these people tend to get ulcers. These tissues also are both acutely and chronically inflamed. As we discuss later, the presence of chronic inflammation per se predisposes the site to cancer. Indeed, people with this disease have a much increased risk of esophageal cancer. The area where the epithelial lining is lost is marked by a box. Similarly, panels C and D show the marked acute and chronic inflammation with partial loss of the epithelial lining of the colon marked by the disease called ulcerative colitis. The area of ulceration is marked also by a box. Unfortunately, besides the horrible symptoms, people with this disease also have an elevated risk of colon cancer.

Since the epithelial cells with ulcers not due to old age are actively trying to repair the damaged lining, they tend to be very metabolically active. This allows us recognize them easily. Specifically, atrophic cells tend to have scant cytoplasm and inactive-appearing (pale, uniform) nuclei. In comparison, cells trying to repair an ulcer tend to have large, open nuclei because they are turning on many RNA and protein synthesis pathways. They also have large structures in the nucleus called *nucleoli*. These structures appear as large dots in the nucleus. The nucleoli is the center of ribosomal RNA synthesis, which, in turn, is important for the many messenger RNAs (mRNAs) the cell will need to make as it tries to repair the area of

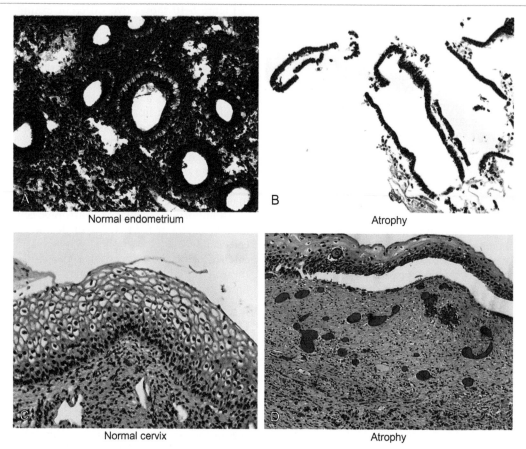

A — Normal endometrium

B — Atrophy

C — Normal cervix

D — Atrophy

Figure 6-10 Epithelial cells and atrophic changes. This figure shows two different examples of atrophy. Panel A is a normal endometrial biopsy from a 25-year-old woman. Note the abundant glands and stroma. Panel B is the endometrial biopsy for an 81-year-old woman. Note that the stroma is so atrophic that it is barely visible and that the epithelia consist of far fewer cells than is evident in panel A. This is the end result of the lack of estrogen stimulation on the endometrium, and reminds us that epithelia are not static but always undergoing cell turnover. Panel C shows a squamous epithelial lining of a cervix in a 32-year-old woman. Compare this to panel D, which is the cervical biopsy of a 30-year-old woman who had a large cervical biopsy taken a few weeks before. The ulceration caused by the biopsy is now re-epithelializing, and the squamous lining is still much thinner than normal. Thus, this is a different cause of atrophy—the loss of the surface epithelia from, in this case, a biopsy and the subsequent re-epithelization of the lining.

injury. Nucleoli are formed in the nucleus around part of the chromosome called NORs (short for nucleolar organizing regions). NORs are composed of many consecutive repeats of rRNAs and are found on several different chromosomes. These are quiescent in the mature epithelial cell (or stromal cell), but can be "reawakened" when the cell revs up its metabolic machinery because of the reparative process. An interesting part of these nucleoli is that they have "roadways" that connect the nucleolar proteins/rRNA to the cytoplasm where they can be exported for mRNA synthesis. Surely, these "roadways" are part of the three-dimensional macromolecule cross-linked network that is held rigidly in place when the cells are fixed. Another interesting part of this "nucleolar to cytoplasm roadway" is the size of the molecules that can easily pass through it. It has been documented that proteins up to 2000 kDa in size can pass through the roadway. Such proteins are huge in comparison to the proteins we use in immunohistochemical and in situ hybridization. As you may recall, the antibodies we use are about 100 kDa in size, streptavidin is 60 kDa, peroxidase is 150 kDa, and so on. Although no doubt formalin fixation, via extensive

cross-linking, creates pores that restrict the size of macromolecules, it is good to remember that the living cell is capable of trafficking molecules much larger than the key reagents we use with immunohistochemistry and in situ hybridization. The bottom line with respect to the histopathology of reparative changes is that it is one of those circumstances (as with cancer cells) in which all the cells start to look the same, regardless of their cell of origin. This is evident from a review of Figure 6-12. Note the similar-looking reactive/reparative cells in the squamous epithelia of the cervix (panel A), the glands of the stomach (panel B), the endothelial cells in the esophagus (panel C), and the endothelial cells/fibroblasts of the stomach by an ulcer (panel D). Also note the strong inflammation associated with each biopsy in the panels where the stroma is evident.

METAPLASIA

The last term to discuss in this section is *metaplasia*. It occurs when one cell type "changes" to a different cell type. There are several places in the body where glandular

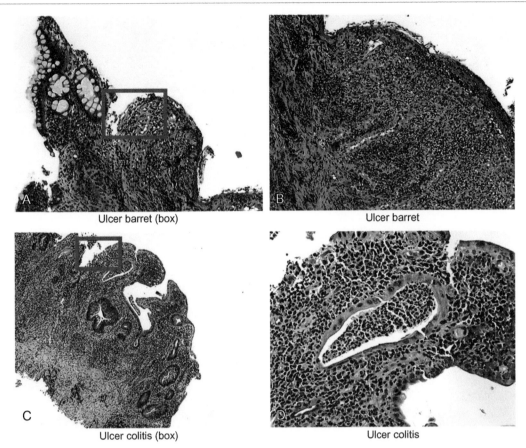

Figure6-11   Epithelial cells and ulceration. Ulceration is defined as the complete loss of the epithelial lining of any tissue; a partial loss is called an erosion. In either case, severe, typically acute, inflammation eventually destroys the epithelial lining but not the stroma. Panels A and B show low and high magnifications of the ulceration found in a person with Barrett's esophagus. This disease is marked by the replacement of the normal thick squamous lining with a much thinner glandular lining. Since the glandular lining often has mucus-producing cells that look like the intestine, it is called intestinal metaplasia. The area where the epithelial lining is lost is marked by a box. Similarly, panels C and D show the marked acute and chronic inflammation with partial loss of the epithelial lining of the colon marked by the disease called ulcerative colitis. The area of ulceration is marked also by a box. Also note the collection of neutrophils in the tubular gland (center of image, panel D). This is called a crypt abscess; an abscess is simply a collection of neutrophils. Also note that the inflammation is spilling over into the stroma. This will lead to the destruction of the small blood vessels. Since the function of the colon epithelia is to absorb water, patients with this horrible disease get bloody diarrhea and cramps.

cells form junctions with squamous cells. Logically, the squamous epithelia are in the part of the body nearer the outside surface, and the glandular epithelia are farther away from the outside surface. Three good examples of this are in the esophagus, the cervix, and at the anal–rectal junction. At these sites, squamous metaplasia is the norm. The term "squamous metaplasia" simply means that the area lined by glandular cells is slowly but surely being replaced by squamous cells. Squamous metaplasia of the cervix is seen in Figure 6-13, panels A and B. Note that the squamous epithelia are overlying the glandular epithelium. This means that, years before, this area was lined by glandular epithelia, but over time it was replaced by squamous epithelia. The process of squamous metaplasia is very important in the cervix, because the metaplastic cells are especially susceptible to infection by human papillomavirus (HPV) infection. HPV infection is an essential co-factor in cervical cancer, as well as in anorectal cancer. Panels C and D show the same metaplasia, only this time

occurring in the esophagus. Metaplasia reminds us how much "plasticity" cells have (as does repair), being able, under the correct conditions, to change into other cell types.

There are other types of metaplasia. There are many different types of glandular cells. Some make mucus, others have cilia, and others are called goblet cells, and so on. Specific types of glandular cells can become other types of glandular cells. For example, in the esophagus the mucus-producing glandular cell can become a goblet cell. Since the goblet cell is commonly found lower in the intestine, this is called *intestinal metaplasia.* An example of intestinal metaplasia is seen in Figure 6-13, panels E and F. The "typical" glandular cells of the esophagus are marked by the small arrow, and the goblet cells by the large arrow. The diagnosis of intestinal metaplasia of the esophagus much increases the risk of adenocarcinoma at this site, again most likely reflecting the chronic inflammation at this site. This disease process (chronic

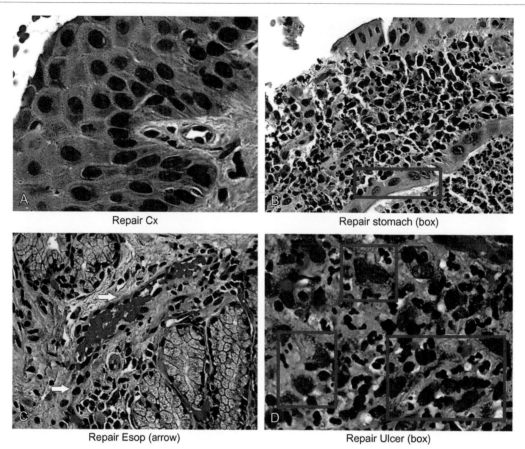

Repair Cx

Repair stomach (box)

Repair Esop (arrow)

Repair Ulcer (box)

Figure 6-12   Epithelial cells and repair. If epithelial cells are exposed to chronic inflammation, they show typical cellular changes called *repair*. The cytologic changes will look very similar to stromal cells undergoing chronic or acute inflammation. The cell's metabolic machinery changes, and this is seen with enlarged, clear nuclei with small nucleoli. The cytoplasm usually remains nondescript. The end result is that the nuclear/cytoplasmic ratio increases. This is seen in panel A for the squamous cells in the cervix, in panel B for the glandular cells of the stomach, in panel C in the endothelial cells in the esophagus, and in panel D in the endothelial cells and fibroblasts around an ulcer.

inflammation with intestinal metaplasia) in the esophagus is called Barrett's esophagus.

SPECIALIZED STROMAL CELLS

Metaplasia reminded me of the amazing ability of cells to change their appearance in response to a variety of factors, such as inflammation. Still, when it comes to cells being able to change their appearance and function, it is hard to beat the stromal cell.

My intention in this chapter is to give you the histologic foundation to be able to identify the cell types that are found in most human and formalin-fixed, paraffin-embedded tissues from other animals. Of course, some of you might be working with specific tissue types such as skeletal muscle or brain. This section deals with some of these other specific tissue types.

The basic stromal cell is the fibroblast. It is long and spindly in appearance. It has relatively little cytoplasm. The reason is that most of its product is secreted into the stroma, where it forms the main skeleton and support of the tissue. Of course, this support protein is collagen.

The endometrium shows how the fibroblast has evolved to serve very different functions. Endometrial stroma is highly cellular and consists of cells that are round with little cytoplasm, at least during the proliferative phase of the menstrual cycle. However, at the second stage of this cycle, the endometrial stroma changes dramatically. The stromal cells show very abundant cytoplasm as they fill with proteins that would help the fertilized egg implant and grow in case of a pregnancy. These enlarged stromal cells are called predecidual cells and can be seen in Figure 6-14, panel B.

Of course, stromal cells can show much more dramatic appearances than predecidua. Panels C and D show stromal cells that are secreting large amounts of protein (mostly collagen), which appears dark pink (large arrow). Note that some of the collagen is blue (small arrow), which reflects partial calcification of the collagen. The latter process greatly strengthens the collagen and is the basic extracellular material of bone. Panels C and D reflect bony tissue present in the region of the mandible. Actually, at times fibroblasts can actually grow bone de novo in a process called *osseous metaplasia*. In panels C and D, we are actually seeing part of the mandibular bone.

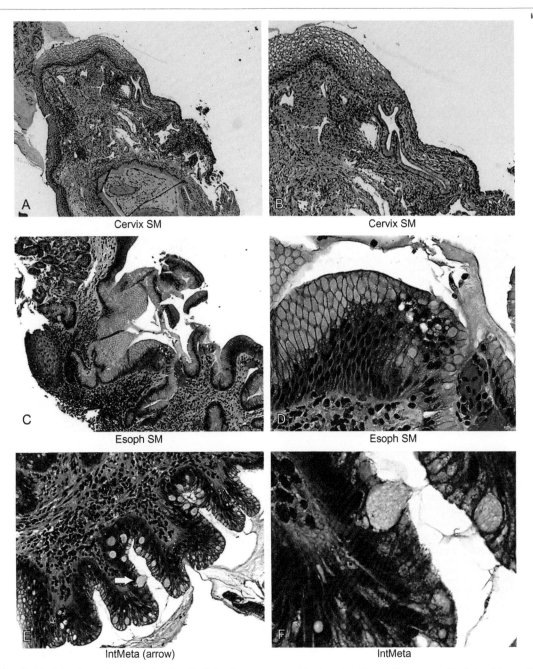

Figure 6-13 Epithelial cells and metaplasia. If epithelial cells are exposed to chronic inflammation, one epithelial cell type often will change to a different cell type. In the cervix and esophagus, in each case the simple glandular epithelia will typically be replaced by stratified squamous epithelia as a "protective" measure. This is seen in panels A and B and, for the esophagus, in panels C and D. Panels E and F show another type of metaplasia of the esophagus called intestinal metaplasia. Here, the lining of the esophagus is replaced by simple columnar epithelia that have large pools of mucus; such cells are called goblet cells and are diagnostic of Barrett's esophagus (or intestinal metaplasia). People with this condition are at increased risk for adenocarcinoma, likely as a reflection of the chronic inflammatory state and associated repair.

Other specialized types of stromal cells include *striated muscle cells*, as seen in skeletal muscle and cardiac tissue, and *smooth muscle cells*, which line all large blood vessels, the uterus, bladder, gallbladder, and the intestine, since each site needs to be able to alter its diameter for blood pressure control and peristalsis, respectively. Stromal cells can also be storage cells; large amounts of fat stored in such cells produces the *adipose* cell. *Cartilage* is made up of cells that secrete a certain type of avascular matrix that helps give tissue both support and the fluidity to move; thus, it is common in joints. In the central nervous system, we see, besides the ubiquitous endothelial cells/blood vessels, three very specialized cell types:

1. The neuron, which shows very large cells with prominent nucleoli. Indeed, these cells are so metabolically active, having to make the neurotransmitters that may have to be able to move down many centimeters

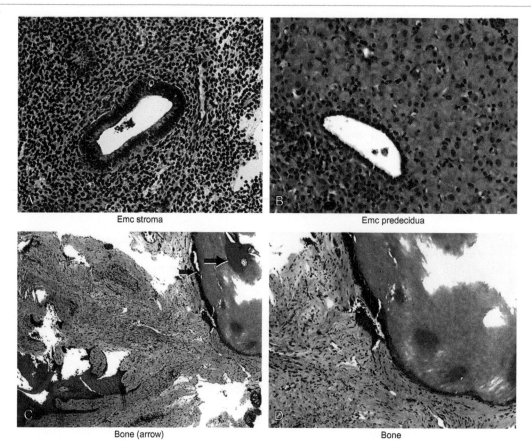

Figure6-14 Stromal cell changes. The cells in the stroma are capable of showing an incredible range of cytologic changes that are much more marked than epithelial cells. Endometrial stroma is highly cellular and consists of cells that are round with little cytoplasm, at least during the proliferative phase of the menstrual cycle. However, at the second stage of this cycle, the endometrial stroma cells show very abundant cytoplasm as they fill with proteins that would help the fertilized egg implant and grow in case of a pregnancy. These enlarged stromal cells are called predecidual cells (panel A, normal; and panel B, predecidua). Panels C and D show stromal cells that are secreting large amounts of protein (mostly collagen), which appears dark pink (large arrow). Note that some of the collagen is blue (small arrow), which reflects partial calcification of the collagen. The latter process greatly strengthens the collagen and is the basic extracellular material of bone. Panels C and D reflect bony tissue present in the region of the mandible.

of axons, that we may see the blue-colored RNA molecules in the cytoplasm, where they are called Nisselb odies.

2. The astrocytes, which are smaller round cells that form the main support of the brain and spinal cord via their interlacing processes.
3. The microglial cells, which are simply the macrophage that is an endogenous cell in the brain.

Final mention should be made of the melanocyte. This specialized cell comes from a specific part of the embryo called the neural crest. This cell produces a specific and unique chemical called melanin, which is derived from the amino acid tyrosine. Melanin is able to absorb UV radiation and transform it into heat. This protects the DNA in the cells from UV-related damage and mutagenesis. Needless to say, this explains why melanin-producing cells are in the skin, and there is a strong inverse relationship between the number of melanocytes in the skin and the susceptibility to skin cancer. Melanin also gives hair its color. Finally, melanin is important when performing immunohistochemistry or in situ hybridization because the brown pigment may give us a false-positive reaction

(unless we recognize the melanin) if we use DAB as the chromogen.

PART TWO: THE DIFFERENTIATION BETWEEN BENIGN AND MALIGNANT CELLS IN FORMALIN-FIXED, PARAFFIN-EMBEDDED TISSUES

Of course, many of you may be doing research on oncogenesis. Thus, you routinely are doing immunohistochemistry and in situ hybridization on tissues that may contain cancer. It is very common for formalin-fixed, paraffin-embedded tissues from surgical biopsies taken from patients with cancer to show both benign and malignant tissues in the same biopsy. It is certainly very common that benign and malignant tissue show differential results when you are looking for a specific target with either immunohistochemistry or in situ hybridization. Thus, it is useful to be able to accurately differentiate benign from malignant tissues in the formalin-fixed, paraffin-embedded tissues we study.

BENIGN TISSUE

The key word as it relates to benign tissue is *orderly.* Benign tissue grows in a very organized, orderly pattern. This is best appreciated at low magnification. Indeed, when I was a resident in anatomic pathology, I was extremely fortunate to have been trained by pathologists such as Dr. Karl Perzin. On many occasions Dr. Perzin would remind us: "A high-powered pathologist uses low-power magnification!" So, the first trick of the trade is to remember to examine the tissues with a 2.5× or 5× ocular magnification.

The orderly, uniform pattern of benign tissues at low magnification is due, of course, to the controlled growth of such tissues. Although formalin-fixed, paraffin-embedded tissues may look static under the microscope, they are constantly undergoing cell turnover. Benign tissue does this in a slow, orderly fashion. Malignant tissue does the same, but in a much more rapid, disorderly manner.

Since they grow and turnover slowly, cells in benign tissue rarely show mitoses. Of course, some benign tissue must show rapid growth. Obvious examples include the endometrium in a young woman and the testicular tissue in a young man. Other common examples include a reactive benign lymph node and the bone marrow. In each case, mitotic figures are easy to find. However, even this more rapid cell growth occurs in a very orderly manner.

Since many of us are studying carcinomas, we will often have to compare the growth pattern of benign glands versus malignant glands. This examination at low magnification is indeed the most accurate and simplest way to diagnose carcinomas under the microscope. Benign glands show very organized patterns. We saw this in the colon with the "wallflower" or "wallpaper" pattern in Figure 6-3. This same orderly pattern in the prostate and breast produce a *lobular pattern* of growth. This pattern makes sense when we consider how these organs grow, as well as their function. These "secreting" organs, like the pancreas, must make a lot of proteins for export. Thus, they need a large, well-defined duct system from the gland to the "outside" world. They also need to produce many nests of glands around a given duct. This produces the lobular pattern of a large simple duct surrounded by an orderly array of smaller acini. These acini branch off the duct, much like a cluster of grapes around a stem. Thus, the aggregate of acini will be sharply defined such that you can easily, in your mind's eye, draw a circle around them when viewing this pattern at low magnification.

So, let's summarize the features of benign tissues:

1. Ano rderlyp atterno fgro wth
2. Ano rderlyse quenceo fc ellm aturation
3. Bland,u niformn uclei

These three features of benign tissue are illustrated in Figures 6-15 and 6-16. The orderly sequence of cell maturation refers mostly to epithelia. There are smaller, often flattened or rounded *basal cells* at the bottom of the epithelia. As you progress toward the surface, the cells mature in a uniform fashion until you reach the surface, where you see *terminal* or *complete* differentiation. For the squamous epithelia, this would be an orderly layer of keratin. For the fallopian tube epithelia, this would be ciliated cells. For the small intestine, this would be goblet cells or cells with terminal brush borders, and so on.

The bland, uniform nuclei of benign tissue refers to the fact that these cells, by the time we see them under the microscope after biopsy, have already done their main job and are now in a steady-state mode of just keeping the benign tissue going during the normal cell turnover cycle. That is to say, these cells have well-defined, usually narrow metabolic profiles and, thus, usually have nuclei that are both very similar one to the next and are bland. We used to call this "carbon copy chromatin," but I am not sure that the readers younger than 30 know what a carbon copy is (or was)!

MALIGNANT TISSUE

Whereas benign tissues epitomize the orderly, well-organized state of being, malignant tissues show us the end result of a highly disorganized, rapid, chaotic pattern of growth. Many pathologists in training focus on the nuclear atypia of cancer cells in trying to make the diagnosis. Certainly, the nuclei of cancer cells are often large and show abundant chromatin clumping and prominent nucleoli. However, many *benign* cells can show the same thing, especially if they are undergoing repair in response to some damage, as we saw in Figure 6-12. Yet, here again we see the difference between benign repair, where the nuclei are large but uniform, and malignant cells, where the atypical nuclei vary in size, shape, and color.

The low magnification pattern of cancer is the best way to make the diagnosis. For epithelial tumors, if you cannot, in your mind's eye, draw tight lines around a group of epithelial cells, but rather they are growing in a haphazard fashion, then this probably is a carcinoma. Of course, a key diagnostic feature of cancer is that it invades local tissue. Thus, if the nests of cells are invading any benign tissues (such as fat, adjacent benign epithelia, etc.), by definition, the tissue is malignant.

So, let's summarize the features of malignant tissues:

1. Ad isorderlyp atterno fgro wth
2. Lacko fc ellmatu ration
3. Nuclei that vary in size, shape, and chromaticity
4. Increasedmito ses;atyp icalmito ses
5. Invasiono flo calb enigntissu es

These diagnostic features of malignant tissues are shownin Figures6-17an d6-18 .

Note from the figures that a low-power view that illustrates either point 1 and/or point 5 is sufficient to make a diagnosis of malignant tissue. Since malignant cells are monoclonal, being derived from a single population of malignant cells (I realize that explanation is oversimplistic, but it works for the histologic description), it makes sense that the cancer cells will not show an orderly pattern of maturation from stem/reserve cell to differentiated cell. Also, of course, since they divide rapidly, they will show increased mitotic figures. Indeed, the diagnosis of some cancers (endocervical adenocarcinoma, leiomyosarcoma) is highly dependent on the mitotic count. Many publications have indicated that atypical mitoses are unique to cancer cells, since they reflect abnormal chromosome numbers that are rarely found in benign cells but are common in cancers. So, of course, points 2, 3, and 4 are useful criteria for the diagnosis of cancer. However, if you train

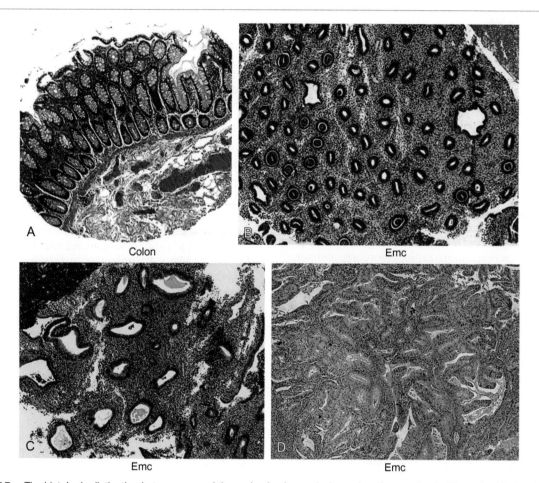

A — Colon
B — Emc
C — Emc
D — Emc

Figure6-15   The histologic distinction between normal tissue, benign hyperplasia, and malignant glands. Normal epithelia will always show an orderly pattern of growth, uniform bland nuclei, and an orderly sequence of cell maturation. This is evident in panels A (colon) and B (endometrium). Compare this pattern to that seen in panel C (endometrium). The glands are still simple glands, which means that they have just one large lumen. Such glands are always benign. However, they show marked variation in size. When benign glands are stimulated to divide too often, some glands grow more rapidly than others. This produces the small lumen/large lumen pattern typical of benign hyperplasia. Panel C is part of an endometrial polyp, the most common form of benign hyperplasia at this site. In contrast, when malignant cells in a gland grow, they do this in a disorganized fashion. This is most easily seen by looking at the lumen. Benign glands always have simple lumens, whereas the disorderly rapidly growing cells of a cancer usually form many glands within one gland (panel D, endometrium). This pattern, called cribiforming, is always malignant.

your eye to focus on points 1 and 5, and do so at the lowest magnification that your microscope will allow, you will soon become very good at rapidly and accurately distinguishing the benign and malignant tissues in the formalin-fixed, paraffin-embedded tissues that are you studying with either immunohistochemistry or in situ hybridization.

PART THREE: PUTTING IT ALL TOGETHER—DETERMINING THE SPECIFIC CELL TYPES THAT CONTAIN YOUR TARGET OF INTEREST

Now that we have discussed the basics of the different cells that are found in pretty much any formalin-fixed, paraffin-embedded tissue and how to distinguish benign from malignant tissues under the microscope, let's put this information together with some immunohistochemical information and then a short quiz.

EPITHELIAL CELLS AND AE 1/3

The fact that epithelial cells serve certain basic functions can be used to our advantage to identify them in formalin-fixed, paraffin-embedded tissues. For example, all epithelial cells contain a type of protein essential for their structural integrity. These are the cytokeratins. Cytokeratins are polymers made from two different intermediate filaments. The polymers are able to form stable supercoils due, in large part, to intra- and intermolecular hydrogen bonds, a process we theorize is also important in the three-dimensional macromolecule cross-linked network. Indeed, in the epithelial cells of formalin-fixed, paraffin-embedded tissues, no doubt keratin forms an essential part of the three-dimensional macromolecule cross-linked cage. The biochemistry of keratin is very similar to that of another, although unrelated structural protein we have already discussed at length: collagen. The triple helix of each gives these proteins their strength, and they are related, in turn, to the high percentage of glycine

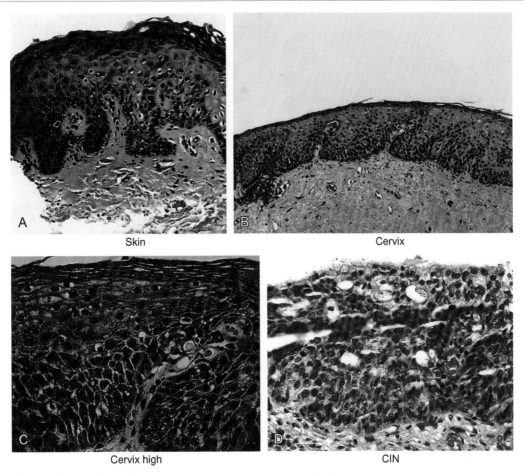

Skin

Cervix

Cervix high

CIN

Figure6-16 The histologic pattern of benign versus malignant squamous cells. Benign squamous cells always show an orderly pattern of growth (panels A–C). At higher magnification, the nuclei are well spaced and the nuclei rarely overlap (panel C). In comparison, malignant squamous cells grow much more rapidly and in an disorganized fashion. This is seen as a much greater cell density and nuclei overlapping(panelD) .

found in these molecules. Indeed, silk is another keratin that can have as much as 80% of glycine and alanine, with many serine amino acids as well. This lends itself to very tightly packed hydrogen bonding, which, again, fits our theoretical construct of one of the essential forces in the three-dimensional macromolecule cross-linked network induced by formalin. Indeed, I can think of no better example of the importance of hydrogen bonding to the biochemistry and structural integrity of proteins than the basic support proteins of the body—keratin, elastin, and collagen.

Figure 6-19 shows the strong staining for keratin at three different sites: the skin, the cervix, and the breast. Note in panel A that no signal is seen without a prior digestion in protease. The serial section (panel B) shows a strong signal if digestion in proteinase K is done prior to immunohistochemistry. Also note that in panel D two different foci are keratin positive. The large arrow shows the classic, well-defined, organized lobule of normal breast. As expected, it is strongly positive for the keratin AE 1/3. The smaller arrow shows a large group of cells. They are also cytokeratin positive, so they must be epithelial cells. Note the two key differences between this group of cells and the benign breast epithelial cells:

1. This group does *not* have a simple lumen as compared to the benign duct cells (panel E) that show small, single, well-defined lumens in each duct/gland.
2. These epithelial cells have a weaker signal for the cytokeratin compared to the benign ductal cells.

Let's re-examine point 1, because it is very important: the group of cells does not have a uniform, single lumen but rather a poorly-defined center where no lumen is clearly identified. For the breast, this one point is enough to confidently state that the nest of cells shown by the small arrow in panel D and at high magnification in panel F must be malignant.

So, in sum, you can use a broadly reactive cytokeratin as a marker of both benign and malignant epithelial cells. There are many different cytokeratins. A list of them is provided at the following website: http://en.wikipedia.org/wiki/List_of_keratins. Although it is true that some cytokeratins are specific for epithelial cells of very specific regions (for example, cytokeratin 20 stains only colonic epithelial cells, and cytokeratin 7 stains epithelial cells from higher up in the gastrointestinal tract and in the lung), I would rather focus on the broadly reactive cytokeratins. If you have done immunohistochemistry or

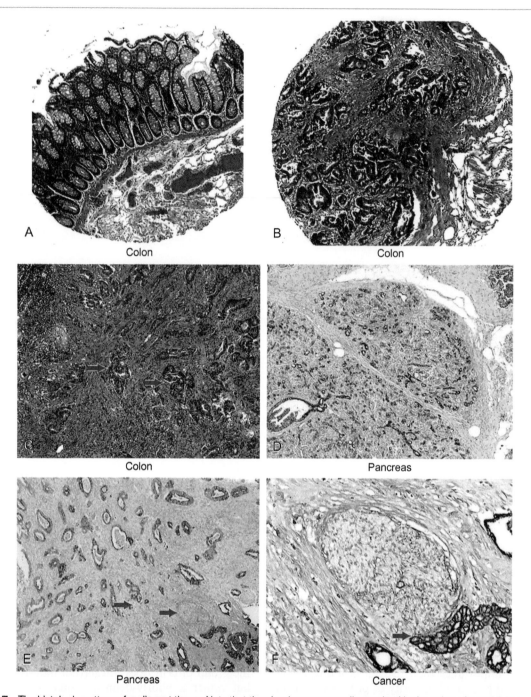

Figure 6-17   The histologic pattern of malignant tissue. Note that the glands are very well organized in the colon of panel A but very disorganized in the colon of panel B. This, by itself, is enough to realize that panel A is from normal colon and B from a person with adenocarcinoma. The same lesion in panel B shows the malignant glands invading the stroma (arrows, panel C, higher magnification). The best way to make a diagnosis of adenocarcinoma is to look at low magnification. Panels D–F are after immunohistochemistry for keratin. Note that the pancreas in panel D is very well-ordered into two defined nodules. Compare this to panel E, where the glands are growing in a haphazard pattern. Panel D is benign and E is malignant. Finally, the keratin in panel F (arrow, also seen in panel E at lower magnification) surrounds a nerve. By definition, when glandular cells directly touch benign tissue, such as a nerve, then the glandular cells must be malignant.

in situ hybridization on a tissue and want to document that the positive cells are epithelial, then I recommend that you use cytokeratin AE 1/3. As a clinical aside, since epithelial cells basically do not belong in lymph nodes (with a few exceptions), then seeing cells in a lymph node that are AE 1/3 positive is highly suggestive of metastatic carcinoma. Indeed, this test is widely used in clinical pathology.

STROMAL CELLS AND VIMENTIN

Do stromal cells contain a unique structural protein in their cytoplasm that is the equivalent to cytokeratin in

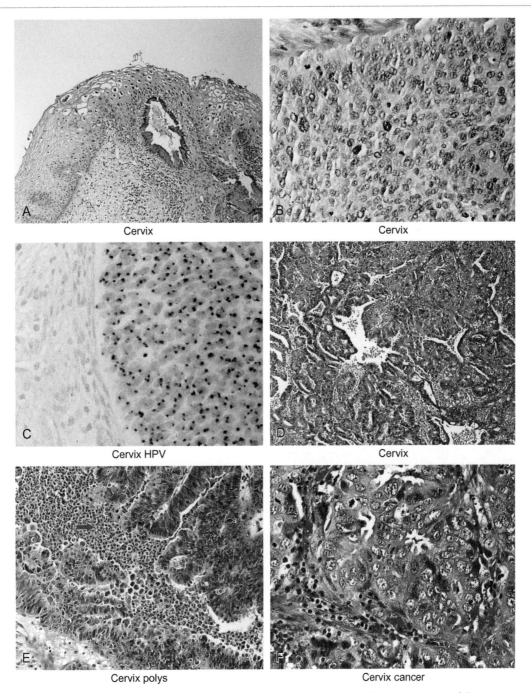

Figure6-18 The histologic pattern of malignant tissue: the cervix. Panel A shows the well-ordered pattern of the squamous and glandular epithelia at the transformation zone. Note in panel B how much thicker is the squamous epithelia layer. Neoplastic cells grow very rapidly and, unlike benign cells that grow rapidly, do not respect their neighbor's borders. Thus, as seen in panel B, the cells are haphazard with the classic cytologic feature of squamous cell carcinoma in situ where the nuclei overlap. Such lesions are invariably HPV DNA positive (panel C); note the punctuate pattern typical of integrated HPV DNA in cancers and that the stroma on the left of the image is negative for the virus and, thus, a good internal negative control. The patterns in panels D, E, and F are all very disorganized. Note the occasional lumens. This is the easiest way, with the disorganized pattern of growth, to recognize an adenocarcinoma. Since cancers grow so rapidly, they tend to outgrow their blood supply. This leads to necrosis after which the neutrophils are sure to follow. A collection of neutrophils is seen in panel E (arrow).

epithelial cells? The answer is yes, and the structural protein is called *vimentin*.

Unlike keratin, which is found only in vertebrates, vimentin is found in all cell types including bacteria,

indicating that it is a more highly conserved protein. Thus, if we were looking for a positive control for animal tissues and we had to decide between antibodies to human keratin or vimentin, the latter would more likely

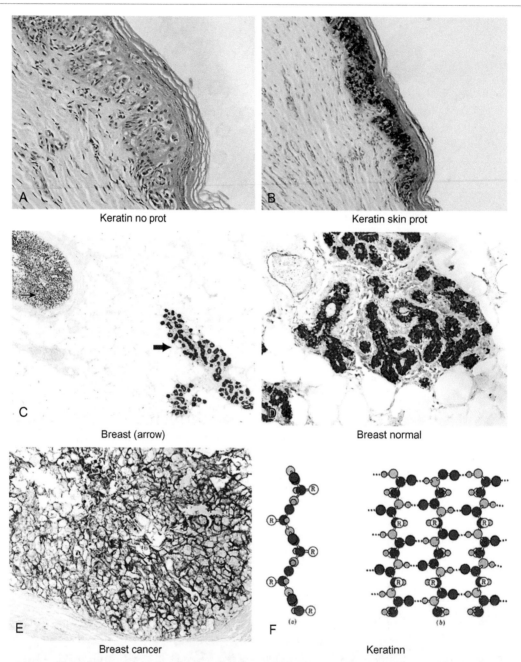

Figure 6-19 Cytokeratin in benign and malignant tissues. Panels A and B show a section of the skin and serve as a reminder that a signal by immunohistochemistry for keratin requires prior protease digestion. Also note that in panel C two different foci are keratin positive. The large arrow shows the classic well-defined, organized lobule of normal breast. As expected, it is strongly positive for the keratin AE 1/3. The smaller arrow shows a large group of cells. They are also cytokeratin positive, so they must be epithelial cells. This group does not have well-defined simple lumens and shows a weaker signal; these are the features of adenocarcinoma of the breast. Higher magnifications of these two areas are shown in panels D and E, respectively. Keratin is an extremely strong protein resistant to both pressure and water damage. It owes this high resistance in large part to its molecular structure (panel F), in which its rigid individual chains are held in a three-dimensional network that allows for maximum interactions of the many hydrogen bonds of its side chains, much like is present in double-stranded DNA.

yield a reaction in the nonhuman tissues. Throughout this book I have referred to the three-dimensional protein/protein cross-linked network. We can now be a bit more specific, since vimentin (or keratin in epithelial cells) typically is strongly associated with other proteins such as tubulin (important in mitoses) and actin (important in cell mobility) to form the cytoskeleton and, thus, form the basis of our three-dimensional protein/protein cross-linked cage. Indeed, it is well established that vimentin is attached to the nucleus, mitochondria, and all cell organelles. An important concept in carcinomas is the transition from epithelial cell to mesenchymal cells (EMT). Not

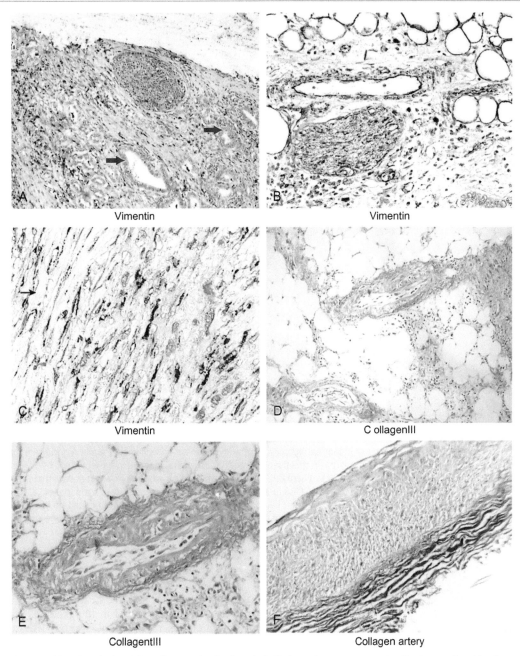

Figure 6-20 Immunohistochemical staining of stromal cells. Panels A–C show that most stromal cells show a signal for the protein vimentin by immunohistochemistry. This tissue is from a pancreatic cancer and note that the glands are negative for vimentin (arrows), serving as an internal negative control. Also note that all cell types, including fibroblasts, adipose cells, and nerve cells are positive. In comparison, note that the immunohistochemistry for type III collagen in panels D–F only shows a signal for the outer wall of the large blood vessels.

surprisingly, vimentin is a marker of such a metaplastic event in cancers.

I have also theorized in this book that the three-dimensional protein/protein cross-linked network is likely responsible for controlling the movement of the key in situ hybridization and immunohistochemistry reagents during the techniques. This has been documented in vivo for vimentin. For example, vimentin can bind to a variety of chemicals and, thus, alter their rate of diffusion in the cell and/or facilitate the processing of such chemicals. Perhaps the clinically most important example with vimentin

is its ability to control the transport of the protein LDL (low-density lipoprotein), given the latter's importance in atherosclerosis.

As seen in Figure 6-20, vimentin is strongly positive in most stromal cells, and either not expressed or only weakly expressed in epithelial or inflammatory cells.

LYMPHOCYTES AND CD45

Lymphocytes are usually easy to recognize in formalin-fixed, paraffin-embedded tissues because of their small

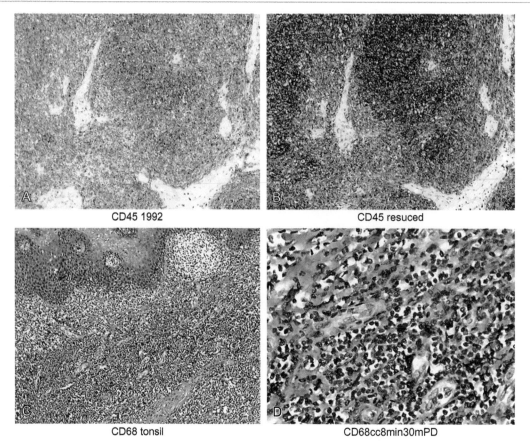

CD45 1992

CD45 resuced

CD68 tonsil

CD68cc8min30mPD

Figure6-21 Immunohistochemical staining of inflammatory cells: lymphocytes and macrophages. Panels A and B are from a lymph node tested for CD45, which is the predominant antigen in this tissue. Note that the signal is weak in panel A (block 20 years old), but was enhanced after "regeneration" with IHCPerfect (panel B). Whereas CD45 is the marker for lymphocytes and shows that they dominate in the lymph node, tonsil, or spleen, the macrophage marker CD68 (panels C and D) will decorate far fewer cells (panel C), which will have much more cytoplasm than lymphocytes (panel D). Thus, if there is a poorly differentiated cancer, keratin would be the logical test to determine if it has an epithelial origin, vimentin to determine if it has a stromal cell origin, and CD45 for a lymphocyte origin.

round shape, lack of cytoplasm, and tendency to aggregate. At times it is useful to document that specific cells that contain a target of interest are lymphocytes. In this case, the protein to use is CD45, because it is present on mostlymp hocytes(Figure6-21).

CD45 is a member of the large protein tyrosine phosphatase family of proteins. It has a key role in cell signaling. Since the main function of lymphocytes is to signal other cells to do things in response to a variety of stimuli, it is not surprising that CD45 is the marker for most lymphocytes. CD45 also inhibits several kinases (such as the JAK kinases), which tend to inhibit cytokine expression. This serves to "quiet down" inflammation, because the bottom line is that inflammation is basically a process of cytokine expression. The most common use of CD45 in clinical pathology is when you are faced with a poorly differentiated cancer in which the cells are so undifferentiated that it is impossible to determine the cell of origin. If the cells are AE 1/3 positive, the origin is a carcinoma; if they are CD45 positive, it is a lymphoma; and if they are vimentin positive (and negative for these other markers); it is a sarcoma.

MACROPHAGES AND CD 68

CD68 is a protein found in the granules of macrophages, which includes not just the standard tissue macrophage but some such cells that are given specific names in certain organs. For example, the macrophage of the liver is called a Kupffer cell, the macrophage of the brain is a microglial cell, the macrophage of the placenta is the Hofbauer cell, and the macrophage of bone is the osteoclast. Here is another example where pathologists tend to complicate matters by using too many names for the same thing. Macrophages play a critical role in the cross-talk between lymphocytes with inflammation and between cancer cells in cancer progression. Thus, we often want to co-label cells positive for a certain protein, DNA, or RNA marker after immunohistochemistry or in situ hybridization to prove they are macrophages. In such cases, CD68 is the best choice.

INFLAMMATION AND TNF ALPHA

A common question in both diagnostic and research pathology is how to assess the degree of inflammation in a tissue. This is especially important when the question concerns possible infectious diseases, autoimmune diseases, or the progression of a given cancer. As we have discussed, the presence of inflammatory cells in a tissue is one indicator of inflammation at that site. However, it is a poor indicator. There are several reasons for this:

1. Some tissues (e.g., the brain, testes, placenta) are "protected" organs that may show substantial inflammation but little inflammatory cell infiltrate. The reason is that native cells are producing the inflammatory cytokines.
2. Mononuclear cells at a given site may be quiescent or may actually be *inhibiting* the inflammatory process by reducing the number of cytokines produced by the lymphocytes and macrophages at the site.

Inflammation is defined by the presence of cytokines. The reason is that it is the cytokines released by the cells at a given site (or from a distal site in the case of sepsis or the cytokine storm in a fetus) that actually mediate the tissue damage. It is the cytokines that cause the leaky blood vessels, the microvessel clotting, the vasoconstriction, and so on, that are the cause of the actual tissue damage with any inflammatory process.

There are many cytokines. Perhaps the most extensively studied is the cytokine called tumor necrosis factor alpha (TNF alpha or TNFα for short).

TNFα is a potent endogenous pyrogen, and thus is responsible for the fever when a person is ill. It can induce death of cells (such as neurons in the neonate who has fetal inflammatory syndrome as a result of an infection). I recommend the use of immunohistochemistry for TNFα if you need to establish the presence and/or degree of inflammation in a given tissue that you are analyzing. It works best with a brief protease digestion, as seen in Figure 6-22.

BENIGN VERSUS MALIGNANT CELLS AND KI-67

The last protein to discuss that can be of general help to use in the interpretation of our immunohistochemistry and in situ hybridization data is Ki-67, as shown in Figure 6-23.

Ki-67 is a nuclear protein that is strongly associated with mitotic activity, probably through its interactions with the rRNAs present in the nucleolus. Pathologists are trained to count the number of mitoses per high power field. A *much* more accurate way to do this is by doing a Ki-67 immunohistochemistry stain and counting the number of positive nuclei per 100 cells. This approach is called the Ki-67 index. By the way, a common commercially available antibody that can detect Ki-67 is called MIB-1.

I previously listed the major criteria for diagnosing a lesion as benign or malignant. These criteria are certainly important, and serve as the basis for the surgical pathologist rendering such a diagnosis on routine hematoxylin and eosin stain. However, there is one other way to make the distinction of benign versus malignant, and this is using immunohistochemistry. If you do immunohistochemistry for Ki-67 and find that the cells in question have an index of 0–15, they are most likely benign. If the index is >25, then they are most likely malignant (realizing, again, that this must be interpreted in the light

of the other histologic findings). Finally, using co-expression with Ki-67 and your target of interest that is present in cancer cells is a simple way to determine whether the protein, DNA, or RNA target is specific for proliferating cells, nonproliferating cells, or neither.

PUTTING IT ALL TOGETHER WITH AN EXAMPLE

Let's look at an example in which some of the information was put to use for a research-related problem. This work reflects collaborations I have done with the laboratory of Dr. Carlo Croce. I have yet to directly work with an investigator who has impressed me more with his knowledge, dedication, and skill than Dr. Croce. His laboratory did a project led by Drs. Muller Fabbri and Federica Calore. The project involved the cross-talk between cancer cells and inflammatory cells. It directly addressed the question I have referred to in this book on many occasions: the documented association between cancer and inflammation.

In sum, Drs. Croce, Fabbri, Calore, and others showed that cancer cells could release certain factors via exosomes that appeared to be going to inflammatory cells and, in this way, altering the metabolism of the cancer cells to facilitate their growth and metastases. The marker for the exosomes is CD9. Thus, we needed to document whether the CD9+ factors from the cancer cells (which were carcinoma cells and, thus, AE 1/3 positive) were taken up by the macrophages, which are CD68 positive. Since some of the work was done in mice, we used the analogue of CD 68 for murine tissues. Some representative data are shown in Figure 6-24. Note the following points:

1. The tumor cells are the main source of the micro RNA (panel A).
2. If you do co-expression analyses for the micro RNA and CD9 (panels B and C), it is clear that the CD9+ exosomes are present only where the tumor meets the benign tissue (referred to as the tumor interface).
3. The exosomes that contain the micro RNA of interest are not evident even 1 mm away in the normal lung (panel D). This is an excellent example illustrating where a sharp eye and a knowledge of surgical pathology of tumors in the lung can give you much insight into the meaning of the immunohistochemistry data.
4. The cells that contain the CD9/micro RNA+ targets are mostly macrophages (panel D), because they co-express CD68.
5. The cells that contain the CD9/micro RNA+ targets at the tumor interface are not cancer cells, since only the cancer cells toward the center of the tumor co-express CD9, not the AE 1/3+ tumor cells at the tumorin terface.
6. In sum, the data show that cancer cells can release factors that can be exported to macrophages. A lot of the supporting data to this extremely interesting point came from the immunohistochemistry, in situ hybridization, co-expression analysis and, perhaps more importantly, from researchers being able to see the histologic distribution of the above markers and know how to interpret the findings due to a solid knowledge of surgical pathology.

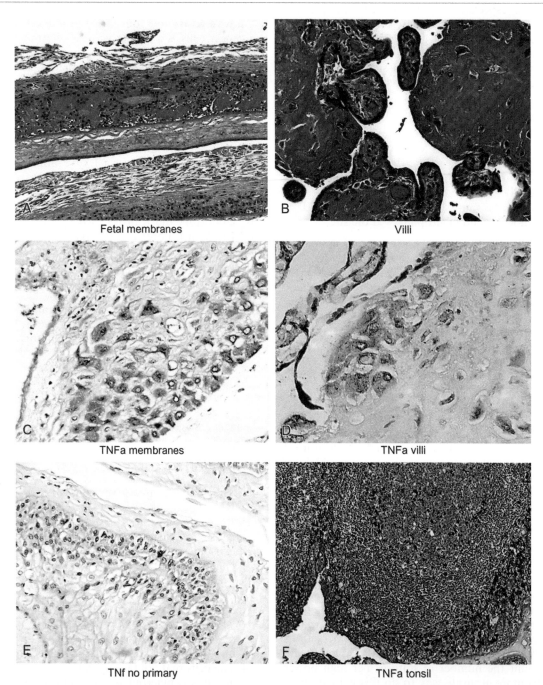

Fetal membranes

Villi

TNFa membranes

TNFa villi

TNf no primary

TNFa tonsil

Figure6-22   The importance of immunohistochemical staining to determine if there is a fetal inflammatory response in the placenta. Panels A and B show hematoxylin and eosin stains from a placenta where the neonate had severe morbidity. Note that there are only nonspecific histologic findings. Since endogenous cells are the primary source of cytokines in such tissues, immunohistochemistry for cytokines such as TNF alpha is the best way to determine if there is a serious ongoing in utero insult that would be responsible for the morbidity. Note that many cells in this case indeed expressed TNF alpha in the placenta and fetal membranes, proving that there was a marked fetal inflammatory response, which, of course, would have caused substantial damage to the neonate's brain. Panel E is one of the negative controls (no primary antibody); and panel F, a positive control; TNF alpha testing of a chronically inflamed tonsil shows a strong signal in scattered cells that have the cytology of macrophages.

A SHORT QUIZ

As promised, let's end this chapter with a brief quiz.

Case 1. Biopsies A and B are each from the cervix and were obtained from women who were 69 and 49 years old, respectively. Choose the correct answer:

A. Each biopsy shows CIN (cervical intraepithelial neoplasia).
B. Eachb iopsysh owsb enignatro phy.
C. Biopsy A shows CIN, and biopsy B shows atrophy.
D. Biopsy A shows atrophy, and biopsy B shows CIN.

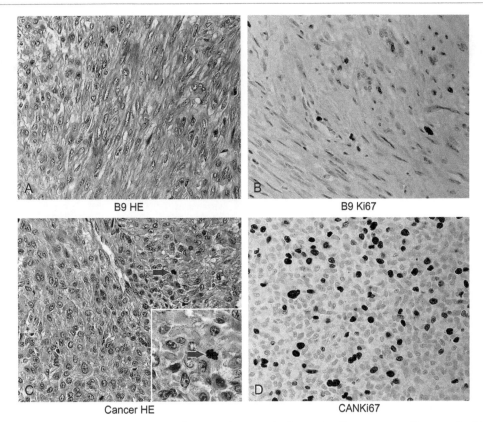

Figure6-23 The importance of the Ki-67 index to differentiate between benign and malignant tumors. Compare the histologic features of panels A and B (each showing tumors of the uterine wall). Note that panel C is more cellular and does show mitotic activity (arrows), but this is not sufficient to differentiate benign versus malignant. Benign tumors show a low Ki-67 index (usually less than 10%), and malignant tumors usually show high indexes, usually much more than 10%. The Ki-67 immunohistochemistry test for the tumor in panel A is very low, consistent with a benign tumor (leiomyoma); whereas the Ki-67 index is much higher for the other lesion (panel D), consistent with the diagnosis of leiomyosarcoma.

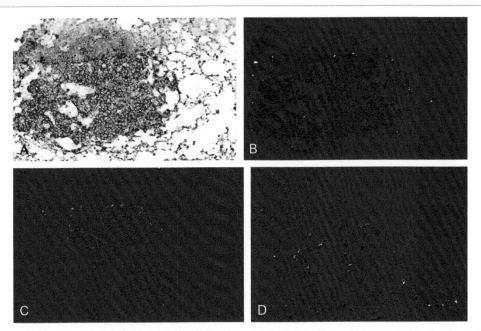

Figure 6-24 Panel A shows a metastatic tumor nodule in the middle of normal lung after analysis for miR-29a (blue) and CD 9 (red). The Nuance-analyzed image shows that co-expression of miR-29a and CD9 (fluorescent yellow) is confined to the perimeter of the tumor (panel B, high magnification and panel C, low magnification); note in the low magnification that the area of normal lung directly adjacent to the tumor is completely negative for co-expression. Panel D shows that the miR-29a also co-expresses with CD68 at the tumor-normal tissue interface.

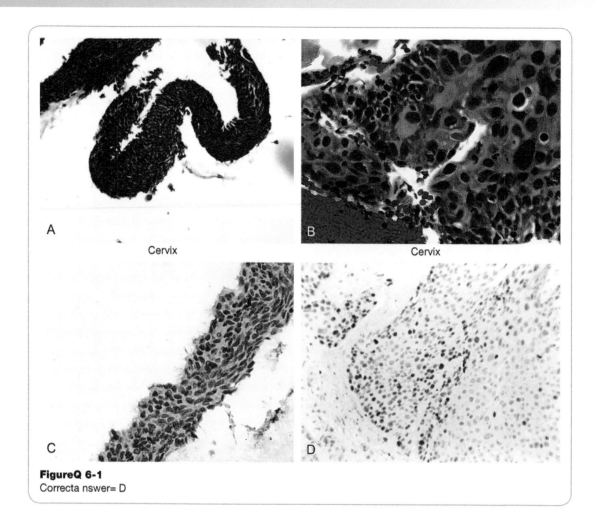

A — Cervix

B — Cervix

C

D

FigureQ 6-1
Correcta nswer= D

LookatF igureQ6-1fo rth ean swer.

Note that biopsy A shows thinned epithelia, whereas biopsy B shows thickened squamous cell lining. Also, note that the nuclei in B vary a great deal in size, shape, and color; whereas the epithelial cells in A are more uniform, though this point is hard to appreciate at low magnification. The proof that A is atrophy is the negative Ki-67 immunohistochemistry stain (panel C). Although the histology of tissue B is classic for a CIN lesion, the irregular staining for the proliferation marker MCM2 (panel D) is also diagnostic of such.

Case 2. Biopsy A was obtained from a 31-year-old woman from the cervix. Choose the correct answer:

A. Theb iopsysh owsatro phy.
B. Theb iopsysh owsCIN.
C. Theb iopsysh owsre parativec hange.
D. Theb iopsysh owsc ervicalc ancer.

Look at Figure Q6-2 for the answer.

Note that biopsy A shows thinned epithelia. We cannot see the epithelia clearly enough at this low magnification to determine whether it shows the open nuclei and prominent nucleoli of repair. However, note the thick band of brown in the submucosa of the cervix. This is not normal. At higher magnification, it is clear that these cells are hemosiderin laden macrophages. The only way these cells would congregate

in the stroma of the cervix is in response to a prior biopsy. Thus, the lesion must represent reparative change. Also, at age 31, the woman is too young to show atrophic changes.

Case 3. Biopsy A shows a skin biopsy from a 59-year-old man. Choose the correct answer:

A. This isn ormals kin.
B. Theb iopsysh owsab enigntu mor.
C. Theb iopsysh owsamalign anttu mor.
D. We can't tell whether the skin is normal or shows a tumor without knowing the site.

Look at Figure Q6-3 for the answer.

The tissue cannot be normal skin for two reasons: (1) it lacks the normal maturation of squamous epithelia; and (2) the squamous epithelia are too thick for any site. Note that there are islands of keratin throughout the lesion. At higher magnifications, note that the cells still show a uniform, orderly maturation. Hence, this is a slowly growing benign tumor. It is also a very common benign tumor called a seborrheic keratosis.

Case 4. Biopsies A, B, and C are from the endometrium of women who were 43, 49, and 50 years old, respectively. Choose the correct answer:

A. Biopsies A and B are benign normal endometrium, and biopsy C is a benign tumor.

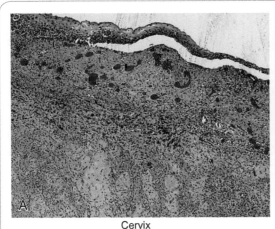
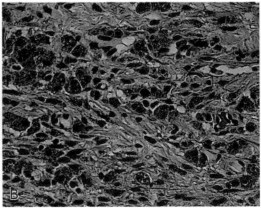

Cervix

FigureQ 6-2
Correcta nswer= C

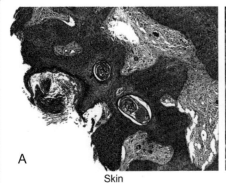
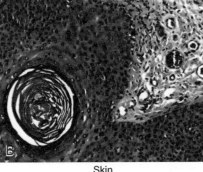
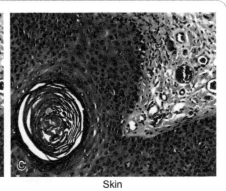

A
Skin

Skin

Skin

FigureQ 6-3
Correcta nswer= B

B. Biopsy A is normal, biopsy B is a benign tumor, and biopsy C is a malignant tumor.
C. Biopsies A and B are benign tumors, and biopsy C is a malignant tumor.
D. Allb iopsiesare b enign.

Look at Figure Q6-4 for the answer.

Note that biopsy A shows a uniform pattern of simple glands. There is a bit of variation in gland size, but this simply reflects the tissue orientation. Biopsy B shows simple glands, each with an obvious lumen, so it must be benign. However, the glands vary a great deal in size and shape. This is a classic finding in *simple hyperplasia* of the endometrium, which is a benign process that reflects unopposed estrogen effect. It is very common in a middle-aged woman. Biopsy B shows another part of this lesion: thick-walled blood vessels. These two features (thick-walled vessels and simple hyperplasia) are diagnostic of the most common benign epithelial tumor of the uterus: an endometrial polyp. Biopsy C shows packed glands that show many lumens inside the main lumen. This complex, disorderly growth pattern is

called cribiforming and is diagnostic of neoplastic epithelial tumors. Cribiformed glands can be found in either complex atypical hyperplasia or adenocarcinoma; this particular biopsy was diagnosed as well-differentiated adenocarcinoma. If you got this answer correct and have no surgical pathology training, you should consider becoming a pathologist! By the way, case D shows tissue present in the ovary. It consists of a simple gland and, as shown by the immunohistochemistry stain, is surrounded by CD10 positive stromal cells. CD10 in this situation is diagnostic of endometrium stroma; this is endometriosis.

Case 5. Biopsies A and B are each from the vulva of two different women who were each 28 years old. Choose the correct answer:

A. BiopsiesAan dB ar en ormals kin.
B. Biopsy A is normal, and biopsy B is simple atrophic.
C. Biopsy A is normal, and biopsy B represents reparativec hange.
D. Biopsy A is normal, and Biopsy B is some inflammatory-based condition where the epithelia becomes atrophic and the stroma has too much collagen.

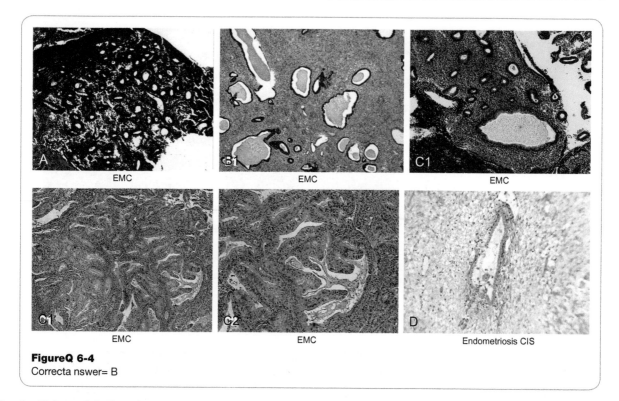

FigureQ 6-4
Correcta nswer= B

LookatF igureQ6-5fo rth ean swer.

Biopsy A is normal skin. Biopsy B shows epithelial atrophy, but this is not expected in a woman who is 28 years old. Note the thick band of inflammatory cells at the base. They are mostly lymphocytes, not macrophages as seen at higher magnification. Also note that the dermis is thickened and pink due to abnormal collagen deposition, and that the collagen gets pale where the epithelial cells meet the dermis. This is called "homogenization" of the papillary (upper) dermis. These are the histologic features of the obscure yet common disease *lichen sclerosus et atrophicus.*

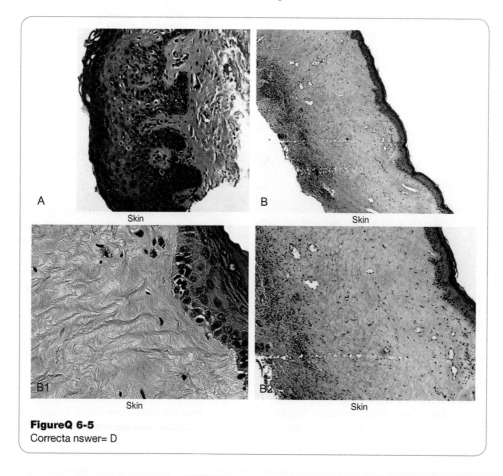

FigureQ 6-5
Correcta nswer= D

The Recommended Protocol for in Situ Hybridization

INTRODUCTION

Up until this chapter, we have focused more on the theoretical aspects of in situ hybridization and immunohistochemistry. We also built a foundation for the histologic interpretations of tissues and discussed at length the biochemistry of the three-dimensional protein/protein cross-linked network, including the remarkable observation that we can not only manipulate it for our benefit, but also can regenerate it, which is especially helpful when using aged blocks.

In this chapter, we do not discuss any of the preceding issues. Rather, I present the more traditional "cookbook" type of protocol I recommend for in situ hybridization. Again, the purpose of this presentation is not to belittle such protocols. However, I am confident that the theoretical discussion we have done prior to this chapter will not only make it clear why I have chosen this protocol, but also that this foundation will help you in troubleshooting your own data.

I would like to add that the company and catalog numbers listed in this chapter were chosen for one reason: I have used them for many years. The fact that I have listed many companies reflects one reason why I have no consulting agreement with any biotechnology company. I do, however, often discuss in situ hybridization and immunohistochemistry with such companies since, of course, they all have investigators with extensive knowledge of these methodologies.

Finally, allow me to stress what I feel is a very important point: I do *not* recommend reading this chapter until you have read all the preceding chapters (with the exception of the prior chapter on histologic interpretation). Although I hope that the protocols in this and the next chapter are helpful, to get the maximum benefit from them, it would be highly advantageous to understand the theoretical basis for each step.

STEP ONE: CUT THE TISSUE SECTIONS ONTO A SILANE-COATED SLIDE

1. Confirm that the blocks are formalin-fixed, paraffin-embeddedtissu es.
2. Determineth eage o fth eb locks.
3. Have the histotechnologist cut successive 4 to 5 micron sections and, depending on the size of the biopsy, put the paraffin ribbons on from 5 to 20 consecutive and labeled silane-coated slides.
4. You do not have to use RNase-free water for this step because the RNA is likely protected from such by the paraffin wax and formalin-induced RNA-proteinc ross-links.
5. Let the water drain from the slides by placing them at an angle on a stand.
6. Put the slides in a rack and bake for 30 minutes at 60°C. *Do not place the slides directly on the hot plate!* Rather, put the slides in a rack and then put the rack in the 60°C oven.
7. Remove the slides and store at RT in a light-tight box or slide holder.

STEP TWO: REMOVE THE PARAFFIN FROM THE SLIDES

1. On the day you wish to begin the in situ hybridization experiment, remove the slides from the box.
2. Incubate the slides for 5 minutes in fresh xylene (the xylene solution should be changed weekly).
3. Incubate the slides for 5 minutes in fresh 100% ethanol (the ethanol solution should also be changed weekly).
4. Air dry the slides at either room temperature or in a 37°Co ven.

STEP THREE: OPTIMIZE THE PROBE/ CHOOSE DIFFERENT PRETREATMENT PROTOCOLS

1. The optimization of the probe presupposes that you have placed at least 2 sections per silane-coated slide.
2. Take four consecutive labeled slides (serial sections) that have been deparaffinized.
3. Check and confirm that the tissue contains the DNA or RNA target of interest.
4. Label the first slide "dilute probe, no protease/protease." Label the second slide "dilute probe, DNA (or RNA) retrieval, no protease/protease." Label the third slide "concentrated probe, no protease/protease." Label the fourth slide "concentrated probe, DNA (or RNA) retrieval, no protease/protease."
5. Make the probe cocktail by diluting the probe with a commercially available DNA or RNA probe diluent

DOI: http://dx.doi.org/10.1016/B978-0-12-415944-0.00007-3

(e.g., Enzo Life Sciences in situ hybridization buffer, catalog #33808; or Exiqon in situ hybridization buffer, catalog #90000, which includes the in situ hybridization buffer as well as proteinase K).

6. For full-length probes (probes made from templates at least 200 base pairs long), the dilute-labeled probe should be diluted to a final concentration of 10 ng/ml, and the concentrated probe should be at 1000 ng/ml.

7. For smaller LNA probes (I do *not* recommend using oligoprobes if they are not LNA modified), the dilute-labeled probe should be diluted to a final concentration of 0.01 picomoles/microliter, and the concentrated probe should be at 1 picomole/microliter.

8. Proceed to the DNA/RNA retrieval and protease steps.

STEP FOUR: PRETREAT THE TISSUES

1. If doing the optimizing protocol, add protease (proteinase K, 1 microgm/ml—Enzo Life Sciences, catalog #33801, diluted in sterile water) to only the *bottom* tissue section to each of the four slides.

2. Incubate the tissue sections in the proteinase K solution for 5 minutes at room temperature.

3. Remove the proteinase K by washing the slides with RNase/DNase-free water for a few seconds. Next, rinse the slides in DNase/RNase-free 100% ethanol for a few seconds. Then air dry in a 37°C incubator.

4. If you are doing the optimization protocol, then take two of the four slides and incubate for 30 minutes at 95°C in any commercially available "antigen retrievalso lution."

5. After the 30 minutes, let the solution cool down for 5 minutes at room temperature; then remove the slides and let air dry.

6. At this stage, if you are doing the optimizing protocol, you have four slides each with at least two sections. The bottom section for each has been digested with protease. Two of the four slides have been exposed to 95°C for 30 minutes in a DNA/RNA retrievalso lution.

STEP FIVE: DENATURE THE PROBE AND TISSUE RNA/DNA

1. For small biopsies (less than 1 cm in size), add 5–10 microliters of the probe cocktail.

2. For larger biopsies (equal to or greater than 1 cm in size), add 10–15 microliters of the probe cocktail.

3. Cover the probe cocktail with a polypropylene coverslip that is cut to be just slightly larger than the tissue section. If doing RNA in situ hybridization, use gloves.

4. For DNA targets, place the slide on a 95°C hot plate for 5 minutes. For RNA targets, place the slide on a 65°C hot plate for 5 minutes.

5. Make sure no bubbles have appeared between the coverslip and the tissue.

6. If you see bubbles, *immediately* and gently remove them using a sterile toothpick.

7. Incubate the tissue in a humidity chamber with water at the base at 37°C for 5–15 hours. For the optimization protocol, I strongly recommend overnight incubation, since it is impossible to know the relative copy number of the target and also the specific activity of the probe.

STEP SIX: THE STRINGENT WASH

1. Remove the coverslips carefully by using fine tweezers. Make certain that the probe cocktail did not dry out. If it did, either the coverslip was too large, there was too little probe cocktail, and/or there was not enough water used for the humidity chamber.

2. Place the slides in a solution of 0.1 XSSC, 2% bovine serum albumin that has been warmed to 50°C in a waterb ath.

3. Incubate the slides in the stringent wash for 10 minutes.

4. Remove the slides *one at a time* for the next step. Do *not* let the slides dry out after the stringent wash.

STEP SEVEN: THE DETECTION PART, THE FIRST STEP

1. Remove the slide from the stringent wash and hold very level. Quickly remove the wash solution from the bottom of the slide and the perimeter of the front of the slide.

2. Using a hydrophobic pen, mark the perimeter of the front of the slide to keep the detection solution over the tissue. Keep the slide level at all times; otherwise, the remaining wash solution will ruin the "hydrophobic dam" you just created.

3. For biotin-labeled probes, add 100 microliters of the streptavidin solution with the reporter enzyme conjugate (Enzo Clinical Labs, catalog # ENZ-32895; this is actually a kit that contains all the necessary ingredients for in situ hybridization using the streptavidin-alkaline phosphatase system and NBT/BCIP as thec hromogen).

4. For digoxigenin-labeled probes, add 100 microliters of the antidigoxigenin/reporter enzyme conjugate. The antidigoxigenin solution should be made fresh at the time it is needed by diluting the antibody 1:200 in a PBS pH 7.0 solution.

5. For either biotin- or digoxigenin-based probes, incubate the streptavidin or antidigoxigenin conjugate at 37°C for 2 hours in a humidity chamber.

STEP EIGHT: THE DETECTION PART, THE LAST STEP

1. If using the alkaline phosphatase conjugate, place the slides in the NBT/BCIP buffer (pH 9.5 solution, Enzo Laboratory, catalog #ENZ-33803)

2. Preheat the NBT/BCIP buffer to 37°C. Confirm that the pH is between 9.0 and 9.5. If it is not, titrate with 0.1 MNaOH.

3. You can add 11 ml of the preheated NBT/BCIP buffer to a five-slide plastic holder.

4. Add 165 microliters of the NBT/BCIP solution (Roche Applied Sciences, catalog #11681451001).

5. Add the slides to the NBT/BCIP solution and incubate at 37°C.

6. If using the peroxidase conjugate, place the slides in the DAB buffer (pH 7.0, PBS solution).
7. Prepare the DAB solution immediately before you are ready to do this step. For manual in situ hybridization using DAB, you may use the Biogenex DAB kit (Super sensitive Link-label Immunohistochemical Detection system, catalog #QD000-5L). This kit has all the reagents you need for immunohistochemistry after the primary antibody step and for in situ hybridization after the probe-target hybridization step if using a biotin-labeled probe. To make the DAB solution, take 2 drops of the DAB chromogen and add to the DAB substrate buffer. Vortex. Then add 1 drop of the DAB substrate. Vortex.
8. Remove the slides from the PBS wash. Wipe the back of the slide and, keeping the slide level, wipe the perimeter of the front of the slide. Place the slide carefully down in a humidity chamber. Do *not* let the slide dry out; otherwise, background will ensue.
9. Add several drops of DAB solution to each tissue section.

STEP NINE: MONITORING THE PRECIPITATE

1. It is *very* important to monitor the precipitation of the chromogen (either DAB or NBT/BCIP) under the microscope, initially every 3 minutes. Do not accept any interruptions at this stage. Stop the reaction when you have determined that the signal has been maximized and/or background is just beginning to develop.
2. Although most NBT/BCIP reactions are finished within 60 minutes, the only determining factor for stopping the reaction is when you have determined that the signal has reached its maximum stage and/or background begins to be evident.
3. Although most DAB reactions are finished within 20 minutes, the only determining factor for stopping the reaction is when you have determined that the signal has reached its maximum stage and/or background begins to be evident.
4. When the signal is at maximum *or* background begins to be evident, immediately place the slide in a jar that contains water. Wash the slides under running tap water for 1 minute to stop the chromogen precipitate.

STEP TEN: THE COUNTERSTAIN

1. If you wish to do co-expression analysis, do *not* do a counterstain at this time.
2. If you wish to use a counterstain, for NBT/BCIP, use Nuclear Fast Red.
3. If you wish to use a counterstain, for DAB, use either Nuclear Fast Red or hematoxylin.
4. For nuclear targets, do a weak counterstain (a few dips in hematoxylin or 30–60 seconds in Nuclear FastRe d).
5. For cytoplasmic targets, use the standard hematoxylin (1 minute) or Nuclear Fast Red (3 minute) counterstain.
6. Rinse the slides in running tap water for 1 minute.
7. Dip the slides in 100% ethanol. Let completely air dry.
8. Placeth es lidesin xyle ne.

9. Coverslip using Permount (Fisher Scientific, catalog #SP15-100).

TROUBLESHOOTING

The most common request I get via email regarding in situ hybridization is to troubleshoot. After working 30 years in the field, I can say with certainty that I have seen every one of the preceding steps have a serious malfunction. Indeed, a major reason that I have included large chapters on the biochemistry and fundamentals of in situ hybridization is to show you the specifics of how each step can have a serious malfunction. By also explaining the theory, I am optimistic that this will not only give you a good guide on how to troubleshoot, but will also allow you to fix the problem.

I can also say with confidence that the two most common problems with in situ hybridization—too much background and no signal—have a few causes that are much more prevalent than other possible malfunctions. Let's look at these points in some detail.

TOO MUCH BACKGROUND

Too much background is the most common problem I see with in situ hybridization in my consulting work. It won't surprise you, from all the data we discussed in previous chapters, that the three most common causes of too much background are

1. Ap robeth atisto oc oncentrated
2. Incorrectp retreatmentc onditions
3. Omission of bovine serum albumin from the wash solution

As we have discussed at length, in situ hybridization is a basic diffusion reaction. The higher the probe concentration, the more likely you will see background. The *primary reason* I recommend always starting the testing of a new tissue and probe by using *two very different* probe concentrations is to use these data to determine which probe concentration will allow you a good signal but with minimal background.

We have also spent a lot of time on the theory of the three-dimensional macromolecule cross-linked cage that surrounds each DNA and RNA target. We have seen on many occasions that poor exposure of the target not only causes a reduced signal, but also may allow background to develop. In my experience, often the reason for this is that the investigator sees little signal and, thus, increases the concentration of the probe. This then leads to the background.

NO (OR VERY WEAK) SIGNAL

No (or very weak) signal is the second most common problem I see with in situ hybridization in my consulting work. By far and away, three causes dominate for this situation:

1. Incorrectp retreatmentc onditions
2. Using a tissue that has very little or none of the target of interest
3. Usingtissu esth atare to oo ld

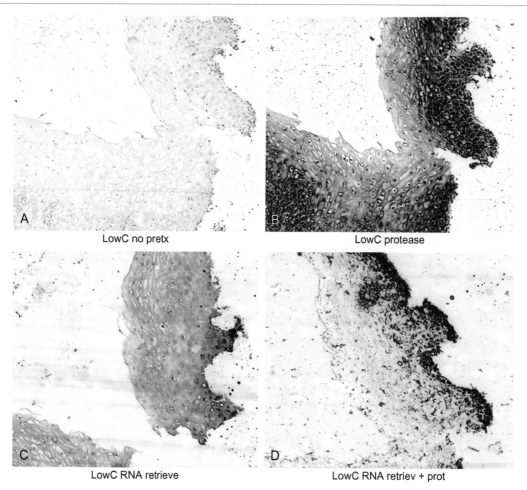

A LowC no pretx

B LowC protease

C LowC RNA retrieve

D LowC RNA retriev + prot

Figure7-1 Optimization protocol for microRNA in a cervical biopsy. Here is an example of an optimization analysis for miR-let-7c in a cervical biopsy. This microRNA was expressed in high copy number in the cervical epithelia. The low probe concentration was 0.1 pmole of probe per microliter of probe cocktail. Note that no expression was evident if no pretreatment step was done (panel A). A strong signal was evident in the epithelial cells if protease digestion was done prior to the in situ hybridization (panel B). A much weaker signal was seen if RNA retrieval was done in place of the protease digestion (panel C). Although a good signal is seen with RNA retrieval and protease digestion (panel D), these pretreatment conditions caused overdigestion with a concomitant loss of the morphologic detail of the tissue.

Please review the extensive data in Chapters 4 and 5, because those chapters remind you how the three-dimensional macromolecule cross-linked network varies from DNA to DNA to RNA to RNA target in formalin-fixed, paraffin-embedded tissues. I realize I am sounding like a broken record again, but unless you optimize the exposure of the specific DNA or RNA molecule with regards to its unique three-dimensional macromolecule cross-linked cage, you will *not* get an optimal signal and may get no signal at all. And, remember, the conditions optimal for one target in a given tissue with in situ hybridization may be *different* for another RNA or DNA target in the same tissue!

The second reason for no signal is quite rare. It is that most investigators have already determined by PCR, qPCR, qRTPCR, or some other method the relative abundance of the nucleic acid in the tissue they are testing.

The third reason is very common. Formalin-fixed, paraffin-embedded tissues that are greater than 5 years old may well have a reasonable copy number of the target of interest. However, the in situ hybridization signal will be negative because the three-dimensional macromolecule cross-linked cage around the target has degenerated too much. As we have discussed at length, this can easily be corrected by the regeneration of that cage.

In my own work, if the tissue block is at least 5 years old, I typically "rejuvenate" the three-dimensional macromolecule cross-linked cage before doing the in situ hybridization experiments. Ultimately, this saves a lot of time and, of course, may be the only way to get useful data.

INTERPRETATION OF DATA

I recommend being rather ritualistic when interpreting the data from the optimizing experiments. For example, you may wish to draw a flow sheet that includes the following:

Probe Concentration	Pretreatment Condition	Signal/ Background
LOW	NONE	
LOW	Protease	
LOW	DNAr etrieval	
LOW	DNAr etrieval + protease	
HIGH	NONE	
HIGH	Protease	
HIGH	DNAr etrieval	
HIGH	DNAr etrieval + protease	

Figure 7-1 shows an example of such an optimizing analysis. Here, the probe was miR-let-7c. The tissue was cervical epithelia. miR-let-7c should be present in high copy number in the squamous epithelia, especially in the more basal aspect of the stratified squamous epithelia. The low probe concentration was 0.1 pmole of probe per microliter of probe cocktail. The data from Figure 7-1 are tabulated below; the high concentration probe (1 pmole/microliter) gave similar results. Thus, for this tissue, where the focus was the relationship of miRNA expression and HPV infection of the cervix, the optimal condition for the detection of miR-let-7c was as follows:

Probe concentration: 0.1 pmole/µl
Pretreatment: Protease digestion

Probe Concentration	Pretreatment Condition	Signal/ Background
LOW	NONE	0/0
LOW	Protease	3/0
LOW	DNAr etrieval	1/0
LOW	DNAr etrieval + protease	3/3 (overdigestion)
HIGH	NONE	0/0
HIGH	Protease	3/weak
HIGH	DNAr etrieval	1.5/0
HIGH	DNAr etrieval + protease	3/3 (overdigestion)

Finally, brief mention should be made of the negative controls we can use with in situ hybridization. Of course, the standard negative control is to omit the probe. This is a useful negative control, because it will tell us if the pretreatment conditions or any part of the detection system is causing background. However, omitting the probe does not address whether the probe per se may be causing background. There are three other negative controls that would address the issue of probe-related background as well as "detection system background":

1. Use a sense probe (for RNA analysis)
2. Use a scrambled or irrelevant sequence probe
3. Use a tissue in which the target is definitely not present

Figure 7-2 shows some examples of these negative controls. Panel A shows the result for in situ hybridization examining a noncoding ultraconserved RNA sequence. Note the strong signal. If the probe was made in the same sequence as the ultraconserved RNA sequence (sense probe), then the signal was eliminated (panel B). Panels C and D show the use of a sample (in this case, cancer cell line) in which we are certain that the cells do not contain the target of interest, in this case reovirus. When this noninfected cell line was analyzed for reovirus RNA using an LNA probe at 10× the normal concentration, background was evident (panel C). This background was eliminated if the probe concentration was brought to 1× (0.1 pmole/microliter). You will recall that we have examined many examples of scrambled probes for microRNA in situ hybridization. These probes have the same number of nucleotides as the miRNA-specific probe, and the same numbers of ATG and C, but the nucleotides are put in a random order, or "scrambled." Of course, Exiqon also makes sure that the scrambled sequence does not have any known homology with any other 20-nucleotide sequence. These scrambled probes are very useful and, since their sequence does not have any known homology with any other 20-nucleotide sequence, it can also serve as an "irrelevant" sequence probe in any microRNA in situ hybridization reaction.

Another useful negative control is to use a tissue sample where you are certain that the target sequence is not present. Perhaps the best example of such a negative control comes from the so-called knockout experiments. In these experiments, which typically involve mice, the genetic makeup of a mouse is manipulated at birth such that a specific sequence or sequences are omitted. Thus, the resultant mice progeny do not have the corresponding DNA, RNA, or protein. An example of such an experiment is seen in Figure 7-3. The mice were "knockout mice" for microRNAs 221 and 222. In the wild-type mouse, you see a strong cytoplasmic-based signal in the epithelia of the mouse breast when using a probe for MiR-221 (panel A). Note that the signal is completely eliminated when analyzing the same probe under the same reaction conditions as breast tissue from a MiR-221/222 knockout mouse (panel B).

The suggested readings [1–157] provide a broad range of different protocols for doing in situ hybridization. I intentionally included papers in which both in situ hybridization and immunohistochemistry were done to show the overlap present in such protocols as used by many different investigators.

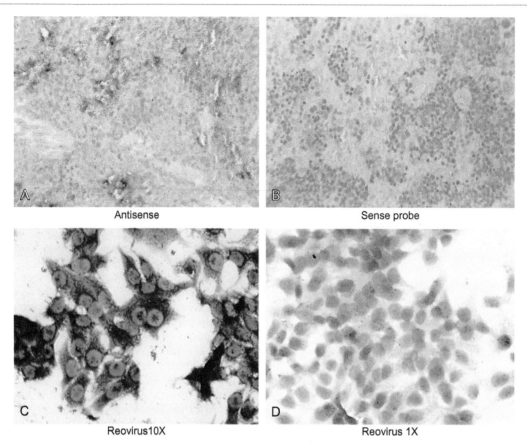

Antisense

Sense probe

Reovirus10X

Reovirus 1X

Figure7-2   Negative control probes for in situ hybridization. This figure shows some examples of these negative controls. Panel A shows the result for in situ hybridization examining a noncoding ultraconserved RNA sequence with a 5-digoxigenin-tagged LNA probe whose sequence was complementary to this RNA sequence (antisense probe). Note the strong signal. If the tagged LNA probe was made in the same sequence as the ultraconserved RNA sequence (sense probe), then the signal was eliminated (panel B). Panels C and D show the use of a sample (in this case, cancer cell line) in which we are certain that the cells do not contain the target of interest, in this case reovirus. When this noninfected cell line was analyzed for reovirus RNA using a LNA probe at 10× the normal concentration, background was evident (panel C). This background was eliminated if the probe concentration was reduced to 1× (0.1 pmole/microliter), panel D.

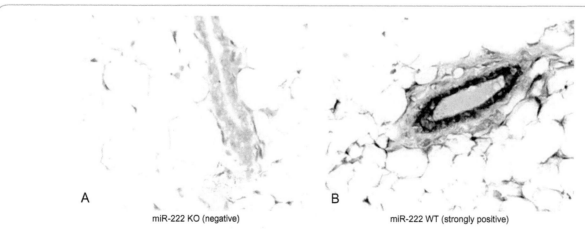

A   B

miR-222 KO (negative)   miR-222 WT (strongly positive)

Figure7-3   These mice were "knockout mice" for microRNAs 221 and 222. In the wild-type mouse, you see a strong cytoplasmic-based signal in the epithelia of the mouse breast when using a probe for MiR-221 (panel A). Note that the signal is completely eliminated when analyzing the same probe under the same reaction conditions as breast tissue from a MiR-221/222 knockout mouse (panel B).

SUGGESTED READINGS

Different published methods on in situ hybridization/immunohistochemistry

[1] Ali AM, Dawson SJ, Blows FM, et al. Comparison of methods for handling missing data on immunohistochemical markers in survival analysis of breast cancer. Br J Cancer 2011;104:693–9.

[2] Angelucci A, Clasca F, Sur M. Anterograde axonal tracing with the subunit B of cholera toxin: a highly sensitive immunohistochemical protocol for revealing fine axonal morphology in adult and neonatal brains. J Neurosci Methods 1996;65:101–12.

[3] Angerer LM, Angerer RC. Localization of mRNAs by in situ hybridization. Methods Cell Biol 1991;35:37–71.

[4] Babic A, Loftin IR, Stanislaw S, et al. The impact of pre-analytical processing on staining quality for H&E, dual hapten, dual color in situ hybridization and fluorescent in situ hybridization assays. Methods 2010;52:287–300.

[5] Bagasra O, Harris T. In situ PCR protocols. Methods Mol Biol 2006;334:61–78.

[6] Ballester M, Galindo-Cardiel I, Gallardo C, et al. Intranuclear detection of African swine fever virus DNA in several cell types from formalin-fixed and paraffin-embedded tissues using a new in situ hybridisation protocol. J Virol Methods 2010;168:38–43.

[7] Barnes DM, Harris WH, Smith P, Millis RR, Rubens RD. Immunohistochemical determination of oestrogen receptor: comparison of different methods of assessment of staining and correlation with clinical outcome of breast cancer patients. Br J Cancer 1996;74:1445–51.

[8] Bashir RM, Harris NL, Hochberg FH, Singer RM. Detection of Epstein-Barr virus in CNS lymphomas by in-situ hybridization. Neurology 1989;39:813–7.

[9] Begum F, Zhu W, Namake M, Frost E. A novel decalcification method for adult rodent bone for histological analysis of peripheral-central nervous system connections. J Neurosci Methods 2010;187:59–66.

[10] Beiske K, Myklebust AT, Aamdal S, et al. Detection of bone marrow metastases in small cell lung cancer patients. Comparison of immunologic and morphologic methods. Am J Pathol 1992;141:531–8.

[11] Bhatt B, Sahinoglu T, Stevens C. Nonisotopic in situ hybridization. Gene mapping and cytogenetics. Methods Mol Biol 1998;80:405–17.

[12] Boorsma DM, Nieboer C, Kalsbeek GL. Cutaneous immunohistochemistry. The direct immunoperoxidase and immunoglobulin-enzyme bridge methods compared with the immunofluorescence method in dermatology. J Cutan Pathol 1975;2:294–301.

[13] Borromeo V, Gaggioli D, Berrini A, Secchi C. Monoclonal antibodies as a probe for the unfolding of porcine growth hormone. J Immunol Methods 2003;272:107–15.

[14] Borzi RM, Piacentini A, Monaco MC, et al. A fluorescent in situ hybridization method in flow cytometry to detect HIV-1 specific RNA. J Immunol Methods 1996;193:167–76.

[15] Broitman-Maduro G, Maduro MF. In situ hybridization of embryos with antisense RNA probes. Methods Cell Biol 2011;106:253–70.

[16] Brown HR, Goller NL, Rudelli RD, Dymecki J, Wisniewski HM. Postmortem detection of measles virus in non-neural tissues in subacute sclerosing panencephalitis. Ann Neurol 1989;26:263–8.

[17] Calábria LK, Teixeira RR, Coelho Gonçalves SM, et al. Comparative analysis of two immunohistochemical methods for antigen retrieval in the optical lobe of the honeybee Apis mellifera: myosin-V assay. Biol Res 2010;43:7–12.

[18] Calmels TP, Mazurais D. In situ hybridization: a technique to study localization of cardiac gene expression. Methods Mol Biol 2007;366:159–80.

[19] Cardoso TC, Gomes DE, Ferrari HF, et al. A novel in situ polymerase chain reaction hybridisation assay for the direct detection of bovine herpesvirus type 3 in formalin-fixed, paraffin-embedded tissues. J Virol Methods 2010;163:509–12.

[20] Carr EL, Eales K, Soddell J, Seviour RJ. Improved permeabilization protocols for fluorescence in situ hybridization (FISH) of mycolic-acid-containing bacteria found in foams. J Microbiol Methods 2005;61:47–54.

[21] Cepica A, Yarson C, Ralling G. The use of ELISA for detection of the antibody-induced conformational change in a viral protein and its intermolecular spread. J Virol Methods 1990;28:1–13.

[22] Chaichana KL, Guerrero-Cazares H, Capilla-Gonzalez V, et al. Intra-operatively obtained human tissue: protocols and techniques for the study of neural stem cells. J Neurosci Methods 2009;180:116–25.

[23] Christensen BB, Sternberg C, Andersen JB, et al. Molecular tools for study of biofilm physiology. Methods Enzymol 1999;310:20–42.

[24] Chuang SS, Li CY. Useful panel of antibodies for the classification of acute leukemia by immunohistochemical methods in bone marrow trephine biopsy specimens. Am J Clin Pathol 1997;107:410–8.

[25] Cmarko D, Koberna K. Electron microscopy in situ hybridization: tracking of DNA and RNA sequences at high resolution. Methods Mol Biol 2007;369:213–18.

[26] Cordes H, Bergstrom AL, Ohm J, Laursen H, Heegaard PM. Characterization of new monoclonal antibodies reacting with prions from both human and animal brain tissues. J Immunol Methods 2008;337:106–20.

[27] Cudahy T, Boeryd B, Fanlund B, Nordenskjold B. A comparison of three different methods for the determination of estrogen receptors in human breast cancer. Am J Clin Pathol 1988;90:588–90.

[28] De Negri F, Campani D, Sarnelli R, et al. Comparison of monoclonal immunocytochemical and immunoenzymatic methods for steroid receptor evaluation in breast cancer. Am J Clin Pathol 1991;96:53–8.

[29] Delage DM, Harris WH, Smith P, Millis RR, Rubens RD. Immunohistochemical determination of oestrogen receptor: comparison of different methods of assessment of staining and correlation with clinical outcome of breast cancer patients. Br J Cancer 1996;74:1445–51.

[30] Del Valle L, White MK, Enam S, et al. Detection of JC virus DNA sequences and expression of viral T antigen and agno-protein in esophageal carcinoma. Cancer 2005;103:516–27.

[31] Delord B, Poveda JD, Astier-Gin T, Gerbaud S, Fleury HJ. Detection of the bunyavirus Germiston in VERO and Aedes albopictus C6/36 cells by in situ hybridization using cDNA and asymmetric RNA probes. J Virol Methods 1989;24:253–64.

[32] Eichele G, Diez-Roux G. High-throughput analysis of gene expression on tissue sections by in situ hybridization. Methods 2011;53:417–23.

[33] Ermert L, Hocke AC, Duncker HR, Seeger W, Ermert M. Comparison of different detection methods in quantitative microdensitometry. Am J Pathol 2001;158:407–17.

[34] Fallert BA, Reinhart T. Improved detection of simian immunodeficiency virus RNA by in situ hybridization in fixed tissue sections: combined effects of temperatures for tissue fixation and probe hybridization. J Virol Methods 2002;99:23–32.

[35] Farr AG, Nakane PK. Immunohistochemistry with enzyme labeled antibodies: a brief review. J Immunol Methods 1981;47:129–44.

[36] Faruharson M, Harvie R, Kennedy A, McNicol A. Detection of mRNA by in-situ hybridization and in northern blot analyses using oligodeoxynucleotide probes labeled with alkaline phosphate. JClin P athol 1992;45:999–1002.

[37] Fenaux M, Halbur P, Haqshenas G,e tal. Cloned genomic DNA of type 2 porcine circovirus is infectious when injected directly into the liver and lymph nodes of pigs characterization of clinical disease, virus distribution, and pathologic lesions. JVir ol 2002;76:541–51.

[38] Fend F, Hittmair A, Rogatsch H,e tal. Seminomas positive for Epstein-Barr virus by the polymerase chain reaction: viral RNA transcripts (Epstein-Barr-encoded small RNAs) are present in intratumoral lymphocytes but absent from the neoplastic cells. ModP athol 1995;8:622–5.

[39] Fernandez-Flores A. Ar eviewo famylo id staining: methods and artifacts. Biotech Histochem 2011;86:293–301.

[40] Filion G, Laflamme C, Turgeon N, Ho J, Duchaine C. Permeabilizationan d hybridization protocols for rapid detection of Bacillus spores using fluorescence in situ hybridization. JM icrobiolM ethods 2009;77:29–36.

[41] Fisher CJ, Gillett CE, Vojtesek B, Barnes DM, Millis RR. Problemswit hp 53 immunohistochemical staining: the effect of fixation and variation in the methods of evaluation. BrJ Can cer 1994;69:26–31.

[42] Fisher E, Anderson S, Dean S,e tal. Solving the dilemma of the immunohistochemical and other methods used for scoring estrogen receptor and progesterone receptor in patients with invasive breast carcinoma. Cancer 2005;103:164–73.

[43] Fong T, DiB isceglie A, Gerber M, Waggoner J, Hoofnagle J. Persistenceo fh epatitisB virus DNA in the liver after loss of HBsAg in chronic hepatitis B. Hepatology 1993;18:1313–8.

[44] Ford C, Faedo M, Rawlinson W. Mouse mammary tumor virus-like RNA transcripts

and DNA are found in affected cells of human breast cancer. Clin Cancer Res 2004;10:7284–9.

[45] Forkert PG. CYP2E1 is preferentially expressed in Clara cells of murine lung: localization by in situ hybridization and immunohistochemical methods. Am J Respir Cell Mol Biol 1995;12:589–96.

[46] Forkert PG, Jackson AC, Parkinson A, Chen S. Diminished expression of CYP1A1 in urethane-induced lung tumors in strain A/J mice: analysis by in situ hybridization and immunohistochemical methods. Am J Respir Cell Mol Biol 1996;14:444–53.

[47] Friedman M, Gentile P, Tarectecan A, Bluchs A. Malignant mesothelioma: immunohistochemistry and DNA ploidy analysis as methods to differentiate mesothelioma from benign reactive mesothelial cell proliferation and adenocarcinoma in pleural and peritoneal effusions. Arch Pathol Lab Med 1996;120:959–66.

[48] Fukuda A, Kumazaki M, Onizuka K. Labeling and identification of living donor cells in brain slices of recipient hemiparkinsonian model rats for physiological recordings: methods for physiological assessments of neural transplantation. Exp Neurol 1996;137:309–17.

[49] Gelmetti D, Grieco V, Rossi C, Capucci L, Lavazza A. Detection of rabbit haemorrhagic disease virus (RHDV) by in situ hybridisation with a digoxigenin labelled RNA probe. J Virol Methods 1998;72:219–26.

[50] Giltane JM, Molinaro A, Cheng H, et al. Comparison of quantitative immunofluorescence with conventional methods for HER2/neu testing with respect to response to trastuzumab therapy in metastatic breast cancer. Arch Pathol Lab Med 2008;132:1635–47.

[51] Gomes SA, Nascimento JP, Siqueira MM, et al. In situ hybridization with biotinylated DNA probes: a rapid diagnostic test for adenovirus upper respiratory infection. J Virol Methods 1985;12:105–10.

[52] Gonzalez-Melendi P, Testillano PS, Mena CG, et al. Histones and DNA ultrastructural distribution in plant cell nucleus: a combination of immunogold and cytochemical methods. Exp Cell Res 1998;242:45–59.

[53] Gosden JR. In: PRINS and in situ PCR protocols. Totowa, N.J.: Humana Press; 1996.

[54] Grantzdorffer I, Yumlu S, Gioeva Z, et al. Comparison of different tissue sampling methods for protein extraction from formalin-fixed and paraffin-embedded tissue specimens. Exp Mol Pathol 2010;88:190–6.

[55] Gruver AM, Peerwani Z, Tubbs RR. Out of the darkness and into the light: bright field in situ hybridisation for delineation of ERBB2 (HER2) status in breast carcinoma. J Clin Pathol 2010;63:210–9.

[56] Groben R, Medlin L. In situ hybridization of phytoplankton using fluorescently labeled rRNA probes. Methods Enzymol 2005:299–310.

[57] Gulley ML, Eagan PA, Quintanilla-Martinez L, et al. Epstein-Barr virus DNA is abundant and monoclonal in the Reed-Sternberg cells of Hodgkin's disease: association with mixed cellularity subtype and Hispanic American ethnicity. Blood 1994;83:1595–602.

[58] Gutstein HB. In situ hybridization in neural tissues. Methods Mol Med 2003:95–105.

[59] Habeeb AF. Comparative studies on radiolabeling of lysozyme by iodination and reductive methylation. J Immunol Methods 1983;65:27–39.

[60] Hagiwara T, Hattori J, Kaneda T. PNA-in situ hybridization method for detection of HIV-1 DNA in virus-infected cells and subsequent detection of cellular and viral proteins. Methods Mol Biol 2006;326:139–49.

[61] Halbhuber KJ, Gossrau R, Moller U, et al. The cerium perhydroxide-diaminobenzidine (Ce-H202-DAB) procedure. New methods for light microscopic phosphatase histochemistry and immunohistochemistry. Histochemistry 1988;90:289–97.

[62] Hannah MJ, Weiss U, Huttner WB. Differential extraction of proteins from paraformaldehyde-fixed cells: lessons from synaptophysin and other membrane proteins. Methods 1998;16:170–81.

[63] Hasson KW, Hasson J, Aubert H, Redman RM, Lightner DV. A new RNA-friendly fixative for the preservation of penaeid shrimp samples for virological detection using cDNA genomic probes. J Virol Methods 1997;66:227–36.

[64] Henke RT, Eun Kim S, Maitra A, Paik S, Wellstein A. Expression analysis of mRNA in formalin-fixed, paraffin-embedded archival tissues by mRNA in situ hybridization. Methods 2006;38:253–62.

[65] Hirsch VM, Zack PM, Vogel AP, Johnson PR. Simian immunodeficiency virus infection of macaques: end-stage disease is characterized by widespread distribution of proviral DNA in tissues. J Infect Dis 1991;163:976–88.

[66] Hoefler H, Childers H, Montminy MR, et al. In situ hybridization methods for the detection of somatostatin mRNA in tissue sections using antisense RNA Probes. Histochem J 1986;18:597–604.

[67] Jackson JB. Detection and quantitation of human immunodeficiency virus type 1 using molecular DNA/RNA technology. Arch Pathol Lab Med 1993;117:473–7.

[68] Jeppesen P. Immunofluorescence techniques applied to mitotic chromosome preparations. Methods Mol Biol 1994:253–85.

[69] Jilbert A. In situ hybridization protocols for detection of viral DNA using radioactive and nonradioactive DNA probes. Methods Mol Biol 2000;123:177–93.

[70] Jiwa M, Steenbergen RD, Zwaan FE, et al. Three sensitive methods for the detection of cytomegalovirus in lung tissue of patients with interstitial pneumonitis. Am J Clin Pathol 1990;93:491–4.

[71] Jones RG, Liu Y, Rigsby P, Sesardic D. An improved method for development of toxoid vaccines and antitoxins. J Immunol Methods 2008;337:42–8.

[72] Jubb AM, Pham TQ, Frantz GD, Peale Jr. FV, Hillan KJ. Quantitative in situ hybridization of tissue microarrays. Methods Mol Biol 2006;326:255–64.

[73] Kadkol SS, Gage WR, Pasternack GR. In situ hybridization—theory and practice. Mol Diagn 1999;4:169–83.

[74] Kerman K, Vestergaard M, Tamiya E. Electrochemical DNA biosensors: protocols for intercalator-based detection of hybridization in solution and at the surface. Methods Mol Biol 2009;504:99–113.

[75] Kasai T, Ikeda H, Tomaru U, et al. A rat model of human T lymphocyte virus type I (HTLV-I) infection: in situ detection of HTLV-I provirus DNA in microglia/macrophages in affected spinal cords of rats with HTLV-I-induced chronic progressive myeloneuropathy. Acta Neuropathol 1999;97:107–12.

[76] Khalidi HS, Lones MA, Zhou Y, Weiss LM, Medeiros LJ. Detection of Epstein-Barr virus in the L & H cells of nodular lymphocyte predominance Hodgkin's disease: report of a case documented by immunohistochemical, in situ hybridization, and polymerase chain reaction methods. Am J Clin Pathol 1997;108:687–92.

[77] Kim DG. Differentially expressed genes associated with hepatitis B virus HBx and MHBs protein function in hepatocellular carcinoma. Methods Mol Biol 2006;317:141–55.

[78] Kim J, Chae C. Differentiation of porcine circovirus 1 and 2 in formalin-fixed, paraffin-wax-embedded tissues from pigs with postweaning multisystemic wasting syndrome by in-situ hybridization. Res Vet Sci 2001;70:265–9.

[79] Kim J, Chae C. Optimized protocols for the detection of porcine circovirus 2 DNA from formalin-fixed paraffin-embedded tissues using nested polymerase chain reaction and comparison of nested PCR with in situ hybridization. JVir olM ethods 2001;92:105–11.

[80] Klinger MH, Kammerer R, Hornei B, Gauss-Muller V. Perinuclearac cumulationo f hepatitis A virus proteins, RNA and particles and ultrastructural alterations in infected cells. ArchVir ol 2001;146:2291–307.

[81] Kohlberger PD, Obermair A, Sliutz G,e tal. Quantitative immunohistochemistry of factor VIII-related antigen in breast carcinoma: a comparison of computer-assisted image analysis with established counting methods. AmJ Clin P athol 1996;105:705–10.

[82] Kristensen T, Engvad B, Nielsen O,e tal. Vacuum sealing and cooling as methods to preserve surgical specimens. Appl Immunohistochem Mol Morphol 2011;19:460–9.

[83] Krywko DM, Gomez HF. Detectiono f Loxosceles species venom in dermal lesions: a comparison of 4 venom recovery methods. AnnE mergM ed 2002;39:475–80.

[84] Krynska B, DelValle L, Croul S,e tal. Detection of human neurotropic JC virus DNA sequence and expression of the viral oncogenic protein in pediatric medulloblastomas. ProcNat lAc adS ciUS A 1999;96:11519–24.

[85] Lanciego JT, Goede PH, Witter MP, Wouterlood FG. Useo fp eroxidaseu bstrate Vector VIP for multiple staining in light microscopy. JNe urosciM ethods 1997;74:1–7.

[86] Lebo RV, Su Y. Positionalc loningan d multicolor in situ hybridization. Principles and protocols. MethodsM olB iol 1994;33:409–38.

[87] Lecuyer E, Parthasarathy N, Krause HM. Fluorescent in situ hybridization protocols in Drosophila embryos and tissues. MethodsM ol Biol 2008;420:289–302.

[88] Leong AS, Leong TY. Standardization in immunohistology. MethodsM olB iol 2011;724:37–68.

[89] Liu CQ, Shan L, Balesar R,e tal. A quantitative in situ hybridization protocol for formalin-fixed paraffin-embedded archival

post-mortem human brain tissue. Methods 2010;52:359Ð66.

[90] Lossi L, Gambino G, Salio C, Merighi A. Direct in situ rt-PCR. Methods Mol Biol 2011;789:111Ð26.

[91] Mahad DJ, Ziabreva I, Campbell G, et al. Detection of cytochrome c oxidase activity and mitochondrial proteins in single cells. J Neurosci Methods 2009;184:310Ð9.

[92] Marks DL, Wiemann JN, Burton KA, et al. Simultaneous visualization of two cellular mRNA species in individual neurons by use of a new double in situ hybridization method. Mol Cell Neurosci 1992;3:395Ð405.

[93] Martinez F, Kiriakdou M, Strauss III JF. Structural and functional changes in mitochondria associated with trophoblast differentiation: methods to isolate enriched preparations of syncytiotrophoblast mitochondria. Endocrinology 1997;138:2172Ð 83.

[94] Maygarden SJ, Strom S, Ware JL. Localization of epidermal growth factor receptor by immunohistochemical methods in human prostatic carcinoma, prostatic intraepithelial neoplasia, and benign hyperplasia. Arch Pathol Lab Med 1992;116:269Ð73.

[95] Meyer PN, Fu K, Greiner TC, et al. Immunohistochemical methods for predicting cell of origin and survival in patients with diffuse large B-cell lymphoma treated with rituximab. J Clin Oncol 2011;29:200Ð7.

[96] Mikulov‡ V, Kološtov‡ K, Zima T. Methods for detection of circulating tumor cells and their clinical value in cancer patients. Folia Biol 2011;57:151Ð61.

[97] Miyanaga T, Hirato J, Nakazato Y. Ampliﬁcation of the epidermal growth factor receptor gene in glioblastoma: an analysis of the relationship between genotype and phenotype by CISH method. Neuropathology 2008;28:116Ð26.

[98] Moench TR, Gendelman HE, Clements JE, Narayan O, Grifﬁn DE. Efﬁciency of in situ hybridization as a function of probe size and ﬁxation technique. J Virol Methods 1985;11:119Ð30.

[99] Mogensen J, Pedersen S, Hindkjaer J, Kolvraa S, Bolund L. Primed in situ (PRINS) labeling of RNA. Methods Mol Biol 1994;33:265Ð75.

[100] Mojsilovic-Petrovic J, Nesic M, Pen A, Zhang W, Stanimirovic D. Development of rapid staining protocols for laser-capture microdissection of brain vessels from human and rat coupled to gene expression analyses. J Neurosci Methods 2004;133:39Ð48.

[101] Monet-Tschudi F, Defaux A, Braissant O, Cagnon L, Zurich MG. Methods to assess neuroinßammation. Curr Protoc Toxicol 2011; Chapter 12, Unit 12.19.

[102] Morasso MI. Detection of gene expression in embryonic tissues and stratiﬁed epidermis by in situ hybridization. Methods Mol Biol 2010;585:253Ð60.

[103] Morrison HW, Downs CA. Immunological methods for nursing research: from cells to systems. Biol Res Nurs 2011;13:227Ð34.

[104] Murai N, Murakami Y, Matsufuji S. Protocols for studying antizyme expression and function. Methods Mol Biol 2011;720:237Ð67.

[105] Mutch DG, Soper JT, Budwit-Novotny DA, et al. Endometrial adenocarcinoma estrogen receptor content: association of clinicopathologic features with

immunohistochemical analysis compared with standard biochemical methods. American Journal of Obstetrics and Gynecology 1987;157:924Ð31.

[106] Myoken Y, Sugata T, Mikami Y, Murayama SY, Fujita Y. Identiﬁcation of Aspergillus species in oval tissue samples of patients with hematologic malignancies by in situ hybridizationÑa preliminary report. J Oral Maxillofac Surg 2008;66:1905Ð12.

[107] Nakamura RM. Overview and principles of in-situ hybridization. Clin Biochem 1990;23:255Ð 9.

[108] Narayanswami S, Dvorkin N, Hamkalo BA. Nucleic acid sequence localization by electron microscopic in situ hybridization. Methods Cell Biol 1991;35:109Ð32.

[109] Nassel DR, Ekstrom P. Detection of neuropeptides by immunocytochemistry. Methods Mol Biol 1997;72:71Ð101.

[110] Navarro B, Daros JA, Flores R. Reverse transcription polymerase chain reaction protocols for cloning small circular RNAs. J Virol Methods 1998;73:1Ð9.

[111] Neely LA, Patel S, Garver J, et al. A single-molecule method for the quantitation of microRNA gene expression. Nat Methods 2006;3:41Ð6.

[112] Newman SJ, Gentleman SM. Microwave antigen retrieval in formaldehyde-ﬁxed human brain tissue. Methods Mol Biol 1997;72:145Ð 52.

[113] Niehusman P, Mittelstaedt T, Bien CG, et al. Presence of human herpes virus 6 DNA exclusively in temporal lobe epilepsy brain tissue of patients with history of encephalitis. Epilepsia 2010;51:2478Ð83.

[114] Nina Hornickel I, Kacza J, Scnapper A, et al. Demonstration of substances of innate immunity in the esophageal epithelium of domesticated mammals. Part IÑMethods and evaluation of comparative ﬁxation. Acta Histochem 2011;113:163Ð74.

[115] Nuovo GJ. In situ PCR: protocols and applications. PCR Methods Appl 1995;4:S151Ð167.

[116] Nuovo GJ. In: PCR in situ hybridization: protocols and applications, 3rd ed. Philadelphia, PA: Lippincott-Raven Publishers; 1996.

[117] Nuovo GJ. In situ detection of precursor and mature microRNAs in paraﬁn embedded, formalin ﬁxed tissues and cell preparations. Methods 2008;44:39Ð46.

[118] Nykanen M, Kuopio T. Protein and gene expression of estrogen receptor alpha and nuclear morphology of two breast cancer cell lines after different ﬁxation methods. Exp Mol Pathol 2010;88:265Ð71.

[119] Oliver C, Jamur MC. Immunocytochemical methods and protocols. Methods Mol Biol 2010;588:ivÐv.

[120] Ota M, Fukushima H, Akamatsu T, et al. Availability of immunostaining methods for identiﬁcation of mixed-up tissue specimens. Am J Clin Pathol 1989;92:665Ð9.

[121] Palop JJ, Roberson ED, Cobos I. Step-by-step in situ hybridization method for localizing gene expression changes in the brain. Methods Mol Biol 2011;670:207Ð30.

[122] Papadopoulos GC, Dori I. DiI labeling combined with conventional immunocytochemical techniques for correlated light and electron microscopic studies. J Neurosci Methods 1993;46:251Ð8.

[123] Perez-Martin E, Rovira A, Calsamiglia M, et al. A new method to identify cell types that support porcine circovirus type 2 replication in formalin-ﬁxed, paraﬁn-embedded swine tissues. J Virol Methods 2007;146:86Ð95.

[124] Petersen KH. Novel horseradish peroxidase substrates for use in immunohistochemistry. J Immunol Methods 2009;340:86Ð9.

[125] Politch JA, Wolff H, Hill JA, Anderson DJ. Comparison of methods to enumerate white blood cells in semen. Fertil Steril 1993;60:372Ð 5.

[126] Porteous DJ. Protocols for chromosome-mediated gene transfer. Selection strategies, transgenome analysis, enrichment cloning, and mapping. Methods Mol Biol 1994;353Ð78.

[127] Presti B, Weidner N. Granulomatous prostatitis and poorly differentiated prostate carcinoma. Their distinction with the use of immunohistochemical methods. Am J Clin Pathol 1991;95(3):330Ð4.

[128] Redi C. Signal transduction immunohistochemistryÑ methods and protocols. Eur J Histochem 2011;5 ejh.2011. br2017.

[129] Rognum TO, Brandtzaeg P, Orjasaeter H, Fausa O. Immunohistochemistryo fe pithelial cell markers in normal and pathological colon mucosa. Comparison of results based on routine formalin- and cold ethanol-ﬁxation methods. Histochemistry 1980;67:7Ð21.

[130] Sandinha MT, Farquharson MA, Roberts F. Identiﬁcation of monosomy 3 in choroidal melanoma by chromosome in situ hybridisation. BrJ Op hthalmol 2004;88:1527Ð 32.

[131] Sansonno D, Lauletta G, Montrone M, et al. Hepatitis C virus RNA and core protein in kidney glomerular and tubular structures isolated with laser capture microdissection. ClinE xpI mmunol 2005;140:498Ð506.

[132] Santangelo KS, Baker SA, Nuovo G, et al. Detectable reporter gene expression following transduction of adenovirus and adeno-associated virus serotype 2 vectors within full-thickness osteoarthritic and unaffected canine cartilage in vitro and unaffected guinea pig cartilage in vivo. JOr thopRe s 2010;28:149Ð 55.

[133] Sato T, Terui T, Kogawa K, et al. A case of true malignant histiocytosis: identiﬁcation of histiocytic origin with use of immunohistochemical and immunocytogenetic methods. AnnHe matol 2002;81:285Ð8.

[134] Sato Y, Shimonohara N, Hanaki K, et al. ImmunoAT method: an initial assessment for the detection of abnormal isoforms of prion protein in formalin-ﬁxed and paraﬁn-embeddedt issues. JVir olM ethods 2010;146:86Ð95.

[135] Scorisa JM, Duobles T, Oliveira GP, Macimino Ther eviewo f theme thods to obtain no-neuronal cells to study glial inßuence on Amyotrophic Lateral Sclerosis pathophysiology at molecular level in vitro. ActaCir B ras 2010;25:281Ð9.

[136] Schmidtmayerova H, Alfano M, Nuovo G, Bukrinsky M. Humanim munodeﬁciency virus type 1T-lymphotropic strains enter macrophages via a CD4- and CXCR4-mediated pathway: replication is restricted at a postentry level. JVir ol 1998;72:4633Ð42.

[137] Seymour L, Meyer K, Esser J, et al. Estimation of PR and ER by immunocytochemistry in breast cancer. Comparison with radioligand

bindingme thods. AmJ Clin P athol 1990;94:S35–40.

[138] Shi SR, Prince JB, Jones CM, Kalra KL, Tandon AK. Useo fmo noclonalan tibodies inimmu nohistochemistry. MethodsM olB iol 1995;89–108.

[139] Smith MD, Ahern M, Coleman M. The use of combined immunohistochemical labeling and in situ hybridization to colocalize mRNA and proteinin tissu ese ctions. MethodsM olB iol 2006;326:235–45.

[140] Sollberg S, Peltonen J, Uitto J. Combined use of in situ hybridization and unlabeled antibody peroxidase anti-peroxidase methods: simultaneous detection of type I procollagen mRNAs and factor VIII-related antigene pitopesin k eloidtissu e. LabIn vest 1991;64:125–9.

[141] Sprickett CM, Wiswedel I, Siems W, Zarkovic K, Zarkovic N. Advancesin me thodsfo r the determination of biologically relevant lipidp eroxidationp roducts. FreeRad icRe s 2010;44:1172–202.

[142] Sundquist SJ, Nisenbaum LK. Fast Fos: rapid protocols for single- and double-labeling c-Fos immunohistochemistry in fresh frozen brain sections. JNe urosciM ethods 2005;141:9–20.

[143] Terenghi G, Polak JM. Preparation of tissue sectionsan dslid esfo rmRNAh ybridization. MethodsM olB iol 1994;28:187–91.

[144] Tse C, Brault D, Gligorov J, et al. Evaluation of the quantitative analytical methods real-time PCR for Her-2 gene quantification and ELISA of serum HER-2 protein and comparison with fluorescence in situ

hybridization and immunohistochemistry for determining HER-2 status in breast cancer patients. ClinCh em 2005;51:1093–101.

[145] Vaney DI. Photochromic intensification of diaminobenzidine reaction product in the presence of tetrazolium salts: applications for intracellular labelling and immunohistochemistry. JNe urosciM ethods 1992;44:217–23.

[146] Vetter DE, Mugnaini E. An evaluation of retrograde tracing methods for the identification of chemically distinct cochlear efferentn eurons. ArchI talB iol 1990;128:331–53.

[147] Vogelaar FJ, Mesker WE, Rijken AM, et al. Clinical impact of different detection methods for disseminated tumor cells in bone marrow of patients undergoing surgical resection of colorectal liver metastases: a prospective follow-ups tudy. BMCCan cer 2010;10:153.

[148] Warner CK, Whitfield SG, Fekadu M, Ho H. Proceduresf orr eproducibled etectiono f rabies virus antigen mRNA and genome in situin f ormalin-fixedt issues. JVir olM ethods 1997;67:5–12.

[149] Wasserman J, Maddox J, Racz M, Petronic-Rosic V. Updateo nimmu nohistochemical methodsr elevantt od ermatopathology. Arch PatholL abM ed 2009;133:1053–61.

[150] Whitworth IH, Dor'e C, Hall S, Green CJ, Terenghi G. Differentmu sclegr aftd enaturing methodsan dt heiru sef orn erver epair. BrJ PlastS urg 1995;48:492–9.

[151] Wrigley EC, McGown AT, Buckley H, Hall A, Crowther D. Glutathione-S-transferase

activity and isoenzyme levels measured by two methods in ovarian cancer, and their value asmar kerso fd iseaseo utcome. BrJ Can cer 1996;73:763–9.

[152] Wu W, Zhang C, Chen Z, Zhang G, Yang J. Differencesin h eatingme thodsmay account for variation in reported effects on gammaH2AXf ocusf ormation. MutatRe s 2009;676:48–53.

[153] Yan F, Wu X, Crawford M, et al. The search for an optimal DNA, RNA, and protein detection by in situ hybridization, immunohistochemistry, and solution-based methods. Methods 2010;52:281–6.

[154] Yong VC, Ong KW, Sidik SM, Rosli R, Chong PP. Amo difiedin s ituRT -PCR method for localizing fungal-specific gene expression in Candida-infected mice renal cells. JM icrobiolM ethods 2009;79:242–5.

[155] Young HM, Kunze WA, Pompolo S, Furness JB, Bornstein JC. Combinedin tracellular injection of Neurobiotin and pre-embedding immunocytochemistry using silver-intensified goldp robesin mye ntericn eurons. JNe urosci Methods 1994;51:39–45.

[156] Zheng XZ, Du LF, Wang HP. An immunohistochemical analysis of a rat model of proliferative vitreoretinopathy and a comparison of the expression of TGF-β2an d PDGFamo ngt hein ductionme thods. BosnJ BasicM edS ci 2010;10:204–9.

[157] Zhou M, Grofova I. The use of peroxidase substrate Vector VIP in electron microscopic singlean dd oublean tigenlo calization. JNe urosciM ethods 1995;62:149–58.

The Recommended Protocol for Immunohistochemistry

8

INTRODUCTION

Now that we have discussed a protocol for in situ hybridization, let's turn our attention to the equivalent protocol for immunohistochemistry. As we have discussed at length, the optimal conditions for immunohistochemistry can vary a great deal from epitope to epitope. Indeed, as we have seen, the same cells in a given formalin-fixed, paraffin-embedded tissue often show marked variation in the optimal conditions for different epitopes, even if the epitopes localize to the same exact subcellular component. We have theorized that this reflects, in part, the hypothesis that each epitope has a biochemically unique three-dimensional macromolecule cross-linked cage that requires different optimal pretreatment conditions to access. We have discussed a lot of data that support this hypothetical model. Some of the strongest supporting data for this theory, at least from an operational standpoint, comes from the observation that if you pretreat tissues with the IHCPerfect reagent, the optimal conditions for immunohistochemistry for different epitopes in a given tissue (or different tissues) tend to be the same. Thus, this gives you the option of pretreating formalin-fixed, paraffin-embedded tissues with IHCPerfect and using the same pretreatment conditions for each epitope, with a good chance of success. At this stage, such a strategy is not readily available. Thus, we discuss a protocol that should allow you to maximize a signal for at least 95% of epitopes in formalin-fixed, paraffin-embedded tissues.

So, with this in mind, let's look at a "cookbook" type of protocol I recommend for immunohistochemistry. Again, the purpose of this presentation is not to belittle such protocols. For simplicity's sake, whenever the protocols for in situ hybridization and immunohistochemistry have similar steps, I have included this information with the current immunohistochemistry protocol so that you avoid having to go back to the previous chapter.

Allow me to repeat that the company and catalog numbers listed in this chapter were chosen for one reason: I have used them for many years. The fact that I have listed many companies reflects one reason why I have no consulting agreement with any biotechnology company, as this allows me to keep my independence. I do, however, often discuss in situ hybridization and immunohistochemistry with such companies, since, of course, they all have investigators with extensive knowledge of these methodologies.

Finally, allow me to stress what I feel is a very important point. I do *not* recommend reading this chapter until you have read all the preceding chapters (with the exception of the previous chapters on histologic interpretation and the in situ hybridization protocol, respectively). Although I hope that the protocols in this chapter are helpful, if you want to get the maximum benefit from them, it would be highly advantageous to understand the theoretical basis for each step.

STEP ONE: CUT THE TISSUE SECTIONS ONTO A SILANE-COATED SLIDE

1. Confirm that the blocks are formalin-fixed, paraffin-embeddedtissu es.
2. Determine the age of the blocks. If the blocks are no more than 2 years old, proceed with the protocol. If the tissue blocks are over 7 years old, realize that the risk of getting a false-negative result *or* a much reduced signal for a high copy protein is about 75%. Consider either getting more recent equivalent biopsies or treating the tissue slides with IHCPerfect.
3. Have the histotechnologist cut successive 4 to 5 micron sections and, depending on the size of the biopsy, put the paraffin ribbons on from 5 to 20 consecutive and labeled silane-coated slides.
4. Let the water drain from the slides by placing them at an angle on a stand.
5. Put the slides in a rack and bake for 30 minutes at 60°C. *Do not place the slides directly on the hot plate.* Rather, put the slides in a rack and then put the rack in the 60°C oven.
6. Remove the slides and store at room temperature (RT) in a light-tight box or slide holder.

STEP TWO: REMOVE THE PARAFFIN FROM THE SLIDES

1. On the day you wish to begin the immunohistochemistry experiment, remove the slides from the box.
2. Incubate the slides for 5 minutes in fresh xylene (the xylene solution should be changed weekly). An

207

DOI: http://dx.doi.org/10.1016/B978-0-12-415944-0.00008-5

alternative to remove the paraffin is to use SubX; in this case, incubate the slides in this solution for 30min utes.

3. Then incubate the slides for 5 minutes in fresh 100% ethanol (the ethanol solution should also be changed weekly).

4. Air dry the slides at either RT or in a 37°C oven.

STEP THREE: OPTIMIZE THE PRIMARY ANTIBODY/CHOOSE DIFFERENT PRETREATMENT PROTOCOLS

1. The optimization of the primary presupposes that you have placed at least two sections per silane-coatedslid e.

2. Take four consecutive labeled slides (serial sections) that have been deparaffinized.

3. Check and confirm that the tissue contains the protein epitope target of interest.

4. Check and confirm that the primary antibody was generated in an animal to which you have the appropriate secondary antibody. If you are using an automated kit, most have secondary antibodies that will recognize only primary antibodies generated in either mice or rabbits.

5. Label the first slide "dilute primary antibody, no protease/protease." Label the second slide "dilute primary antibody, antigen retrieval, no protease/protease." Label the third slide "concentrated primary antibody, no protease/protease." Label the fourth slide "concentrated primary antibody, antigen retrieval, no protease/protease."

6. Dilute the primary antibody with a commercially available primary antibody diluent (e.g., Enzo Life Sciences diluent, catalog #33808; DAKO diluent buffer, catalog #90000; or Ventana primary antibody diluent, catalog #760-219).

7. For the concentrated primary antibody, I recommend the following:
   For an RTU antibody (ready to use, already diluted), use at a 1:1 dilution.
   For a primary antibody that is provided as a stock solution, use at a 1:50 dilution.

8. For the diluted primary antibody, I recommend the following:
   For an RTU antibody (ready to use, already diluted), use at a 1:10 dilution.
   For a primary antibody that is provided as a stock solution, use at a 1:500 dilution.

9. Remember to use one of the tissues on each slide for protease digestion, and do not put any reagent on the other section during the pretreatment steps.

10. Proceedto t hean tigenre trievalan dp roteases teps.

STEP FOUR: PRETREAT THE TISSUES

1. If doing the optimizing protocol, add protease (proteinase K, 1 microgm/ml, Enzo Life Sciences, catalog #33801, diluted in sterile water) to only the *bottom* tissue section of each of the four slides.

2. Incubate the tissue sections in the proteinase K solution for 5 minutes at room temperature.

3. Remove the proteinase K by washing the slides with sterile water for a few seconds. Next, rinse slides in DNase/RNase-free 100% ethanol for a few seconds. Then air dry in a 37°C incubator.

4. If you are doing the optimization protocol, take two of the four slides and incubate for 30 minutes at 95°C in any commercially available "antigen retrieval solution."

5. After the 30 minutes, let the solution cool down for 5 minutes at room temperature. Then remove the slides and let air dry.

6. At this stage, if doing the optimizing protocol, you have four slides each with at least two sections. The bottom section for each has been digested with protease. Two of the four slides have been exposed to 95°C for 30 minutes in a DNA/RNA retrieval solution.

7. If you are using a DAB-based system, the slides should be incubated at this stage in a solution of 3% H_2O_2 for 10 minutes at RT.

STEP FIVE: ADD THE PRIMARY ANTIBODY

1. For small biopsies (less than 1cm in size), manually add 5–10 microliters of the primary antibody mixture (*concentrated* form) to two of the slides (one no protease/protease and the other antigen retrieval with no protease/protease). Also add 5–10 microliters of the primary antibody mixture (*diluted* form) to the remaining two slides (one no protease/protease and the other antigen retrieval with no protease/protease).

2. For larger biopsies (equal to or greater than 1cm in size), manually add 20–35 microliters of the primary antibody mixture (*concentrated* form) to two of the slides (one no protease/protease and the other antigen retrieval with no protease/protease). Also add 20–35 microliters of the primary antibody mixture (*diluted* form) to the remaining two slides (one no protease/protease and the other antigen retrieval with no protease/protease) of the probe cocktail.

3. If you are using an automated system, you need to add 100 microliters of the primary antibody to each pretreatedslid e.

4. Cover the primary antibody mixture with a polypropylene coverslip that is cut to be just slightly larger than the tissue section (for manual method only).

5. Incubate the slides in a humidity chamber with water at the base at 37°C for 3–5 hours. The primary probe cocktails are much more likely to dry out than the primary probe cocktail used with in situ hybridization. Thus, make certain to add enough water to the base of the humidity chamber.

STEP SIX: THE STRINGENT WASH

1. Remove the coverslips carefully by using fine tweezers. Make certain that the primary antibody mixture did not dry out. If it did, either the coverslip was too large, there was too little of the primary antibody cocktail, and/or there was not enough water used for the humidity chamber.

2. Place the slides in a solution of PBS that contains 1% bovine serum albumin that has equilibrated to room temperature.

3. Incubate the slides in the stringent wash for 5 minutes.
4. Remove the slides *one at a time* for the next step. Do *not* let the slides dry out after the stringent wash.

STEP SEVEN: THE DETECTION PART, THE FIRST STEP

1. Remove the slide from the stringent wash and hold very level. Quickly remove the wash solution from the bottom of the slide and the perimeter of the front of the slide.
2. Using a hydrophobic pen, mark the perimeter of the front of the slide to keep the detection solution over the tissue. Keep the slide level at all times; otherwise, the remaining wash solution will ruin the "hydrophobic dam" you have just created.
3. Reconfirm the animal source of the primary antibody. If the primary antibody was generated in a mouse or rabbit (which is by far the most common scenario), add 100 microliters of the appropriate secondary antibody. For this step, you can use commercially available anti-rabbit (Biogenex, catalog #HK336-9R), anti-mouse (Biogenex, catalog #HK335-9M), as well as multilink secondary antibodies (Biogenex, catalog #HK340-9K), that are able to detect any primary antibody generated in a mouse, rabbit, rat, guinea pig, or goat. These commercially available secondary antibodies are conjugated with biotin.
4. Although you can use the Multilink secondary antibody, I strongly recommend for the secondary antibody using *only* the one specific for the animal used to generate the primary antibody.
5. If you are working with mouse tissues, I strongly recommend using only primary antibodies generated in rabbits. If this is not possible, primary antibodies generated in goats will also work.
6. If you are using an automated system (e.g., Ventana Medical Systems), it is important to realize that it is *not* based on biotin-conjugated secondary antibodies. It will detect primary antibodies generated only in a mouse or rabbit. If using such a system, you can still use a primary antibody directed in an animal other than a mouse or rabbit (for the purposes of this example, a goat). The protocol would be to add the primary antibody, incubate for 15 minutes, and then add a rabbit-derived antibody directed against any goat primary antibody.
7. Whatever secondary antibody system you are using, incubate the tissue in the secondary antibody for 1–2 hours.

If you are doing the immunohistochemistry by hand, you have the option of adding 100 microliters of the secondary antibody to each slide and putting the slides in the humidity chamber. With this large a volume, the solution should not dry out, and, thus, there is no need for polypropylene coverslips. However, if you wish to use smaller volumes of the secondary antibody, cut the polypropylene coverslips to a size that should allow you to use 25–50 microliters of the secondary antibody per slide.

STEP EIGHT: THE DETECTION PART, THE SECOND STEP

1. Place the slides in a solution of PBS that contains 1% bovine serum albumin and 3% H_2O_2 at room temperature.
2. Incubatefo r5min utesin th ewash so lution.
3. Remove the slide from the stringent wash and hold very level. Quickly remove the wash solution from the bottom of the slide and the perimeter of the front of the slide.
4. There should be no need to use the hydrophobic pen again because it should still be present. Keep the slide level at all times; otherwise, the remaining wash solution will ruin the "hydrophobic dam" you have justc reated.
5. If using a peroxidase-based system, add the streptavidin-peroxidase conjugate (e.g., Biogenex, catalog #HK330-9K; or Enzo Life Sciences, catalog #ENZ-32840).
6. If using the alkaline phosphatase-based system, add the streptavidin-peroxidase conjugate (e.g., Enzo Life Sciences, catalog #ENZ-32870).

Incubate the slides for 1–2 hours in the streptavidin conjugate. If you add 100 microliters of the streptavidin conjugate to each slide and put the slides in the humidity chamber, the solution should not dry out, and thus there is no need for polypropylene coverslips. However, if you wish to use smaller volumes of the streptavidin conjugate, cut the polypropylene coverslips to a size that should allow you to use 25–50 microliters of the streptavidin conjugate per slide.

STEP NINE: THE DETECTION PART, THE LAST STEP

1. If using a peroxidase-based system, prepare the DAB solution immediately before you are ready to do this step. For manual immunohistochemistry using DAB, you may use the Biogenex DAB kit (Super sensitive Link-label Immunohistochemical Detection system, catalog #QD000-5L). This kit has all the reagents you need for immunohistochemistry after the primary antibody step. To make the DAB solution, take 2 drops of the DAB chromogen and add to the DAB substrate buffer. Vortex. Then add 1 drop of the DAB substrate. Vortex.
2. If using the *Fast Red* system, prepare the Fast Red chromogen as per the directions of the manufacturer. Be careful! Some Fast Red precipitates are soluble in ethanol/xylene, whereas others are not. If you are using a Fast Red chromogen that is soluble in ethanol/xylene, you must use a water-based mounting media such as Aquamount (catalog #HK079-7K).
3. Remove the slides from the PBS wash. Wipe the back of the slide and, keeping the slide level, wipe the perimeter of the front of the slide. Place the slide carefully down in a humidity chamber. Do *not* let the slide dry out; otherwise, background will ensue.
4. Add several drops of DAB solution to each tissue section (for a peroxidase-based assay).
5. Add several drops of Fast Red solution to each tissue section (for a alkaline phosphatase-based assay).

STEP TEN: MONITOR THE PRECIPITATE

1. It is *very* important to monitor the precipitation of the chromogen (either DAB or Fast Red) under the microscope, initially every 3 minutes. Do not accept any interruptions at this stage. Stop the reaction when you have determined that the signal has been maximized and/or background is just beginning to develop.
2. Although most Fast Red reactions are finished within 20 minutes, the only determining factor for stopping the reaction is when you have determined that the signal has reached its maximum stage and/or background begins to be evident.
3. Although most DAB reactions are finished within 20 minutes, the only determining factor for stopping the reaction is when you have determined that the signal has reached its maximum stage and/or background begins to be evident.
4. When the signal is at maximum *or* background begins to be evident, immediately place the slide in a jar that contains water. Wash the slides under running tap water for 1 minute to stop the chromogen precipitate.

STEP ELEVEN: THE COUNTERSTAIN

1. If you wish to do co-expression analysis, do *not* do a counterstain at this time.
2. If you wish to use a counterstain, for Fast Red, use hematoxylin.
3. If you wish to use a counterstain, for DAB, use either Nuclear Fast Red or hematoxylin.
4. For nuclear targets, do a weak counterstain (a few dips in hematoxylin or 30–60 seconds in Nuclear Fast Red).
5. For cytoplasmic targets, use the standard hematoxylin (1 minute) or Nuclear Fast Red (3 minute) counterstain.
6. Rinse the slides in running tap water for 1 minute.
7. Dip the slides in 100% ethanol. Let completely air dry. (Again, if you are using Fast Red as the chromogen, make certain that the product you are using is insoluble in ethanol/xylene. If it *is* soluble in ethanol/xylene, then use Aquamount for putting the coverslip on the slide.)
8. Placeth eslid esin xyle ne.
9. Coverslip using Permount (Fisher Scientific, catalog #SP15-100).

IMPORTANT NOTES

1. I purposely made the times of incubation listed in the preceding steps longer than many other protocols. This is based on the many different time-dependent data we discussed in previous chapters. Specifically, we saw that if all conditions were not optimal and/or we were dealing with a low copy target, then increasing the times of incubations at each step could enhance the signal. I recommend starting with the *shortest* times that are listed in this and the in situ hybridization protocol. If the signal is weak under these conditions, you can increase the incubation times of the different steps.

2. The DAB-based peroxidase system is by far the most popular for immunohistochemistry. I recommend you begin using this system. A trick of the trade when using the DAB system is to add 3% H_2O_2 to the primary antibody and the secondary antibody, because this will decrease the risk of background.

TROUBLESHOOTING

As with in situ hybridization, I think that the key to troubleshooting problems with immunohistochemistry is to have a solid base of knowledge regarding the biochemistry of the process. Similar to in situ hybridization, a few issues make up the majority of the problems with immunohistochemistry and require you to figure out how to solve the problem.

Let's look at the issues that are most likely to cause problems when doing immunohistochemistry.

TOO MUCH BACKGROUND

As with in situ hybridization, too much background is the most common problem that comes my way via emails and phone calls from people doing immunohistochemistry. Given the many similarities at a biochemical level, I think you will not be surprised that the causes of background with immunohistochemistry are very similar to the causes of background with in situ hybridization:

1. A primary antibody that is too concentrated
2. Incorrectp retreatmentc onditions
3. Omission of bovine serum albumin from the wash solution

As we have discussed at length, immunohistochemistry is a basic diffusion reaction in which various R side chains of amino acids either inhibit the movement of the key immunohistochemistry reagents and/or facilitate their docking with the target epitope and other secondary detection reagents. The higher the concentration of the primary antibody, the more likely you will see background. The *primary reason* I recommend always starting the testing of a new tissue and primary antibody by using *two very different* primary antibody concentrations is to use these data to determine which primary antibody concentration will allow you a good signal, but with minimal background.

We have also spent a lot of time on the theory of the three-dimensional macromolecule cross-linked cage that surrounds each DNA and RNA target. We have seen on many occasions that poor exposure of the target not only causes a reduced signal, but also may allow background to develop. In my experience, often the reason is that the investigator sees little signal and, thus, increases the concentration of the probe. This then leads to the background. Figures 8-1 and 8-2 show representative data for the optimization of the protein human chorionic gonadotropin (hCG). The tissue is placenta that contains extremely high copy numbers of this protein. The hCG localizes to only one cell type in the placenta, which are called trophoblasts. These are the cells that line the villi of the placenta and do most of the metabolic work for the

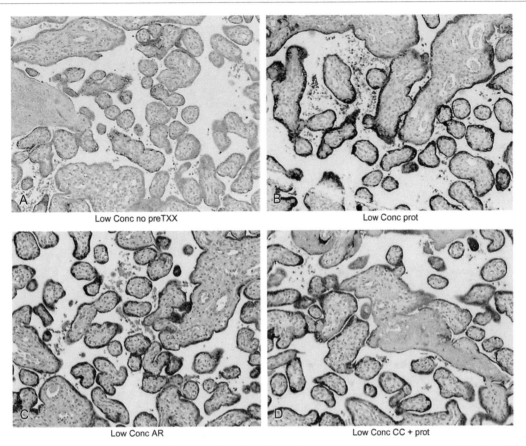

Figure8-1 The association of background with immunohistochemistry to the concentration of the primary antibody and the pretreatment conditions. Panels A–D show the results for the testing of placenta tissue for hCG after different pretreatment conditions and at a low concentration of the primary antibody. Note that the signal in each panel localized to just the trophoblasts, and thus, no background was evident. The signal is most intense with protease digestion, strong with antigen retrieval or antigen retrieval plus protease digestion, and weak with no pretreatment.

fetus. Thus, the placenta and hCG serve as a good assay system to study the effects of pretreatment and primary antibody concentration on signal and background.

Note in Figures 8-1 and 8-2 that if we use the *concentrated* primary antibody and antigen retrieval, then there is intense background that completely masks the signal. However, the *concentrated* primary antibody produces a strong signal present only in the trophoblasts and no background if protease digestion alone was done as the pretreatment. Clearly, the epitope must be hidden to some degree because no pretreatment produces a weak signal. However, subjecting the tissue to 95°C for 30 minutes in the antigen retrieval solution apparently changes the biochemical microenvironment to the point that the primary hCG antibody sticks nonspecifically to many sites. The latter process is strongly dependent on the concentration of the primary antibody. This is evident in Figure 8-2 where the *diluted* primary antibody (a 1:10 dilution of the concentrated form) caused *no* background, and an intense signal, after antigen retrieval.

NO (OR VERY WEAK) SIGNAL

No (or very weak) signal is also a common problem. By far and away, three causes dominate for this situation:

1. Incorrectp retreatmentc onditions
2. Using the primary antibody at too dilute a concentration
3. Using tissue blocks that are too old

For those who say "Using a tissue that has very little or none of the target of interest," is another cause of no signal, my comment is, "That is certainly a valid point." However, most investigators will have some information about the relative copy number of their protein target of interest, either from their own Western blot data and/or from the published literature. Thus, I did not include this point in the list.

Let's look at some examples of point 1, as this is a very common reason that there is no signal with immunohistochemistry when, in fact, the tissue contains many cells with high copy number of the protein of interest.

Figures 8-3 and 8-4 show representative examples for IL-28 and amylase, respectively. In each case, a colleague reported that he had not been able to generate a signal with immunohistochemistry. This colleague had already done Western blots and shown that the tissues he was studying had high copy numbers of the protein of interest. As seen in panel A of Figure 8-3, there was no signal for IL-28 when antigen retrieval was done. However, note what happens if the same concentration

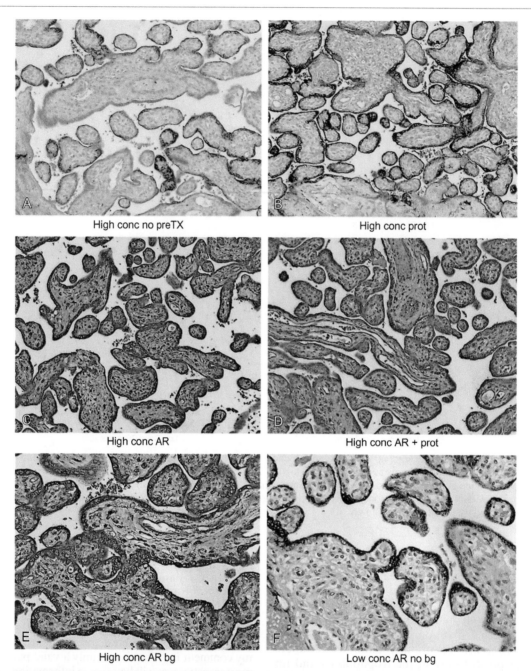

High conc no preTX

High conc prot

High conc AR

High conc AR + prot

High conc AR bg

Low conc AR no bg

Figure8-2   The association of background with immunohistochemistry to the concentration of the primary antibody and the pretreatment conditions. The signal for the high concentration is DAB and for the low antibody concentration is Fast Red. The experiments were the same as in Figure 8-1, but a 10× higher concentration of the hCG primary antibody was used. Note that the results with no pretreatment (panel A) and with protease digestion (panel B) are similar to that noted for the low concentration primary antibody; a signal is evident (weak, no pretreatment, and strong, protease digestion). However, if antigen retrieval or antigen retrieval plus protease digestion was used with the high concentration of the primary antibody, then most cells showed the DAB precipitate. Since many of these cells were not trophoblasts, this must represent background (panels C–E). Compare these results to that for the low concentration of the primary antibody and antigen retrieval (panel F) where the signal was strong and no background was evident.

of the primary antibody was used, but either no pretreatment (panel B) or protease digestion prior to immunohistochemistry (panels C and D). A signal is noted with no pretreatment or protease digestion. Note that, at high magnification (panel D), it is clear that the signal localizes to the cytoplasm of elongate cells that are concentrated in the stroma of the tissue. These spindly cells have the

cytologic appearance of either myofibroblasts or fibroblasts; co-expression analyses would have to be done to determine the phenotype of the positive cells.

Now compare the data in Figure 8-3 to that in Figure 8-4. The initial experiments were done with no protease digestion (panel A). My colleague was correctly concerned that, since he was studying pancreas tissue, which,

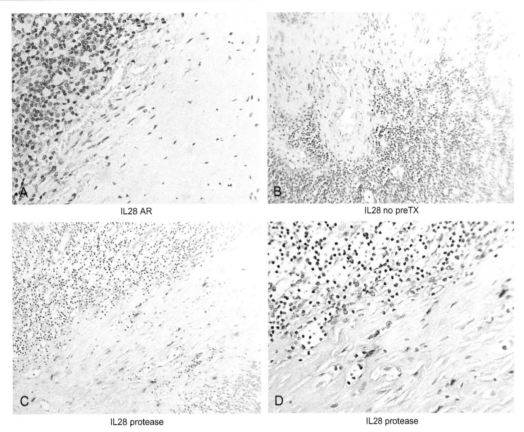

IL28 AR

IL28 no preTX

IL28 protease

IL28 protease

Figure8-3   Optimizing experiments for IL-28 detection by immunohistochemistry. As seen in panel A, there was no signal for IL-28 when antigen retrieval was done. However, note what happens if the same concentration of the primary antibody was used, but either no pretreatment (panel B) or protease digestion prior to immunohistochemistry (panels C and D). A signal is noted with no pretreatment or protease digestion. Note that, at high magnification (panel D), it is clear that the signal localizes to the cytoplasm of elongate cells that are concentrated in the stroma of the tissue. Thus, IL-28 is an example of an epitope where antigen retrieval is actually detrimental to its detectionb yi mmunohistochemistry.

of course, is very rich in proteases, any additional protease digestion would probably destroy the signal and/or the tissue. We then tried protease digestion (panel B and, at higher magnification, panel C). Note that we now see a signal that localizes to the cells that certainly would be maximizing amylase, the acinar cells of the pancreas. However, we would expect the signal to be even stronger, because the pancreas should be the tissue with the highest concentrations of amylase. As seen in panel D, an intense signal was indeed seen if the tissue was pretreated with antigen retrieval. These data also remind us how capable formalin fixation is in terms of inactivating the endogenous proteases, as well as DNases and RNases.

You may recall from previous chapters that you can prestain the tissue with either hematoxylin, eosin, or both and still get a strong signal with immunohistochemistry (or in situ hybridization). Indeed, this simple point that pre-existing nuclear and/or cytoplasmic signals will not interfere with immunohistochemistry or in situ hybridization is the basis of co-expression analyses, which we discuss in the next chapter. We can use this observation to our advantage. Say that you have done an immunohistochemistry

experiment and got no signal. This scenario is illustrated in Figure 8-5 for IL-18. The conditions were antigen retrieval with a primary antibody concentration of 1:100. You realize that when the tissue was optimized, you found the correct primary antibody concentration to be 1:25. All you needed to do (and what was done in this experiment) was to remove the coverslip and redo the immunohistochemistry. Since antigen retrieval had already been done, there was no reason to repeat that part of the protocol. Note that when the same tissue was reused, but with the higher concentration of the primary antibody, a strong signal was evident (panels C and D). Remembering what you read in Chapter 6 on histologic interpretation, look at the histologic pattern of the signal for IL-18. It is "talking to us," telling us a lot about its function in this tissue (tonsil).

Here is what I would list as the salient features of the signal for IL-18:

1. It is cytoplasmic based and present in large cells.
2. It is present in a reticulated pattern.
3. The positive cells are interconnected by the reticulatedp attern.

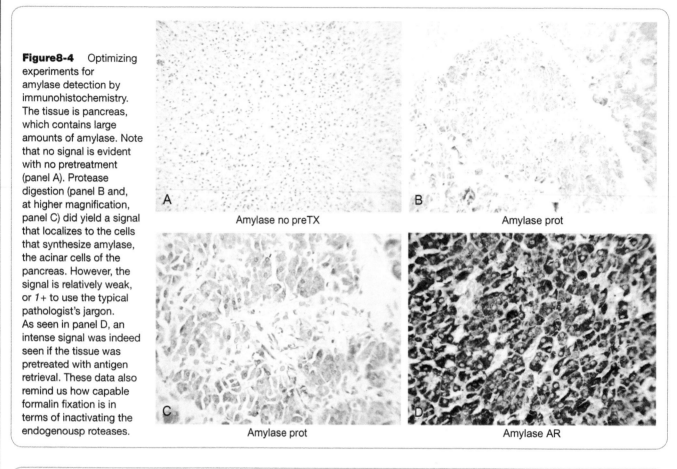

Figure8-4 Optimizing experiments for amylase detection by immunohistochemistry. The tissue is pancreas, which contains large amounts of amylase. Note that no signal is evident with no pretreatment (panel A). Protease digestion (panel B and, at higher magnification, panel C) did yield a signal that localizes to the cells that synthesize amylase, the acinar cells of the pancreas. However, the signal is relatively weak, or *1*+ to use the typical pathologist's jargon. As seen in panel D, an intense signal was indeed seen if the tissue was pretreated with antigen retrieval. These data also remind us how capable formalin fixation is in terms of inactivating the endogenousp roteases.

Amylase no preTX

Amylase prot

Amylase prot

Amylase AR

Figure8-5 Prior testing by immunohistochemistry does not preclude successful subsequent immunohistochemistry. The tissue is tonsil, tested for IL-18 by antigen retrieval at a concentration of 1:100 (panel A). If the test was repeated at a higher concentration (1:25), then a strong signal was seen (panel D) in cells that have the cytologic appearance and geographic distribution of dendritic cells. Similarly, the IL-28 immunohistochemistry test was negative (panel B), and the reason was that, by error, the primary antibody was too dilute. The coverslip was removed, and the tissue retested at the proper primary antibody concentration where a strong signal was now evident(panelC).

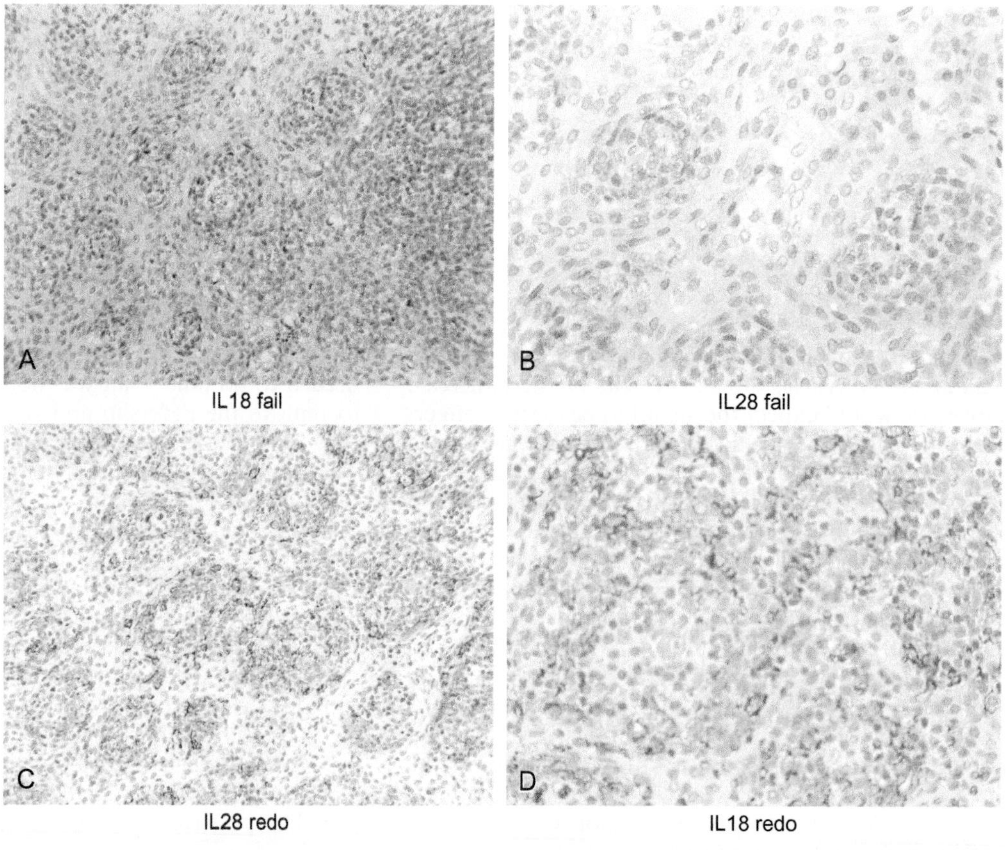

IL18 fail

IL28 fail

IL28 redo

IL18 redo

For the nonpathologists, look again at Figure 8-5. Do you see this pattern? If so, what type of cell in the immune system would have such a pattern? The answer is the follicular dendritic cell. These are cells that are also called "antigen-presenting cells." Their job is to be on the front lines and be the first to recognize some foreign antigen. When they do recognize a foreign antigen, they communicate among themselves and other immune-modulating cells to decide what should be the next step. This is why they have this complex interconnected pathway connecting them to each other and to other immune cell modulators. This pattern is unique for the follicular dendritic cell in the lymph node and tonsil.

Let's look at one last example of how signal and background can be strongly correlated to the pretreatment conditions. The tissue is tonsil, and it had been fixed for an extended period of time (at least several weeks). Thus, it was logical to assume that the tissue would probably need antigen retrieval and protease digestion when it was being used to test for TUSC-1. This protein is thought to be a tumor suppressor in a variety of cancers. Representative data for the optimization protocol are shown in Figure 8-6. Note the moderately strong signal with no pretreatment (panel A). The signal is cytoplasmic

and in larger cells in the sinusoids. This is certainly consistent with macrophages. The signal became intense after protease digestion (panel B) and then much weaker with clearcut background if antigen retrieval was used (panel C). Finally, antigen retrieval and protease destroyed the signal and intensified the background (panel D). These data continue to remind us that protease digestion and antigen retrieval certainly must cause quite different biochemical changes for both in situ hybridization and immunohistochemistry.

Point 3 is very common for no signal with an immunohistochemistry reaction. Formalin-fixed, paraffin-embedded tissues that are greater than 5 years old may well have a reasonable copy number of the target of interest. However, as we saw with the in situ hybridization signal, the immunohistochemistry signal may be negative because the three-dimensional macromolecule cross-linked cage around the target has degenerated too much. As we discussed at length, this can easily be corrected by the regeneration of that cage. Another example of this is provided in Figure 8-7. I purposely used tissue blocks that were not too old. This is testis tissue from an older man; the tissue was fixed and embedded in paraffin in 2005. Note that the signal for smooth muscle actin (SMA) and Ki-67 is present

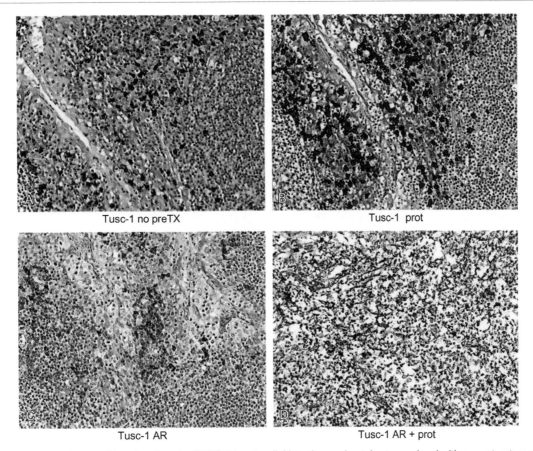

Tusc-1 no preTX

Tusc-1 prot

Tusc-1 AR

Tusc-1 AR + prot

Figure8-6 Optimizing immunohistochemistry for TUSC-1 in a tonsil. Note the moderately strong signal with no pretreatment (panel A). The signal is cytoplasmic and in larger cells in the sinusoids. Thus, these are probably macrophages. The signal became intense after protease digestion (panel B) and then much weaker with clearcut background if antigen retrieval was used (panel C). Finally, antigen retrieval and protease destroyed the signal and intensified the background (panel D). These data show very well the inverse relationship between signal and background, and remind us not to assume that antigen retrieval will be the optimal pretreatment for all new primary antibodies.

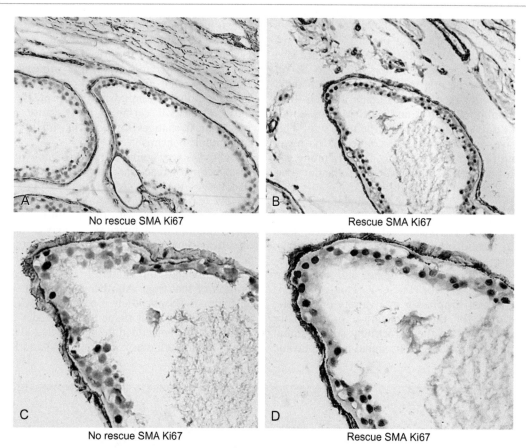

A   No rescue SMA Ki67   B   Rescue SMA Ki67

C   No rescue SMA Ki67   D   Rescue SMA Ki67

Figure 8-7   Co-expression analysis for smooth muscle actin and Ki-67 in a testis processed in 2007. The signal for smooth muscle actin and Ki-67 should each be strong in the testis. However, in this case, the signal was weak (panel A for actin and panel C for Ki-67). After IHCPerfect pretreatment, the actin and Ki-67 were both much enhanced (panel B and, for Ki-67, panel D). It is important to note that for each primary antibody, the optimizing experiments showed that antigen retrieval was optimal. This is the reason that we can do co-expression analyses with one section—plus, of course, the fact that Ki-67 and smooth muscle actin are found in histologically distinct cell types.

but weak. SMA is always found as an intense signal in any large-walled vessel in any organ in the body.

In my own work, if the tissue block is at least 10 years old, I typically "rejuvenate" the three-dimensional macro-molecule cross-linked cage before doing the immunohisto-chemistry experiments. Ultimately, this saves a lot of time and, of course, may be the only way to get useful data.

INTERPRETATION OF DATA

As with in situ hybridization, I recommend being a bit compulsive when interpreting the data from the optimiz-ing immunohistochemistry experiments. For example, you may wish to draw a flow sheet that includes the following:

Primary Concentration	PretreatmentC ondition	Signal/ Background
LOW	NONE	
LOW	Protease	
LOW	Antigenr etrieval	
LOW	Antigenr etrieval +p rotease	
HIGH	NONE	
HIGH	Protease	
HIGH	Antigenr etrieval	
HIGH	Antigenr etrieval +p rotease	

Figures 8-1 and 8-2 showed some raw data of such an optimizing analysis for the protein hCG in placenta. The tabulated data follows:

PrimaryA ntibody Concentration	Pretreatment Condition	Signal/ Background
LOW	NONE	1/0
LOW	Protease	3/0
LOW	Antigenr etrieval	3/0
LOW	Antigenr etrieval + protease	3/weak (due to overdigestion)
HIGH	NONE	2/0
HIGH	Protease	3/0
HIGH	Antigenr etrieval	1/3
HIGH	Antigenr etrieval + protease	0/3

These data provide a good illustration of how a given tissue can have more than one optimal pretreatment con-dition that is dependent on the concentration of the pri-mary antibody. If you use the low concentration primary antibody for hCG, then either protease digestion or anti-gen retrieval will yield superior results. However, if you use the high concentration of the hCG primary antibody, then you must use protease digestion in order to maxi-mize the signal.

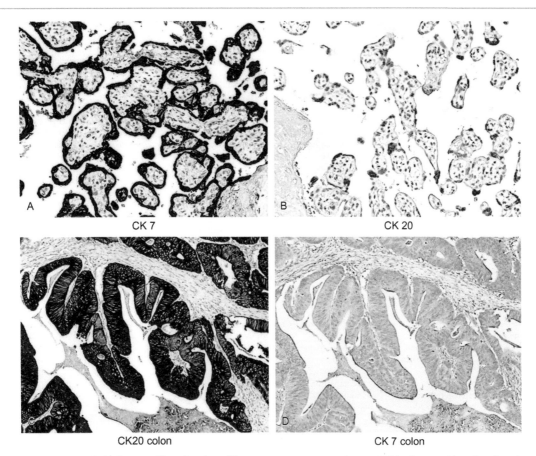

CK 7

CK 20

CK20 colon

CK 7 colon

Figure 8-8   Negative controls for immunohistochemistry. The most common negative control for immunohistochemistry is omission of the primary antibody. This test is useful, but it does not indicate whether the primary antibody per se can cause background. To address this point, you can choose a tissue that is known to not have the epitope of interest. This is illustrated for cytokeratins 7 and 20. Cytokeratin 7 is found in most keratin-positive tissues, whereas cytokeratin 20 is found in the epithelia of the lower gastrointestinal tract. Note that there is an intense signal for cytokeratin 7 in the placenta (panel A), whereas cytokeratin 20 is not evident in the placenta (panel B). In comparison, cytokeratin 20 is intensely positive in the colon (panel C) and negative in the placenta (panel D). Hence, the placenta can serve as a negative control for cytokeratin 20, and the colon can be a negative control tissue for cytokeratin 7.

Finally, Figure 8-8 shows the value of another type of negative control. In this negative control, you uses tissue known *not* to have the target of interest.

The suggested readings [1–67] provide a broad range of different protocols for doing immunohistochemistry.

Again, I have intentionally included papers in which both in situ hybridization and immunohistochemistry were done to show the overlap present in such protocols as used by many different investigators.

SUGGESTED READINGS

Different published protocols for immunohistochemistry/in situ hybridization

[1] Ahmed A, Gilbert-Barness E, Lacson A. Expression of c-kit in Ewing family of tumors: a comparison of different immunohistochemical protocols. PediatrDe vP athol 2004;7:342–7.

[2] Angelucci A, Clasca F, Sur M. Anterograde axonal tracing with the subunit B of cholera toxin: a highly sensitive immunohistochemical protocol for revealing fine axonal morphology in adultan dn eonatalb rains. JNe urosciM ethods 1996;65:101–12.

[3] Bagasra O. Protocols for the in situ PCR-amplification and detection of mRNA and DNA sequences. NatP rotoc 2007;2:2782–95.

[4] Bagasra O, Harris T. In situ PCR protocols. MethodsM olB iol 2006;334:61–78.

[5] Ballester M, Galindo-Cardiel I, Gallardo C, et al. Intranuclear detection of African swine fever virus DNA in several cell types from formalin-fixed and paraffin-embedded tissues using a new ins ituh ybridisationp rotocol. JVir olM ethods 2010;168:38–43.

[6] Beier K. Light microscopic morphometric analysis of peroxisomes by automatic image analysis: advantages of immunostaining over the alkalineDAB me thod. JHis tochemCyt ochem 1992;42:49–56.

[7] Bittermann AG, Knoll G, Nemeth A, Plattner H. Quantitative immuno-gold labelling and ultrastructural preservation after cryofixation

(combined with different freeze-substitution and embedding protocols) and after chemical fixation and cryosectioning. Analysis of the secretory organelle matrix of Paramecium trichocysts. Histochemistry 1992;97:421–9.

[8] Bzorek Sr. M, Petersen BL, Hansen L. Simultaneous phenotyping and genotyping (FICTION-methodology) on paraffin sections and cytologic specimens: a comparison of 2 differentp rotocols. AppII mmunohistochem MolM orphol 2008;16:279–86.

[9] Carr EL, Eales K, Soddell J, Seviour RJ. Improved permeabilization protocols for fluorescence in situ hybridization (FISH) of mycolic-acid-containing bacteria found in foams. JM icrobiolM ethods 2005;61:47–54.

[10] Chaichana KL, Guerrero-Cazares H, Capilla-Gonzalez V, et al. Intra-operatively obtained human tissue: protocols and techniques for the study of neural stem cells. J Neurosci Methods 2009;180:116–25.

[11] Chang CC, Huang TY, Shih CL, et al. Whole-mount identification of gene transcripts in aphids: protocols and evaluation of probe accessibility. Arch Insect Biochem Physiol 2008;68:186–96.

[12] Chen L, Liu F, Fan X, et al. Detection of hepatitis B surface antigen, hepatitis B core antigen, and hepatitis B virus DNA in parotid tissues. Int J Infect Dis 2009;13:20–3.

[13] Chen WL, Huang MT, Liu HC, Li CW, Mao SJ. Distinction between dry and raw milk using monoclonal antibodies prepared against dry milk proteins. J Dairy Sci 2004;87:2720–9.

[14] Cserni G. Complete sectioning of axillary sentinel nodes in patients with breast cancer. Analysis of two different step sectioning and immunohistochemistry protocols in 246 patients. J Clin Pathol 2002;55:926–31.

[15] Dabbs DJ. Immunohistochemical protocols: back to the future. Am J Clin Pathol 2008;129:355–6.

[16] de Jong D, Koster A, Hagenbeek A, et al. Impact of the tumor microenvironment on prognosis in follicular lymphoma is dependent on specific treatment protocols. Haematologica 2009;94:70–7.

[17] Derradji H, Bekaert S, Van Oostveldt P, Baatout S. Comparison of different protocols for telomere length estimation by combination of quantitative fluorescence in situ hybridization (Q-FISH) and flow cytometry in human cancer cell lines. Anticancer Res 2005;25:1039–50.

[18] Epple S, Mittmann M, Reichrath J. Immunohistochemical detection of retinoic acid receptor alpha in human skin: a comparison of different fixation protocols. Histochem J 1996;28:657–60.

[19] Ertault-Daneshpouy M, Deschaumes C, Leboeuf C, et al. Histochemical and immunohistochemical protocols for routine biopsies embedded in Lowicryl resin. Biotech Histochem 2003;78:35–42.

[20] Familiari G, Verlengia C, Nottola S, et al. Heterogeneous distribution of fibronectin, tenascin-C, and laminin immunoreactive material in the cumulus-corona cells surrounding mature human oocytes from IVF-ET protocols—evidence that they are composed of different subpopulations: an immunohistochemical study using scanning confocal laser and fluorescence microscopy. Mol Reprod Dev 1996;43:392–402.

[21] Filion G, Laflamme C, Turgeon N, Ho J, Duchaine C. Permeabilization and hybridization protocols for rapid detection of Bacillus spores using fluorescence in situ hybridization. J Microbiol Methods 2009;77:29–36.

[22] Fisher, E., Colangelo, L., Wieand, S., Fisher, B., Wolmark, N., 2003. Lack of influence of cytokeratin-positive mini micrometastases in "Negative Node" patients with colorectal cancer: findings from the National Surgical Adjuvant Breast and Bowel Projects protocols R-01 and C-01. Dis Colon Rectum 46, 1021–1025, discussion 1025–1026.

[23] Frederiks WM, Mook OR. Metabolic mapping of proteinase activity with emphasis on in situ zymography of gelatinases: review and protocols. J Histochem Cytochem 2004;52:711–22.

[24] Fried J, Ludwig W, Psenner R, Schleifer KH. Improvement of ciliate identification and quantification: a new protocol for fluorescence in situ hybridization (FISH) in combination with silver stain techniques. Syst Appl Microbiol 2002;25:555–71.

[25] Gosden JR. In: PRINS and in situ PCR protocols. Totowa, N.J.: Humana Press; 1996.

[26] Grabau D, Ryden L, Fernö M, Ingvar C. Analysis of sentinel node biopsy—a single-institution experience supporting the use of serial sectioning and immunohistochemistry for detection of micrometastases by comparing four different histopathological laboratory protocols. Histopathology 2011;59:129–38.

[27] Gu J. In: Analytical morphology: theory, applications, and protocols. Cambridge, MA: Birkhauser Boston; 1997.

[28] Guidi AJ, Berry DA, Broadwater G, et al. Association of angiogenesis and disease outcome in node-positive breast cancer patients treated with adjuvant cyclophosphamide, doxorubicin, and fluorouracil: a Cancer and Leukemia Group B correlative science study from protocols 8541/8869. J Clin Oncol 2002;20:732–42.

[29] Ha SK, Choi C, Chae C. Development of an optimized protocol for the detection of classical swine fever virus in formalin-fixed, paraffin-embedded tissues by seminested reverse transcription-polymerase chain reaction and comparison with in situ hybridization. Res Vet Sci 2004;77:163–9.

[30] Hayward PA, Bell JE, Ironside JW. Prion protein immunocytochemistry: reliable protocols for the investigation of Creutzfeldt-Jakob disease. Neuropathol Appl Neurobiol 1994;20:375–83.

[31] Hopman AH, Kamps MA, Smedts F, et al. HPV in situ hybridization: impact of different protocols on the detection of integrated HPV. Int J Cancer 2005;115:419–28.

[32] Ishii T, Omura M, Mombaerts P. Protocols for two- and three-color fluorescent RNA in situ hybridization of the main and accessory olfactory epithelia in mouse. J Neurocytol 2004;33:657–69.

[33] Jilbert A. In situ hybridization protocols for detection of viral DNA using radioactive and nonradioactive DNA probes. Methods Mol Biol 2000;123:177–93.

[34] Kerman K, Vestergaard M, Tamiya E. Electrochemical DNA biosensors: protocols for intercalator-based detection of hybridization in solution and at the surface. Methods Mol Biol 2009;504:99–113.

[35] Khalil SH. Book review: PCR in situ hybridization: protocols and applications. Ann Saudi Med 1997;17:491.

[36] Kim J, Chae C. Optimized protocols for the detection of porcine circovirus 2 DNA from formalin-fixed paraffin-embedded tissues using nested polymerase chain reaction and comparison of nested PCR with in situ hybridization. J Virol Methods 2001;92:105–11.

[37] Kim O, Yi SJ. Optimization of in situ hybridization protocols for detection of feline herpesvirus 1. J Vet Med Sci 2003;65:1031–2.

[38] Kita-Matsuo H, Barcova M, Prigozhina N, et al. Lentiviral vectors and protocols for creation of stable hESC lines for fluorescent tracking and drug resistance selection of cardiomyocytes. PLoS One 2009;4:e5046.

[39] Kraft O, Sevcik L, Klat J, et al. Detection of sentinel lymph nodes in cervical cancer. A comparison of two protocols. Nucl Med Rev Cent East Eur 2006;9:65–8.

[40] Lebo RV, Su Y. Positional cloning and multicolor in situ hybridization. Principles and protocols. Methods Mol Biol 1994;33:409–38.

[41] Lecuyer E, Parthasarathy N, Krause HM. Fluorescent in situ hybridization protocols in Drosophila embryos and tissues. Methods Mol Biol 2008;420:289–302.

[42] Lichter P, Fischer K, Joos S, et al. Efficacy of current molecular cytogenetic protocols for the diagnosis of chromosome aberrations in tumor specimens. Cytokines Mol Ther 1996;2:163–9.

[43] Liu CQ, Shan L, Balesar R, et al. A quantitative in situ hybridization protocol for formalin-fixed paraffin-embedded archival post-mortem human brain tissue. Methods 2010;52:359–66.

[44] Lourenco HM, Pereira TP, Fonseca RR, Cardoso PM. HER2/neu detection by immunohistochemistry: optimization of in-house protocols. Appl Immunohistochem Mol Morphol 2009;17:151–7.

[45] Manavis J, Gilham P, Davies R, Ruszkiewicz A. The immunohistochemical detection of mismatch repair gene proteins (MLH1, MSH2, MSH6, and PMS2): practical aspects in antigen retrieval and biotin blocking protocols. Appl Immunohistochem Mol Morphol 2003;11:73–7.

[46] McQuaid S, Allan GM. Detection protocols for biotinylated probes: optimization using multistep techniques. J Histochem Cytochem 1992;40:569–74.

[47] Mojsilovic-Petrovic J, Nesic M, Pen A, Zhang W, Stanimirovic D. Development of rapid staining protocols for laser-capture microdissection of brain vessels from human and rat coupled to gene expression analyses. J Neurosci Methods 2004;133:39–48.

[48] Motohashi T, Tabara H, Kohara Y. Protocols for large scale in situ hybridization on C. elegans larvae. WormBook 2006:1–8.

[49] Murai N, Murakami Y, Matsufuji S. Protocols for studying antizyme expression and function. Methods Mol Biol 2011;720:237–67.

[50] Navarro B, Daros JA, Flores R. Reverse transcription polymerase chain reaction protocols for cloning small circular RNAs. J Virol Methods 1998;73:1–9.

[51] Norberg L, Stratis M, Dardick I. Quantitation and localization of cycling tumor cells in pleomorphic adenomas and myoepitheliomas: an immunohistochemical analysis. J Oral Pathol Med 1997;26:124–8.

[52] Oberman F, Yisraeli JK. Two protocols for nonradioactive in situ hybridization to Xenopus oocytes. Trends Genet 1995;11:83–4.

[53] Oliver C, Jamur MC. Immunocytochemical methods and protocols. Methods Mol Biol 2010;588:iv–v.

[54] Paul P, Rouas-Freiss N, Moreau P, et al. HLA-G, -E, -F preworkshop: tools and protocols for analysis of non-classical class I genes transcription and protein expression. Hum Immunol 2000;61:1177–95.

[55] Peng EC, Chen S, Bentivoglio M. A sensitive double immunostaining protocol for Fos-immunoreactive neurons. Brain Res Bull 1995;37:101–5.

[56] Porteous DJ. Protocols for chromosome-mediated gene transfer. Selection strategies, transgenome analysis, enrichment cloning, and mapping. Methods Mol Biol 1994:353–78.

[57] Privat N, Sazdovitch V, Seilhean D, LaPlanche JL, Hauw JJ. PrP immunohistochemistry: different protocols, including a procedure for long formalin fixation, and a proposed

schematic classification for deposits in sporadic Creutzfeldt-Jakobd isease. MicroscRe sT ech 2000;50:26–31.

[58] Redi C. Signal transduction immunohistochemistry—methods and protocols. EurJ Histo chem 2011;5 ejh.2011. br2017.

[59] Salguero FJ, Diaz-San SF, Brun A, Cano MJ, Torres JM. Comparisono fth reemo noclonal antibodies for use in immunohistochemical detection of bovine spongiform encephalopathy protease-resistantp rionp rotein. JVe tDiagn Invest 2006;18:106–9.

[60] Schutz A, Tannapfel A, Wittekind C. Comparison of different double immunostaining protocols for paraffin embeddedlive rtissu e. AnalCe llP athol 1999;18:227–33.

[61] Shibata Y, Fujita S, Takahashi H, Yamaguchi A, Koji T. Assessment of decalcifying protocols for detection of specific RNA by non-radioactive in situ hybridization in calcified tissues. Histochem Cell Biol 2000;113:153–9.

[62] Sundquist SJ, Nisenbaum LK. Fast Fos: rapid protocols for single- and double-labeling c-Fos immunohistochemistry in fresh frozen brain sections. JNe urosciM ethods 2005;141:9–20.

[63] Syrjanen S, Partanen P, Syrjanen K. Comparison of in situ DNA hybridization protocols using 35S-labeled and biotin-labeled probes in detection of human papillomavirus DNAs equences. CancerCe lls 1987:329–36.

[64] Vekemans K, Rosseel L, Wisse E, Braet F. Immuno-localization of Fas and FasL in rat hepatic endothelial cells: influence of different fixationp rotocols. Micron 2004;35:303–6.

[65] Waiser J, Schwaar S, Bohler T, et al. Immunohistochemical double-staining of renal allograft tissue: critical assessment of threed ifferentp rotocols. VirchowsAr ch 2002;440:648–54.

[66] Warren KC, Coyne KJ, Waite JH, Cary SC. Use of methacrylate de-embedding protocols for in situ hybridization on semithin plastic sections withmu ltipled etections trategies. JHis tochem Cytochem 1998;46:149–55.

[67] Willmore-Payne C, Metzger K, Layfield LJ. Effects of fixative and fixation protocols on assessment of Her-2/neu oncogene amplification statusb yfl uorescencein s ituh ybridization. Appl Immunohistochem Mol Morphol 2007;15:84–7.

Co-Expression Analyses

INTRODUCTION

One of the most exciting developments with in situ hybridization and immunohistochemistry over the past 5 years is the rapid development of co-expression analyses. As a pathologist, I always find it exciting to see a well-done co-expression analysis slide, because it provides so much information about the interplay of the two targets of interest. The fuel for the development of the field in the past few years has certainly been the explosion in the interest in the biological properties of microRNA (miRNA) molecules.

If you are my age or older, you may remember a term we heard a lot in molecular biology classes 30–35 years ago. That term was "junk DNA." Most of us may recall when the professor indicated that a high percentage of human DNA had no coding function and, thus, was "junk DNA" because it was not doing anything "important." In this era of investments called "high quality" that in reality are junk, it is nice to see that nature reversed that situation and made "junk DNA" into one the most important and valuable discoveries of the past 10 years—noncoding RNA—although, admittedly, this is just one part of "junk DNA."

As the name implies, noncoding RNAs do not generate proteins per se. However, they can have profound effects on the function of key mRNAs and proteins. One of the most important families in the overall class of noncoding RNAs is the miRNAs. As of the writing of this book, 1921 mature miRNAs have been identified in humans. It is estimated that the majority of all metabolic pathways in normal physiology, as well as oncogenesis, are directly impacted by microRNAs.

miRNAs function mainly by sharing sufficient homology with the 3-untranslated region (UTR) of a given mRNA. By binding to the UTR, the miRNA can either stop (or slow down) the transcription of the mRNA to a protein, or it can lead to the mRNA's complete degradation. Either way, the miRNA can effectively eliminate the mRNA from the cell as effectively as a mutation. miRNAs are, as their name implies, very small molecules of typically 20 nucleotides in size. Since typically only 7 to 8 of their nucleotides participate in the binding of the UTR of the target mRNA, we can easily see from our discussions of the melting curve (Tm) that such hybridizations may be fraught with a strong propensity to denature. To "compensate" for this relatively weak force of hybridization, miRNAs have seemed to develop two other features that are important relative to their detection with in situ hybridization:

1. They tend to be in very high copy number when physiologicallyac tive.
2. They tend to be strongly associated with multiple different proteins, including the Argonaute proteins that are part of the RISC (RNA-induced silencing complex)c omplex.

Of course, point 1 works very much to our advantage when doing in situ hybridization. This is the same reason that it usually is very easy to detect DNA viruses by in situ hybridization or immunohistochemistry. The reason is that DNA viruses such as herpes simplex, human papillomavirus, Epstein-Barr virus, and cytomegalovirus produce large amounts of their DNA, RNA, and proteins. Since miRNAs are so small and, thus, would have a low melting temperature in nature, it seems logical to assume that this "primitive" system of mRNA control evolved to be associated with very high copy number when physiologically active. Whatever the reason, this will be enormously advantageous to us when we do in situ hybridization for miRNA.

Point 2 is an interesting point. It has been well established that miRNAs are strongly bound to a variety of proteins that play key roles in "presenting" and "docking" the miRNA to the mRNA target. The multiple protein "cage" associated with microRNAs is called the RNA-induced silencing complex (or RISC for short). It uses the mature miRNA as the template for recognizing the mRNA to which the miRNA has homology in the 3-UTR region. When this happens, as indicated previously, either RNase is activated and the mRNA is destroyed, or the transcription process slows down dramatically, effectively eliminating the physiologic role of this mRNA. It is interesting to speculate that the dense protein cage that surrounds miRNAs in the RISC may help compensate for the relatively weak homology between the miRNA and the mRNA 3-UTR region by providing an environment that tends to stabilize the miRNA/mRNA complex. A key player in the protein scaffold of the RISC is the Argonaute protein family. These proteins represent the catalytic part of the RISC. Indeed, they possess endonuclease activity that may be important in mRNA destruction, as well as in the activation of a precursor miRNA to the active, mature

DOI: http://dx.doi.org/10.1016/B978-0-12-415944-0.00009-7

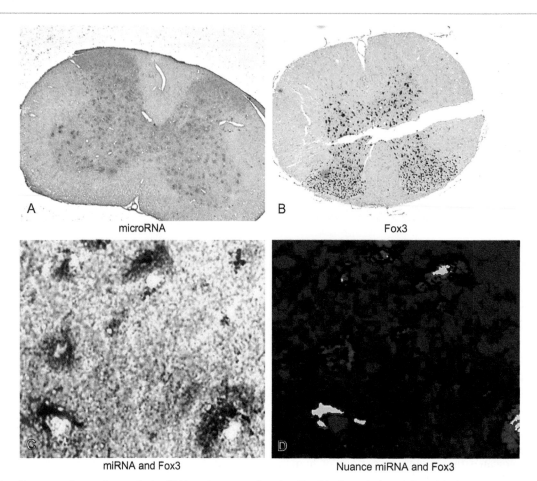

Figure9-1   Co-expression analyses of microRNA and a neuronal marker Fox 3 in the spinal cord. Panel A shows a strong expression of a microRNA in the spinal cord of a mouse that was paralyzed after induced ischemia to the spinal cord. Based on the in situ controls and solution phase data, it was clear that this microRNA was upregulated in the paralyzed mouse. Note the distribution of the microRNA; it is present in the ventral and dorsal horns of the spinal cord. This is the exact region where the larger motor neurons are present in the spinal cord. Of course, the large motor neurons would be the cells that are directly responsible for paralysis. Thus, our knowledge of brain histopathology allows us to have a high index of suspicion that most of the microRNA positive cells are neurons. To further document this point, immunohistochemistry analysis was done on the serial section for the neuron-specific protein Fox 3. As you recall, the use of serial sections allows us to study the same cell populations. Note that Fox 3 has the same distribution as the miRNA (panel B). To get further verification that the miRNA was expressed in neurons, the sections were treated with the miRNA first and then the protein Fox 3. These data are presented in panel C. The miRNA is blue, and the Fox 3 protein is brown. When these co-expression images were analyzed by the Nuance system, the miRNA signal was converted to a fluorescent blue and the Fox 3 signal to a fluorescent red. Cells with both targets are evident as fluorescent yellow (panel D). These data remind us that serial section analyses and a good knowledge of histopathology provide the foundation to co-expression analyses.

form. Interestingly, it has been established that the structure of the Argonaute protein contains a highly conserved basic pocket that is able to bind to the 5-region of the miRNA. The fact that parts of the Argonaute proteins are strongly conserved throughout living forms and found even in bacteria points to their role having developed early in the history of evolution.

We have seen data in previous chapters suggesting that miRNAs may be surrounded by much more developed three-dimensional protein/protein cross-linked cages than other DNA or RNA molecules in the cells in a given formalin-fixed, paraffin-embedded tissue. This may partly explain why, in general, microRNAs tend to need stronger pretreatment conditions than do DNA or RNA molecules in the tissue of interest. The information just

discussed relative to the RISC proteins in general, and the Argonaute proteins in particular, provides us with a more solid biochemical basis to speculate that miRNAs may be surrounded by dense, large protein coats that would, of course, have a strong influence on the ability to detect these very small miRNAs by in situ hybridization.

As indicated, miRNA research has fueled enormous interest in co-expression analyses. There are several ways you can determine whether a given miRNA may be inhibiting a specific mRNA:

1. Do a computer-based scan that matches the so-called seed region of the miRNA with the complementary sequence in the 3-UTR of the mRNA of interest. This can be done via some websites such as TargetScan.

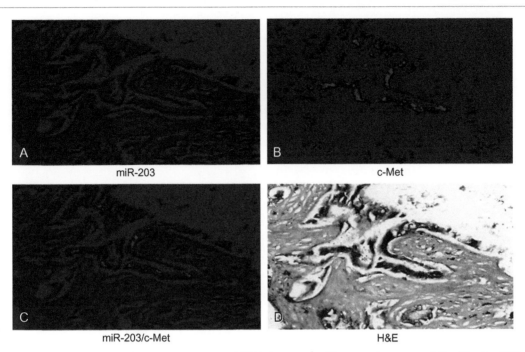

Figure9-2 Co-expression analyses of microRNA-203 and cMET in lung cancer. There are times, especially with microRNA and its putative target, when serial section analysis of the two targets show that in a given nest of cancer cells some of the cancer cells are making the microRNA and some of the cancer cells are making the putative target. In such cases, serial section analyses cannot tell us if these are two distinct populations of cells. To determine whether the cancer cells that express the miRNA are expressing or not expressing the putative protein target, we must do co-expression analyses. The routine microscopy analysis (panel D) shows a nest of cancer cells in which the blue of the tumor suppressor microRNA-203 and the red of the oncoprotein cMET are intermingled. The Nuance system converted the microRNA signal to fluorescent blue (panel A) and the cMET signal to fluorescent red (panel B). Merging the two images (panel C) shows that miR-203 positive cells do not express cMET, and vice versa. This bit of data supports the large amount of solution phase data that suggests that the miR-203 can regulate cMET expression.

2. Use the luciferase assay and, by making vectors that include the luciferase molecule downstream of the 3-UTR sequence, document that the miRNA can inhibit the expression of luciferase, which would happen if the upstream mRNA was not being transcribed due to the presence of the miRNA.
3. Show that by mutational analysis to the 3-UTR region noted in point 2, earlier in the chapter, that the inhibition of expression of the mRNA via the miRNA is lost. That is, you can change the sequence of the 3-UTR such that the miRNA is no longer able to bind to it. Thus, the luciferase is normally expressed.

Although each of the three methods detailed here is useful, all are done in the test tube and, thus, none directly examine the physiologic effect of the miRNA on the putative mRNA target. This can only be done with co-expression analysis for the miRNA and target in question. Co-expression analyses also allow us to readily answer a question that cannot be answered if the miRNA analysis is done by solution phase methods, such as qRTPCR, since the tissue is ground up and destroyed in such analysis. Specifically, the co-expression analysis of the microRNA and phenotypic marker will allow us to determine the specific phenotype of the cell that is expressing the miRNA. These two key points are illustrated in Figures 9-1 and 9-2. In Figure 9-1, panel A shows a strong expression of a microRNA in the spinal cord of a mouse that

was paralyzed after induced ischemia to the spinal cord. When compared to the control, this microRNA was much increased in the mouse with the ischemic cord. Note the distribution of the microRNA; it is present in the ventral and dorsal horns of the spinal cord. This is the exact region that the larger motor neurons are present in the spinal cord. Of course, the large motor neurons would be the cells that are directly responsible for paralysis. Indeed, these are the target cells of the polio virus, which, of course, causes paralysis in its victims. To further document that these cells that are positive for the miRNA upregulated in induced spinal cord paralysis in the mouse are neurons, immunohistochemistry analysis was done on the serial section for the neuron-specific protein Fox 3. As you recall, the use of serial sections allows us to study the same cell populations. Note that Fox 3 has the same distribution as the miRNA (panel B). To get further verification that the miRNA was expressing cells, the sections were treated with the miRNA first and then the protein Fox 3. These data are presented in panel C. The miRNA is blue, and the Fox 3 protein is brown. When these co-expression images were analyzed by the Nuance system, the miRNA signal was converted to a fluorescent blue and the Fox 3 signal to a fluorescent red. Cells with both targets are evident as fluorescent yellow. As is evident from Figure 9-1, panel D, many cells that express the miRNA are neurons, since they co-express the neuronal marker Fox 3.

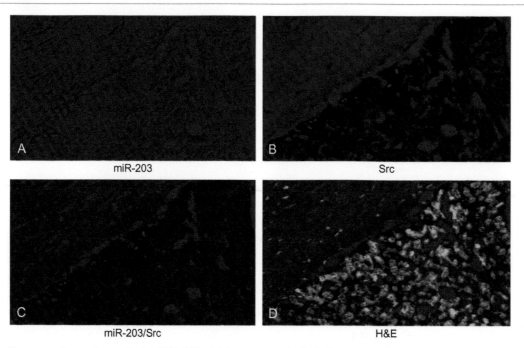

Figure9-3 Co-expression analyses of microRNA-203 and the oncoprotein Src in lung cancer. In this tumor the slide was analyzed for miR-203 and Src after the optimal conditions were detected for each target in this particular tissue. It is clear that the tumor cells in this area do not express miR-203 (panel A, fluorescent blue), but that basically all express Src (panel B, fluorescent red). Thus, co-expression analyses (panel C) showed no tumor cells expressing both. This figure also shows the value of adding a counterstain (methylene green), because it allows us to see the cancer cells (lower part of image) and adjacent stromal cells (upper part of image, panel D). Still, I prefer to use the regular RGB image (panel D, Figure 9-2) to show the histologic features of the tissue when using the Nuance system.

Figures 9-2 and 9-3 show two different examples of how co-expression analysis can be very useful for the study of putative targets of microRNAs. Work done by Dr. Michela Garofalo in the laboratory of Dr. Carlo Croce showed that, in lung cancers, certain critical oncoproteins may be directly regulated by specific microRNAs. Two such oncoproteins that are increased in lung cancer, and associated with a worsened prognosis, are cMET and Src. These researchers had strong data that indicated miR-203 downregulation could be responsible, in large part, for the increased expression of cMET and Src in lung cancers. Clearly, this would give clinicians another possible route to treat lung cancers, because the introduction of miR-203 would simultaneously reduce the expression of these two (and probably other) oncoproteins. When we did serial section analysis of miR-203, cMET, and Src in the lung cancers, it was obvious that some cancer cells were expressing each of the three targets. The best way to examine their physiologic interdependence was to do co-expression analysis. As seen in Figures 9-2 and 9-3, the cancer cells that are expressing miR-203 are clearly not expressing cMET or Src, and vice versa. This gives a powerful additional layer to the foundation of the hypotheses that miR-203 may be regulating these two important oncoproteins.

Let's examine Figures 9-4 and 9-5 for the "reverse image" of the downregulation experiments seen in Figures 9-2 and 9-3. Figure 9-4 shows a series of cancer cell lines that were either exposed to the oncolytic virus called reovirus or given a "sham" infection. It is known that reovirus can kill cancer cells, and that its entry into

malignant cells is associated with proteins such as caspase-3 (which can lead to the direct apoptosis of the cancer cells) and p38 (which enhances reovirus entry into the cancer cell). Finally, it is known that PKR inhibits the growth of the reovirus in the cancer cells, and indeed, the ability of p38 to enhance reoviral growth probably reflects its ability to inactivate PKR. Note in panel A we did a co-expression of reovirus and caspase-3 in a reovirus-treated cell line. Of course, you cannot examine serial sections of cell preparations, because you simply add the cells directly to the silane coated slide. (Of course, you could put the cells in a paraffin block and then cut sections, but I do not see any advantage to adding this step.) Again, it is simply not possible for our eyes to discriminate the red reovirus signal from the brown caspase-3 signal from the blue hematoxylin signal. But, of course, a computer-based system can do this easily. First, note in panel B that the different colors can be easily separated out by the Nuance system (caspase is fluorescent green, and the insert shows the fluorescent red signal of the reovirus). When the signals are mixed, you can see the fluorescent yellow that indicates most of the cancer cells with the reoviral protein also express caspase-3. As a negative control, reovirus and PKR were analyzed by co-expression (panel D). Note the lack of co-expression, which serves as a good negative control of the reovirus/caspase-3 data. Panel E shows that the reovirus positive cells also express p38 in the reovirus-infected cell line. Panel F is the corresponding negative control of reoviral/p38 co-expression analyses in a cell line not exposed to the virus.

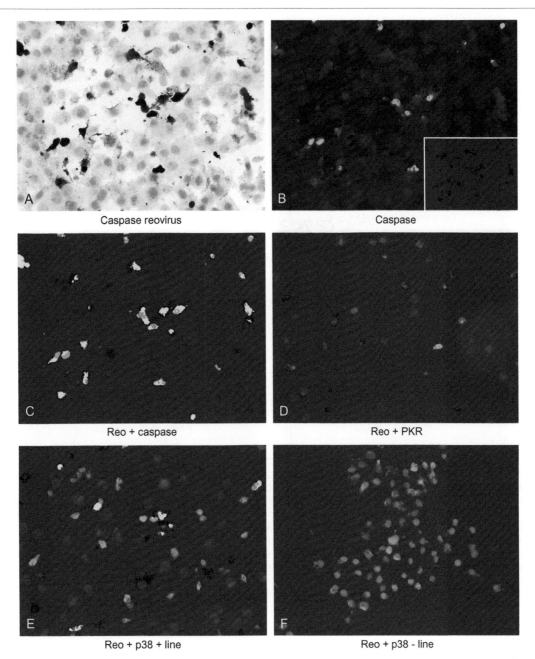

A
Caspase reovirus

B
Caspase

C
Reo + caspase

D
Reo + PKR

E
Reo + p38 + line

F
Reo + p38 - line

Figure9-4 Molecular correlates of reovirus infection in cancer cells. Panel A shows the regular light microscopy image for the co-expression of reovirus and caspase-3 in a reovirus-treated cell line. Of course, you cannot examine serial sections of cell preps because you simply add the cells directly to the silane-coated slide. Plus, it not possible for the naked eye to discriminate the red reovirus signal from the brown caspase-3 signal from the blue hematoxylin counterstain—hence, the value of the computer based co-expression imaging. Note in panel B the isolated caspase signal (fluorescent green); the insert shows the fluorescent red signal of the reovirus. When the signals are mixed, you can see the fluorescent yellow that indicates most of the cancer cells with the reoviral protein also express caspase-3 (panel C). As a negative control, reovirus and PKR were analyzed by co-expression (panel D). Note the lack of co-expression, which serves as a good negative control of the reovirus/caspase-3 data. Panel E shows that the reovirus positive cells also express p38 in the reovirus-infected cell line. Panel F is the corresponding negative control of reoviral/p38 co-expression analyses in a cell line not exposed to the virus. These images show the value of computer-based analyses when doing co-expression analyses.

Figure 9-5 shows the equivalent data in reoviral-infected tissue. The tissue came from a patient with colon cancer who had a liver metastasis. The patient was given reovirus before the removal of the liver metastasis. We did co-expression of reovirus with tubulin, p38, and cas-pase-3 in serial sections. The purpose of showing this data is three-fold: (1) it highlights the value of analyzing serial sections for co-expression analyses, since we are examining the same sets of cells; (2) it shows that we cannot determine with regular microscopy whether cells are co-expressing two different proteins if the two are in the same cellular compartment (each protein in this

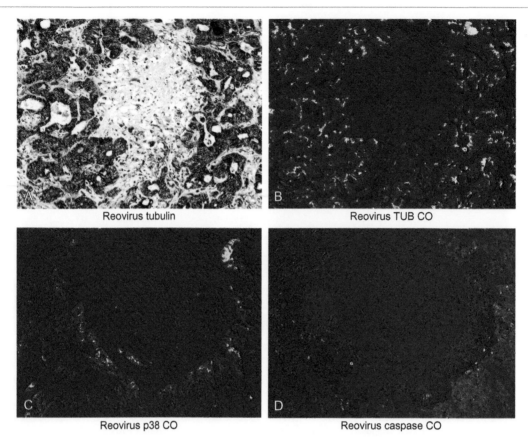

Reovirus tubulin

Reovirus TUB CO

Reovirus p38 CO

Reovirus caspase CO

Figure 9-5   Molecular correlates of reovirus infection in tissue. Unlike the experiments in Figure 9-4, when using tissue, we can combine co-expression analysis with serial section testing, which increases the power of the assays by doing the analysis in the same groups of cancer cells. The tissue is an adenocarcinoma from a person treated with reovirus prior to the surgical removal of the tumor. Note that the routine microscopy (panel A) shows the tumor cells and surrounding stromal cells very clearly after the immunohistochemical analysis of reovirus and tubulin. Each epitope required 30 minutes of antigen retrieval. Thus, the reovirus analysis (Fast Red signal) was done first using antigen retrieval, followed by tubulin analysis (brown signal) in which no additional pretreatment was done. A common mistake made at this stage is to redo the antigen retrieval after the reovirus for the tubulin. This would amount to a total of 1 hour of antigen retrieval; most antigens will not yield good signals under such conditions. Note that indeed most of the cells with detectable reovirus were productively infected based on the co-expression with tubulin (panel B, seen as fluorescent yellow). Serial section analysis showed most of the cancer cells were p38 positive (panel C), which makes sense because expression of this antigen facilitates the replication of the virus in the cell. Also note that the reovirus protein was found only in the cancer cells. Finally, note that a minority of the productively infected cells expressed caspase-3, which parallels the cancer cells undergoing apoptotic death (panel D).

set is cytoplasmic based); (3) we can clearly see perhaps the strongest feature of in situ hybridization and immunohistochemistry: the ability to document that the reovirus is specifically targeting cancer cells. Such information could not be obtained from PCR or Western blot-based data.

Reovirus needs to attach to the microtubulin scaffold when productively infecting a cell. As seen in Figure 9-5, panels A (regular image) and definitively in panel B (Nuance-based image), most of the reovirus-infected cells indeed co-express microtubulin. Thus, the virus has not been simply passively absorbed by the cells but, rather, is productively replicating in the cancer cells. The serial section shows that the reovirus positive cells co-express p38 (panel C), and that a smaller percentage of these cancer cells are also expressing the protein caspase-3 that likely signals their death via apoptosis (panel D).

Now let's look at different ways we can do co-expression analyses.

DIFFERENT METHODOLOGIES FOR CO-EXPRESSION ANALYSES

USE EITHER TWO PROBES OR TWO ANTIBODIES THAT CAN DETECT THEIR RESPECTIVE TARGETS IN COMPLETELY DISTINCT CELL POPULATIONS

Using either two probes or two antibodies is certainly the easiest way to do co-expression analyses. The only two criteria for using this simple form of co-expression analyses are:

1. The two different targets must be in two completely different cell populations that are morphologically distinct.
2. The two different targets need to have similar immunohistochemistry or in situ hybridization optimizationp rofiles.

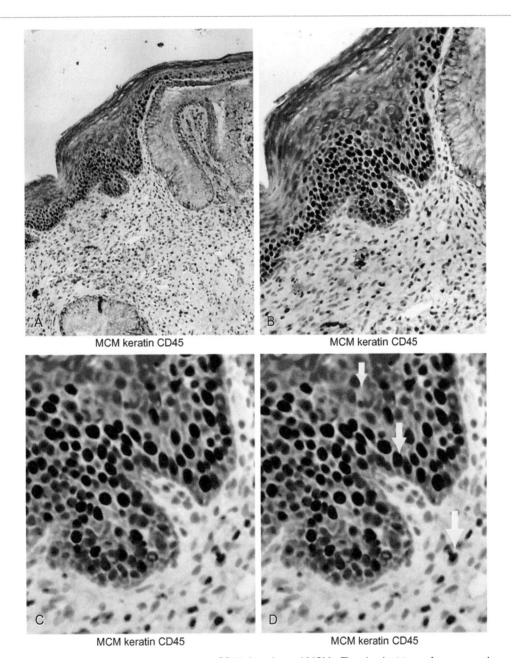

MCM keratin CD45

MCM keratin CD45

MCM keratin CD45

MCM keratin CD45

Figure9-6 Co-expression analysis with one chromogen: CD45; keratin; and MCM₂. The simplest type of co-expression analysis is when two or more targets are present in different cell types and/or cell compartments that are easily differentiated on cytologic grounds. Keratin is found in the cytoplasm of squamous cells, MCM₂ is a nuclear epitope present in rapidly dividing cells, and CD45 is present in the cytoplasm of lymphocytes. Thus, we can analyze a CIN biopsy for all three targets and get the same exact results as if three serial sections were used, one for each target. This saves reagents and time, and reminds us that the generation of a given signal with immunohistochemistry or in situ hybridization will not interfere with the simultaneous development of another signal. Note in panel A that the entire squamous epithelia clearly shows the cytoplasmic signal that corresponds to cytokeratin AE1/3, whereas the more basal cells of the squamous cell layer show the intense nuclear signal of MCM₂. Many cells in the stroma show the cytoplasmic signal of CD45; T- and B-cells are invariably present in the stroma of the cervix. These three distinct regions/cellular localization patterns are seen in panel D as *small* arrow (squamous cells), *middle*-sized arrow (MCM₂), and *large* arrow (lymphocytes).

In this book, we have already seen many examples in which two different protein epitopes were detected simultaneously by doing immunohistochemistry with one chromogen and two separate primary antibodies. Figure 9-6 shows one such example. The tissue is a cervical biopsy from a woman who had cervical intraepithelial neoplasia (CIN). This is the precursor lesion to cervical cancer, and it is invariably associated with human papillomavirus (HPV) infection. The lesion was analyzed *at the same time* with cytokeratin (epithelial marker), MCM₂ (mitotic activity marker), and CD45 (lymphocyte marker). Of course, we can easily differentiate the cytoplasmic signal of cytokeratin from that of CD45. The reason is that the cytokeratin is in the very large, stacked squamous cells, whereas the CD45 signal is in the much smaller lymphocytes that dominate in the submucosa. The MCM$_2$ signal is easily differentiated from the other two signals because it is nuclear and the other two are cytoplasmic.

Note in Figure 9-6, panel A, that the entire squamous epithelia clearly shows the cytoplasmic signal that corresponds to cytokeratin AE1/3, whereas the more basal cells of the squamous cell layer show the intense nuclear signal of MCM$_2$. Finally, the stroma, as expected, shows the CD45 positive cells, because the lymphocytes will predominate in this area. These three distinct regions/cellular localization patterns are seen in panel D as *small* arrow (squamous cells), *middle*-sized arrow (MCM$_2$), and *large* arrow (lymphocytes).

As you are probably aware, several companies offer excellent co-labeling kits. These kits allow you to detect two (or more) antigens in a given immunohistochemical experiment. These commercial kits use one color for one antigen and another color for the other antigen, which typically are located in completely different cell populations. The results I have seen from such products are excellent. However, it is clear that we can do the exact same experiments right now using the same chromogen under the guidelines outlined previously. I see three solid advantages to this simplest of co-expression analysis:

1. It strengthens our surgical pathology/histopathology knowledge because it requires us to be able to differentiate different cell types and the cytoplasm from then ucleus.
2. It strengthens our immunohistochemical and in situ hybridization knowledge because it requires us to have a thorough knowledge of the optimization profiles of the two or more targets.
3. It allows us to save money on reagents and generate much more data with fewer slides and experiments. I suppose this may reflect my Vermont upbringing, since Vermonters are known for their frugality!

Let's look at one more example of doing multiple analyses (again for three distinct targets) at the same time, using the same chromogen. The tissue is breast cancer. The three targets are: (1) the cancer cells (of course, easily differentiated by the disorganized growth pattern and the variation in nuclear size, shape, and color. Do you remember the name of the classic pattern of adenocarcinoma where glands form within the larger gland? See below); (2) smooth muscle cells (easily distinguished since they line large blood vessels); and (3) lymphocytes (small round cells in the stroma).

As seen in Figure 9-7, panel A, the cancer cells are positive for AKT (nuclear-based signal after this oncogene is activated—small arrow), the lymphocytes are positive for CD45 (intermediate arrow), and the smooth muscle cells are positive for SMA (large arrow). Panel B shows the adenocarcinoma and lymphocytes at higher magnification, and panel C shows the smooth muscle cells at higher magnification. By the way, this tissue was 12 years old, and the AKT signal was not even evident in the original material. However, when the tissue was pretreated with IHCPerfect, the AKT signal became evident (panels A and B), and the CD45 and SMA signals became stronger (panel C, after rescue of the signal, and panel D, the

original data without pretreatment). These data remind us that making a stronger three-dimensional macromolecule cross-linked network, of course, not only makes invisible signals evident, but can also make signals that are 1+ or 2+ become stronger. By the way, the adenocarcinoma pattern in panel A is called the *cribiform* pattern.

Of course, there will be many times when we want to analyze two or more targets in the same tissue sections in cases where the different targets will be present in either the same cell types and/or the same subcellular compartments. In such circumstances, we will need to use other strategies, which we discuss next.

ANALYZE THE SAME SECTION MORE THAN ONCE

Let's go over two points before we discuss doing co-expression analyses by analyzing the same tissue section two or more times with immunohistochemistry, in situ hybridization, or both:

- **Point 1.** Although serial section analysis is very helpful with co-expression analyses, it cannot reliably tell us whether a given cell is making each target.
- **Point 2.** Under a regular light microscope, you cannot reliably differentiate two different colored signals if they are in the same cellular compartment of the same cell.

Let's discuss each point in turn.

Regarding point 1, let's examine Figure 9-8, panels A and B. Note that these are clearly serial sections, with panel A showing the histologic distribution of cKIT (CD 117) in the tonsil and panel B showing the distribution of IL22 in the serial section. Note the cluster of adipose cells on the right of each image and the solitary fat cell on the left. Clearly, these two targets show a similar distribution pattern in the interfollicular region of this tonsil. But we cannot reliably say whether a given cell has both targets, even though the sections are only 4 microns apart. Compare this to Figure 9-1. Here, we could clearly say that the miRNA and the protein target Fox3 had the same *population* distribution (larger motor neurons of the spinal cord), which is strong evidence that this cell type is expressing both. But in the tonsil, the cells are smaller and not as cytologically distinct. So, the best plan of action with the tonsil is to go to computer-based co-expression analysis. The results of such an experiment are seen in Figure 9-8, panel C (light microscopy), where the same tissue was analyzed using Fast Red for cKIT and then using DAB for IL22. The Nuance system converted the cKIT image to fluorescent red (panel D) and the IL22 signal to fluorescent green (panel E). By merging the images, the Nuance system clearly shows that cells making cKIT (a marker of natural killer cells) also express IL22 (seen as fluorescent yellow in panel F). Indeed, basically all the cKIT+ cells are also IL22+ in this image.

It follows from such data that a computer-based co-expression system will be very helpful when doing co-expression analyses if both targets are in the same cell types and in the same subcellular compartments. I want to stress that the analysis of *serial sections* of the different targets is still an *essential* part of the co-expression analyses in such cases.

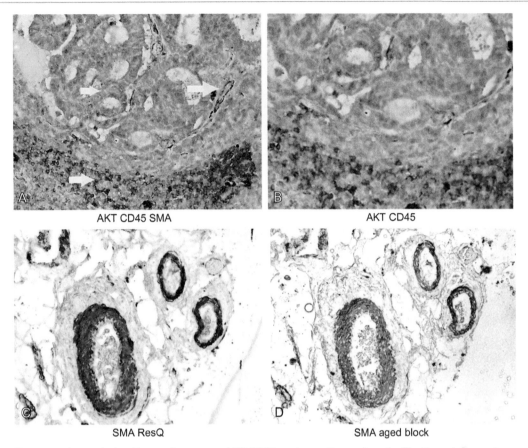

AKT CD45 SMA

AKT CD45

SMA ResQ

SMA aged block

Figure9-7 Co-expression analysis with one chromogen: AKT; CD45; and smooth muscle actin. The tissue is from a lung cancer biopsied 12 years prior to this testing. Note the large complex gland in the middle of panel A. Do you recall what this pattern (many lumens inside a larger gland) is called? It is called cribiforming, and is diagnostic of adenocarcinoma. Note that the tumor cells are positive for the oncogene AKT (small arrow), the lymphocytes are positive for CD45 (intermediate arrow), and the smooth muscle cells of a blood vessel that is supplying blood to the tumor are positive for SMA (large arrow). Panel B shows the adenocarcinoma and lymphocytes at higher magnification, and panel C shows the smooth muscle cells at higher magnification. The AKT signal was not even evident in the original material. However, when the tissue was pretreated with IHCPerfect, the AKT signal became evident (panels A and B), and the CD45 and SMA signals became stronger (panel C, after rescue of the signal, and panel D, the original data without pretreatment). These data remind us that making a stronger three-dimensional macromolecule cross-linked network, of course, not only makes invisible signals evident, but can also make signals that are 1+ or 2+ become 3+, to use the typical pathology jargon.

Now let's review some observations that are relevant to co-expression analyses that we have discussed in previous chapters before we go on to point 2:

A. You can still get good immunohistochemical or in situ hybridization results even if the tissue has been stained with hematoxylin, eosin, or hematoxylin and eosin.

B. You can still get good immunohistochemical or in situ hybridization results even if the tissue has already been analyzed by immunohistochemical or in situ hybridization.

C. Point 2 assumes one of two conditions:

  a. The second target has the *same* pretreatment optimizing regime as the first target. (This is by far the most common scenario where, for example, each target needs either protease digestion or antigen retrieval/DNA/RNA retrieval for optimald etection.)

  b. The pretreatment of the first target does not interfere with the pretreatment regime of the second target. (A good example of this is where no pretreatment is needed to detect HPV in situ followed by immunohistochemical for a protein that needs antigen retrieval for optimal detection.)

D. It is *not* possible to reliably remove the precipitate when using DAB, NBT/BCIP, or Fast Red (Ventana) as the chromogen.

E. It *is* possible, however, to use a chromogen that is soluble in ethanol or xylene such that you can do immunohistochemistry, photograph the results, and then remove the chromogen and stain the exact same tissue for another target using a different colored chromogen.

Points A–C remind us that we can readily stain a given tissue for two or more targets under the conditions listed in these points, and then use a computer-based system to tabulate and analyze our data. Point E reminds us that we do not have to do this if we use a chromogen (such as Fast Red from Biogenex) that is ethanol or xylene soluble.

Regarding point 2—not being able to reliably differentiate two different colored signals if they are in the same cellular compartment of the same cell under a regular light

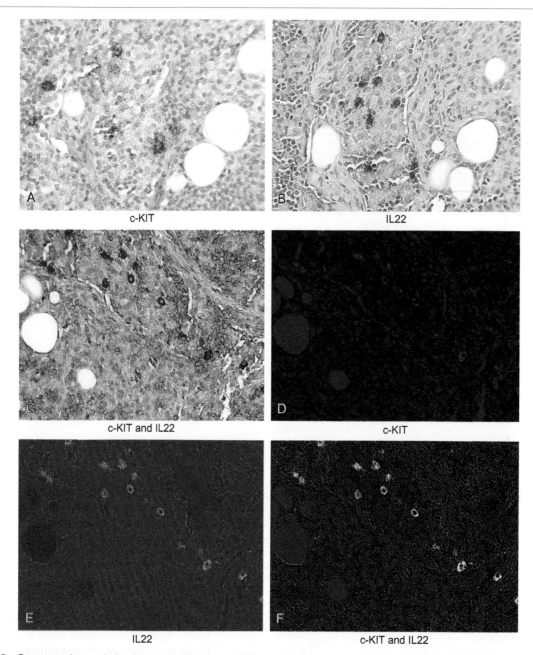

Figure 9-8 Co-expression analysis with two chromogens: cKIT and IL22. Panels A and B show serial sections from a tonsil showing the histologic distribution of cKIT (CD 117, panel A) and IL22 in the serial section. Groups of fat cells were used as the landmarks to illustrate that these two targets show a similar distribution pattern in the interfollicular region of this tonsil. Also note that the positive cells for each have a very similar cytologic appearance, and the signal is in both instances in the cytoplasm. Hence, we need to use two different chromogens to do the co-expression analysis. This is seen in panel C (light microscopy), where the same tissue was analyzed using Fast Red for cKIT and then using DAB for IL22. The Nuance system converted the cKIT image to fluorescent red (panel D) and the IL22 signal to fluorescent green (panel E). By merging the images, the Nuance system clearly shows that cells making cKIT (a marker of Natural Killer cells) also express IL22 (seen as fluorescent yellow in panel F). IL22 production in NK cells defines a specific stage of NK cell activation.

microscope—this is evident by re-examining panel D of Figure 9-8. These points remind us that, if two or more signals are present in the same cells in the same cellular compartments, then we need to follow one of two protocols:

- **Protocol 1.** Do the first test with Fast Red as the chromogen (soluble in ethanol), aquamount the slide, photograph the regions of interest, remove the coverslip, remove the chromogen with an ethanol wash, then do the second immunohistochemical reaction, and photograph the same cells.

- **Protocol 2.** Do the two separate immunohistochemical and/or in situ hybridization reactions with two separate chromogens. Then analyze with a computer-based system that is capable of isolating the different chromogens, mixing them, and thus performing the co-expression analyses.

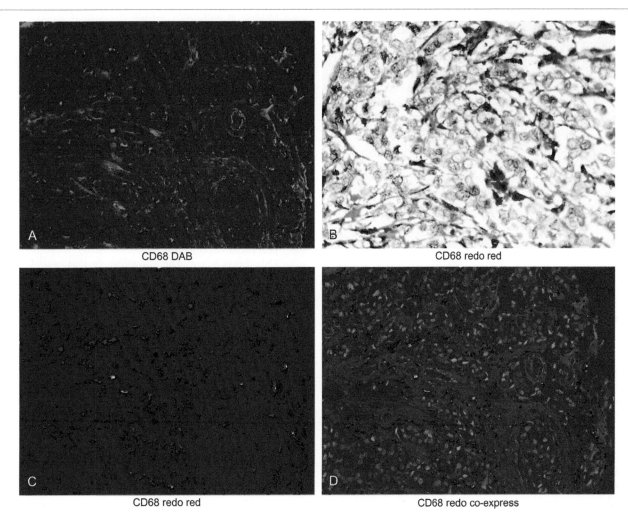

CD68 DAB

CD68 redo red

CD68 redo red

CD68 redo co-express

Figure9-9   Co-expression analysis for the same antigen as a quality control tool. Panel A shows the immunohistochemistry using CD68 as the chromogen in a lung cancer. Note that very few cells were positive (panel A shows the results after Nuance conversion of the DAB image to fluorescent green). The suspicion was that the CD68 DAB signal was not optimized, as the laboratory used too low a concentration of the primary antibody and the incorrect pretreatment. To confirm this, we did CD68 immunohistochemistry again, but this time including protease digestion and using Fast Red as the chromogen. Panel B shows many more positive cells (regular light microscopy image). Panel C shows that by then doing co-expression for CD68 (DAB and Fast Red) and using fluorescent yellow as the co-expression indicator, the protease digestion and higher concentration of the antibody produced many more positive cells (fluorescent red alone). This again reminds us that a prior immunohistochemistry reaction should not interfere with a subsequent reaction, unless the pretreatment for the first antigen eliminates the possibility of detecting the second antigen. Panel D shows the same image, but the fluorescent blue of the hematoxylin counterstain was added to show the nuclei of the cells.

Having done many experiments using Protocol 1, I can attest that it is not too difficult. Of course, it is time consuming. Protocol 2 is much quicker, and you have a permanent record of both target visualizations. Of course, Protocol 2 requires investment in a computer-based co-expression analysis system (see Appendix).

Before we leave this section, I would like to go over the use of co-expression analyses as a quality control tool for immunohistochemistry. Figure 9-9 shows the results for immunohistochemistry using CD68 as the chromogen in a lung cancer. Very few cells were positive (panel A shows the results after Nuance conversion of the DAB image to fluorescent green). The suspicion was that the CD68 DAB signal was not optimized. To confirm this, we did CD68 immunohistochemistry again, but this time including protease digestion and using Fast Red as the chromogen.

Panel B shows many more positive cells (regular light microscopy image). Panel C shows that by then doing co-expression for CD68 (DAB and Fast Red) and using fluorescent yellow as the co-expression indicator, the CD68 assay for this tissue needed the protease digestion step. Panel D shows the same image, but the fluorescent blue of the hematoxylin counterstain was added to better appreciate the tissue morphology.

USE COMPUTER-BASED CO-EXPRESSION

The rest of this chapter is devoted to doing co-expression analyses of two or three targets using a computer-based co-expression analysis system.

Let me start by presenting the protocol I recommend for co-expression analysis where it is assumed that the two (or more) targets in question are located, at least in part, in the same subcellular compartment of the same cells. I begin by assuming that a large percentage of the readers interested in doing co-expression analyses wish to do so for a micro-RNA and a putative target. Thus, the protocols use micro-RNA by in situ hybridization and putative protein target by immunohistochemistry as the template system. If you are interested in doing protein/protein co-expression, then simply substitute the word "protein" for the word "micro-RNA." If you wish to do RNA/DNA in situ hybridization co-expression, just substitute the letters "RNA" for "micro-RNA" and the letters "DNA" for "protein," and so on.

PROTOCOL FOR CO-EXPRESSION ANALYSIS IN FORMALIN-FIXED, PARAFFIN-EMBEDDED TISSUES

STEP ONE: OPTIMIZING STEP

Determine the optimal conditions, including necessary pretreatment (if any) and probe concentration for the microRNA of interest using the protocol in Chapter 7. Simultaneously, determine the optimal conditions including necessary pretreatment conditions (most likely some pretreatment will be needed) and primary antibody concentration for the epitope of interest using the protocol in Chapter 8. *Remember to test both the microRNA and protein with no pretreatment, protease alone, "RNA or antigen retrieval" alone, and RNA/antigen retrieval plus protease digestion*, because this will be important information.

STEP TWO: DETERMINE THE ORDER OF CO-EXPRESSION ANALYSIS

Write down the results for the optimizing experiments for the tissue of interest for the detection of the microRNA and the protein. Remember to pay close attention to the results of *all* experimental conditions. If antigen retrieval yielded a 3+ signal for immunohistochemistry, but protease digestion alone and antigen retrieval + protease digestion each yielded a 2+ signal with no background, the latter two conditions may be acceptable for co-expression analyses. This is especially true if using the Nuance-based computer system for analyses, because it is much more sensitive to the signal than are our naked eyes.

Choose one of the following sets of conditions:

A. miRNA – no pretreatment optimal and protein, protease digestion or antigen retrieval optimal
   The microRNA in situ hybridization must go first (recall, this is very rare).
B. miRNA – protease digestion optimal and protein, protease digestion optimal
   Either the microRNA in situ hybridization or protein immunohistochemistry can go first. If the latter is done first, remember to follow RNase precautions for the microRNA in situ hybridization.
C. miRNA – protease digestion optimal and protein, antigen retrieval optimal

This is a common situation. Under such circumstances, examine the scores for the miRNA in situ hybridization if RNA retrieval was used and the scores for immunohistochemical if protease digestion alone was used. Remember that antigen retrieval alone (or protease digestion alone) may yield too much background for immunohistochemistry with one concentration, but each may yield excellent results for both if the concentration of the primary antibody is reduced. In my experience, in more than 90% of these cases (miRNA –protease digestion is optimal, and protein – antigen retrieval is optimal), you can find conditions that will allow both the miRNA and protein to be detected simultaneously.

D. miRNA – RNA retrieval optimal and protein, antigen retrieval optimal
   This too is a common situation. In this case, I recommend doing the microRNA in situ hybridization first and then the protein detection by immunohistochemistry. Remember to *not* redo the antigen retrieval for the protein or the signal will probably be diminished.
E. miRNA – RNA retrieval optimal and protein, protease digestion optimal
   This is an unusual situation. When it does occur, in my experience, the protein can usually be detected satisfactorily with antigen retrieval if the primary antibody concentration is adjusted.

To restate this important point, in most cases scenario C or D here will apply. This is another reason I have stressed in the optimization protocols for micro-RNA/RNA and DNA in situ hybridization and protein detection by immunohistochemistry that you test different concentrations *and* different pretreatment conditions (nothing, protease digestion, antigen retrieval, and antigen retrieval + protease digestion), because this will allow you to see the range of possible pretreatments for co-expression analyses. Finally, I purposely left out some possibilities, such as both microRNA and protein need no pretreatment, because I have never seen this and, if it happened, clearly you could test each in sequence.

STEP THREE: TEST SERIAL SECTIONS FOR THE microRNA BY IN SITU HYBRIDIZATION AND THE PROTEIN BY IMMUNOHISTOCHEMISTRY

I consider Step Three to be the key step when doing co-expression analyses. In this step, you take serial sections and analyze them for the microRNA and protein of interest. You then carefully record and interpret the data.

In this step, you bring together the two critical fields of expertise needed for optimal interpretation of immunohistochemistry and in situ hybridization. These are, of course, an in-depth knowledge of molecular pathology *and* histopathology. I like to photodocument the data at this stage, taking photos of the same areas of the serial sections so as to study the same groups of cells for the two different targets.

The reason I consider Step Three to be the most important step of co-expression analyses is that its goal is to

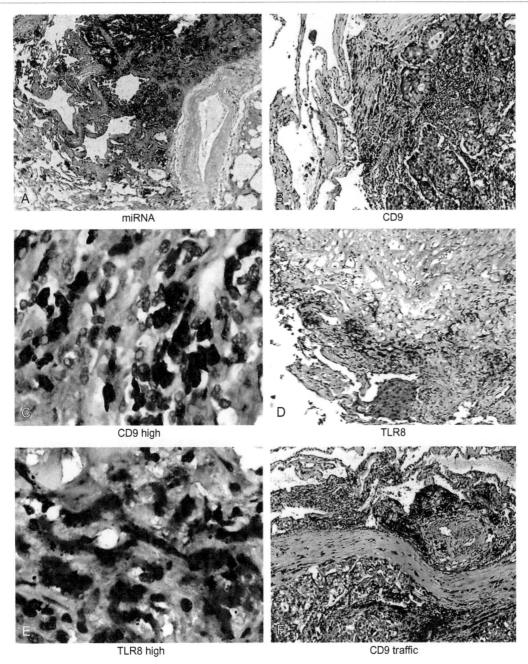

Figure9-10 Use of multiple co-expression analyses to dissect the molecular "cross-talk" between cancer and inflammatory cells in lung cancer: Part A: miR-29a was much increased in the cancer cells (upper and mid part of panel A), but not in the cells of the adjacent normal lung present at the tumor interface. In the serial sections of this lung cancer, CD9 was found in the cancer cells and in some inflammatory cells, but only at the interface of the cancer (panel B; CD9 is a marker of exosomes, which are the vehicle that cells can use to transport molecules to other cells). A higher magnification of the tumor interface shows that the CD9 positive cells had the cytologic features of macrophages (panel C), reminding us again of how invaluable a strong foundation in histopathology can be in such interpretations. Analysis of the serial section for TLR-8 shows that the positive cells were found mostly at the tumor interface and in cells with the cytologic features of macrophages (panel D and, at higher magnification, panel E). Using the fact that macrophages routinely traffic from the cancer to the regional lymphoid accumulations/lymph nodes, we obtained additional evidence that the CD9 positive cells were probably macrophages, as they were concentrating in the lymphoid aggregates around the tumor (panel F).

basically let you determine the interplay between the two (or more) targets *before* you do the computer-based analyses. In other words, I prefer to use the computer-based analysis system as the *corroborative* test and not the *primary* test for co-expression analysis.

Let's look at a couple of examples in Figures 9-10 and 9-11. This work came from the laboratory of Dr. Carlo Croce and included several of his team members, including Dr. Muller Fabbri and Federica Calore. They had extensive solution phase evidence suggesting that certain

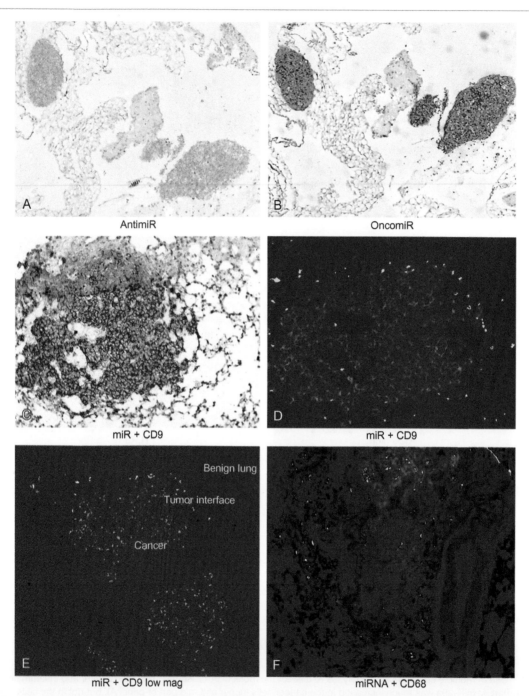

AntimiR

OncomiR

miR + CD9

miR + CD9

miR + CD9 low mag

miRNA + CD68

Figure9-11   Use of multiple co-expression analyses to dissect the molecular "cross-talk" between cancer and inflammatory cells in lung cancer: Part B: In these experiments, the cancer cells were transfected with anti-miRNAs, injected into the mouse, and then the lung metastases were examined to determine whether the anti-miRNA had rendered the cancer cells incapable of metastasis. These experiments are based on the observation that not all cancer cells transfected with a given anti-miRNA will acquire the anti-miRNA. Indeed, we did document that the cancer cells that metastasized did not have the sequence of the "anti-miRNA" used to silence the oncomiRNA (panel A), but did still have the sequence of the oncomiRNA (panel B). We then showed that the metastases in the mice showed the same geographic pattern of the miRNA and CD9 (concentrating at the tumor interface; panel C) as we noted in the human tumor (Figure 9-10). The co-expression analyses did indeed show that the miRNA and CD9 were concentrating in a thin rim specifically at the tumor interface (panel D), but not in the alveolar macrophages 1 mm away in the normal lung (panel E). The co-expression analyses also showed that some of the miRNA+ cells at the tumor interface were CD68+ macrophages (panel F) which, since they were not making the miRNA, must have acquired it via the exosomal release from the cancer cell, in a strong example of how cancer cells can influence the surrounding benign inflammatory cells via "cross-talk" mediated by the microRNAs.

microRNAs could be exported by cancer cells and, in turn, modulate inflammatory cell function. They provided me with extensive mouse tissue and human tissue with a lung cancer model. They wanted me to address the issue of whether co-expression analysis could provide additional evidence that there was direct trafficking between the cancer and inflammatory cells. Some representative data are shown in Figure 9-10. Let's summarize what this figure shows:

1. In the human cancer, the microRNA was found mostly in cancer cells (panel A).
2. In the serial sections of the human cancer, CD9 was found in the cancer cells and in some inflammatory cells, but only at the interface of the cancer and the benign tissue (panel B; CD9 is a marker of exosomes, which are the vehicle that cells can use to transport molecules to other cells).
3. A higher magnification of the tumor interface shows that the CD9 positive cells had the cytologic features of macrophages (panel C).
4. Analysis of the serial section for TLR-8 shows that the positive cells were found mostly at the tumor interface and in cells with the cytologic features of macrophages (panel D and, at higher magnification, panelE).
5. Since macrophages routinely traffic from the cancer to the regional lymphoid accumulations/lymph nodes, we looked to see whether the CD9+ cells were found in the regional lymphoid accumulations. As you can see in panel F, CD9+ cells were indeed concentrating in these areas.

So, to summarize, before doing any co-expression experiments, we obtained a lot of information from the serial section experiments for the different targets of interest. It seems clear that the miRNA is associated with an exosomal marker, but in a very specific geographic locale—specifically, where the tumor meets the adjacent benign tissue, commonly called the tumor interface. It is also clear that the tumor interface is the place where macrophages that contain the exosomal marker CD9 are congregating. Finally, the TLR-8 accumulation at this site would be consistent with evidence that the cancer cells may indeed be modulating the macrophage's behavior.

Dr. Croce's laboratory staff then documented that if they partially knocked out the miRNA in the mouse model, then far fewer lung metastases were evident. Of course, partial knockout via transfection will allow some cancer cells to still express the microRNA. Here is a good example of the power of in situ hybridization. We analyzed the lung metastases in the mice with the partial knockout of the miRNA. Our hypothesis was that these tumors would still express the miRNA, and not the "anti-miRNA." As seen in Figure 9-11, this indeed was the case. The cancer cells did not have the sequence of the "anti-miRNA" used to silence the oncomiRNA (panel A), but did still have the sequence of the oncomiRNA (panel B). We when showed that the metastases in the mice showed the same geographic pattern of the miRNA and CD9 (concentrating at the tumor interface; panel C). Similar patterns were seen with the macrophage mouse marker and TLR-7 (data not shown).

It was only after having all of this data that we did the co-expression analyses. These analyses did indeed show that the miRNA and CD9 were concentrating in a thin rim specifically at the tumor interface (panel D), but not in the alveolar macrophages 1 mm away in the normal lung (panel E). The co-expression analyses also showed that some of the miRNA+ cells at the tumor interface were CD68+ macrophages (panel F), which, since they were not making the miRNA, must have acquired it via the exosomal release from the cancer cell. The final point to make about Figures 9-10 and 9-11 is that they show that the different targets of interest are being expressed in the same subcellular compartment (primarily, the cytoplasm), and that the cells expressing these different targets overlap (cancer cells and inflammatory cells). This is why we needed to do the co-expression analyses. If the targets were in the same cell populations but different subcellular compartments, there may not be a need to do computer-based co-expression analysis, but, rather, we could rely on the serial section and colorimetric co-expression data.

The point of showing Figures 9-10 and 9-11 is not to review the data of cancer cell and macrophage cross-talk, although it is, to put it mildly, fascinating. It is, however, to show that our molecular pathology and surgical pathology expertise can give us a great deal of information with serial section analyses before we do the co-expression analysis.

STEP FOUR: DO AT LEAST TWO DIFFERENT OPTIMIZING PROTOCOLS WHEN DOING THE CO-EXPRESSION ANALYSIS

The final important technical point makes perfect sense considering what we have learned about optimizing signals for in situ hybridization and immunohistochemistry. Let's summarize the main points here, as they are the basis of Step Four:

1. For in situ hybridization, signal and background are dependent on the probe concentration and the pretreatmentc onditions.
2. For immunohistochemistry, signal and background are also dependent on the primary antibody concentration and the pretreatment conditions.
3. The optimal conditions for two different targets, even if in the same subcellular compartment, can differ a great deal in a given tissue.
4. The optimal conditions for a given DNA, RNA, or protein target can vary from tissue to tissue depending on fixation conditions, over which we usually have no control and have no specific knowledge about.

To get right to the point, the preceding four points are going to make our co-expression analyses more challenging from a technical standpoint! Even if we have optimized a microRNA by in situ hybridization and its putative protein target using immunohistochemistry on a given tissue, and found that in each case protease digestion was optimal at a specific concentration, it does not mean that we can be certain that the same optimal conditions may apply to a different tissue. I also want to stress that this problem seems to be more important for microRNA in situ hybridization. As we discussed at length in previous chapters, the optimal conditions for HPV in situ hybridization (using this as a model of DNA in situ hybridization) are usually rather straightforward. You typically need to do either no pretreatment or protease digestion to optimize the HPV in situ hybridization for most tissues. Yes, in some tissues, DNA retrieval with or without protease digestion is needed for successful HPV in situ hybridization, but these are the exception, not the rule. Not so with microRNA in situ hybridization. You may recall that we theorized that this is the case because microRNAs are typically associated with a much denser, more complex three-dimensional protein/protein cross-linked network than other DNA or RNA sequences. Whatever the explanation, if you are doing co-expression analyses with microRNA in situ hybridization and your putative target, you will face the challenge of needing to reoptimize the optimal conditions for each additional tissue.

It follows that if your interest is primarily co-expression with two distinct proteins using immunohistochemistry, then Step Four will not be as relevant for you. In most cases, the optimal conditions for a given protein for immunohistochemistry are relatively constant and, as we have stressed throughout this book, in most cases they will involve antigen retrieval. This is why co-expression with two or more proteins that exist in separate cell compartments and/or different cell types allows for such simple co-expression analysis. You need only to add the two (or more) primary antibodies, do the reaction with one chromogen, and be very confident that the reaction will work very well because you have already determined that each primary antibody requires the same pretreatment conditions.

But, since I am assuming that many of you who want to do co-expression are interested in examining a microRNA by in situ hybridization and its putative protein by immunohistochemistry, let's discuss how to address the previous four points in a protocol. Here is what I recommend:

Extended Protocol for microRNA In Situ Hybridization and Putative Protein Target Co-expression Analyses

1. Have your technician place serial sections on 10 unstained silane-coated slides, labeling the slides 1 to 10. As always, make certain that there are from two to three sections per slide, depending on the size of the formalin-fixed, paraffin-embedded tissue.
2. Use slides 1 and 2 to determine the optimal conditions for microRNA in situ hybridization on this tissue by comparing the following variables:
No pretreatment and protease digestion (slide 1)
RNA retrieval and RNA retrieval + protease digestion (slide 2)
(Use the same concentration of the microRNA probe that you determined was optimal when you initially optimized the microRNA in situ hybridization; see Step Three of Chapter 7.)
3. Use slides 3 and 4 to determine the optimal conditions for putative protein target by immunohistochemistry on this tissue by comparing the following variables:
No pretreatment and protease digestion (slide 3)
Antigen retrieval and antigen retrieval + protease digestion (slide 4)

(Use the same concentration of the primary antibody that you determined was optimal when you initially optimized the epitope detection; see Step Three of Chapter 8.)
4. Do *not* counterstain slides 1–4. Rather, coverslip them and then photograph the sections which represent the specific condition that was optimal for the detection of the microRNA and the specific condition that was optimal for the detection of the putative protein target.
5. After photographing the data, write down your observations as to which specific cell type or types are expressing the microRNA of interest and what specific cell type or types are expressing the putative protein target of interest. Also note what specific cellular compartments contain the microRNA and protein of interest.
6. Use slide 5 to perform the microRNA in situ hybridization reaction using the optimal conditions you documented above in step 4 of this protocol. However, since you have at least two sections per slide, use one section as the negative control (either omission of the probe or use of a scrambled or irrelevantp robe).
7. Similarly, use slide 6 to perform the immunohistochemistry reaction using the optimal conditions that you documented in step 4 of this protocol. However, since you have at least two sections per slide, use one section as the negative control (typically omission of the primary antibody).
8. Confirm under the microscope that slides 5 and 6 performed as expected.
9. Use slide 5 to do the immunohistochemistry reaction for the putative protein target of the microRNA.
10. Use slide 6 to do the in situ hybridization for the microRNA of interest.
(Of course, the order may be dictated by the optimal conditions for the microRNA and protein target detection; see "Step Two: Determine the Order Of Co-expression Analysis" earlier in this chapter.
11. If you are using NBT/BCIP as the microRNA counterstain and DAB as the chromogen for the putative protein target, you can use Nuclear Fast Red as the counterstain. Do *not* use hematoxylin for a counterstain with NBT/BCIP as the chromogen. For the Nuance co-expression analyses, you do not need a counterstain. However, you do need a counterstain if you are using the InForm quantification system for co-expression analysis. If you are using NBT/BCIP and either Fast Red or DAB, methylene green can serve as an effective counterstain.

Let's look at an example of the preceding protocol. Before we do, there is one more point to make. Assuming that you are using a computer-based system to document the co-expression data, it is important to realize that the computer is doing the equivalent of a three-dimensional "CT" scan of the cell. But instead of measuring differences in density of tissue, it is measuring differences in the color staining of the cell. Also, the computer will not discriminate between cytoplasm and nucleus when doing this analysis. Rather, it will use a plane of about 150 nm,

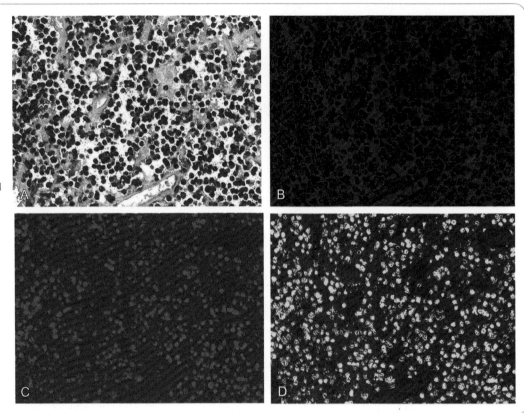

Figure 9-12 Co-expression of EBV RNA and protein in lymphoma cells. Panel A shows a diffuse large B-cell lymphoma that was analyzed for both EBER-1 and -2 by in situ hybridization (blue signal; separated in panel C), and the EBV protein bzlf-1 by immunohistochemistry (red signal, separated in panel B). Each target is nuclear-based, and thus, you must do co-expression analyses to determine which cancer cells are expressing both targets. The Nuance system (panel D) clearly shows that the cancer cells are making both EBER-1 and -2, and bzlf-1, which is indicative of lytici nfection.

and if the signal is present in the nucleus *and* the other signal is present directly above the nucleus in the cytoplasm (that is, in the same plane), then the computer will score the cell as "positive for co-expression." So this is not the same as co-localization in which only two signals present in the exact same subcellular compartment are scored as positive. Rather, this is truly "cellular co-expression analysis" in which we are trying to determine whether a given cell does or does not have our two targets of interest. Of course, that is what we are interested in with such experiments, and not true "co-localization."

Let's look at Figure 9-12 to illustrate this important point. The RNA that was detected is EBER-1 and 2. This is an RNA made in abundance by the Epstein-Barr virus (EBV), which is also called human herpesvirus 4 (HHV-4), and it is nuclear based. The same tissue was also analyzed for the EBV protein bzlf-1 by immuno-histochemistry with DAB as the chromogen (panel A shows the co-expression as seen under light microscopy, and panel C shows the Nuance-isolated EBER-1 and -2 signal). As is evident, the bzlf-1 protein is also nuclear based (panel C, shown as a fluorescent red signal by the Nuance system). If cells infected with HHV-4 show both EBER-1 and -2 as well as bzlf-1, then this indicates that the infection is lytic as opposed to latent. It is not possible to determine if the EBER-1 and -2 RNA positive cells also contain bzlf-1 protein, although it is certainly clear many cells are expressing at least one of these targets. The Nuance system (Figure 9-12, panel D) clearly shows this is a lytic viral infection, because many of the cells show the fluorescent yellow color indicative of co-expression.

Now let's go to Figure 9-13. These series of experiments were done in collaboration with Dr. Croce's laboratory and were headed by Dr. Michela Garofalo. She has done extensive and excellent work on the microRNA modulators of various cancers, including lung. In this set of experiments, we were interested in the effect of various microRNAs on the tumor suppressor gene LATS2. Figure 9-13 shows some of the data from the optimizing experiments for LATS2 and its putative miRNA regulator using the protocol outlined previously. As is evident, no signal was generated for LATS2 in the normal lung if no pretreatment (panel A) or protease digestion (panel B) was used. A strong signal was evident with antigen retrieval with no detectable background (panel C). Note that the signal stays sharply localized to the mononuclear cells. If antigen retrieval and protease digestion were used, the same cells in the serial section showed a strong signal, but background was now evident. The background is evident as the relatively weak red staining over the stroma (arrows). Thus, we can conclude that, in the co-expression experiments, LATS2 detection will be optimized by antigen retrieval. Now look at the data for the miRNA after in situ hybridization optimization. A strong signal is noted with protease digestion (panel E). Note the typical features of a strong signal and minimal background with miRNA in situ hybridization detection; the intense signal localizes to a specific cell type (cancer cells, large arrow) and is not evident in an adjacent different cell type (normal lung, small arrow). Compare this to panel F, where the miRNA was detected after RNA retrieval. The signal is evident but is less distinct, being

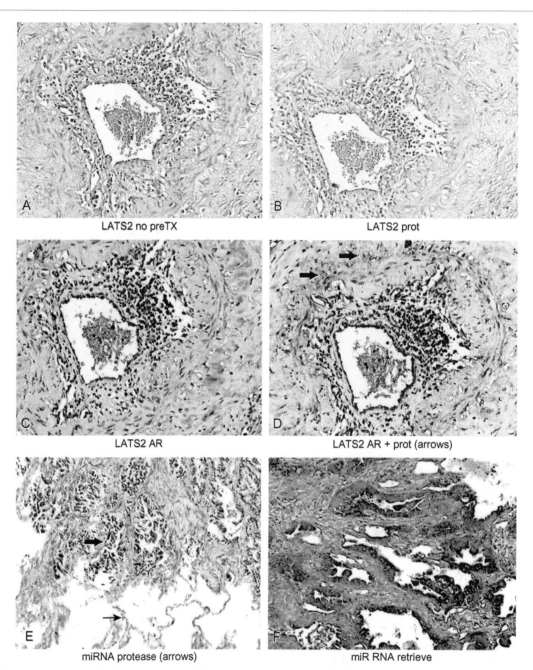

Figure9-13 Co-expression of LATS2 and a microRNA: importance of optimizing pretreatment conditions with serials sections: Part A. Panel A shows that no signal was generated for LATS2 in the normal lung if no pretreatment (panel A) or protease digestion (panel B) was used. A strong signal was evident with antigen retrieval with no detectable background (panel C). Note that the signal stays sharply localized to the mononuclear cells. If antigen retrieval and protease digestion was used, then the same cells in the serial section showed a strong signal, but background was now evident. The background is evident as the relatively weak red staining over the stroma (arrows). Compare this to the optimizing data for the miRNA that may regulate LATS2. A strong signal is noted with protease digestion (panel E). Note the typical features of a strong signal and minimal background with miRNA in situ hybridization detection; the intense signal localizes to a specific cell type (cancer cells, large arrow) and is not evident in an adjacent different cell type (normal lung, small arrow). Compare this to panel F where the miRNA was detected after RNA retrieval. The signal is much less distinct. Thus, the co-expression experiment would be best done with the miRNA first (protease digestion) followed by LATS2 after antigen retrieval; remember to use a lower concentration of the primary antibody as discussed in Chapter 5, because this will reduce the background seen in panel D.

present in the cancer cells and surrounding stromal cells. If RNA retrieval was followed by protease digestion, then background became very high (data not shown). Thus, the optimizing experiments indicated that the microRNA would be best detected with protease digestion.

We then analyzed serial sections for LATS2 (antigen retrieval) and the miRNA (protease digestion). The slide tested for LATS2 first by antigen retrieval was then tested for the miRNA by in situ hybridization with no additional pretreatment. The slide tested for the microRNA

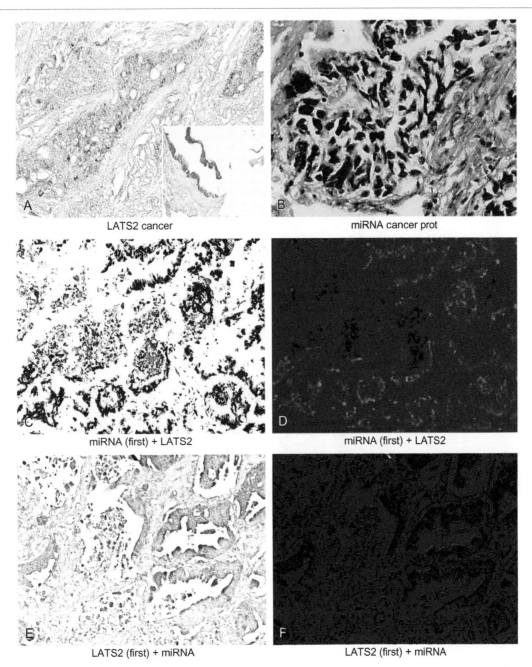

LATS2 cancer

miRNA cancer prot

miRNA (first) + LATS2

miRNA (first) + LATS2

LATS2 (first) + miRNA

LATS2 (first) + miRNA

Figure 9-14   Co-expression of LATS2 and a microRNA: Part B. As evident in Figure 9-14, rare cancer cells showed LATS2 signal (panel A). Note the strong signal for LATS2 in the same tissue for a benign bronchiole (insert), which serves as an internal positive control. In comparison, many more cancer cells were positive for the miRNA (panel B), where, again, a nuclear and cytoplasmic signal were each evident. When the microRNA in situ hybridization was done first, followed by the LATS2 detection (panel C), each signal could easily be recognized by routine microscopy. This is a situation in which the computer-based system is invaluable; there are nests of cancer cells where both the microRNA and its putative target are present. Note that the co-expression analysis (panel D) did not demonstrate any co-expression, which, thus, is consistent with the theory that the microRNA may be regulating the LATS2 protein. Look at the data if the immunohistochemistry for LATS2 was done first, followed by the microRNA (panel E). The signal for the LATS2 is still evident, but the microRNA signal is no longer evident, as the RNA retrieval was simply not the correct pretreatment to optimize the signal, as we saw in Figure 9-13. Thus, the co-expression data is not useful because only the LATS2 signal is present (panel F).

by in situ hybridization after protease digestion was then tested for LATS2 using antigen retrieval. In each case, the miRNA signal was NBT/BCIP (blue), and the LATS2 signal was red due to Fast Red.

As evident in Figure 9-14, rare cancer cells showed LATS2 signal (panel A). Note the strong signal for LATS2 in the same tissue for a benign bronchiole (insert), which serves as an internal positive control. The LATS2 signal is seen in both the cytoplasm and the nucleus. In comparison, many more cancer cells were positive for the miRNA (panel B) where, again, a nuclear and cytoplasmic signal were each evident. When the microRNA in situ hybridization was done first, followed by the LATS2 detection (panel C), each signal could easily be recognized by routine microscopy. This is a situation in which the computer-based system is invaluable; there are nests of cancer cells where both the microRNA and its putative target are present. Note that the co-expression analysis

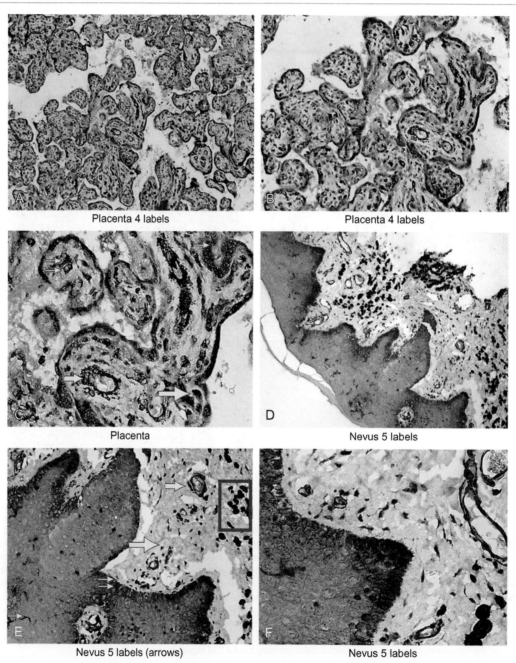

Placenta 4 labels

Placenta 4 labels

Placenta

Nevus 5 labels

Nevus 5 labels (arrows)

Nevus 5 labels

Figure9-15   Co-expression analysis of four and five different antigens using one chromogen. A placental tissue was analyzed for four targets at the same time with immunohistochemistry using DAB as the chromogen (panels A–C). Note that you can differentiate the signal for hCG (small arrow, trophoblasts), CD31 (double arrow, endothelial cells of capillaries), smooth muscle actin (intermediate arrow, larger blood vessel), and CD68 (large arrow, macrophage). A nevus was analyzed for five targets at the same time with immunohistochemistry using Fast Red as the chromogen (panels D–F). Note that you can differentiate the signal for S100 (small arrow, Langerhans cell that is present in epidermis), keratin (double arrow, squamous cells), smooth muscle actin (intermediate arrow, larger blood vessel), CD45 (large arrow, lymphocytes), and melanin-A (box, benign melanocytes).

(panel D) did not demonstrate any co-expression, which, thus, is consistent with the theory that the microRNA may be regulating the LATS2 protein. Look at the data if the immunohistochemistry for LATS2 was done first, followed by the microRNA (panel E). The signal for the LATS2 is still evident, but the microRNA signal is no longer evident, as the RNA retrieval was simply not the correct pretreatment to optimize the signal, as we saw in Figure 9-13. Thus, the co-expression data is not useful because only the LATS2 signal is present (panel F).

Remember that, with a good knowledge of the cytopathology and histopathology and the subcellular localization of the targets, you can detect even four (Figure 9-15, panels A–C) or five (panels D–F) antigens in a given tissue using just one chromogen. Of course, in this case, each target had the same optimization profile. I never

cease to be amazed as to how "clean" the in situ hybridization and immunohistochemistry is at the biochemical level, in that you can easily run many separate reactions simultaneously.

Figure 9-16 shows the value of performing co-expression analyses on serial sections. The tissue is a lung cancer. Serial section 1 was tested for miR-34a; note that most of the cancer cells are negative (panel A). Panel B is the next serial section tested for PDGFR beta, and panel C is the subsequent serial section tested for PDGFR alpha. It is clear from these data that miR-34a is not being expressed in the cancer cells, but that most are expressing both forms of PDGFR. This is confirmed in panel D, where there is no evident co-expression when the Nuance system analyzed miR-34a and PDGFR beta in the same tissue section.

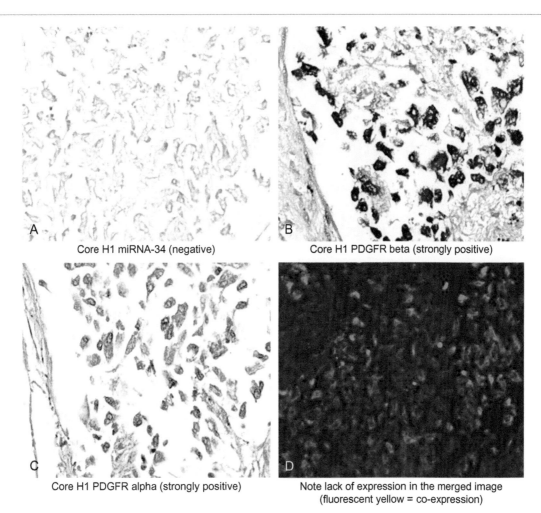

Core H1 miRNA-34 (negative)

Core H1 PDGFR beta (strongly positive)

Core H1 PDGFR alpha (strongly positive)

Note lack of expression in the merged image
(fluorescent yellow = co-expression)

Figure 9-16   The value of serial sections in co-expression analyses. Serial sections for this lung cancer were tested for miR-34a (panel A), PDGFR beta (panel B), and PDGFR alpha (panel C). It is clear that the cancer cells are expressing abundant PDGFR alpha and beta but not the miR. Indeed, extensive additional analyses including the luciferase assay and mutational analyses (by Dr. M Garofalo) showed that the miRNA was regulating these two forms of PDGFR. Panel D confirms the lack of co-expression of miR-34a and PDGFR beta (panel D) (the miRNA is fluorescent blue, the PDGFR beta is fluoresent green, and the co-expression would be seen as fluorescent yellow).

PCR In Situ Hybridization and RT In Situ PCR

10

INTRODUCTION

In this chapter we discuss PCR in situ hybridization for DNA targets and RT in situ PCR for RNA targets.

In our discussions about in situ hybridization and immunohistochemistry throughout the book, we have stressed that the reactions are diffusion reactions in which the three-dimensional protein/protein cross-linked network creates the equivalent of ion exchange and "hydrogen bond resins," and pore size and hydrophobic and/or hydrophilic pockets dictate both the movement of the key in situ hybridization and immunohistochemistry reagents, including assisting them in docking with the target of interest. Of course, since we are using just colorimetric systems in our in situ hybridization and immunohistochemistry protocols, ultimately the success of the reaction depends on an enzyme (peroxidase or alkaline phosphatase) working within this three-dimensional protein/protein cross-linked network. We have also stressed that the "high throughput" PCR or ELISA platforms pale next to in situ hybridization or immunohistochemistry in terms of the numbers of reactions at a given time. Since most 10 mm biopsies would contain about 100,000 cells that are exposed to the in situ hybridization or immunohistochemistry reagents, when we do either in situ-based process, we are simultaneously performing about 100,000 reactions! Based on this information, it will not come as a surprise that we can do PCR within intact tissues. This approach is just like in situ hybridization in the sense that we need to get key reagents, including the primers; nucleotides, including the ones with the label; and the polymerase to enter the cell, find the DNA or RNA target, and then for the polymerase to synthesize the new DNA (or cDNA to DNA). The primers are the same size as the LNA probes we use for in situ hybridization, and of course, the individual nucleotides are much smaller. We routinely put the reporter molecule (digoxigenin or biotin) on much larger molecules than the nucleotides when we do in situ hybridization or immunohistochemistry. Finally, the Taq polymerase (or rTth polymerase for RT in situ PCR) is actually smaller than the peroxidase or alkaline phosphatase conjugates we use with in situ hybridization or immunohistochemistry. So, when you get right down to it, in situ PCR is simply the equivalent of doing in situ hybridization within formalin-fixed, paraffin-embedded tissues with only one key difference: with in situ PCR, we are employing a polymerase to synthesize either DNA or cDNA/DNA for our target of interest. Of course, this amplification step will serve to increase the copy number of the target of interest. Clearly, that is the main point with in situ PCR. Indeed, the fuel that fired up such interest with in situ PCR and its successful application by many laboratories was the in situ search for the one integrated copy of the provirus of the AIDS virus (HIV-1). To this day, the crowning achievement of PCR in situ was the simultaneous demonstration by several laboratories that HIV-1 massively infects the CD4 population (and CD21 population) of cells *before* the person shows any diagnostic features of AIDS. This, of course, led to the strategy of treating people recently infected with anti-HIV-1 with antiretroviral therapy *prior* to developing AIDS, an approach that has saved countless lives over the past 15 years.

Since in situ PCR and in situ hybridization share so much in common, it will also be no surprise that success with in situ PCR is highly dependent on determining the correct pretreatment conditions. Unless we optimize the pretreatment conditions, either the reagents will not be able to access the target of interest and no DNA synthesis will occur, or the reagents may access the target but the permeability is so increased that the amplicons diffuse out of the cell. It is also clear that formalin-fixed, paraffin-embedded tissues will be the optimal types of tissues to use for in situ PCR, because the denaturing fixatives are much too likely to show diffusion of the amplicons as well as poor entry of the key reagents during the in situ amplification. Finally, if we are using old formalin-fixed, paraffin-embedded tissues, we will need to rejuvenate the three-dimensional protein/protein cross-linked network that surrounds the DNA or RNA target of interest; otherwise, the reaction will likely not succeed.

Thus, in a way, this last chapter on in situ PCR is simply a review of the discussions we have already had with in situ hybridization. The discussions we had in Chapters 3, 4, and 7 regarding in situ hybridization—such as diffusion kinetics of key reagents, the need to optimize the pretreatment conditions in each case, the observation that the three-dimensional protein/protein cross-linked cage surrounding each potential target of interest will vary, and so on—apply equally well to in situ PCR.

Since RT in situ PCR is much more commonly used than PCR in situ hybridization, let's begin our discussion with the detection of RNA targets after reverse transcriptase of the RNA to cDNA and then the amplification of the cDNA sequence into detectable amplicons.

243

DOI: http://dx.doi.org/10.1016/B978-0-12-415944-0.00010-3

RT IN SITU PCR

mRNAs and viral RNAs are often present in relatively low copy number. Also, 15–20 years ago, it was a much more tedious process to make probes for mRNA or viral RNA targets. Typically, we had to clone the cDNA sequence of interest into a vector that was "bidirectional." That is, the plasmid vector would have a T7 and SP6 promoter to start the synthesis of the RNA probe in the sense (negative control) and antisense (test probe) directions.

Of course, the dynamics of DNA-DNA and DNA-RNA hybridization are basically the same after we denature the DNA target of interest. Indeed, assuming an identical GC content and a probe/target of the same size, the melting temperature of the DNA-RNA hybrids will be slightly higher than the DNA-DNA hybrids.

It is important to remember the context of molecular pathology when PCR in situ hybridization and RT in situ PCR were first invented in the early 1990s. Immunohistochemistry had been in wide use for about 5 years, and in situ hybridization was still in its infancy. So, if we were looking for a low copy DNA or RNA target, and if the RNA made relatively small amounts of protein, the odds were that neither in situ hybridization nor immunohistochemistry would be successful in detecting the nucleic acid sequence or epitope. So much has changed in the past 20 years in the fields of immunohistochemistry and in situ hybridization! The sensitivity of immunohistochemistry has increased at least 10-fold due to improvements in the detection systems that basically allow us to get many more reporter molecules and reporter enzymes in any given primary antibody-epitope complex.

The sensitivity of in situ hybridization has increased for two reasons:

1. Improvements of the detection system that, again, allow us to get many more reporter enzymes in any given DNA-DNA or DNA-RNA hybrid complex.
2. The discovery of the LNA probes about 5 years ago.

I realize that, as one of the first groups associated with in situ PCR, my laboratory has developed a reputation for considering RT in situ PCR a very useful method and therefore uses it frequently. It may, thus, surprise you to hear that my laboratory rarely uses RT in situ PCR or PCR in situ hybridization at this time. The primary reason can be stated in three letters: LNA.

I have done a series of experiments comparing the signal (both in terms of intensity of signal as well as the percentage of positive cells) between LNA probes and RT in situ PCR. As you might expect, HIV-1 RNA was one of the targets that was tested, given the fact that in situ detection of HIV-1 nucleic acids was *the* reason that PCR in situ hybridization came to be in the first place. The case involved a young man who had died unexpectedly, and the autopsy documented one of the rare, yet well-documented, complications of HIV-1 infection: myocarditis. As you know, HIV-1 in such instances will localize to the CD4 cells in the spleen and lymph nodes and to other cell types in the testes. It will also localize directly to cells in the myocardium, including CD 68 positive macrophages. If you were to examine the tissues just for HIV-1 DNA by in situ hybridization using an oligoprobe that was not LNA modified, no signal would be evident; similarly, no signal would be evident for the viral RNA if an oligoprobe that was not LNA modified was used (data not shown). However, as is evident in Figures 10-1 and 10-2, an intense signal was evident in many of the CD4+ cells in the spleen and testes when in situ hybridization was done with the LNA probe.

My laboratory recently did an extensive side-by-side comparison of the detection of reovirus by RT in situ PCR, LNA-based in situ hybridization, and immunohistochemistry. Reovirus is a double-stranded RNA virus that has been getting a great deal of attention in the past few years. The reason for the attention is that reovirus specifically can target cancer cells. Reovirus is a common virus that can cause nonspecific infections that typically resolve. The proliferation of the virus in cells is much increased if

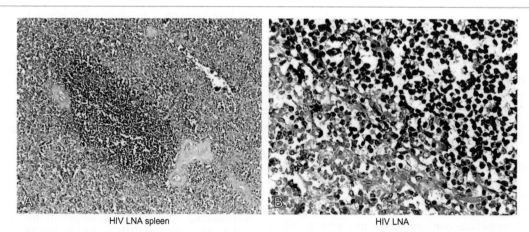

HIV LNA spleen   HIV LNA

Figure 10-1 LNA probes are as sensitive as RT in situ PCR for HIV-1 RNA. The tissues are from the spleen and a lymph node of a person recently diagnosed with HIV-1 infection. When in situ hybridization was done with the tagged SK38 and SK39 probes (specific for the gag region), no signal was evident (data not shown). Typically, RT in situ PCR would be needed to see the distribution of HIV-1 in these tissues in the asymptomatic state. However, when the 5′ tagged SK38 and SK39 probes were modified with LNA nucleotides, an intense signal was seen in both tissues (panel A and B).

the cell expresses the protein p38. The presence of p38, in turn, is much increased if the RAS pathway has been activated. As you know, the RAS pathway is a multi-armed system that is typical of cancer cells.

When the reovirus is given IV in either an animal or a human who has a malignant tumor, it has been very well established that the virus will hone in on the cancer cells. Since viral proliferation is associated with ample production of some capsid proteins, immunohistochemistry is a good and reliable assay to detect the proliferating virus. As seen in Figures 10-3 and 10-4, viral protein is readily abundant in the cancer cells, yet totally undetectable in the surrounding stromal or inflammatory cells. It is also well documented that treating tumors with reovirus, either in the animal models or in patients with cancers, when done in combination with chemotherapy, will in a synergistic fashion increase the death of cancer cells.

How is it that reovirus, when included with chemotherapy, will result in the death of more cancer cells than with reovirus or chemotherapy alone? This is a good example of where co-expression analyses can give us an enormous amount of insight and information into this very important clinical question. Solution phase-based experiments clearly showed that chemotherapy was augmenting the ability of reovirus to productively infect cancer cells. The chemotherapy appeared to be doing this through several mechanisms:

1. Chemotherapy changed the molecular profile of the endothelial cells in the blood vessels of the tumor, including the increased synthesis of VEGF receptor 2, that facilitated the exit of the virus from the vascular system into the tumor.
2. Chemotherapy increased the expression of RAS and p38 proteins on the cancer cells, which, in turn, facilitated the proliferation of the virus.
3. The anti-inflammatory effect of the chemotherapy may have allowed more reovirus to persist in the tumor.

Although the preceding information was all very useful, since the tissue was pulverized and the DNA/RNA/proteins extracted prior to analysis, it could not show us the cell-by-cell interaction of the virus with some of these molecular markers such as RAS or p38, nor could it allow us to correlate viral expression with indicators of cancer cell death. Thus, a series of co-expression analyses was done to examine the relationship of productive reoviral infection to the molecular biology of the cancer cells.

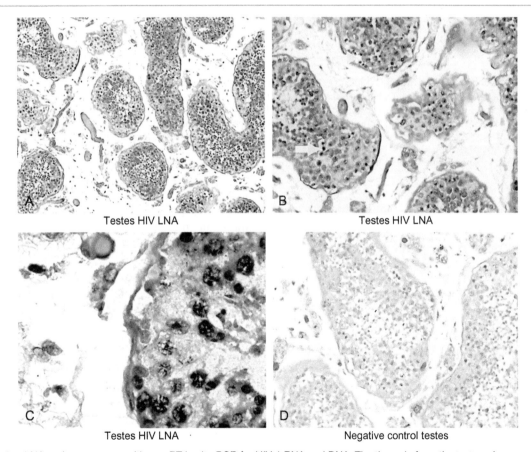

Testes HIV LNA

Testes HIV LNA

Testes HIV LNA

Negative control testes

Figure 10-2 LNA probes are as sensitive as RT in situ PCR for HIV-1 RNA and DNA. The tissue is from the testes of a man recently diagnosed with HIV-1 who had died. It is well documented that HIV-1 can infect cells in the testes, which, in turn, is related to the frequent occurrence of testicular atrophy found in men with AIDS. Many HIV-1 infected cells were evident after in situ hybridization with the LNA-modified probes. Note the extent of the infection at low magnification (panel A). Also note that the signal is evident in the progenitor spermatogonia (panel B, arrow; and panel C, high magnification). Panel D is one of the negative controls, which shows the use of a scrambledp robe.

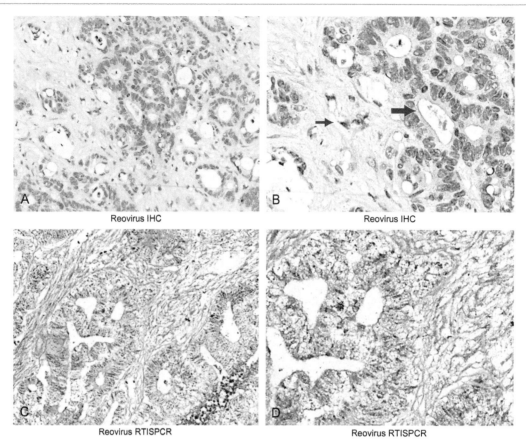

Figure 10-3 Detection of reoviral RNA by RT in situ PCR; comparison with the detection of reoviral protein by immunohistochemistry. The tissue is from a cancer removed from a patient soon after he had received a cycle of reovirus IV. Note the disorganized pattern of growth of the glands consistent with adenocarcinoma (panel A) and the cribiform pattern seen at higher magnification in panels C and D. Reoviral protein (panels A and B) was detected in a minority of the cancer cells as determined by immunohistochemistry (panel B, large arrow depicts cancer cells positive for reovirus with their red signal; the small arrow depicts viral negative stromal cells that show blue color due to the hematoxylin counterstain). RT in situ PCR was done, and it showed that in this area basically all the cancer cells had detectable viral RNA (panels C and D); note the lack of signal in the stromal areas that are stained light pink.

These data are provided in Figure 10-4; also note the following:

1. The reoviral protein was evident in the cytoplasm of a variable percentage of cancer cells.
2. The reoviral protein strongly co-localized with p38 and other protein markers of the activated RAS system, such as pI3 kinase.
3. Reoviral proliferation requires the virus to attach to the cancer cells' microtubulin scaffold. Co-expression of tubulin and reovirus showed an intense co-expression pattern, and thus provided another piece of evidence of productive viral infection (as compared to passive viral absorption by the cancer cells).
4. The cancer cells showed variable apoptotic death, which was associated, as expected, with the expression of the activated form of caspase-3.
5. The reoviral protein was strongly co-expressed with the activated form of caspase-3.

Thus, when we put together the in situ-based and solution phase-based data, the picture emerges of the strong specificity of the reovirus to the cancer cells being a consequence of the p38 protein to facilitate productive viral infection, which, in turn, somehow activates caspase-3 expression and, ultimately, the death of the cancer cells.

Two important questions follow:

1. Is viral RNA present in cells (cancer or benign) that lack detectable viral protein?
2. What is the molecular link between productive reoviral infection and the apoptotic death of the infected cancerc ells?

Each question can be addressed by in situ hybridization-based experiments. Let's look at question 1 first. This is a perfect example in which RT in situ PCR can be a very valuable tool. Since the question basically examines whether the viral genome can enter the cell and be metabolically inactive, we could easily see the value of RT in situ PCR, because there would be little to no protein expression from the viral genome, nor would there be much in the way of a viral copy number. We, thus, did RT in situ PCR on a variety of tumors from patients who were either exposed or not exposed to the virus. These were done in the typical blinded fashion. To be honest, we found the results surprising; they are shown in Figure 10-4. Note that the virus was indeed present in many

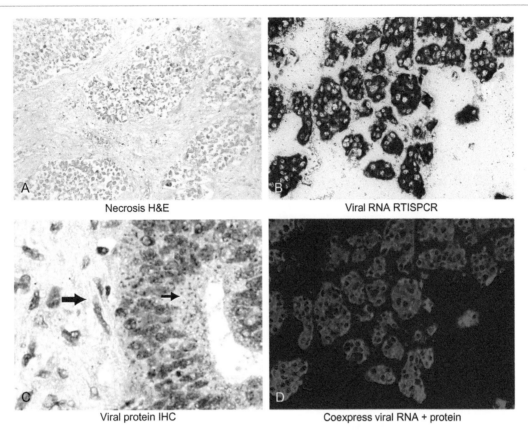

Necrosis H&E

Viral RNA RTISPCR

Viral protein IHC

Coexpress viral RNA + protein

Figure 10-4   Co-expression analysis of reoviral protein and RNA by immunohistochemistry and RT in situ PCR, respectively. Since RT in situ PCR has an extremely low detection threshold, which may indeed be one copy per cell, it is well suited to the analysis of viral infections where detection of a viral proliferation marker, such as reoviral capsid protein, can be used in conjunction with it to determine latent versus active infection. Panel A shows the hematoxylin and eosin stain from a tumor of a patient recently treated with reovirus. Note that RT in situ PCR shows that basically all of the cancer cells, but none of the surrounding stromal cells, have the viral genome (panel B; note the cytoplasmic localization of the viral RNA). Also note that reoviral protein was found in some cancer cells, also in the cytoplasm (panel C, small arrow). The large arrow indicates that the viral protein is only evident in the cancer cells and not the stromal cells. Finally, RT in situ PCR for reoviral RNA was done with immunohistochemistry for reoviral protein as a co-expression experiment and analyzed with the Nuance system (panel D). Fluorescent yellow marks cancer cells that contain viral RNA and protein. Clearly, only a small percentage of the viral genome positive cells also show a detectable amount of viral proteins. This suggests that reoviral-induced killing of cancer cells could be augmented by increasing the percentage of productively infected cells in the cancer.

cells, far more than contained the viral protein when the serial section was analyzed by immunohistochemistry. Also note that the viral genome, even after RT in situ PCR, was strictly confined to the cancer cells. When we do co-expression analyses of reoviral RNA and protein, as seen in Figure 10-5, it is clear that a relatively low percentage of the cancer cells that contain the viral genome are also expressing the viral capsid protein. Thus, the RT in situ PCR data, in conjunction with the immuno-histochemistry data, give us some useful insight into the dynamics of reoviral infection of cancer cells:

1. The viral genome avidly enters cancer cells.
2. The viral genome is not able to enter the surrounding benign cells (or, at least, does this very poorly).
3. In a certain variable percentage of the cancer cells that contain reoviral RNA, the infection becomes productive as marked by the co-expression of the viral RNA with the protein (or by the viral protein and the cell's microtubulin protein).

4. It is in the cells with productive viral infection that caspase-3 is activated and the cancer cell then undergoes an apoptotic death.

Now let's look at point 2. Since microRNAs are a major regulator of cancer-associated proteins, we analyzed a series of cancer cell lines that were infected and not infected with reovirus. The infected cancer cells showed a significant increase in the percentage of cells undergoing apoptotic death. The infected and noninfected cells were then analyzed for global changes in the microRNA profile. The microRNA most downregulated in the cancer cells infected with reovirus as microRNA-let-7d. By using the luciferase assay, we showed that microRNA-let-7d was able to downregulate the active form of caspase-3. Further molecular analyses showed that the reoviral genome had sequences that had sufficient homology with microRNA-let-7d that it could inactive it. This, in turn, would upregulate caspase-3 expression. The last piece of the puzzle was to show that

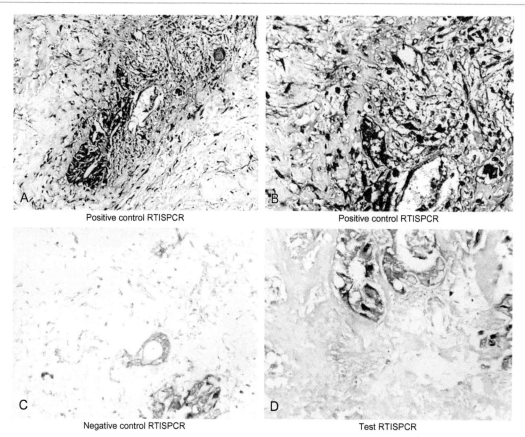

Figure10-5   The internal positive and negative controls for RT in situ PCR. To demonstrate that the pretreatment conditions were optimal for RT in situ PCR, we must take one tissue and not digest it with DNase, and then take another tissue and digest with DNase; then we must do RT and PCR with either no primers or with nonsense primers. It is only when the no DNase control (positive control) shows a majority of cells with a strong positive nuclear-based signal (panel A and, at higher magnification, in B) *and* the DNase/no primers or nonsense primers show no signal that the test can be considered valid. A successful positive control is seen in panels A and B; note the strong nuclear signal in all cell types. A successful negative control is seen in panel C, as no signal is evident; only the pink counterstain can be visualized. With these data for the negative and positive controls, we are confident that the signal in the cancer cells in panel D (RT in situ PCR for reoviral RNA, the signal is blue) is target specific. The localization of the signal to just the cancer cells gives us another layer of confidence in the specificity of the signal.

the cancer cells with productive reoviral infection were negative for microRNA-let-7d by co-expression analysis and positive for caspase-3. This, indeed, was documented by co-expression analyses, as shown in Figure 10-5.

The preceding data regarding reovirus and the molecular events in the infected cancer cells provide a good example of the power of combining solution phase data and in situ-based data, especially when the latter includes co-expression data. As we get to the end of the book, there were two reasons I wanted to go over these data in detail:

1. As indicated, these data show the power of combining in situ-based data with solution phase data in generating a much more complete understanding of this very important phenomenon of reoviral-induced cancer cell death.
2. They also show the high sensitivity of LNA probes when compared to RT in situ PCR.

The data in Figures 10-4 and 10-5 for the reoviral genome were generated by RT in situ PCR. Let's compare these data to that generated with LNA probes that correspond to the same primers used in RT in situ PCR. As seen in

Figure 10-6, the digoxigenin-tagged LNA probes, when used for in situ hybridization, performed just as well as the corresponding sequences did when used for RT in situ PCR! We have now seen this for HIV-1 and reovirus. I could show other examples of this as well, but I think the point has been made. Figure 10-7 reminds us that the pretreatment conditions when using LNA probes as well as the probe concentration are the primary determinants of background.

An obvious question regarding these reovirus data is this: what is the mechanism whereby productive reoviral infection actually causes the apoptotic death of the cancer cells? Through a combination of in situ hybridization with immunohistochemistry and extensive co-expression analyses plus much solution phase data, it was determined that part of the reoviral genome is complementary to the microRNA-let-7d. This microRNA, in turn, regulates caspase-3 expression. Thus, this acute inactivation of microRNA-let-7d led to the increased caspase-3 production that, in turn, led to the apoptotic death of the cancer cells. A schema that puts these data together is shown in Figure 10-8 and reminds us of the power of combining the in situ-based data with the solution phase data.

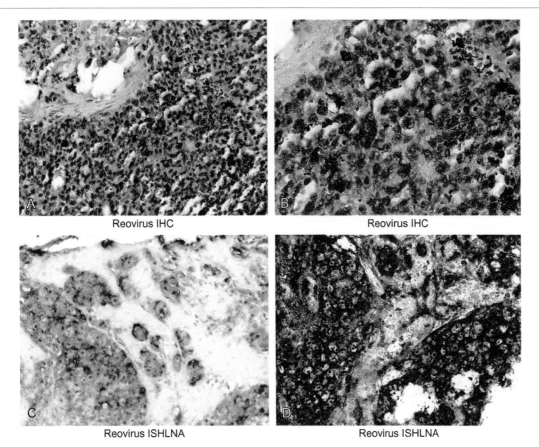

A Reovirus IHC

B Reovirus IHC

C Reovirus ISHLNA

D Reovirus ISHLNA

Figure10-6 LNA probes have sensitivity equivalent to RT in situ PCR: reovirus. These experiments involved mice that were injected with cancer cells exposed to the reovirus. The immunohistochemistry (panels A and, at higher magnification, B) showed scattered cancer cells positive for reoviral protein (DAB signal). In situ hybridization was then done on serial sections for reoviral RNA using LNA probes. Note that the results are equivalent to RT in situ PCR detection of reoviral RNA (e.g., Figure 10-5). The viral genome localizes to only the cancer cells (panel C), and most of the cancer cells show a strong cytoplasmic signal (panel D). The RT in situ PCR/reoviral protein co-expression data showed us that most of the cancer cells were latently infected by the virus. Such latent infection is typically associated with nonproliferating viral RNA. Thus, the LNA probes show a strong sensitivity in this direct comparison with RT in situ PCR.

Many claims have been made for a variety of in situ hybridization systems that are capable of detecting one copy of the target per cell. These claims have been made over the past 15 years or so. In my experience, such claims usually do not stand the test of actually using them in your own lab. Yes, you can detect the one copy of HPV DNA in SiHa cells with different HPV in situ hybridization systems—however, in my experience, usually not in the majority of the cells. I believe the LNA probes have brought us to the point with in situ hybridization that very low copy targets are now detectable in a routine manner. Still, LNA probes are expensive, whereas the primers for RT in situ PCR (or PCR in situ hybridization) are very inexpensive. So, for a variety of reasons, you may wish to do RT in situ PCR or PCR in situ hybridization to detect a low copy RNA or DNA sequence. Since RT in situ PCR is so much more commonly used, let me now present an in-depth protocol for doing RT in situ PCR on formalin-fixed, paraffin-embedded tissues. To make it easier for the investigator, I include steps also used for in situ hybridization so that you do not have to keep referring to Chapter 7.

A PROTOCOL FOR RT IN SITU PCR: FIRST DETERMINE THE OPTIMAL PRETREATMENT CONDITIONS

STEP ONE: CUT THE TISSUE SECTIONS ONTO A SILANE-COATED SLIDE

1. Confirm that the blocks are formalin-fixed, paraffin-embeddedtissu es.
2. Determineth eage o fth eb locks.
3. Have the histotechnologist cut successive 4 to 5 micron sections and, depending on the size of the biopsy, put the paraffin ribbons on from 5 to 20 consecutive and labeled silane-coated slides. Although this step is important for in situ hybridization, it is very important for RT in situ PCR, since this will be the simplest and most direct way of generating the method's positive and negative controls that are essential for ensuring the specificity of the data.
4. You do not have to use RNase-free water for this step, because the RNA is likely protected from such by the paraffin wax and formalin-induced RNA-proteinc ross-links.

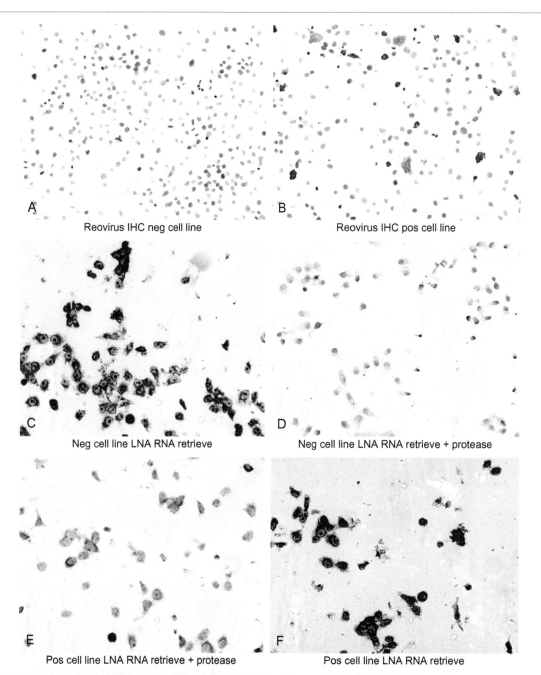

Figure 10-7 Importance of pretreatment conditions for signal and background with LNA probes. Of course, we stressed in Chapter 7 that the signal-to-background ratio with in situ hybridization was typically inversely related and that each, in turn, could be affected by the pretreatment conditions and the probe concentration. In these experiments, cancer cell lines either exposed or not treated with reovirus were analyzed with LNA probes with different pretreatment conditions. Note that no signal was present after immunohistochemical analysis in the cell line shown in panel A (negative control line) and that scattered cancer cells were positive for viral protein in the positive control line (panel B). The negative control cell line showed a signal with the reovirus LNA probe if RNA retrieval was done (panel C), but not if RNA retrieval was combined with protease digestion (panel D). Under these pretreatment conditions, most of the cancer cells exposed to reovirus showed a cytoplasmic signal for reoviral RNA using the LNA probe (panel E). Compare this signal to the background evident in the positive cell line if only RNA retrieval was done (panel F). This again underscores the value of testing multiple pretreatment conditions on the same slide when doing in situ hybridization.

5. Let the water drain from the slides by placing the slides at an angle on a stand.
6. Put the slides in a rack and bake for 30 minutes at 60°C. *Do not place the slides directly on the hot plate;* rather, put the slides in a rack and then put the rack in the 60°C oven.
7. Remove the slides and store at RT in a light-tight box or slide holder.

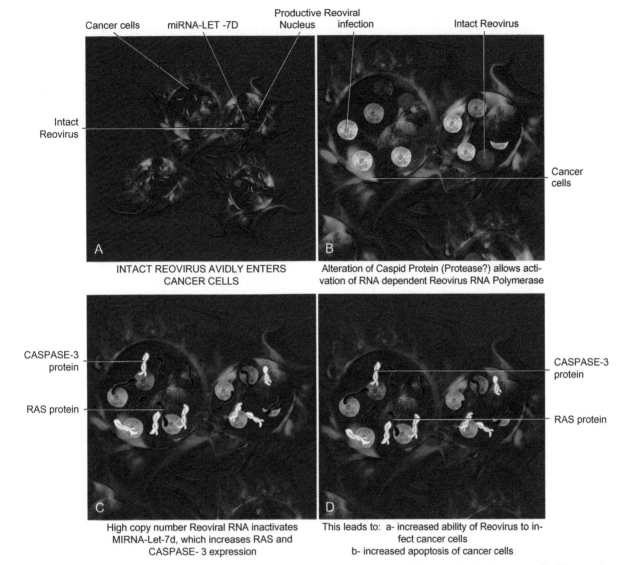

Cancer cells   miRNA-LET -7D   Productive Reoviral Nucleus   infection   Intact Reovirus

Intact Reovirus

INTACT REOVIRUS AVIDLY ENTERS CANCER CELLS

Cancer cells

Alteration of Caspid Protein (Protease?) allows activation of RNA dependent Reovirus RNA Polymerase

CASPASE-3 protein

RAS protein

CASPASE-3 protein

RAS protein

High copy number Reoviral RNA inactivates MIRNA-Let-7d, which increases RAS and CASPASE- 3 expression

This leads to: a- increased ability of Reovirus to infect cancer cells
b- increased apoptosis of cancer cells

Figure 10-8 Schematic representation of the key molecular events surrounding reoviral-induced death of cancer cells. The panels represent an artist's rendition of the key events that surround reoviral infection of cancer cells, including the avid entry of the viral genome into the cancer cells; the variable induction of productive viral infection; the upregulation of RAS, which, in turn, facilitates viral productive infection; and the inactivation by parts of the viral genome of microRNA-let-7d, which, in turn, upregulates caspase-3 and RAS production.

STEP TWO: REMOVE THE PARAFFIN FROM THE SLIDES

1. On the day you wish to begin the in situ hybridization experiment, remove the slides from the box.
2. Incubate the slides for 5 minutes in fresh xylene (the xylene solution should be changed weekly).
3. Incubate the slides for 5 minutes in fresh 100% ethanol (the ethanol solution should also be changed weekly).
4. Air dry the slides at either room temperature or in a 37°Co ven.

STEP THREE: OPTIMIZE THE PROTEASE DIGESTION PROTOCOL

1. This is the *key* step when doing RT in situ PCR. By optimizing the protease digestion, you can do direct target-specific incorporation of the reporter nucleotide during the PCR part of the reaction. This obviates the need to do a hybridization step, although you still have that option if you prefer.
2. Take two consecutive labeled slides (serial sections) that have been deparaffinized.
3. Check and confirm that the tissue contains the DNA or RNA target of interest. (I will assume that each slide has three sections per slide.)
4. Do *not* protease digest the top section. Protease digest the second section for 10 minutes and protease digest the last section furthest from the frosted end for 45 minutes.
5. After the protease digestion is done, briefly wash the slides in RNase-free water, followed by a quick rinse in 100% ethanol. Air dry.

6. Store one slide in a light-tight box at room temperature. Take the other slide and prepare a 100 μl DNase solution. You can make the solution by diluting the DNase buffer 1:10 in sterile RNase-free water to which you have added 100 units of DNase. Add 30 μl of the DNase solution per section and cover with a polypropylene coverslip that was cut tosiz e.

7. Incubate the tissues in the DNase solution overnight at 37°C.

8. Remove the coverslips and wash the DNase solution away with sterile DNase/RNase-free water followed by a brief rinse in 100% ethanol. Then air dry.

9. Prepare the "DNA synthesis solution" as per the following recipe:

   DNA Synthesis Solution
   10 μlo fth eE ZrT thb uffer[1]
   1.6 μl *each* of dATP, dCTP, dGTP, and dTTP (10 mMsto ck[2])
   1.6 μlo f2% (w/v)b ovinese rumalb umin
   1.0 μlo fRN asin[3]
   16 μlDE PCwate r
   12.4 μlo f10 mMM nCl
   0.6 μlo fd igoxigenind UTP(1 mMsto ck)
   2.0 μlo fth erT th
   (Reagents available in the EZ RT PCR kit from AppliedB iosystems)

10. Add 15 ml of the "DNA synthesis solution" to each tissue section *of each slide* and cover with a polypropylene coverslip cut to size.

11. Incubate the two slides at 37°C for 1 hour; overlay the slide with mineral oil to prevent evaporation of the DNA synthesis solution.

12. Remove the coverslip and incubate the slides in xylene for 5 minutes to remove the mineral oil and then for 5 minutes in 100% ethanol to remove the xylene. Air dry.

STEP FOUR: THE STRINGENT WASH

1. Place the slides in a solution of 0.1 XSSC, 2% bovine serum albumin that has been warmed to 50°C in a waterb ath.

2. Incubate the slides in the stringent wash for 10 minutes.

3. Remove the slides *one at a time* for the next step. Do *not* let the slides dry out after the stringent wash.

STEP FIVE: THE DETECTION PART—THE FIRST STEP

1. Remove the slide from the stringent wash and hold very level. Quickly remove the wash solution from the bottom of the slide and the perimeter of the front of the slide.

[1] Omit the primers for the negative control; substitute DEPC water or use irrelevant primers.
[2] Omit the primers for the negative control; substitute DEPC water or use irrelevant primers.
[3] Omit the primers for the negative control; substitute DEPC water or use irrelevant primers.

2. Using a hydrophobic pen, mark the perimeter of the front of the slide to keep the detection solution over the tissue. Keep the slide level at all times; otherwise, the remaining wash solution will ruin the "hydrophobic dam" you have just created.

3. To detect the digoxigenin tag, add 100 microliters of the antidigoxigenin/reporter enzyme conjugate. The antidigoxigenin solution should be made fresh at the time you need it by diluting the antibody 1:200 in a PBS pH 7.0 solution.

4. Incubate the antidigoxigenin conjugate at 37°C for 2 hours in a humidity chamber.

STEP SIX: THE DETECTION PART—THE LAST STEP

1. If using the alkaline phosphatase conjugate, place the slides in the NBT/BCIP buffer (pH 9.5 solution, Enzo Laboratory, catalog #ENZ-33802).

2. Preheat the NBT/BCIP buffer to 37°C. Confirm that the pH is between 9.0 and 9.5. If it is not, titrate with 0.1 MNaOH.

3. You can add 11 ml of the preheated NBT/BCIP buffer to a five-slide plastic holder.

4. Add 165 microliters of the NBT/BCIP solution (Roche Applied Sciences, catalog #11681451001).

5. Add the slides to the NBT/BCIP solution and incubate at 37°C.

6. If using the peroxidase conjugate, place the slides in the DAB buffer (pH 7.0, PBS solution).

7. Prepare the DAB solution immediately before you are ready to do this step. For manual in situ hybridization using DAB, you may use the Biogenex DAB kit (Super Sensitive Link-label Immunohistochemical Detection system, catalog #QD000-5L). This kit has all the reagents you need for immunohistochemistry after the primary antibody step, and for in situ hybridization after the probe-target hybridization step if using a biotin-labeled probe. To make the DAB solution, take 2 drops of the DAB chromogen and add to the DAB substrate buffer. Vortex. Then add 1 drop of the DAB substrate. Vortex.

8. Remove the slides from the PBS wash. Wipe the back of the slide and, keeping the slide level, wipe the perimeter of the front of the slide. Place the slide carefully down in a humidity chamber. Do *not* let the slide dry out; otherwise, background will ensue.

9. Add several drops of DAB solution to each tissue section.

STEP SIX: MONITOR THE PRECIPITATE

1. It is *very* important to monitor the precipitation of the chromogen (either DAB or NBT/BCIP) under the microscope, initially every 3 minutes. Do not accept any interruptions at this stage. Stop the reaction when you have determined that the signal has been maximized and/or background is just beginning tod evelop.

2. For RT in situ PCR, although most NBT/BCIP reactions are finished within 20 minutes, the only determining factor for stopping the reaction is when you

have determined that the signal has reached its maximum stage and/or background begins to be evident.

3. Although most DAB reactions for RT in situ PCR are finished within 20 minutes, the only determining factor for stopping the reaction is when you have determined that the signal has reached its maximum stage and/or background begins to be evident.

4. When the signal is at maximum *or* background begins to be evident, immediately place the slide in a jar that contains water. Wash the slides under running tap water for 1 minute to stop the chromogen precipitate.

STEP SEVEN: THE COUNTERSTAIN

1. If you wish to do co-expression analysis, do *not* do a counterstain at this time.

2. If you wish to use a counterstain for NBT/BCIP, use Nuclear Fast Red.

3. If you wish to use a counterstain for DAB, use either Nuclear Fast Red or hematoxylin.

4. For nuclear targets, do a weak counterstain (a few dips in hematoxylin or 30–60 seconds in Nuclear FastRe d).

5. For cytoplasmic targets, use the standard hematoxylin (1 minute) or Nuclear Fast Red (3 minute) counterstain.

6. Rinse the slides in running tap water for 1 minute.

7. Dip the slides in 100% ethanol. Let completely air dry.

8. Placeth es lidesin xyle ne.

9. Coverslip using Permount (Fisher Scientific, catalog #SP15-100).

STEP EIGHT: INTERPRETATION OF THE OPTIMIZING PROTOCOL FOR RT IN SITU PCR

At this stage, you have two slides, each of which has three tissue sections. The three tissue sections have been pretreated with nothing, 10 minutes of protease, and 60 minutes of protease. One slide has had each tissue section treated with DNase overnight. The key to the interpretation is to focus on the *nuclei* of the cells. This is the place where the nonspecific DNA synthesis pathway occurs. It is this pathway that will tell you the optimal conditions for RT in situ PCR. Specifically, this is what you need to see to document that this condition (no pretreatment, 10 minutes of protease, or 60 minutes of protease digestion) is optimal for the *specific* tissue of interest:

1. An intense nuclear-based signal in at least 50% of the cells with no DNase.

2. The complete elimination of the nuclear-based signal with the same pretreatment conditions after DNase digestion.

Figures 10-9 and 10-10 show examples of some of these controls and the associated test-related data for another RNA virus, the rabies virus.

Here are the interpretations if you do not have optimal pretreatment conditions:

A. *Signal with DNase digestion.* When you see a signal with the DNase digestion, it means that the pretreatment did *not* allow the DNase complete access to the genomic DNA. This is the equivalent to in situ hybridization or immunohistochemistry when the three-dimensional macromolecule cross-linked network was too dense and, thus, did not permit the probe/primary antibody or other essential reagents access to the target, as discussed at length in Chapter 4. The solution is to increase the protease digestion and/or add RNA retrieval.

B. *Overdigestion with no/weak signal.* This, of course, is the same problem we have seen with in situ hybridization and immunohistochemistry and discussed at length in previous chapters. The solution is reducing the protease digestion time.

C. *No signal with no DNase digestion sections.* This is the equivalent to seeing no signal with in situ hybridization or immunohistochemistry with one big difference: by definition, every cell in formalin-fixed, paraffin-embedded tissues can generate DNA synthesis from the nonspecific primer independent pathway. The reason is that every such cell will have nicks in its DNA that serve as the starting points of the DNA synthesis due to the "proofreading" exonuclease ability of Taq polymerase or rTth. So if you see no signal with the no DNase digestion sections, here are the possibilities:

1. *Inactive protease.* Remember that if the three-dimensional macromolecule cross-linked network is dense, and the protease does not have much activity, the key reagents simply cannot access the nicks in the DNA to generate the DNA synthesis that is needed to see the darkly stained nuclei after this "optimizing protocol."

2. *Inactive polymerase.* Although my colleagues at ABI insist this is extremely unusual, if you do not add the correct buffer or the correct ratio of nucleotides or digoxigenin-tagged nucleotide, etc., then this is a possibility. It would be helpful in this regard to verify that the polymerase is functioning by doing a solution phase PCR or RTPCRre action.

3. *Inactive antidigoxigenin conjugate.* My recommendation to rule out this possibility is to do a standard in situ hybridization reaction with a high copy number target (e.g., HPV 6/11 and vulvar condyloma) to verify that each of the detection reagents is operating well.

After you have determined the optimal pretreatment conditions for the specific tissue, then you are ready to do the RT in situ PCR reaction. The protocol is described next.

A PROTOCOL FOR RT IN SITU PCR: USING THE ONE-STEP RTTH SYSTEM

STEP ONE: CUT THE TISSUE SECTIONS ONTO A SILANE-COATED SLIDE

1. Get two more of the serial sections from the group prepared for the optimizing protocol.

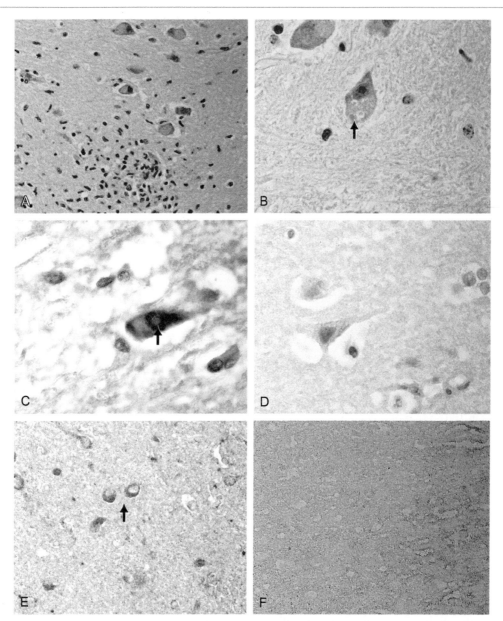

Figure 10-9   RT in situ PCR and immunohistochemistry for the detection of the rabies virus. Panel A shows an area where a microglial nodule is evident, although such inflammatory-type changes are rare in rabies encephalitis. Panel B is a higher magnification of a large neuron in this area where a cytoplasmic inclusion (Negri body) is evident (arrow). PCR-amplified rabies cDNA was evident in the cytoplasm of many neurons in this area; panel C shows a high magnification so that the Negri body (arrow) is also visible. Viruses usually are associated with specific cell types and, when they have inclusions such as the rabies virus, provide another cytologic marker for the specificity of the reaction. The negative controls each showed loss of signal and included pretreatment in RNase (panel D), omission of rabies-specific primers, substitution with irrelevant (HPV-specific) primers, and testing of noninfected brain tissue (not shown). Panel E shows detection of rabies protein by immunohistochemistry; note the cytoplasmic-based red signal in the larger neurons. The negative controls each showed loss of signal and included omission of the primary antibody (panel F) and testing of noninfected brain tissue (not shown).

STEP TWO: REMOVE THE PARAFFIN FROM THE SLIDES

1. Incubate the slides for 5 minutes in fresh xylene (the xylene solution should be changed weekly).
2. Incubate the slides for 5 minutes in fresh 100% ethanol (the ethanol solution should also be changed weekly).
3. Air dry the slides either at room temperature or in a 37°Co ven.

STEP THREE: PRETREAT THE SLIDES WITH THE OPTIMAL PROTEASE DIGESTION CONDITIONS

1. Protease digest each of the three sections for the optimal protease digestion time.
2. After the protease digestion is done, briefly wash the slides in RNase-free water, followed by a quick rinse in 100% ethanol. Air dry.

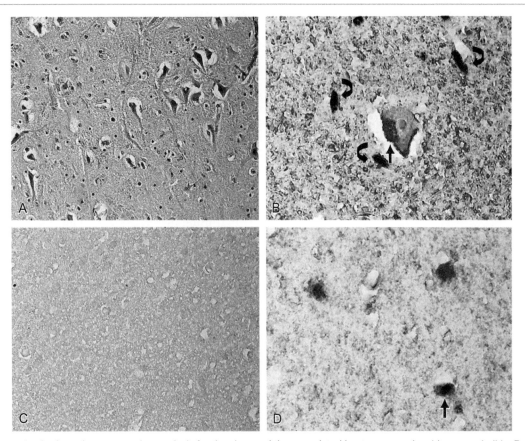

Figure 10-10   In situ-based co-expression analysis for the virus and the associated host response in rabies encephalitis. Panel A shows an area of the brain in which the histologic changes are minimal and nonspecific. The bland histology is typical of rabies encephalitis and, of course, is in stark contrast to the clinical state. In such situations, you often find that the endogenous cells are expressing high levels of cytokines, which, in turn, are in large part responsible for the marked symptomatology. Many rabies-infected cells were evident in this area with other cells expressing TNFα and iNOS. Co-labeling experiments with rabies protein by immunohistochemistry (red signal, small arrow) and TNFα mRNA by RT in situ PCR (blue signal, large arrows) showed that the cells expressing TNF were usually not infected by the rabies virus but, rather, surrounded rabies-infected neurons (panel B). Note that the cells expressing TNFα have the elongate phenotype characteristic of microglial cells. The signals were lost in the serial section if the tissue was pretreated with RNase, tested for TNFα mRNA expression, and then tested for rabies protein with omission of the primary antibody (panel C). Although most of the rabies-infected cells had the phenotype of neurons, occasional cells may have been astrocytes; co-labeling experiments with GFAP (red signal) and rabies RNA (blue signal) documented rare infected astrocytes (panel D, arrow).

3. Take each slide and prepare a 100 μl DNase solution. The solution can be made by diluting the DNase buffer 1:10 in sterile RNase-free water to which you have added 100 units of DNase. Add 30 μl of the DNase solution per section and cover with a polypropylene coverslip that was cut to size. *For one slide, do not DNase digest one of the sections;* this will serve as the positive control (label this slide as the positive and negative control slide).

4. Incubate the tissues overnight in the DNase solution at 37°C.

5. Remove the coverslips and wash the DNase solution away with sterile DNase/RNase-free water followed by a brief rinse in 100% ethanol. Air dry.

6. Prepare the "RT in situ PCR solution" as per the following recipe:
   RT in situ PCR Solution
   10 μl of the EZ rT buffer[4]

1.6 μl *each* of dATP, dCTP, dGTP, and dTTP (10 mM stock[5])
1.6 μl of 2% (w/v) bovine serum albumin
1.0 μl of RNasin[6]
3.0 μl of primer 1 and primer 2 (20 μM stock of each primer [*test* slides])
13 μl DEPC water
12.4 μl of 10 mM MnCl

---

[4]Omit the primers for the negative control; substitute DEPC water or use irrelevant primers.

[5]Omit the primers for the negative control; substitute DEPC water or use irrelevant primers.

[6]Omit the primers for the negative control; substitute DEPC water or use irrelevant primers.

0.6 μl o fd igoxigenind UTP(1 mMsto ck)

2.0 μl o fth erT th

(Reagents available in the EZ RT PCR kit from AppliedB iosystems)

7. Add 15 ml of the ÒRT in situ PCR solutionÓ to the slide that has had each section treated with DNase and cover with a polypropylene coverslip cut to size. This will be the *test* slide, and you will have tested three serial sections for the same RNA of interest.

8. To the other slide, add the RT in situ PCR solution but *omit* the primers. This slide can serve as the negative control (no primers) and two separate positive controls. One of the positive controls will be the no DNase section. This is the positive control for the PCR in situ reaction. The other positive control will be the DNase-treated section to which you add to the solution Òhousekeeper primersÓ (like actin). This will be the mRNA positive control slide.

9. Incubate the two slides at 60¡C for 30 minutes (RT step). Remember to overlay the slide with mineral oil to prevent evaporation of the RT in situ PCR solution.

10. Denature the slides at 95¡C for 3 minutes; then do 25 cycles at 60¡C for 1½ minutes and 95¡C for 1 minute. Bring the cycler to 4¡C.

11. Remove the coverslip and incubate the slides in xylene for 5 minutes to remove the mineral oil and then for 5 minutes in 100% ethanol to remove the xylene. Air dry.

STEP FOUR: THE STRINGENT WASH

1. Place the slides in a solution of 0.1 XSSC, 2% bovine serum albumin that has been warmed to 50¡C in a waterb ath.

2. Incubate the slides in the stringent wash for 10 minutes.

3. Remove the slides *one at a time* for the next step. Do *not* let the slides dry out after the stringent wash.

STEP FIVE: THE DETECTION PART—THE FIRST STEP

1. Remove the slide from the stringent wash and hold very level. Quickly remove the wash solution from the bottom of the slide and the perimeter of the front of the slide.

2. Using a hydrophobic pen, mark the perimeter of the front of the slide to keep the detection solution over the tissue. Keep the slide level at all times; otherwise, the remaining wash solution will ruin the Òhydrophobic damÓ you have just created.

3. To detect the digoxigenin tag, add 100 microliters of the antidigoxigenin/reporter enzyme conjugate. The antidigoxigenin solution should be made fresh at the time you need it by diluting the antibody 1:200 in a PBS pH 7.0 solution.

4. Incubate the anti-digoxigenin conjugate at 37¡C for 2 hours in a humidity chamber.

STEP SIX: THE DETECTION PART—THE LAST STEP

1. If using the alkaline phosphatase conjugate, place the slides in the NBT/BCIP buffer (pH 9.5 solutionÑ Enzo LaboratoryÑ catalog #ENZ-33802).

2. Preheat the NBT/BCIP buffer to 37¡C. ConÞrm that the pH is between 9.0 and 9.5. If it is not, titrate with 0.1 MNaOH.

3. You can add 11 ml of the preheated NBT/BCIP buffer to a Þve-slide plastic holder.

4. Add 165 microliters of the NBT/BCIP solution (Roche Applied Sciences, catalog #11681451001).

5. Add the slides to the NBT/BCIP solution and incubate at 37¡C.

6. If using the peroxidase conjugate, place the slides in the DAB buffer (pH 7.0, PBS solution).

7. Prepare the DAB solution immediately before you are ready to do this step. For manual in situ hybridization using DAB, you may use the Biogenex DAB kit (Super Sensitive Link-label Immunohistochemical Detection systemÑ catalog #QD000-5L). This kit has all the reagents you need for immunohistochemistry after the primary antibody step, and for in situ hybridization after the probe-target hybridization step if using a biotin-labeled probe. To make the DAB solution, take 2 drops of the DAB chromogen and add to the DAB substrate buffer. Vortex. Then add 1 drop of the DAB substrate. Vortex.

8. Remove the slides from the PBS wash. Wipe the back of the slide and, keeping the slide level, wipe the perimeter of the front of the slide. Place the slide carefully down in a humidity chamber. Do *not* let the slide dry out; otherwise, background will ensue.

9. Add several drops of DAB solution to each tissue section.

STEP SIX: MONITOR THE PRECIPITATE

1. It is *very* important to monitor the precipitation of the chromogen (either DAB or NBT/BCIP) under the microscope, initially every minute. Do not accept any interruptions at this stage. Since the RT in situ PCR reaction can develop so quickly, I monitor the reaction at every minute. Stop the reaction when you have determined that the signal has been maximized and/or background is just beginning to develop.

2. For RT in situ PCR, although most NBT/BCIP reactions are Þnished within 20 minutes, the only determining factor for stopping the reaction is when you have determined that the signal has reached its maximum stage and/or background begins to be evident.

3. Although most DAB reactions for RT in situ PCR are Þnished within 20 minutes, the only determining factor for stopping the reaction is when you have determined that the signal has reached its maximum stage and/or background begins to be evident.

4. When the signal is at maximum *or* background begins to be evident, immediately place the slide in a jar that contains water. Wash the slides under running tap water for 1 minute to stop the chromogen precipitate.

STEP SEVEN: THE COUNTERSTAIN

1. If you wish to do co-expression analysis, do *not* do a counterstain at this time.
2. If you wish to use a counterstain for NBT/BCIP, use Nuclear Fast Red.
3. If you wish to use a counterstain for DAB, use either Nuclear Fast Red or hematoxylin.
4. For nuclear targets, do a weak counterstain (a few dips in hematoxylin or 30–60 seconds in Nuclear Fast Red).
5. For cytoplasmic targets, use the standard hematoxylin (1 minute) or Nuclear Fast Red (3 minute) counterstain.
6. Rinse the slides in running tap water for 1 minute.
7. Dip the slides in 100% ethanol. Let completely air dry.
8. Placeth es lidesin xyle ne.
9. Coverslip using Permount (Fisher Scientific, catalog #SP15-100).

Confirm that the negative control and PCR positive control worked satisfactorily. If so, then the RT in situ PCR reaction for the DNase-treated sections and the test primers should yield a target-specific reaction.

If you follow the protocol outlined here, although the reaction will be target-specific, you still have the option of omitting the digoxigenin from the RT in situ PCR reaction and, instead, using a digoxigenin-tagged LNA probe after the RT and in situ PCR steps as per the standard in situ hybridization protocol presented in Chapter 7.

PCR IN SITU HYBRIDIZATION

PCR in situ hybridization is similar to the RT in situ PCR protocol listed in the preceding section. As I have already indicated, I rarely use this protocol because direct head-to-head comparisons have indicated that the LNA probes give equivalent results to PCR in situ hybridization and RT in situ PCR (see Figures 10-1 and 10-2).

If you wish to do PCR in situ hybridization, there are three key differences with RT in situ PCR:

1. Ofc ourse,o mitth eDNase ste p.
2. Do not use the digoxigenin-tagged nucleotide, but rather use a labeled probe after the PCR in situ reaction to detect the amplicon.
3. Use the hot-start maneuver to optimize the number of amplicons generated during PCR in situ hybridization.

Hot-start PCR is well documented to improve the yield of amplification. It works because the high temperature of the hot-start reaction is above the melting temperature of the nonspecific pathways (primer dimerization, mispriming) and below the melting temperature of the target specific pathway, where the homology will be 100%.

As with any in situ hybridization reaction, you first have to determine the optimal pretreatment conditions for the target in the specific tissue. Hence, test at least four slides as detailed next.

STEP ONE: OPTIMIZE THE PROBE/CHOOSE DIFFERENT PRETREATMENT PROTOCOLS

1. The optimization of the probe presupposes that you have placed at least two sections per silane-coated slide.
2. Take four consecutive labeled slides (serial sections) that have been deparaffinized.
3. Check and confirm that the tissue contains the DNA or RNA target of interest.
4. Label the first slide "Dilute probe, no protease/protease." Label the second slide "Dilute probe, DNA (or RNA) retrieval, no protease/protease." Label the third slide "Concentrated probe, no protease/protease." Label the fourth slide "Concentrated probe, DNA (or RNA) retrieval, no protease/protease."
5. Make the probe cocktail by diluting the probe with a commercially available DNA or RNA probe diluent (e.g., Enzo Life Sciences in situ hybridization buffer, catalog #33808; or Exiqon in situ hybridization buffer, catalog #90000, which includes the in situ hybridization buffer as well as proteinase K).
6. For full-length probes (probes made from templates at least 200 base pairs long), the dilute-labeled probe should be diluted to a final concentration of 10 ng/ml, and the concentrated probe should be at 1000 ng/ml.
7. For smaller LNA probes (I do *not* recommend using oligoprobes if they are not LNA modified), the dilute-labeled probe should be diluted to a final concentration of 0.01 picomole/microliter, and the concentrated probe should be at 1 picomole/microliter.
8. Proceed to the DNA/RNA retrieval and protease steps.

STEP TWO: PRETREAT THE TISSUES

1. Add protease (proteinase K, 1 microgm/ml, Enzo Life Sciences, catalog #33801, diluted in sterile water) to only the *bottom* tissue section to each of the fourslid es.
2. Incubate the tissue sections in the proteinase K solution for 15 minutes at room temperature.
3. Remove the proteinase K by washing the slides with RNase- and DNase-free water for a few seconds; then rinse slides in DNase/RNase-free 100% ethanol for a few seconds. Then air dry in a 37°C incubator.
4. Take two of the four slides and incubate for 30 minutes at 95°C in any commercially available "antigen retrievalso lution."
5. After the 30 minutes, let the solution cool down for 5 minutes at room temperature. Then remove the slides and let air dry.
6. At this stage, you have four slides, each with at least two sections. The bottom section for each has been digested with protease. Two of the four slides have been exposed to 95°C for 30 minutes in a DNA/RNA retrieval solution.

STEP THREE: THE HOT-START PCR IN SITU REACTION

1. Prepareth efo llowingso lution:
   PCR In Situ Hybridization Synthesis Solution
   5 μlo fth eT aqp olymeraseb uffer

1.6 μl *each* of dATP, dCTP, dGTP, and dTTP, (10 mMsto ck)

1.6 μlo f2%(w/ v)b ovinese rumalb umin

3.0 μl of primer 1 and primer 2 (20 μM stock of each primer [*test*slid es])

14 μlDE PCwate r

9.0 μlo f10 mMM gCl

2.0 μlo fth eT aqp olymerase

(Reagents available in the PCR kit from Applied Biosystems)

2. Heat the slide to 60¡C before adding the PCR in situ hybridization synthesis solution. Then add the solution under a polypropylene coverslip. Anchor with a few small drops of nail polish; overlay with mineral oil.

3. Denatureat9 5¡Cfo r3min utes.

4. Cycle for 35 cycles at 60¡C for 1½ minutes and 95¡C for 1 minute.

5. Bringto 4¡C.

6. Remove the mineral oil with xylene and 100% ethanol washes. Air dry.

STEP FOUR: DENATURE THE PROBE AND TISSUE RNA/DNA

1. For small biopsies (less than 1 cm in size), add 5Đ10 microliters of the probe cocktail.

2. For larger biopsies (equal to or greater than 1 cm in size), add 10Đ15 microliters of the probe cocktail.

3. Cover the probe cocktail with a polypropylene coverslip that is cut to be just slightly larger than the tissue section. If doing RNA in situ hybridization, useglo ves.

4. For DNA targets, place the slide on a 95¡C hot plate for 5 minutes. Make sure no bubbles have appeared between the coverslip and the tissue.

5. If you see bubbles, *immediately* and gently remove them using a sterile toothpick.

6. Incubate the tissue in a humidity chamber with water at the base at 37¡C for 5Đ15 hours. For the optimization protocol, I strongly recommend overnight incubation because it is impossible to know the relative copy number of the target and also the speciĐc activity of the probe.

STEP FIVE: THE STRINGENT WASH

1. Remove the coverslips carefully by using Þne tweezers. Make certain that the probe cocktail did not dry out. If it did, either the coverslip was too large, there was too little of the probe cocktail and/or there was not enough water used for the humidity chamber.

2. Place the slides in a solution of 0.1 XSSC, 2% bovine serum albumin that has been warmed to 50¡C in a waterb ath.

3. Incubate the slides in the stringent wash for 10 minutes.

4. Remove the slides *one at a time* for the next step. Do *not* let the slides dry out after the stringent wash.

STEP SIX: THE DETECTION PART—THE FIRST STEP

1. Remove the slide from the stringent wash and hold very level. Quickly remove the wash solution from the bottom of the slide and the perimeter of the front of the slide.

2. Using a hydrophobic pen, mark the perimeter of the front of the slide to keep the detection solution over the tissue. Keep the slide level at all times; otherwise, the remaining wash solution will ruin the Þhydrophobic damÓyou have just created.

3. For biotin-labeled probes, add 100 microliters of the streptavidin solution with the reporter enzyme conjugate (Enzo Clinical Labs, catalog #ENZ-32895; this is actually a kit that contains all the necessary ingredients for in situ hybridization using the streptavidin-alkaline phosphatase system and NBT/BCIP as thec hromogen).

4. For digoxigenin-labeled probes, add 100 microliters of the antidigoxigenin/reporter enzyme conjugate. The antidigoxigenin solution should be made fresh at the time you need it by diluting the antibody 1:200 in a PBS pH 7.0 solution.

5. For either biotin- or digoxigenin-based probes, incubate the streptavidin or anti-digoxigenin conjugate at 37¡C for 2 hours in a humidity chamber.

STEP SEVEN: THE DETECTION PART—THE LAST STEP

1. If using the alkaline phosphatase conjugate, place the slides in the NBT/BCIP buffer (pH 9.5 solution, Enzo Laboratory, catalog #ENZ-33802).

2. Preheat the NBT/BCIP buffer to 37¡C. ConÞrm that the pH is between 9.0 and 9.5; if it is not, titrate with 0.1 MNaOH.

3. You can add 11 ml of the preheated NBT/BCIP buffer to a Þve-slide plastic holder.

4. Add 165 microliters of the NBT/BCIP solution (Roche Applied Sciences, catalog #11681451001).

5. Add the slides to the NBT/BCIP solution and incubate at 37¡C.

6. If using the peroxidase conjugate, place the slides in the DAB buffer (pH 7.0, PBS solution).

7. Prepare the DAB solution immediately before you are ready to do this step. For manual in situ hybridization using DAB, you may use the Biogenex DAB kit (Super Sensitive Link-label Immunohistochemical Detection system, catalog #QD000-5L). This kit has all the reagents you need for immunohistochemistry after the primary antibody step, and for in situ hybridization after the probe-target hybridization step if using a biotin-labeled probe. To make the DAB solution, take 2 drops of the DAB chromogen and add to the DAB substrate buffer. Vortex. Then add 1 drop of the DAB substrate. Vortex.

8. Remove the slides from the PBS wash. Wipe the back of the slide and, keeping the slide level, wipe the perimeter of the front of the slide. Place the slide carefully down in a humidity chamber. Do *not* let the slide dry out; otherwise, background will ensue.

9. Add several drops of DAB solution to each tissue section.

STEP EIGHT: MONITOR THE PRECIPITATE

1. It is *very* important to monitor the precipitation of the chromogen (either DAB or NBT/BCIP) under

the microscope, initially every 3 minutes. Do not accept any interruptions at this stage. Stop the reaction when you have determined that signal has been maximized and/or background is just beginning to develop.

2. Although most NBT/BCIP reactions with PCR in situ hybridization are finished within 30 minutes, the only determining factor for stopping the reaction is when you have determined that the signal has reached its maximum stage and/or background begins to be evident.

3. Although most DAB reactions are finished within 10 minutes, the only determining factor for stopping the reaction is when you have determined that the signal has reached its maximum stage and/or background begins to be evident.

4. When the signal is at a maximum *or* background begins to be evident, immediately place the slide in a jar that contains water. Wash the slides under running tap water for 1 minute to stop the chromogen precipitate.

STEP NINE: THE COUNTERSTAIN

1. If you wish to do co-expression analysis, do *not* do a counterstain at this time.
2. If you wish to use a counterstain for NBT/BCIP, use Nuclear Fast Red.
3. If you wish to use a counterstain for DAB, use either Nuclear Fast Red or hematoxylin.
4. For nuclear targets, do a weak counterstain (a few dips in hematoxylin or 30–60 seconds in Nuclear FastRe d).
5. For cytoplasmic targets, use the standard hematoxylin (1 minute) or Nuclear Fast Red (3 minute) counterstain.
6. Rinse the slides in running tap water for 1 minute.
7. Dip the slides in 100% ethanol. Let completely air dry.
8. Placeth es lidesin xyle ne.
9. Coverslip using Permount (Fisher Scientific, catalog #SP15-100).

To troubleshoot PCR in situ hybridization, use the same rationale as with in situ hybridization. This makes sense because the two processes are basically identical.

THE THEORY BEHIND RT IN SITU PCR

Since, in my experience, many more investigators are interested in RT in situ PCR than PCR in situ hybridization, let me end this section by discussing the theory of RT in situ PCR.

You cannot do target-specific direct incorporation of the reporter molecule (e.g., digoxigenin dUTP) for DNA targets in paraffin-embedded tissue sections, which I refer to as *in situ* PCR. The reason is that the nonspecific DNA synthesis, primarily in the form of DNA repair, invariably would cause a false-positive signal with paraffin-embedded tissue sections. Although DNA repair can be blocked by pretreatment in a solution that contains a

dideoxy nucleotide, the exacting requirements of correct protease digestion time and the preincubation time with the dideoxy nucleotide made this "dideoxy-assisted *in situ* PCR" prone to residual nonspecific false-positive results. Another way to avoid this nonspecific DNA synthesis pathway is to use cryostat sections fixed in 10% buffered formalin and *never* exposed to high dry heat. The DNA in such samples will not have nicks, which are the entire basis of the primer-independent DNA synthesis pathway during in situ PCR. In my experience, this is the best and most reliable way to detect DNA targets using the hot-start maneuver and direct incorporation of the reporter nucleotide.

The nonspecific primer independent signal, as indicated previously, can reliably and completely be eliminated with an overnight digestion in RNase-free DNase if (and only if) the sample was first optimally digested with a protease. Clearly, this approach would not be of use for the detection of either viral or native DNAs. However, these observations lend themselves very well to the in situ detection of RNA viruses or mRNAs using the direct incorporation of the reporter molecule. Let's now discuss each possible pathway that can be operative during in situ amplification of DNA.

THE POTENTIAL SOURCES OF THE SIGNAL WITH IN SITU PCR/RT IN SITU PCR

There are four potential DNA synthesis pathways that may be operative inside the cell during in situ PCR. They are

1. Target-specificamp lification
2. Nontarget primer-dependent amplification (mispriming)
3. Primero ligomerization
4. The primer-independent DNA synthesis pathway (in situ nick translation)

Of these, DNA repair is invariably present in paraffin-embedded tissues or any other cell or tissue preparation that has been exposed to dry heat of at least 55°C for ≥ 1 hour. Indeed, this primer-independent signal may be evident even after several minutes of exposure to such dry heat. DNA repair will usually involve all of the different cell types in a tissue section. In comparison, the target-specific cDNA-based signal for RT in situ PCR or the target-specific signal with PCR in situ hybridization will rarely involve all cell types unless the target chosen is a "housekeeper" transcript. Thus, the presence of a signal in all cell types after RT in situ PCR or PCR in situ hybridization suggests a nonspecific signal. Alternatively, a signal that is restricted to only one or a few cell types after RT in situ PCR suggests a target-specific signal. This simple but important point can be easily demonstrated with viral infections, given the fastidious tropism exhibited by many viruses (e.g., Figures 10-1 and 10-2).

Mispriming is another source of DNA synthesis that is operative during in situ PCR if the hot-start maneuver is not used. In comparison, the hot-start method will not inhibit the DNA repair pathway. You can demonstrate the existence of mispriming by using frozen, fixed tissues that lack the primer-independent pathway and primers that do not correspond to any target in the cells in the tissue sample. Mispriming in situ is the equivalent that you can see with solution phase PCR.

In comparison, in solution phase PCR, a very common source of background is primer oligomerization. This is the place where primers bind to themselves and serve as the nidus of DNA synthesis via the polymerase. Primer oligomerization does not appear to induce any signal with in situ PCR. This can be demonstrated by using paraffin-embedded tissues in which DNA repair and mispriming have been blocked by DNase digestion after optimal protease digestion. If you then do in situ PCR using primers that do not correspond to any target in the cells, no nuclear signal is evident.

Target-specific DNA synthesis is, of course, operative during in situ PCR if the primers correspond to a target in the cells in the sample.

Finally, it is important to remember, and an appropriate way to end this section, that the three-dimensional macromolecule cross-linked network is every bit as important to PCR in situ hybridization and RT in situ PCR as it is for in situ hybridization and immunohisto-chemistry. This makes sense because, at their basis, each process involves diffusion of similarly sized key reagents through a labyrinth of cross-linked proteins and needs the protein cage to dock with the target and not to diffuse from the site of the target.

Appendix 1: Locked Nucleic Acids—Properties and Applications

Peter Mouritzen, Jesper Wengel, Niels Tolstrup, Søren MorgentalerEchwa ld, Johan Wahlin, and InaK. Dahlsveen
Exiqon A/S, Skelstedet 16, 2950 Vedbæk, Denmark

INTRODUCTION TO LOCKED NUCLEIC ACIDS (LNA™)

What Is LNA™?

Locked nucleic acids (LNA™) are a class of high-affinity RNA analogues in which the ribose ring is "locked" in the ideal conformation for Watson–Crick binding (Figure A-1). As a result, LNA™ oligonucleotides exhibit unprecedented thermal stability when hybridized to a complementary DNA or RNA strand. For each incorporated LNA monomer, the melting temperature (Tm) of the duplex increases by 2°C–8°C (Figure A-2). In addition, LNA™ oligonucleotides can be made shorter than traditional DNA or RNA oligonucleotides and still retain a high Tm. This is important when the oligonucleotide is used to detect small or highly similar targets.

Nucleic acid duplexes fall into two major conformational types, the A-type and the B-type, which are dictated by the puckering of the single nucleotides, namely a C3′-endo (N-type) conformation in the A-type and a C2′-endo (S-type) conformation in the B-type (Figure A-1) [1]. The A-type is adopted by RNA when dsRNA duplex regions are found, whereas the dsDNA in the genome adopts a B-type. RNA duplexes have a higher binding affinity than DNA duplexes due to the difference in conformation. Restricting nucleotides to the N-type conformation can therefore increase the affinity of binding to DNA and RNA by generating highly stable A-type duplexes.

Locked nucleic acids (LNA™) are nucleotides in which the O2′ and the C4′ atoms are linked by a methylene group, thereby introducing a conformational lock of the molecule into a near perfect N-type conformation. A major structural characteristic of LNA is its close resemblance to the natural nucleic acids [2]. In addition, LNA shares many physical properties with DNA and RNA, including water solubility, which means that LNA oligonucleotides can be used in many standard experimental applications. Furthermore, LNA-containing oligonucleotides are synthesized by conventional phosphoramidite chemistry, allowing automated synthesis of fully modified LNA-sequences, as well as chimeras with DNA, RNA, modified monomers, or labels.

LNA oligonucleotides are usually defined as sequences containing one or more of the 2′-O,4′-C-methylene-β-D-ribofuranosyl nucleosides called LNA monomers (Figure A-1). A number of LNA:RNA and LNA:DNA hybrids have been structurally characterized by NMR spectroscopy and X-ray crystallography. In general, the

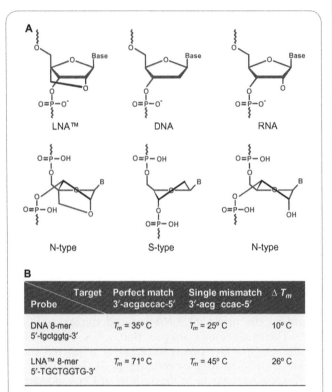

FigureA -1 Structure and conformations of nucleotides (A) and the effect of LNA™ on melting temperature of duplexes (B).

Probe	Target	Perfect match 3′-acgaccac-5′	Single mismatch 3′-acg ccac-5′	Δ T_m
DNA 8-mer 5′-tgctggtg-3′		$T_m = 35°$ C	$T_m = 25°$ C	10° C
LNA™ 8-mer 5′-TGCTGGTG-3′		$T_m = 71°$ C	$T_m = 45°$ C	26° C

duplexes retain the features common for native nucleic acid duplexes, i.e., usual Watson–Crick base pairing, nucleobases in the anti-orientation, base stacking, and a right-handed helical conformation. Importantly, LNA monomers conformationally tune the flanking monomers, especially the 3′-flanking monomer, to adopt an N-type furanose conformation [3].

LNA™ Improves Hybridization to DNA and RNA

The most important feature of LNA is the unprecedented hybridization to complementary nucleic acids. The introduction of one or more LNA monomers increases duplex stability for both complementary RNA and DNA. The affinity-enhancing effect of incorporation of LNA™ is demonstrated by an increase in the duplex melting temperature (T_m) per LNA monomer, and this effect is generally more pronounced against complementary RNA than against complementary DNA (normally from +3°C to +7°C against RNA and from +2°C to +5°C against

Same length, higher T_m						Shorter length, similar T_m			
DNA	23-mer	tcgatcgattagctacgtacgta	T_m: 60°C			23-mer	tcgatcgattagctacgtacgta	T_m: 60°C	DNA
DNA/LNA™	23-mer	tcgatcgattAgctacgtacgta	T_m: 64°C	+ LNA™		16-mer	---atcgattAgctAcgta----	T_m: 60°C	DNA/LNA™
DNA/LNA™	23-mer	tcgatcGattAgctaCgtaCgta	T_m: 78°C			8-mer	----------aGCtacGT----	T_m: 61°C	DNA/LNA™

DNA: atcg   LNA: ATCG

FigureA -2   On the left, substitution of DNA nucleotides with LNA™ increases the melting temperature of the oligonucleotide. On the right, LNA™ substitutions allow shortening of the probe while maintaining the same Tm.

DNA). Furthermore, LNA monomers with pyrimidine bases tend to induce larger thermal stability increases than LNA monomers with purine bases [4]. Figure A-2 illustrates the effect of introducing LNA to a DNA oligonucleotide when hybridized to RNA.

The position of LNA monomers in an oligonucleotide has an impact on the affinity-enhancing effects [3]. The relative increase in thermal stability per LNA monomer is largest for a single or a few nonconsecutive incorporations in DNA, RNA, or 2′ O methyl-RNA-sequences (mixmers). LNA monomers incorporated in the 3′- or 5′-termini induce a smaller increase in thermal stability than do centrally placed LNA monomers. In addition, the relative increase in thermal stability per LNA monomer is larger in shorter sequences than in longer sequences. Other chemical modifications are also stabilized by the incorporation of LNA, including phosphorothioate sequences and triplex forming molecules (TFOs).

There have been conflicting reports in the literature regarding the origin of LNA-induced duplex stabilization. Thermodynamics suggest that LNA can stabilize the duplex by either preorganization or improved stacking in terms of entropy or enthalpy [4–7]. The reason for the discrepancy may lie in the number of LNA monomers used. The stabilizing effect of introducing a single LNA monomer is more entropic in origin (pre-organizing the nucleotide into the optimal conformation), whereas additional adjacent LNA monomers stabilize the duplex by improved stacking and favorable enthalpic changes [8]. In addition to the effects on entropy and enthalpy, LNA duplexes could also be further stabilized by effects on counterion and water molecule uptake [9].

LNA™ Improves Mismatch Discrimination

> LNA™ improves mismatch discrimination by increasing the difference in binding between the perfect match and mismatch target. Single nucleotide discrimination is made possible in a variety of applications by intelligent placement of LNA™ monomers. The superior mismatch discrimination also improves the specificity of LNA™ containing oligonucleotides in hybridization-baseda pplications.

The increased affinity of LNA for its complementary sequence also increases the discriminatory power between perfect match and mismatched sequences. Discrimination between binding to a complementary sequence and a noncomplementary sequence (or mismatch sequence) is often described in terms of the difference in melting temperature (ΔT_m) between the two binding events. For most types of mismatch, LNA monomers placed at the position of the mismatch lead to an increase in ΔT_m and better discrimination compared to DNA (Figure A-2b). Different patterns of LNA modifications have been studied in terms of mismatch discrimination. Placing LNA monomers at bases neighboring a mismatch has been shown to improve discrimination. In addition, three LNA bases centered around the mismatch provide the best discriminatory power in many cases [10].

The effect on mismatch discrimination is dependent on the length of the sequence and position of the mismatch, as well as the actual sequence surrounding the mismatch [10]. Shorter oligonucleotides confer the best possibilities for mismatch discrimination, because discrimination decreases as sequence length increases. As LNA monomers increase duplex stability, LNA-containing oligonucleotides can be designed to be shorter than unmodified DNA oligonucleotides, thereby enhancing mismatch discrimination. The position of the mismatch is also important, because mismatches located at the center of a sequence show slightly improved discrimination compared to mismatches located closer to the ends of a duplex.

The thermodynamic basis for the increase in mismatch discrimination achieved by the use of LNA is not fully understood. Fluorescence experiments suggest that LNA modifications enhance base stacking of perfectly matched base pairs and decrease stabilizing stacking interactions of mismatched base pairs [10]. It has been hypothesized that the thermodynamic differences originate from the shift of the duplex conformation from B-form to A-form, with the level of mismatch discrimination increasing as the conformational equilibrium is shifted toward the A-form [8]. This is probably the result of energetic changes in stacking interactions, H-bonding of base pairs, and the hydration envelope when the duplex turns to the A-like conformation.

LNA™ Enables Tm Normalization

Because addition of LNA monomers to an oligonucleotide sequence increases the melting temperature, it is possible to control the Tm of a nucleotide duplex by varying the LNA content. This feature can be used to normalize the Tm across a population of short sequences with varying GC content. For AT-rich nucleotides, which give low melting temperatures, more LNA is incorporated into the LNA™ oligonucleotide to raise the Tm of the duplex. This enables the design of numerous LNA oligonucleotides with a narrow Tm range, which is beneficial in many

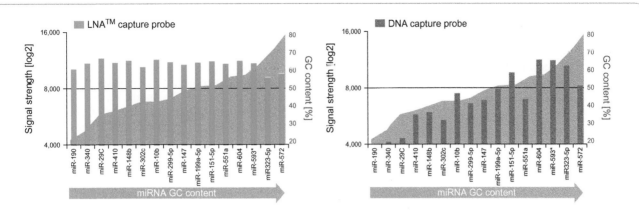

FigureA -3 The signal from DNA-based capture probes varies with the GC content of the target whereas LNA™-probes give similar signals regardless of target GC content. Signal intensity from microarray experiments using LNA™-enhanced (blue) or DNA based (gray) capture probes. MicroRNA targets with varying GC content were added at 100 amol each.

research applications such as microarray, PCR, and other applications in which sensitive and specific binding to many different targets must occur under the same conditions simultaneously.

The powerful effect of Tm normalization is clearly illustrated by the use of LNA-enhanced tools for the analysis of microRNA, which are short (20–22 nt) noncoding RNAs with widely varying GC content (5%–95%). The large sequence variability is problematic in applications such as microarray profiling, where many microRNA targets are analyzed under the same experimental conditions. However, when the LNA content is varied, oligonucleotides with specific duplex melting temperatures can be designed, regardless of the GC content of the microRNA. The power of Tm-normalization is demonstrated by the comparison of DNA and LNA probes for microarray-based detection of microRNA targets with a range of GC content (Figure A-3). The signal from DNA-based capture probes varies with GC content due to the differences in Tm among the different probes and microRNA targets. However, Tm-normalized LNA probes result in similar signals from all microRNAs, regardless of GC content. Tm normalization is also useful for simultaneous detection of microRNA in situ, microRNA inhibition, and design of qPCR primers for microRNA expression profiling.

LNA™ Improves in Vivo Stability of Oligonucleotides

An additional advantage of using LNA in oligonucleotides is the high in vivo stability. The issue of physiological stability mostly concerns the potential degradation of oligonucleotides by nucleases. A fully modified LNA sequence has been reported to be fully resistant toward the 3′-exonuclease SVPDE [11], whereas only minor protection against the same enzyme is obtained with one LNA monomer in the 3′-end or in the middle of a sequence. Endblocked sequences, i.e., LNA-DNA-LNA gapmers that contain DNA flanked by LNA, display a high stability in human serum compared to similar 2′-OMe modified sequences [12]. Only three LNA monomers at each end were sufficient to increase the half-life of the oligonucleotides 10-fold in human serum. Another study showed that two terminal LNA monomers

DNA	ncRNA
Real-time/quantitative PCR	Real-time/quantitative PCR
SNP detection/allele specific PCR	Microarray analysis
Methylation analysis	In situ hybridization
Bead-based applications	Northern blotting
Chromosomal FISH	Fluorescence activated cell
Comparative genome hybridization	sorting
Proteomics of isolated chromatin	Inhibition of RNA function
segments (PICh)	RNA modification (frame
Antigene inhibition	shifting/exon skipping)
Mutagenesis	

mRNA	miRNA
Real-time/quantitative PCR	Real-time/quantitative PCR
Microarray analysis	Microarray analysis
In situ hybridization	In situ hybridization
Northern blotting	Northern blotting
Bead-based applications	Bead-based applications
Fluorescence activated cell sorting	Inhibition of RNA function
Isolation	
Inhibition of RNA function	
RNA modification (frame	PCR based approaches
shifting/exon skipping	Hybridization based approaches
DNAzymes	In vivo based approaches

FigureA -4 Overviewo fd ifferenta pplicationsf orLN A™

provided significant protection against a Bal-31 exonucleolytic degradation [13]. The high in vivo stability of LNA oligonucleotides is fundamental in their use as antisense agents and in therapeutics.

APPLICATIONS FOR LOCKED NUCLEIC ACIDS

The unique characteristics of LNA make it a powerful tool for the detection of low abundance, short, or highly similar targets in a number of applications (Figure A-4). Since the physical properties (e.g., water solubility) of LNA are very similar to those of RNA and DNA, conventional experimental protocols can easily be adjusted to their use. LNA has been successfully used to overcome the difficulties of studying very short sequences and has greatly improved, and in many cases enabled, specific and sensitive detection of noncoding RNA and other small RNA molecules. The unique ability of LNA oligonucleotides to discriminate between highly similar sequences has further

263

been exploited in a number of applications targeting longer RNA sequences such as mRNA. The high in vivo stability combined with the increased binding affinity, specificity, and possibilities for Tm normalization make LNA oligonucleotides highly attractive for in vivo applications.

LNA™ in microRNA Research

The short length and varying GC content (5%–95%) of microRNAs make them challenging to analyze using traditional methods. DNA- or RNA-based technologies for microRNA analysis can result in poor data quality because the melting temperature (Tm) of the oligonucleotide/microRNA duplex will vary greatly depending on the GC content of the sequences. This is especially problematic in applications such as microarray profiling and high throughput experiments, where many microRNA targets are analyzed under the same experimental conditions. Another challenge of studying microRNAs is the high degree of similarity between the sequences. Some microRNA family members vary by a single nucleotide.

These challenges in microRNA analysis can be overcome by using LNA-enhanced oligonucleotides. The use of LNA in probes and primers enables highly sensitive detection and analysis of the short microRNA sequences. When the LNA content is varied, oligonucleotides with specific duplex melting temperatures can be designed, regardless of the GC content of the microRNA. Exiqon has used the LNA technology to Tm-normalize primers, probes, and inhibitors to ensure that they all perform well under the same experimental conditions. Finally, LNA™ can be used to enhance the discriminatory power of primers and probes to allow excellent discrimination of closely related microRNA sequences.

LNA™ Tools for microRNA Research

LNA™ offers significant improvement in sensitivity and specificity and ensures optimal performance for all microRNA targets. Exiqon offers a complete product portfolio for microRNA research including RNA isolation, microRNA profiling and expression analysis by microarray and qPCR, microRNA localization by ISH, and microRNA functional analysis using in vitro and in vivo inhibitors. Visit www.exiqon.com to learn more.

LNA™ in Hybridization-Based Approaches

Due to their unique hybridization properties, LNA probes are the ideal choice for the detection of RNA and DNA, especially when the target sequence is short or AT rich. LNA probes have been used successfully to detect a variety of RNA species in situ including mRNA and long noncoding RNA. LNA is particularly useful for discrimination between different mRNA splice variants or highly similar isoforms. In addition to providing unmatched specificity, these probes give strong signals even with short hybridization times. LNA™ probes can also be used for chromosomal FISH, giving strong signals even after hybridization times of less than 1 hour.

LNA probes were fundamental in the seminal work describing the tissue-specific microRNA expression patterns in Zebrafish [14]. This was the first indication that

these short noncoding RNAs may have wide-ranging roles in animal development. Since then, LNA probes have been used to investigate the precise localization of microRNA expression in a variety of animals and plants [15–17]. LNA-based detection of microRNAs can also be used for clinical molecular diagnostics [18]. Researchers from Vejle Hospital in Denmark have recently presented results from a large clinical study of miR-21 levels in FFPE samples from 520 patients [19]. The level of miR-21 as quantified by LNA in situ hybridization was found to be an independent predictive marker of poor recurrence-free cancer-specific survival.

LNA probes can also be used with great advantage for fast and sensitive Northern blot analysis [20,21]. Due to the increased affinity of LNA containing Northern blot probes, it is possible to significantly reduce the amount of sample needed and reduce exposure times to just a few hours. Mismatch discrimination is also greatly improved.

Gene expression analysis using microarrays is another hybridization-based application that can benefit greatly from the use of LNA [22]. Hybridization of hundreds of different targets to a microarray of relatively short capture probes has to occur under specific hybridization conditions. The possibility to Tm normalize the capture probes so that they all work efficiently to capture their targets under those conditions is essential. In addition, the use of LNA also allows the generation of short capture probes with high affinities, which improves specificity and increases mismatch discrimination. These principles have been fundamental in the development of LNA enhanced microarrays for microRNA expression profiling [23].

Other hybridization-based applications for the capture and detection of nucleic acids that can be improved by LNA include bead-based approaches in which capture probes are attached to either magnetic or fluorescently labeled beads. Examples include purification of mRNA using an LNA oligoT capture probe [24], and purification of protein complexes associated with specific genomic sequences [25]. LNA has also been used in combination with FACS to monitor viral RNA in infected cells using LNA flow FISH [26].

LNA™ in PCR

LNA has been shown to improve the performance of both standard and real-time PCR assays, and can be included in both probes and primers as well as PCR blockers. LNA is widely applicable in various types of PCR probes, both displacement type and hydrolysis Taqman™-type probes. The key advantage of using LNA in probes is the ability to increase their affinity and specificity toward their complementary DNA or RNA targets and the increased local specificity by design of shorter probes. Short LNA-containing PCR probes also increase the flexibility for design, can lead to better signal-to-noise ratios, and have been shown to improve PCR efficiency [10,27,28]. Increased specificity at specific SNP sites can be achieved by spiking in LNA bases at, or very close to, a specific site of interrogation in a probe [10].

LNA-enhanced PCR primers offer several advantages over conventional DNA primers [29,30]. They increase the flexibility in primer positioning, which is particularly important for AT-rich targets. The use of LNA also allows precise adjustments of primer melting temperatures and facilitates the design of multiple primer sets that can be used under the same amplification conditions, even if target sequences are highly variable. In addition, LNA-enhanced primers can be made shorter than conventional primers, which is advantageous when working with very short and/or low abundant targets or targets from degraded/highly complex samples. Allele-specific primers have been used for many applications including SNP typing and methylation analysis; and adding LNA modifications to the 3′ end of primers, primarily at the point of discrimination either at the last or penultimate nucleotide, can greatly enhance specificity [31,32].

Clamping or blocking probes can be used in PCR to enhance the detection of low-level mutations or rare alleles by preventing amplification of the wild-type or undesired allele. Clamping probes can be designed to bind very specifically to unwanted nontarget sequences, and prevent detection either by interfering with the PCR primer sites or detection probe binding sites, or by inhibiting polymerase transcription of PCR templates by binding between primers [33–35]. The use of LNA in clamping probes can greatly improve the specificity of the probe and has been shown to enable detection of rare targets in a high background of molecules with close resemblance to the target.

LNA also confers a great advantage for gene expression analysis of microRNAs using qPCR. The use of LNA has allowed the development of a microRNA qPCR system that uses two short microRNA-specific primers in the PCR amplification step. The LNA-enhanced primers make the system the most sensitive microRNA detection method available [36] and enables accurate and robust quantification of microRNAs from challenging clinical samples such as LCM material and biofluids.

LNA™ for Inhibition of RNA

The superior affinity of LNA oligonucleotides to their complementary sequences, combined with the high specificity of short LNA oligonucleotides, makes them ideal for antisense knockdown or RNA inhibition applications. Inhibition of RNA can be achieved through different approaches such as steric block antisense, RNase H-mediated antisense, LNAzymes, siLNA, and LNA aptamers [37]. In addition, cellular delivery, physiological stability, as well as toxicity LNA inhibitors have been studied in detail [3,38]. LNA oligonucleotides can be delivered into cells using standard cationic transfection agents [3,39].

When used in antisense constructs, LNA has been reported to combine substantial increased potency in vitro and in vivo with minimal toxicity [39]. mRNA inhibition by RNase H recruitment can be enhanced by the use of LNA/DNA mixmers that are extremely nuclease resistant. A DNA stretch at the center of the mixmer targets the nuclease activity to the desired sequence [12]. LNA is also substantially compatible with the siRNA machinery, and siLNAs exhibit greatly improved biostability and show enhanced inhibition at certain RNA

targets [40]. It has been shown that LNA can be used to reduce sequence-related off-target effects by either lowering incorporation of the siRNA sense-strand and/or by reducing the ability of inappropriately loaded sense-strands to cleave the target RNA.

LNA-enhanced microRNA inhibitors are ideal for use as specific suppressors of microRNA activity. Other applications include identification and validation of microRNA targets, the study of microRNA function in cellular processes and pathological pathways, and microRNA regulation of gene expression. The high affinity of the LNA-enhanced microRNA inhibitors makes them highly effective at physiological temperatures even when used in low concentrations. Intelligent probe design minimizes potential secondary and off-target effects. Since the LNA-enhanced microRNA inhibitors can be Tm-normalized, they are ideal for use in multiplexing and screening assays. The modification of the oligonucleotide backbone with phosphorothioate has allowed the development of microRNA inhibitors with increased potency. The LNA technology has been successfully applied in vivo, combining low concentrations of the microRNA inhibitor with high serum stability and low toxicity [41].

DESIGN OF LNA™ OLIGONUCLEOTIDES
General Design Guidelines

Exiqon Is the Home of LNA™

Exiqon's scientists have the expertise to design LNA™-enhanced oligonucleotides with high melting temperatures (Tm), optimal mismatch discrimination, and high binding specificity, while keeping secondary structure and self-complementarity to a minimum. More than 10 years of experience in working with LNA™ applications has allowed us to develop advanced algorithms and innovative design tools for a variety of applications. Visit http://www.exiqon.com/oligo-toolst ol earnmo re.

The use of LNA in oligonucleotides is a powerful tool to increase binding affinity, optimize mismatch discrimination, and adjust melting temperature of oligonucleotides. It is, however, important to consider the placement of LNA monomers within a sequence carefully to optimize the positive effects and reduce any possibilities of negative effects. LNA:LNA duplexes are inherently very stable, so to avoid application-limiting self-hybridization, this should be carefully considered when designing LNA sequences. First, unless using very short probes (8–10 bp) in qPCR applications, it is recommended to keep the number of LNA monomers to between 30% and 60% of the sequence to avoid "sticky" probes. To ensure best possible synthesis yields and purity, avoid stretches of 4 or more LNA (expected when using very short probes). It is also best to avoid placing LNA in stretches of 3 or more Cs or Gs.

In the design of hybridization probes, it is essential to examine candidate sequences for their tendency to fold into hairpins, homodimers, and other undesirable secondary structures. If secondary structures cannot be avoided, their stability needs to be minimized to prevent competition with desired probe-target duplex formation, which means avoiding LNA at positions that could stabilize them. Therefore, no LNA should be placed in palindromic sequences. In this respect, it is also recommended to keep the length of the LNA oligonucleotides to below 30 nt for optimal performance. Longer stretches of sequence also tend to have higher risk of forming secondary structures.

In the design of allele-specific primers or probes for SNP detection, vary the length and LNA positioning to obtain comparable melting temperatures (T_m) for the alleles, while keeping the difference in melting temperatures (ΔT_m) between the perfect match and mismatch binding as high as possible. To optimize mismatch discrimination, a triplet of LNAs can be centered on the mismatch site. However, LNA modification of the guanine nucleotide or either of its nearest-neighbor bases should be avoided in a G/T mismatch site. To maximize the effect of the mismatch, shorter probes are advisable. It is also advantageous to make sure the mismatch site is close to the center of the sequence to achieve maximum discriminatory power.

GeneralD esign Guidelines

- LNA™ will bind very tightly to other LNA™ residues. Avoid self-complementarity and cross-hybridization to other LNA™-containingo ligonucleotides.
- Keep the GC content between 30% and 60%.
- Avoid stretches of more than 4 LNA™ bases, except when verysho rt(9–10 nt)o ligonucleotidesa red esigned.
- Avoid LNA in stretches of 3 or more Gs or Cs.
- No LNA™ bases should be placed in palindromes (G-C base pairs are more critical than A-T base pairs).
- For most hybridization-based approaches, aim for a DNA Tm of around 75°C.
- The optimal length for LNA detection probes and capture probesi s20–25 nt.
- For SNP detection, pay specific attention to the position of the SNP and placement of LNAs to optimize discrimination.
- For novel applications, design guidelines may have to be establishede mpirically.
- Useful LNA™ oligonucleotide design guidelines and design tools are available at exiqon.com/oligo-tools.

Design of LNA™ Detection Probes for ISH

For detection of microRNA in situ, Exiqon provides pre-designed detection probes for most microRNAs in miR-base and can design custom probes for any small RNA. In the design of detection probes for microRNAs, there is limited flexibility in the selection of sequence, so most probes tend to cover the full length of the microRNA.

Tm can be adjusted by varying the number and position of LNA monomers, and the DNA Tm should ideally be between 75°C and 85°C. The same principles apply for the detection of other short RNAs.

In the design of detection probes for mRNA ISH, there are a lot more parameters to consider due to the length of the target. In order to simplify and improve the process for design, Exiqon has developed the Custom LNA mRNA Detection Probe design tool, which uses a sophisticated algorithm to quickly identify optimal LNA-enhanced probes for target mRNAs. The software very quickly evaluates more than 5,000 probe designs based on more than 20 design criteria. This process ensures that high-quality probes can be designed for any mRNA sequence.

The design criteria include:

- Optimization of probe length and the number and positions of LNA bases within the probe. This ensures that the probes have the correct melting temperatures and that any potential self-hybridization is avoided.
- A detailed analysis of the LNA pattern of the probe based on Exiqon's extensive knowledge of oligonucleotide design rules. This ensures that LNA bases are not placed at positions where they could negatively affect the performance of the probe.
- Calculations of the target's secondary structure to ensure that the probes target only accessible regions of themRNA.
- BLAST searches to ensure that the probes will target only the relevant mRNA sequence.

LNA detection probes can be labeled with a variety of labels such as DIG, biotin, and FITC. The addition of a DIG moiety at both the 3′ and 5′ end has been shown to increase sensitivity and improve signal-to-noise ratios [42]. A double FTIC label could also have a similar effect (unpublished in-house data).

Exiqon has years of experience in the design of detection probes for a variety of RNA species and is happy to offer assistance during the design process.

For more guidelines on LNA oligonucleotides design for other applications, please visit http://www.exiqon.com/oligo-tools.

To learn more about microRNA *in situ* hybridization using LNA™ detection probes and to find protocols, please visit http://www.exiqon.com/microrna-in-situ-Hybridization.

For more information about mRNA ISH, please go to http://www.exiqon.com/mrna-in-situ-hybridization.

Get expert advice and technical assistance by contacting us at http://www.exiqon.com/contact.

REFERENCES

[1] Saenger W. In: Principles of Nucleic Acid Structure. New York: Springer; 1984.

[2] Petersen M, Nielsen CB, Nielsen KE, Jensen GA, Bondensgaard K, Singh SK, et al. The conformations of locked nucleic acids (LNA). J Mol Retard 2000;13:44–53.

[3] Petersen M, Wengel J. LNA: a versatile tool for therapeutics and genomics. Trends Biotechnol 2003;21:74–81.

[4] McTigue PM, Peterson RJ, Kahn JD. Sequence-dependent thermodynamic parameters for locked nucleic acid (LNA)-DNA duplex formation. Biochemistry 2004;43:5388–405.

[5] Doi T, Imanishi T, Obika S, Nanbu D, Hari Y, Andoh J, Morio K. pleacs Stability and structural features of the duplexes containing nucleoside analogues with a fixed N-type conformation, 2′-O,4′-Cmethyleneribonucleosides. Tetrahedron Lett. 1998;39:5401–4.

[6] Christensen U, Jacobsen N, Rajwanshi VK, Wengel J, Koch T. Stopped-flow kinetics of locked nucleic acid (LNA)-oligonucleotide duplex formation: studies of LNA-DNA and DNA-DNA interactions. Biochem J 2001;354:481–4.

[7] Koshkin AA, Nielsen P, Meldgaard M, Rajwanshi VK, Singh SK, Wengel J. LNA (locked nucleic acid): an RNA mimic forming exceedingly stable LNA:LNA duplexes. J Am Chem Soc 1998;120:13252–53.

[8] Owczarzy R, You Y, Groth CL, Tataurov AV. Stability and mismatch discrimination of locked nucleic acid-DNA duplexes. Biochemistry 2011;50:9352–67.

[9] Kaur H, Wengel J, Maiti S. Thermodynamics of DNA-RNA heteroduplex formation: effects of locked nucleic acid nucleotides incorporated into the DNA strand. Biochemistry 2008;47:1218–27.

[10] You Y, Moreira BG, Behlke MA, Owczarzy R. Design of LNA probes that improve mismatch discrimination. Nucleic Acids Res. 2006;34:e60.

[11] Frieden M, Hansen HF, Koch T. Nuclease stability of LNA oligonucleotides and LNA-DNA chimeras, Nucleosides Nucleotides Nucleic Acids 2003;22:1041–3.

[12] Kurreck J, Wyszko E, Gillen C, Erdmann VA. Design of antisense oligonucleotides stabilized by locked nucleic acids. Nucleic Acids Res 2002;30:1911–8.

[13] Crinelli R, Bianchi M, Gentilini L, Magnani M. Design and characterization of decoy oligonucleotides containing locked nucleic acids. Nucleic Acids Res 2002;30:2435–43.

[14] Wienholds E, Kloosterman WP, Miska E, Alvarez-Saavedra E, Berezikov E, de Bruijn E, et al. MicroRNA expression in zebrafish embryonic development. Science 2005;309:310–1.

[15] Wheeler G, Valoczi A, Havelda Z, Dalmay T. In situ detection of animal and plant microRNAs. DNA and Cell Biology 2007;26:251–5.

[16] Nuovo GJ, Elton TS, Nana-Sinkam P, Volinia S, Croce CM, Schmittgen TD. A methodology for the combined in situ analyses of the precursor and mature forms of microRNAs and correlation with their putative targets. Nat Protoc 2009;4:107–15.

[17] Nuovo GJ. In situ detection of precursor and mature microRNAs in paraffin embedded, formalin fixed tissues and cell preparations. Methods 2008;44:39–46.

[18] Nielsen BS, Jorgensen S, Fog JU, Sokilde R, Christensen IJ, Hansen U, et al. High levels of microRNA-21 in the stroma of colorectal cancers predict short disease-free survival in stage II colon cancer patients. Clin Exp Metastasis 2011;28:27–38.

[19] Kjaer-Frifeldt S, Hansen TF, Nielsen BS, Joergensen S, Lindebjerg J, Soerensen FB, Depont Christensen R, Jakobsen A. The prognostic importance of miRNA-21 in stage II colon cancer: A population based study. Br J Cancer 2012;107(7):1169–74.

[20] Kim SW, Li Z, Moore PS, Monaghan AP, Chang Y, Nichols M, et al. A sensitive non-radioactive northern blot method to detect small RNAs. Nucleic Acids Res 2010;38:e98.

[21] Varallyay E, Burgyan J, Havelda Z. MicroRNA detection by northern blotting using locked nucleic acid probes. Nat Protoc 2008;3:190–6.

[22] Tolstrup N, Nielsen PS, Kolberg JG, Frankel AM, Vissing H, Kauppinen S. OligoDesign: optimal design of LNA (locked nucleic acid) oligonucleotide capture probes for gene expression profiling. Nucleic Acids Res 2003;31:3758–62.

[23] Castoldi M, Schmidt S, Benes V, Noerholm M, Kulozik AE, Hentze MW, et al. A sensitive array for microRNA expression profiling (miChip) based on locked nucleic acids (LNA). RNA 2006;12:913–20.

[24] Jacobsen N, Eriksen J, Nielsen PS. Efficient poly(A)+ RNA selection using LNA oligo(T) capture. Methods Mol Biol 2011;703:43–51.

[25] Dejardin J, Kingston RE. Purification of proteins associated with specific genomic loci. Cell 2009;136:175–86.

[26] Robertson KL, Verhoeven AB, Thach DC, Chang EL. Monitoring viral RNA in infected cells with LNA flow-FIS. RNA 2010;16:1679–85.

[27] Ugozzoli LA, Latorra D, Puckett R, Pucket R, Arar K, Hamby K. Real-time genotyping with oligonucleotide probes containing locked nucleic acids. Anal Biochem 2004;324:143–52.

[28] Costa JM, Ernault P, Olivi M, Gaillon T, Arar K. Chimeric LNA/DNA probes as a detection system for real-time PCR. Clin Biochem 2004;37:930–2.

[29] Ballantyne KN, van Oorschot RA, Mitchell RJ. Locked nucleic acids in PCR primers increase sensitivity and performance. Genomics 2008;91:301–5.

[30] Malgoyre A, Banzet S, Mouret C, Bigard AX, Peinnequin A. Quantification of low-expressed mRNA using 5′ LNA-containing real-time PCR primers. Biochem Biophys Res Commun 2007;354:246–52.

[31] Gustafson KS. Locked nucleic acids can enhance the analytical performance of quantitative methylation-specific polymerase chain reaction. J Mol Diagn 2008;10:33–42.

[32] Nakitandwe J, Trognitz F, Trognitz B. Reliable allele detection using SNP-based PCR primers containing locked nucleic acid: application in genetic mapping. Plant Methods 2007;3:2.

[33] Senescau A, Berry A, Benoit-Vical F, Landt O, Fabre R, Lelièvre J, et al. Use of a locked-nucleic-acid oligomer in the clamped-probe assay for detection of a minority Pfcrt K76T mutant population of Plasmodium falciparum. J Clin Microbiol 2005;43:3304–8.

[34] Oldenburg RP, Liu MS, Kolodney MS. Selective amplification of rare mutations using locked nucleic acid oligonucleotides that competitively inhibit primer binding to wild-type DNA. J Invest Dermatol 2008;128:398–402.

[35] Ren XD, Lin SY, Wang X, Zhou T, Block TM, Su Y-H. Rapid and sensitive detection of hepatitis B virus 1762T/1764A double mutation from hepatocellular carcinomas using LNA-mediated PCR clamping and hybridization probes. J Virol Methods 2009;158:24–9.

[36] Jensen SG, Lamy P, Rasmussen MH, Ostenfeld MS, Dyrskjot L, Orntoft TF, et al. Evaluation of two commercial global miRNA expression profiling platforms for detection of less abundant miRNAs. BMC Genomics 2011;12:435.

[37] Veedu RN, Wengel J. Locked nucleic acid as a novel class of therapeutic agents. RNAB iology 2009;6:321–3.

[38] Jepsen JS, Wengel J. LNA-antisense rivals siRNA for gene silencing. CurrOp inDr ug Discov Devel 2004;7:188–94.

[39] Wahlestedt C, Salmi P, Good L, Kela J, Johnsson T, Hökfelt T,e tal. Potentan d nontoxic antisense oligonucleotides containing locked nucleic acids. ProcN atlAc adS ciUS A 2000;97:5633–8.

[40] Elmén J, Thonberg H, Ljungberg K, Frieden M, Westergaard M, Xu Y,e tal. Lockedn ucleic acid (LNA) mediated improvements in siRNA stability and functionality. NucleicAc idsRe s 2005;33:439–47.

[41] Stenvang J, Silahtaroglu AN, Lindow M, Elmen J, Kauppinen S. Theu tilityo fL NA in microRNA-based cancer diagnostics and therapeutics. SeminCan cerB iol. 2008;18:89–102.

[42] Sweetman D, Rathjen T, Jefferson M, Wheeler G, Smith TG, Wheeler GN, et al. FGF-4 signaling is involved in mir-206 expression in developing somites of chicken embryos. Dev Dyn 2006;235:2185–91.

267

Appendix 2

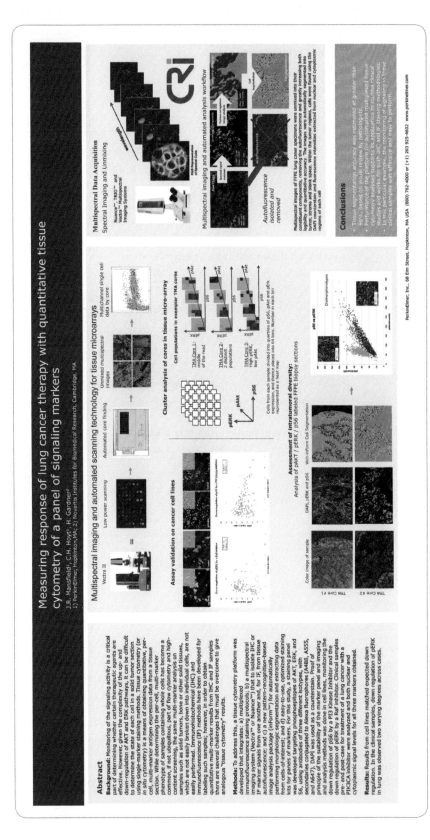

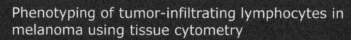

Phenotyping of tumor-infiltrating lymphocytes in melanoma using tissue cytometry

J.R. Mansfield,[1] K. Lane,[1] C.M. van der Loos,[2] C.C. Hoyt[1]
1) PerkinElmer, Hopkinton, MA; 2) Academic Medical Center, Amsterdam, Netherlands.

Abstract

In many cancers, tumor-infiltrating lymphocytes (TILs) indicate levels of tumor immunogenicity and are a strong predictor of survival. An understanding of the phenotype and spatial distribution of TILs within tumor regions would be advantageous for characterizing host response. However, visual TIL assessment is prone to error and multimarker quantitation is difficult with standard methods. Here we present a multi-marker, computer-aided event-counting method for the determining the phenotypes of lymphocytes in melanoma sections using a novel multispectral imaging (MSI) approach. A section of a tissue microarray containing 120 melanoma cores was stained for CD3, S100, Foxp3 and hematoxylin. This was imaged using MSI and the individual staining of each marker separated from each other using spectral unmixing. The images were analyzed using software which had been trained to recognize the tumor area based on the S100 staining pattern. Then the Foxp3 status of each CD3+ TIL was then determined. Results indicate that machine-learning software can be trained to accurately recognize tumor regions within each core. MSI enabled the accurate quantitation of three immunostains in the sample without crosstalk. Within the tumor region of each core it was possible to count the CD3+ TILs and then determine the Foxp3 status of each. This multimarker phenotyping and counting approach shows the potential for broad applicability in the assessment of TILs in many solid tumors.

Multispectral imaging of quadruplex stained melanomas

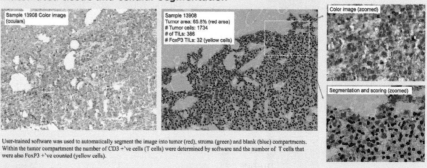

Multispectral imaging and spectral unmixing enables the separation of the 5 colors in the sample each into their own image. These unmixed images (bottom right) can be recombined and recolored in many ways to allow for easy visualization of the markers. At left are simulated single stain IHC images, at right are simulated fluorescence images.

Automated tissue and cellular segmentation

User-trained software was used to automatically segment the image into tumor (red), stroma (green) and blank (blue) compartments. Within the tumor compartment the number of CD3 +'ve cells (T cells) were determined by software and the number of T cells that were also FoxP3 +'ve counted (yellow cells).

Multispectral imaging technology

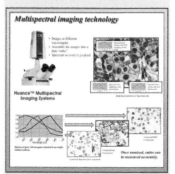

Nuance™ Multispectral Imaging Systems

Morphologic and cellular segmentation

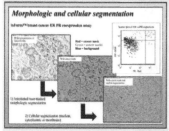

Quantitation and tissue cytometry display

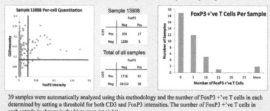

39 samples were automatically analyzed using this methodology and the number of FoxP3 +'ve cells in each determined by setting a threshold for both CD3 and FoxP3 intensities. The number of FoxP3 +'ve T cells in each sample is shown in the histogram (at right).

Conclusions

- Multispectral imaging enabled the quantitation of three immunostains (CD3, FoxP3 and S100) in the presence of hematoxylin and melanin.
- Tissue segmentation accuracy and cell counting was estimated at greater than 90%, based on visual review by pathologists.
- Automated multiplexed tissue cytometry analyses is feasible for routine clinical studies and works for both IHC and IF staining methodologies.
- In this particular example, the enumeration of FoxP3 +'ve T cells in these clinical samples was effective and easy to perform.

PerkinElmer, Inc., 68 Elm Street, Hopkinton, MA USA (800) 762-4000 or (+1) 203 925-4602
www.perkinelmer.com

INCREASING SENSITIVITY AND ACCURACY OF QUANTITATIVE IMMUNOFLUORESCENCE IN FFPE TISSUE WITH SPECTRAL IMAGING

Introduction

Understanding cancer and its complexity can be advanced significantly with better tools for measuring proteins in situ in formalin-fixed, paraffin-embedded tissue sections. Separation technologies, such as Western blot, microarrays, and mass spectrometry are used widely, but lose key architecturally-specific signals that reside at the cellular level, and blend signals from multiple cell types. Laser capture micro-dissection attempts to address this issue, but is expensive and laborious to perform. IHC has the significant advantage of retaining tissue architecture and heterogeneity, and presents views of protein expression, but is non-quantitative and variable. Most separation approaches disaggregate tissues, blendingsign alsfro mman yc ellsan dtissu es.

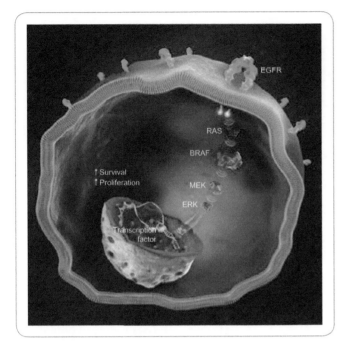

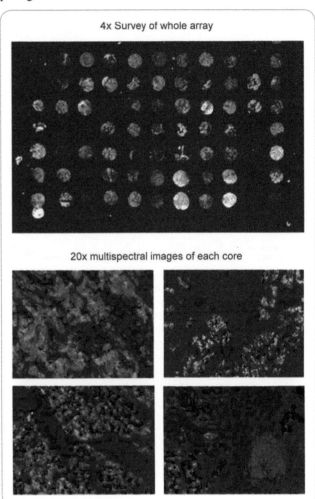

4x Survey of whole array

20x multispectral images of each core

Fluorescence microscopy is becoming increasingly important for these endeavors, compared to chromogenic immunohistochemistry, due to higher multiplexing capability, larger and more linear signal range, and less interference among labels. However, immunofluorescence poses challenges, including the presence and effects of autofluorescence and the inability to distinguish overlapping signals due to cross-talk. Multispectral imaging eliminates these issues through the use of spectral unmixing, which enables isolation of individual biomarker signals, even when signals are substantially overlapping spatially and spectrally, and are obscured by autofluorescence signals. Sensitivity and accuracy can be increased several-fold. The purpose of this application note is to provide an example of the quantitative advantage of multispectral imaging.

Methods

A tissue microarray of lung cancer tissue was labeled with a multicolor immunofluorescence cocktail commercially available from Cell Signaling Technology (Pathscan Node Kit #8999). The kit comprises two tubes of mixed reagents, one of primaries and one of conjugated secondaries. Phospho-AKT is labeled with AlexaFluor™ 488, phospho-ERK with AlexaFluor™ 555, and phospho-S6 with AlexaFluor™ 647. DAPI nuclear counterstain is included. Multispectral images were acquired with a Caliper Nuance multispectral imaging system and spectrally unmixed into label and autofluorescence signals. For comparison, conventional immunofluorescence images were acquired with fluorescence emission filters centered on the emission peak of the respective fluorophores.

For this analysis, we compared the signals for the phospho-ERK (AlexaFluor™ 555) signal in typical microarray cores, as indicated in the spectrally unmixed component image and in the conventional fluorescence emission

monochrome image, acquired with emission filtering centered on the AlexaFluor™555 emission peak. Three locations in the sample were used as reference points: in cytoplasm; in a nucleus; and off the sample. A signal-to-background ratio and autofluorescence contribution were calculated for each image from the cytoplasm and off-sample signal counts.

Results

Measured signals in the conventional immunofluorescence images were significantly higher than in the spectrally unmixed component images, due to the presence of a high autofluorescence signal present, in addition to the pERK signal. In the spectrally unmixed component image, the autofluorescence is removed, and thus the pERK signal is pure and more accurate. A further benefit of spectral unmixing is the removal of cross-talk from overlapping fluorophores spectra, in multi-label assays. Signal-to-background in the spectrally unmixed images is 50 and in the conventional monochrome image is 3.9, a 13× improvement. More importantly, the data suggest that 34% of the signal measured with conventional epifluorescence was actually autofluorescence or cross-talk from another fluorophores labels ((2991−1959)/2991).

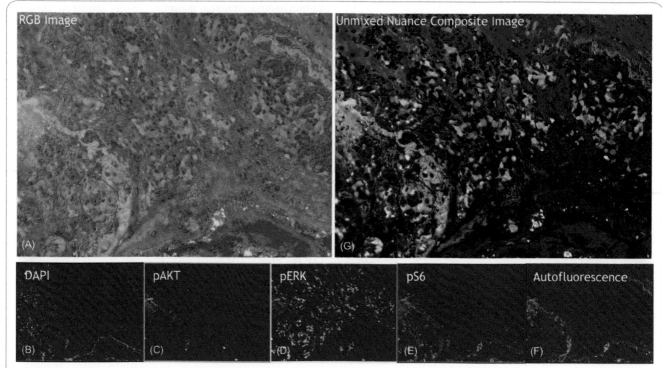

Figure 1 Lung tissue labeled with AlexaFluor™ 488, 555, and 647 and a DAPI counterstain. Panel A shows an RGB image of the sample with pAKT immunolabeled with AF488, pERK immunolabeled with AF555, and pS6 immunolabeled with AF647. Panels B through E display these unmixed component images along with the autofluorescence component in Panel F, whose border colors correspond to the pseudocolors used to form the composite image, G. The autofluorescence spectrum is unmixed in the black channel so it is not visible in the unmixed composite image.

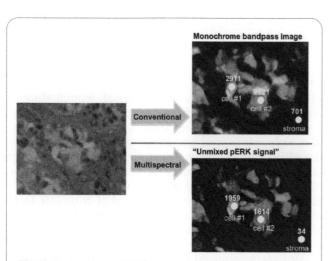

Figure 2 Images of pERK signal generated with conventional fixed filters (above) and with spectral unmixing (below). Signal counts were found from the same three pixel locations in both images, in cytoplasm where pERK is likely expressed, in the nucleus, and a third off the sample to provide a background signal.

Table 1 Signal counts found from the two pERK images in Figure 2. The signal-to-background ratio of each image was calculated by dividing the signal from the cytoplasm by the off-sampleba ckgrounds ignal

	Signal(counts)	
	Conventional	Multispectral
Cell#1	2991	1959
Cell#2	2521	1614
Stroma	701	34
Signal/Background	3.9	50

Conclusion

Spectral imaging enables reliable and accurate assessments of weakly expressing proteins in FFPE sections, even when fluorescence emissions are substantially weaker than background autofluorescence. The approach also automatically corrects for cross-talk between labels in multiplexed assays, thus giving pure signals of biomarkers and greatly increasing the signal-to-noise ratio compared to conventional methods.

RECEPTOR TRAFFICKING ANALYSIS WITH MULTI-COLOR FLUORESCENCE LABELING AND MULTISPECTRAL IMAGING

Example: EGFR Activation

Introduction

Analysis of signaling pathway activity in FFPE tissue is essential in oncology research. For example: the epidermal growth factor receptor (EGFR) is a transmembrane receptor tyrosine kinase that belongs to the HER/ErbB protein family. Dysregulation of EGFR and other receptor signaling pathways through activating mutations or gene amplification has been implicated in the pathogenesis of many human carcinomas, leading to extensive clinical study. Immunohistochemical studies have limitations due to the difficulty in interpreting the results, as well as specificity of staining. Multiplexing allows simultaneous detection of multiple biomarkers, revealing specific signaling configurations and tumor heterogeneity. In this example, we demonstrate multiplex labeling of total EGFR epitopes and Phospho-EGFR epitopes to determine percent of activated receptors within tumor-positive cells. The antibodies recognize total EGFR and phosphorylated-EGFR specifically upon activation of tyrosine residues. The phosphorylation EGFR is rapidly endocytosed, and either recycled back to the plasma membrane or targeted for lysosomal degradation. The ratio of total EGFR versus Phospho-EGFR reveals the percentage of receptor being recycled by tumor cells. Thus, Caliper's Vectra™ multispectral imaging,

automated tumor and cell finding software analysis offers precise detection of individual proteins through spectral unmixing based on their spectral signatures. inForm™ analysis software segments the individual tumor cells and provides quantitation for each cellular compartment.

Methods

The tissues used in this analysis are mouse xenografts of lung cancer cell lines. EGFR signaling has been found to be dysregulated in several types of this cancer. Expression and/or activation have been linked to therapeutic response, angiogenesis, and metastasis. Labeling is done with a commercially available kit from Cell Signaling Technologies (Catalog #7967), used to detect simultaneously expression, localization, and activation state of EGFR, and additionally the downstream signaling through Erk1/2.

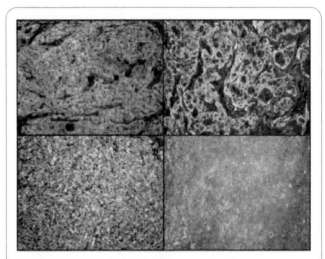

Figure 1 Multispectral 20x images acquired from multiplexed non-small cell lung carcinoma samples with pEGFR immunolabeled with AlexaFluor™ 488, total EGFR immunolabeled with AlexaFluor™ 555, pERK with AlexaFluor™ 647, and a DAPI counterstain. **(A)** Amplified H3255 cell line. **(B)** Amplified Kyse450 cell line. **(C)** Unamplified H1975 cell line. **(D)** Unamplified H1703c ell ine.

EFG, TGF-alpha, etc

P13-K
RAS
GRB2
SOS
RAF
AKT
STAT
MEK
ERK
mTOR

Gene Transcription
Cell Cycle Progression

Cell proliferation | Inhibition of Apoptosis
Angiogenesis | Migration, Adhesion, Invasion

Sample:

- Four human tumor xenografts of non-small cell lung carcinoma cell lines.
- Twoamp lified:H3255/ L858Ran dKyse 450/wild-type.
- Twou namplified:H1975/ L858Ran dH1703/ wild-type.

Immunofluorescence Staining:

- Cell Signaling Technology PathScan EGF Receptor Activation Multiplex IF Kit #7967.
- Primaryan tibodies:p EGFR;to talE GFR;p ERK
- Secondary antibodies: AlexaFluor™ 488; AlexaFluor™ 555; and AlexaFluor™ 647.
- DAPIc ounterstain.

Imaging and Analysis:

- Caliper's Vectra automated imaging system and inForm image analysis software.
- Acquiredfo ur20 × multispectral images per sample.

273

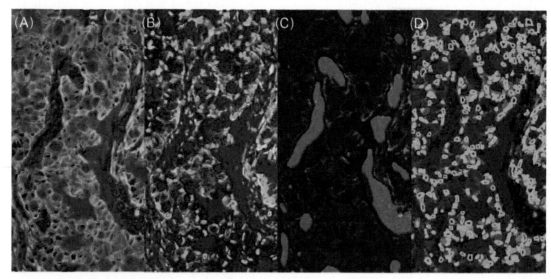

Figure 2 Image cubes of amplified H3255 cells. **(A)** RGB representation of multispectral cube. **(B)** Composite image of unmixed signals with pEGFR represented in green, EGFR represented in red, and co-localization shown in yellow. **(C)** RGB image overlayed with an inForm tissue segmentation map, where red areas are tumor, green is stroma, and the background is in blue. **(D)** RGB image overlayed with an inForm cell segmentation map, where each cell in the tumor area has been segmented into nucleus, cytoplasm, and membrane compartments.

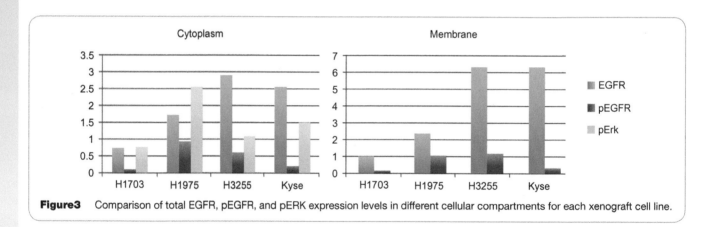

Figure3 Comparison of total EGFR, pEGFR, and pERK expression levels in different cellular compartments for each xenograft cell line.

- Automatedmac hine-learning-basedimage an alysisto :
  - Detectare aso ftu mor
  - Segmentth ec ellsin tosu bcellularc ompartments
  - Extract IF signals on a per-cell basis

Results

Evaluation of individual biomarkers following multispectral unmixing revealed changes in pEGFR expression compared to total EGFR in xenograft expression specific tyrosine kinase mutations. Since cytoplasmic pEGFR represents an activated form of surface EGFR, it was not observed within the proximity of total EGFR. Data analysis revealed tyrosine kinase EGFR mutant xenografts showed lower expression of pEGFR, while wildtype/Kyse450 xenograft samples showed increase in total EGFR, pERK, and a decrease in cytoplasmic pEGFR expression. The data demonstrates that the ratio of pEGFR versus total EGFR is consistent with cytoplasmic to membrane receptor recycling. Activation of pERK did not correlate with EGFR expression, which is not unexpected as it could be activated by RAS/MEK signaling pathway.

Conclusion

Multiplexing provides individual biomarker signals whose expression correlates with dynamic changes in receptor trafficking, activation of specific signaling pathways, and modifications in tumor cell morphology. In conclusion, high-throughput imaging platform along with image analysis software and a panel of phosphorylation-specific antibodies could reveal molecular subtypes of tumors for effective personalized therapy.

AUTOMATED IMAGE ANALYSIS OF TUMOR-INFILTRATING LYMPHOCYTES

Introduction

Tumors contain variable numbers of CD3+, CD4+, CD8+ and FoxP3+ lymphocytes referred to as tumor-infiltrating lymphocytes (TILs). Initially, TILs were thought to reflect the origin of cancer at sites of chronic inflammation. Recent studies revealed a relationship between the intensity as well as the ratio of CD3+, CD4+ CD8+, and FoxP3 lymphocytes with prognostic outcome

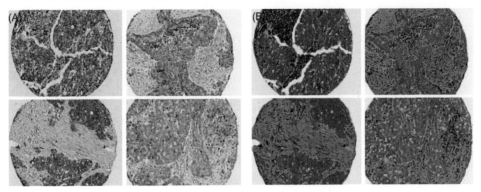

Figure1 **(A)** Multispectral 20x images from four example cores acquired from a TMA of ovarian cancer stained with hematoxylin, Vector Red labeling cytokeratin, and DAB labeling anti-CD3 antibody. **(B)** The same four cores with tissue and cell segmentation maps overlaid with red areas representing tumor, blue areas representing other tissue, and green outlines indicating DAB+T ILs.

of different solid tumors. Potentially, tests that characterize TIL phenotype, location, and density in tumor and surrounding stroma could aid the selection of immunotherapy and define targets for individualized treatment. However, TIL assessment today, based on human visual perception, is prone to inter- and intra-observer variability, due to complicated tissue architectures, ambiguous histology revealed through hematoxylin counterstain, and other human factors. Also, visual TIL assessment is tedious and time consuming. In this study, performed in collaboration with Dr.Michael Feldman and Dr. Ian Hagemann at University of Pennsylvania, we investigate computer-aided histologic event counting, using as our sample use-case the counting of lymphocytes in serous ovarian carcinoma specimens. Samples were prepared as a tissue microarray (TMA) consisting of 618 cases, with clinical follow-up. Caliper's inForm™ image analysis software automatically detected regions of tumor, and counted TILs in tumor regions, thus determining TIL density. Automated results were compared with visual assessments. Kaplan-Meier survival curves were generated for both automated and visual scores.

Methods

Sample:

- Tissue microarray (TMA) containing 618 ovarian cancersamp les.
- The TMA samples were stained with anti-CD3 antibody to identify T-lymphocytes.
- The samples were further stained with epithelial cell specific marker (cytokeratin) to assist in automated segmentation of tumor and stroma.

Automated Scoring:
The TMA slides were scanned using Caliper's Vectra™ multispectral imaging system.

- The scanned images were processed using Caliper's inForm™. A machine-learning algorithm was trained to segment tumor from stroma and identify CD3 cells labeled with DAB, indicating T-lymphocytes.
- Density of TILs within the tumor areas was calculated.

Visual Scoring:

- Ap athologistrate dlymp hocyted ensity.
- A semi-quantitative scale of 0 to 3 was assigned to each core.

Analysis:

- Automated counts were compared to visual scores.
- Manual and automated scoring were compared with survival.

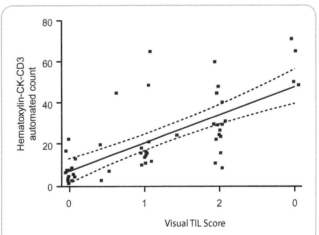

Figure2 Correlation of visual scoring and automated scoring yieldeda nR- valueo f0.6455.

Results

The machine-learning algorithm determined tumor TIL density for 70% of cores and stromal TIL density for 42%. With triplicate representation of each tumor on the array, 93% of tumors had at least one core informative for intratumoral TILs, and 71% had at least one core informative for stromal TILs. There was a significant strong positive correlation between total visual and machine counts (r = 0.6581, p<0.0001 by Spearman's nonparametric test). Kaplan-Meier analysis shows equivalent P values (~0.03) for visual and automated

275

Table 1 Automateds egmentationr esultss howingt hat 436 of 618 TMA cores successfully segmented. Errors were due to issues with the tissue, over- or under-segmentation of tissue, and over- or under-segmentation of lymphocytes.

Total histospots evaluated	618
Pre-algorithmic failures	
Spot fell off	37
Unsuitable tissue (e.g., colon or fat only)	77
Tissue segmentation failures	
Tumor interpreted as stroma	26
Stroma interpreted as tumor	49
Cell segmentation failures	
Overdetection of lymphocytes	9
Underdetection of lymphocytes	3
Spots successfully segmented	436

CD8+/FoxP3+), enabled by Vectra's multispectral capability.

- TMAs, although useful for research investigations, do not support routine clinical work. Future investigations will involve whole biopsy sections.
- These results demonstrate the feasibility of a practical and viable clinical workflow, in which TIL counting is automated by computer and results are reviewed by pathologists to assure data quality.

AUTOMATED QUANTITATIVE IMAGE ANALYSIS OF DIABETIC NEPHROPATHY

Introduction

Late stage diabetic nephropathy is histologically characterized by either diffuse or nodular expansion of the glomerular matrix. Biochemical studies have provided evidence that the microfibrillar collagen type IV is increased in diabetic nephropathy, which is presumed to correlate with functional impairment of the kidney. High-throughput image acquisition and analysis along with

scoring approaches. Although correlation between visual and automated scoring was high, automated scoring consistently determined larger numbers than visual. Larger numbers were due primarily to over-splitting of segmented lymphocytes, inclusion of lymphocytes lacking nuclear counterstain which are ignored during visual scores, and inclusion of lymphocytes at the periphery of tumor areas. Additionally, we found that a pathologist's involvement was essential, to review segmentation results and assure data quality by rejecting data from areas improperly stained, out-of-focus, folded, or otherwise inaccurately segmented.

Conclusions

- Preliminary results indicate machine scoring can meaningfully capture TIL status of tumors and yield quantitative, normalized feature count using consistent rules.
- The prognostic power of the test can be extended by adding labels for lymphocyte phenotyping (e.g., CD3+/

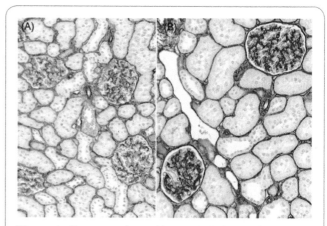

Figure 1 Representative multispectral 20x images from PAS and Collagen IV stained **(A)** negative and **(B)** positive control rats.

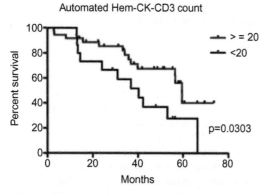
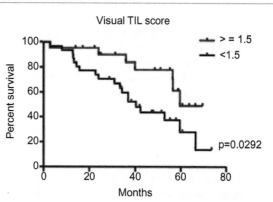

Figure 3 Kaplan-Meier curves for visual and semi-automated scoring compared with survival. These curves have essentially equivalent P-values.

immunohistochemical assessment was designed to evaluate the extent and exact morphologic location of Collagen IV deposition at various stages of diabetic glomerulosclerosis (GS). In late stage nephropathy, intrinsic basement membrane components are no longer produced. Instead, massive accumulation of Collagen IV occurs. This application note describes work performed by Dr. Bruce Homer at Pfizer. It demonstrates the capabilities of the Vectra imaging system and inForm image analysis software (Caliper Life Sciences, Inc.) for segmenting nodular glomeruli and quantitating expression of Collagen IV in Steptozotocin (STZ) induced diabetic mice. The study also compares automated results with data generated with stereological software (Stereologer 2000).

Figure 2 4x mosaic of entire scanned kidney section. Red boxes indicate sections automatically chosen for high-power 20x imaging based on an algorithm created to locate the cortex.

Methods

Sample:

- Male Wistar rats were injected with Steptozotocin (STZ) to induce diabetes.
- Animals with FBG (3 days post-STZ) > 200 mg/dl were included in the study.
- All treatment was initiated 3 days after STZ injection.
- FFPE kidney sections were stained with immunohistochemistry for Collagen IV.

Multispectral Imaging:

- 144 slides from a diabetic nephropathy study were scanned with Caliper's Vectra imaging platform.
- An algorithm was created in Caliper's inForm image analysis software to automatically identify the cortex of the kidney.
- A threshold was set so that regions containing > 40% of cortex region were selected for high-power 20× imaging and further analysis.

Image Analysis:

- 1440 20× multispectral images (10 per kidney) were acquired by Vectra and analyzed with inForm.
- Caliper's inForm image analysis software was used to automatically identify glomerular tufts within the cortical region of each image.
- Once the glomeruli were identified, the mesangial matrix within each was segmented.
- Mesangial matrix volume was calculated by multiplying the kidney tuft volume by the mean percent mesangial matrix area.
- Percent mesangial matrix and mesangial matrix volume determined from the automated platform and from stereological assessment were compared.

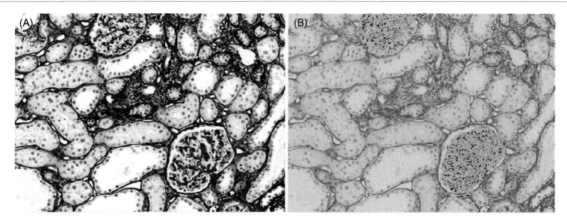

Figure3 (A) RGB representation of a 20x multispectral image acquired with Vectra imaging system. (B) The same image with tissue and cell segmentation masks created by inForm image analysis software. The software was trained to segment the tissue into glomeruli (red) and other tissue (green). inForm was then able segment the mesangial matrix (green outlines) and output quantitative data from thesese gmentedr egions.

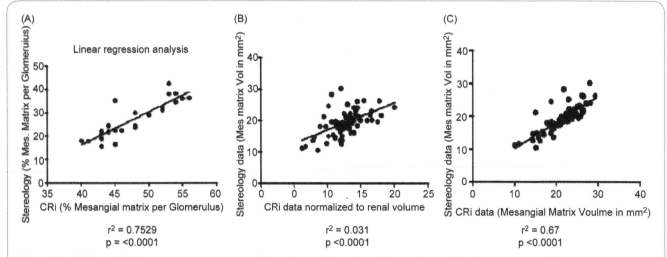

Figure4 **(A)** Correlation between percent mesangial matrix per glomerulus when analyzed with the automated platform versus stereological assessment. **(B)** Correlation between mesangial matrix volume normalized to renal volume when analyzed with the automated platform versus stereological assessment. **(C)** Correlation between mesangial matrix volume normalized to tuft volume when analyzed with the automated platform versus stereological assessment.

Results

Conclusion

- Results indicate a significant correlation in the percentage of mesangial matrix volume analyzed by Caliper's imaging system compared to a stereological analyzer for STZ treated versus non-STZ vehicle samples.
- There was a significant correlation in the matrix volume and glomerular tuft volume analyzed by Caliper's imaging system compared to a stereological analyzer.
- An increase in Collagen type IV expression was observed within glomerular tufts following STZ treated samples.
- High-throughput imaging with Vectra and automated image analysis with inForm are capable of automating analysis of diabetic nephropathy, thus accelerating studies and reducing manual analysis significantly.

STUDIES OF RECEPTOR SIGNALING AND MUTATIONS IN ARCHIVAL TISSUE USING TISSUE MICROARRAYS AND MULTISPECTRAL IMAGING

Example: Androgen receptor analysis in prostate cancer

Introduction

The analysis of immunofluorescence in tissue microarrays offers an efficient and precise method to explore correlations between mutation-driven receptor expression patterns and clinical outcome. Analysis using automated multispectral imaging (with the Caliper Vectra™) enables precise and linear quantitation of receptor, in anatomically appropriate cellular structures, and in archival clinical tissues. Sensitive and precise quantitation is possible even if archival FFPE tissues exhibit strong autofluorescences, and receptor expression levels are significantly less than autofluorescence level. TMAs rather than whole sections offer substantial increases in throughput for image acquisition and analysis, consistency of staining among samples, and efficient use of reagents.

The example case is the assessment of androgen receptor in prostate cancer. Androgen receptor (AR) has a central role in normal growth, in carcinogenesis, and in progression. Androgens function predominantly through their action on the androgen receptor (AR), which belongs to the nuclear receptor family. In addition, 10–30% of prostate carcinomas treated by anti-androgens acquire somatic point-mutation in the AR gene. These genetic changes lead to AR over-expression and hypersensitivity and promiscuous mutant AR proteins activated by non-androgenic ligands or growth modulators. The AR molecule plays a major part in the regulation of androgen-AR complex, and is a potential marker in the prognosis and hormonal responsiveness in PCa. IHC studies have shown variability in the expression of ARs in cancer and the ability to predict clinical progression and survival[1].

Methods

The data presented in this note was generated by Dr. Wei Huang of the University of Wisconsin. She is investigating the expression of AR as a potential marker to augment Gleason score for PCa staging. This study involved tissue microarrays (TMAs) built from formalin-fixed, paraffin-embedded (FFPE) prostatic adenocarcinoma tissue blocks from 183 patients (174 hormone-naïve, 9 castration resistant) and benign prostatic tissue with five year follow-up information.

- **Immunofluorescence**: The TMA sections were immuno-labeled with rat anti-AR monoclonal antibody (mAb) (Abcam, 1:200) and mouse anti-e-cadherin monoclonal antibody (Dako, 1:50). AR was detected with the rat-AR mAb and was labeled with AlexaFluor™ 647, while the mouse anti-E-cadherin was labeled with AlexaFluor™ 488. DAPI was added as a nuclear counterstain.

- **Multispectral Imaging**: Multispectral fluorescence images from the center of each duplicate core of the TMAs was automatically acquired using the Caliper Vectra™ imaging platform. A spectral library for unmixing label and autofluorescence signals was generated from single-stain slides
- **Image Analysis**: inForm™ image analysis software was applied to quantify the expression of AR in epithelial cells, in a per-cell and per-cell-compartment basis. inForm's machine-learning pattern-recognition algorithm was trained by a pathologist to identify epithelial regions automatically. Individual cells within epithelial regions were segmented into associated subcellular compartments (nucleus, cytoplasm, and membrane). Spectrally unmixed AR protein expression level was then extracted from each cell for analysis.
- **Data Analysis**: the AR expression levels were correlated with the Gleason scores (GS 5, 6, 7, 8, and 9), prostatic adenocarcinoma pathological stages (pT2, pT3a, and pT3b), and recurrence status (free, biochemical, and cancer).

RESULTS

Multispectral imaging revealed AR expression in the nucleus, cytoplasm, and membrane compartments of PCa, BPT, and CRecur patients. AR exits in the cytoplasm, bound to heat shock proteins. Upon ligand interaction, the AR homodimerizes, undergoes phosphorylation and translocation to the nucleus, where it binds to androgen response elements and induces transcription of genes. inForm™ analysis comparing AR proteins levels in PCa and BPT patients revealed elevated nuclear expression and lower expression in the cytoplasm. However, AR protein level in all the cellular compartments was lower in the PCa compared to the BPT patients. In addition, analysis of nAR and mAR levels in patients with cancer recurrence (CRecur) was significantly lower compared to recurrence-free (RF) patients. No significant difference in AR levels was observed between RF patients versus biochemical recurrence (BRecur), or between BRecurr and CRecur patients.

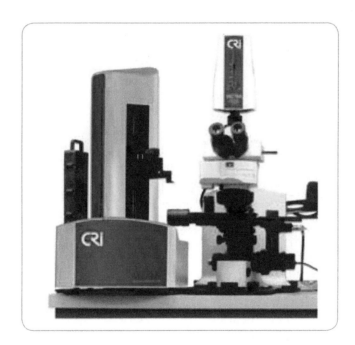

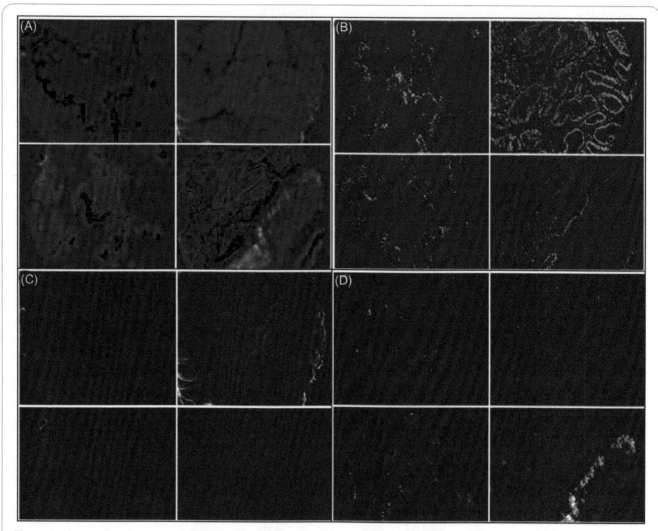

Figure1   Images from four example TMA cores. **(A)** RGB representations of multispectral images. **(B)** Unmixed DAPI component images. **(C)** Unmixed androgen component images. **(D)** Unmixed background and autofluorescence component images.

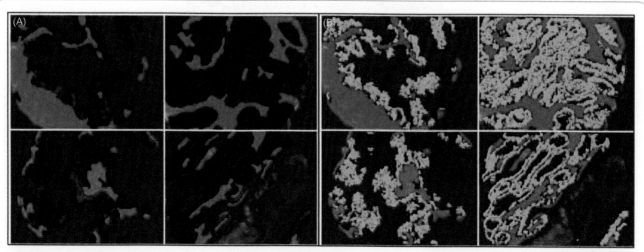

Figure2   RGB representations from four example TMA cores displaying inForm™ epithelial tissue segmentation and epithelial cell segmentation overlays. **(A)** The tissue was segmented into three classes: epithelial (the target tissue); stroma; and background. **(B)**T he individual cells inside the target epithelial tissue were segmented into nuclei, cytoplasm, and membrane.

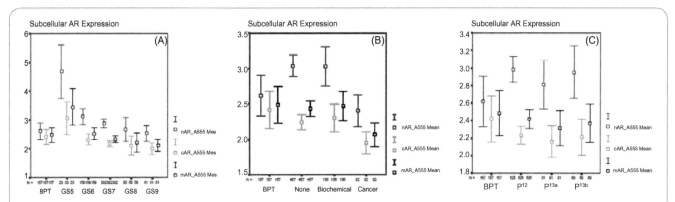

Figure 3 Correlation of subcellular androgen receptor expression with: **(A)** Gleason scores; **(B)** prostate cancer outcomes in terms of recurrence; and **(C)** prostate cancer pathological stages. These correlations are made from expression in the nucleus (nAR), cytoplasm (cAR), and membrane (mAR) of epithelial cells. Each graph displays the correlation with benign prostatic tissue (BPT) as a control.

Conclusion

Automated multispectral imaging of tissue microarrays allows ability to identify multiple proteins and pathways involved in the progression of prostate cancer. inForm™ image analysis revealed the translocation of AR to the nucleus from the cytoplasm that is consistent with previous studies (Waltering et al., 2009; Qiu et al., 2008). In our study low AR expression was observed in patients with metastatic PCa that is consistent with data reported by Takeda et al. 1996 and Segawa et al., 2001. Although the changes in AR levels in all the three compartments were not significantly correlated with PCa stages or

Gleason Score, they are consistent with other reports. This could be due to the heterogeneity of AR immunostaining within a tumor.

Thus, this application demonstrates the ability of automated multispectral imaging system in aiding pathologists, as well as researchers to identify multiple biomarkers involved in the prognosis of prostate cancer. In addition, inForm™ software with its machine-learning abilities is capable of segmenting tissue based on morphological features, performing cell segmentation, and biomarker expression on a per cell basis on large data sets forq uantitativean alysis.

Appendix 3

Folio Biosciences (Columbus, OH) is the leading commercial provider of annotated human tissue samples for research studies. Having over two million human tissue samples, with up to 20 years' follow-up clinical data, Folio is dedicated to advancing biomedical research by providing easy and affordable access to high quality biospecimens. Folio's board-certified pathologists and experienced scientists carry out state of the art analytical methods, including immunohistochemistry (IHC) and cytogenetics (FISH), in its CLIA certified labs. Its proprietary method enables restoration of protein, thereby dramatically improving the signal and RNA and DNA expression levels in FFPE tissue archival specimens.

Phylogeny, Inc. is a functional genomics research services company focused on tissue-based assays. It works in a wide range of formats including formalin-fixed paraffin-embedded (FFPE) cells and tissues performing in situ hybridization (ISH), for measurement of gene expression and miRNA detection. Phylogeny works in all the biological systems of its clients including: human, mouse and other mammal; normal versus disease state; developmental studies; and plant studies. Phylogeny also provides cytogenetic services within its CLIA certified facility. The Ph.D. scientists provide all levels of experimental design and results interpretation (including for large studies), and can also prepare probes for studies. Additional services, such as immunohistochemistry (IHC) and laser capture microdissection, are also available. Phylogeny delivers publication quality data to its clients to streamline their research projects.

Index

Note: Page numbers followed by "*f*" and "*t*" refer to figures and tables respectively.

Printed and bound by CPI Group (UK) Ltd, Croydon, CR0 4YY

08/05/2025

01865026-0001